Lecture Notes in Computer Science 12016

More information about this series at http://www.springer.com/series/7407

Manoj Changat · Sandip Das (Eds.)

Algorithms and Discrete Applied Mathematics

6th International Conference, CALDAM 2020
Hyderabad, India, February 13–15, 2020
Proceedings

 Springer

Editors
Manoj Changat
University of Kerala
Thiruvananthapuram, India

Sandip Das
Indian Statistical Institute
Kolkata, India

ISSN 0302-9743   ISSN 1611-3349   (electronic)
Lecture Notes in Computer Science
ISBN 978-3-030-39218-5   ISBN 978-3-030-39219-2   (eBook)
https://doi.org/10.1007/978-3-030-39219-2

LNCS Sublibrary: SL1 – Theoretical Computer Science and General Issues

This Springer imprint is published by the registered company Springer Nature Switzerland AG
The registered company address is: Gewerbestrasse 11, 6330 Cham, Switzerland

Preface

This volume contains the papers presented at CALDAM 2020: the 6th International Conference on Algorithms and Discrete Applied Mathematics held during February 13–15, 2020, in Hyderabad, India. CALDAM 2020 was organized by the Department of Computer Science and Engineering, Indian Institute of Technology, Hyderabad, and the Association for Computer Science and Discrete Mathematics (ACSDM).

The conference had papers in the areas of graph algorithms, graph theory, combinatorial optimization, combinatorial algorithms, computational geometry, and computational complexity. We received 102 submissions with authors from 15 countries.

The reputation of the conference was enhanced by its Program Committee and the invited talks. We were able to get highly respected researchers to serve on our Program Committee. Each submission received at least one detailed review and nearly all were reviewed by three Program Committee members. The Program Committee had to take tough decisions as the quality of papers were very much competitive. The committee decided to accept 38 papers. The program also included two invited talks by Sergio Cabello and Douglas West, who are prominent figures in discrete mathematics, graph theory, and geometry.

As volume editors, we would like to thank all the authors for contributing high-quality research papers to the conference. We are very much indebted to the Program Committee members and the external reviewers for reviewing the papers at a high standard within a very short period of time. We thank Springer for publishing the proceedings in the *Lecture Notes in Computer Science* series. Our sincerest thanks are due to the invited speakers Sergio Cabello (University of Ljubljana, Slovenia) and Douglas West (Zhejiang Normal University, China, and University of Illinois, USA) for accepting our invitation. We thank the Organizing Committee, chaired by Subramaniyam Kalyanasundaram from Indian Institute of Technology, Hyderabad, for the smooth functioning of CALDAM 2020 and Indian Institute of Technology, Hyderabad, for proving the facilities for the smooth conduct of the conference. We are immensely grateful to the chair of the Steering Committee, Subir Ghosh, for his active help, support, and guidance throughout. We thank our sponsors Google Inc., the National Board for Higher Mathematics (NBHM), the Department of Atomic Energy, Government of India, for their financial support. We also thank Springer for its support of the Best Paper Presentation Awards. We thank the EasyChair conference management system, which was very effective in handling the entire review process.

Our sincerest thanks were also due to French speakers and IFCAM for arranging a relevant pre-conference school (Indo-French School) on Algorithms and Combinatorics. It provided an excellent overview to the participants about the research on emerging graph theory and discrete mathematics techniques.

February 2020

Manoj Changat
Sandip Das

Organization

Program Committee

Amitabha Bagchi	Indian Institute of Technology Delhi, India
Aritra Banik	National Institute of Science Education and Research, India
Niranjan Balachandran	Indian Institute of Technology Bombay, India
Boštjan Brešar	University of Maribor, Slovenia
Tiziana Calamoneri	Sapienza University of Rome, Italy
Manoj Changat (Chair)	University of Kerala, India
Victor Chepoi	Aix-Marseille University, France
Sandip Das (Chair)	ISI Kolkata, India
Josep Diaz	Polytechnic University of Catalonia, Spain
Sumit Ganguly	Indian Institute of Technology Kanpur, India
Daya Gaur	University of Lethbridge, Canada
Sathish Govindarajan	Indian Institute of Science Bangalore, India
Pavol Hell	Simon Fraser University, Canada
Christos Kaklamanis	University of Patras, Greece
Ramesh Krishnamurti	Simon Fraser University, Canada
Van Bang Le	University of Rostock, Germany
Andrzej Lingas	Lund University, Sweden
Anil Maheshwari	Carleton University, Canada
Bodo Manthey	University of Twente, The Netherlands
Bojan Mohar	Simon Fraser University, Canada
Apurva Mudgal	Indian Institute of Technology Ropar, India
N. S. Narayanaswamy	Indian Institute of Technology Madras, India
Sudebkumar Pal	Indian Institute of Technology Kharagpur, India
David Peleg	Weizmann Institute of Science, Israel
Iztok Peterin	University of Maribor, Slovenia
Abhiram Ranade	Indian Institute of Technology Bombay, India
M. V. Panduranga Rao	Indian Institute of Technology Hyderabad, India
André Raspaud	University of Bordeaux, France
Sagnik Sen	Indian Institute of Technology Dharwad, India
Michiel Smid	Carleton University, Canada
Éric Sopena	University of Bordeaux, France
Joachim Spoerhase	Aalto University, Finland
C. R. Subramanian	Institute of Mathematical Science Chennai, India

Organizing Committee

N. R. Aravind	Indian Institute of Technology Hyderabad, India
Swami Dhyanagamyananda	Ramakrishna Mission Vivekananda Educational and Research Institute, India
Partha P. Goswami	Association for Computer Science and Discrete Mathematics, India
Subrahmanyam Kalyanasundaram (Chair)	Indian Institute of Technology Hyderabad, India
Pritee Khanna	Indian Institute of Technology Design and Manufacturing Jabalpur, India
Rogers Mathew	Indian Institute of Technology Hyderabad, India
Tathagata Ray	BITS Pilani Hyderabad, India
Tarkeshwar Singh	BITS Pilani Goa, India
Karteek Sreenivasaiah	Indian Institute of Technology Hyderabad, India

Steering Committee

Subir Kumar Ghosh (Chair)	Ramakrishna Mission Vivekananda Educational and Research Institute, India
Gyula O. H. Katona	Alfréd Rényi Institute of Mathematics, Hungarian Academy of Sciences, Hungary
János Pach	École Polytechnique Fédérale de Lausanne (EPFL), Switzerland
Nicola Santoro	School of Computer Science, Carleton University, Canada
Swami Sarvattomananda	Ramakrishna Mission Vivekananda Educational and Research Institute, India
Chee Yap	Courant Institute of Mathematical Sciences, New York University, USA

Additional Reviewers

Agrawal, Ravi
Alikhani, Saeid
Ashok, Pradeesha
Babu, Sobhan
Balakrishnan, Kannan
Bandyapadhyay, Sayan
Banerjee, Subhashis
Baswana, Surender
Beaudou, Laurent
Benkoczi, Robert
Bessy, Stéphane

Bhattacharya, Srimanta
Bhore, Sujoy
Božović, Dragana
Cabrera Martinez, Abel
Chakraborty, Dibyayan
Chellali, Mustapha
Choudhary, Keerti
Couetoux, Basile
Danda, Sravan
Das, Arun Kumar
de Graaf, Maurits

Dev, Subhadeep
Dey, Hiranya
Diner, Oznur Yasar
Dorbec, Paul
Dourado, Mitre
Fertin, Guillaume
Gahlawat, Harmender
Gologranc, Tanja
González Yero, Ismael
Groshaus, Marina
Gupta, Sushmita
Gurski, Frank
Hernández-Cruz, César
Hoeksma, Ruben
Iranmanesh, Ehsan
Jakovac, Marko
Kalinowski, Rafal
Kalyanasundaram, Subrahmanyam
Kare, Anjeneya Swami
Kelenc, Aleksander
Kern, Walter
Khan, Muhammad Ali
Khare, Niraj
Klasing, Ralf
Knauer, Kolja
Kowaluk, Miroslaw
Kraner Šumenjak
Krithika, R.
Kumar, Neeraj
Lee, Chia-Wei
Levcopoulos, Christos
Macgillivray, Gary
Madireddy, Raghunath Reddy
Misra, Neeldhara
Mittal, Rajat

Mulzer, Wolfgang
N. R., Aravind
Nandakumar, Satyadev
Nandi, Soumen
Nasre, Meghana
Natarajan, Aravind
Nederlof, Jesper
Ordyniak, Sebastian
P. D., Pavan
Padinhatteeri, Sajith
Pal, Sagartanu
Parter, Merav
Pattanayak, Debasish
Paulusma, Daniel
Protti, Fabio
Ramamoorthi, Vijayaragunathan
Ray, Chiranjit
S., Taruni
Sahoo, Uma Kant
Saivasan, Prakash
Samadi, Babak
Sandeep, R. B.
Saxena, Nitin
Serna, Maria
Sharma, Roohani
Singh, Nitin
Singh, Rishi Ranjan
Sivaraman, Vaidy
Tewari, Raghunath
Thangadurai, Ravindran
Togni, Olivier
Valicov, Petru
Venkitesh, S.
Zylinski, Pawel
Štesl, Daša

Abstracts of Invited Talks

Interactions Between Geometry, Graphs and Algorithms

Sergio Cabello[1,2]

[1] Faculty of Mathematics and Physics, University of Ljubljana, Slovenia
[2] Institute of Mathematics, Physics and Mechanics,
Slovenia
sergio.cabello@fmf.uni-lj.si

I will describe some of the interactions between graphs and geometry, with an algorithmic slant. A part of the talk will be devoted to explain how tools from computational geometry become useful to efficiently manipulate distances in planar graphs and graphs of bounded treewidth. In particular, I will explain the main ideas to compute in subquadratic time the sum of the distances and the sum of the inverse of the distances between all pairs of vertices in those graphs.

Another part of the talk will be devoted to the problem of finding a maximum matching in a geometric intersection graph. In this part we will provide the main ideas to compute a maximum matching in the intersection graph of n unit disks in the plane in $O(n^{\omega/2})$ time with high probability, where ω is any constant, $\omega > 2$, such that two $n \times n$ matrices can be multiplied in $O(n^{\omega})$ time.

The second part of the talk is based on joint work with Édouard Bonnet and Wolfgang Mulzer available at https://arxiv.org/abs/1910.02123.

The Slow-Coloring Game on a Graph

Douglas West[1,2]

[1] Departments of Mathematics, Zhejiang Normal University, Jinhua, China
[2] University of Illinois, Urbana IL, USA
dwest@math.uiuc.edu

The *slow-coloring game* is played by Lister and Painter on a graph G. Initially, all vertices of G are uncolored. In each round, Lister marks a non-empty set M of uncolored vertices, and Painter colors a subset of M that is independent in G. The game ends when all vertices are colored. The score of the game is the sum of the sizes of all sets marked by Lister. The goal of Painter is to minimize the score, while Lister tries to maximize it. The score under optimal play is the *sum-color cost* or simply *cost* of the graph, written $\mathring{s}(G)$. The game was introduced by Mahoney, Puleo, and West [6], who obtained various basic results on the problem. This talk presents results from this and subsequent papers.

From the definition of the game, the cost satisfies a recursive formula:

$$\mathring{s}(G) = \max_{\emptyset \neq M \subseteq V(G)} (|M| + \min \mathring{s}(G - I)),$$

where the minimum is taken over subsets I of M that are independent in G.

A general lower bound is given by the *chromatic sum* of the graph, written $\sum(G)$, which is the minimum of $\sum_{v \in V(G)} c(v)$ over all proper colorings c that color the vertices using positive integers (introduced by Kubicka [4, 5]). The chromatic sum is the outcome of the game when Lister follows the strategy of always marking all uncolored vertices.

A greedy strategy for Painter keeps the cost of G to at most $\chi(G)n$ when G has n vertices, with strict inequality unless G has no edges. Wu [8] showed that this is asymptotically sharp for Turán graphs, obtaining more generally $\mathring{s}(K_{r_1,\ldots,r_k}) \geq n + \sum_{i<j} u_{r_i} u_{r_j}$, where u_r is the maximum t such that $\binom{t+1}{2} \leq r$ (approximately $\sqrt{2r}$).

On sparser families with bounded chromatic number, Painter can do better. We begin with n-vertex trees. Mahoney, Puleo, and West [6] showed that the maximum cost is $\lfloor 3n/2 \rfloor$, achieved by trees, and the minimum cost is $n + u_{n-1}$, achieved by stars. Puleo and West [7] obtained a linear-time algorithm and inductive formula to compute $\mathring{s}(G)$ on trees, and they also determined all the extremal n-vertex trees. A *triangular number* is a number having the form $\binom{k}{2}$ for $k \in \mathbb{N}$. With u_r as defined above, these results are the following.

Research supported by National Natural Science Foundation of China grants NNSFC 11871439 and 11971439.

Theorem 1 ([7]). *Let T be a forest. If T has no edges, then $\mathring{s}(T) = |V(T)|$. If v is is a non-leaf vertex in T with at most one non-leaf neighbor, and R is the set of leaf neighbors of v, with $r = |R|$, then*

$$\mathring{s}(T) = \begin{cases} \mathring{s}(T - R - v) + r + 1 + u_r, & \text{if } r+1 \text{ is not a triangular number}, \\ \mathring{s}(T - R) + r + u_r, & \text{if } r+1 \text{ is a triangular number}. \end{cases}$$

Theorem 2 ([7]). *The maximum of $\mathring{s}$ over n-vertex trees is $\lfloor 3n/2 \rfloor$, achieved precisely for trees containing a spanning forest in which every vertex has degree 1 or 3, except for one vertex of degree 0 or 6 when n is odd. The minimum equals $n + u_{n-1}$ and is achieved only by stars, except for a few other star-like trees when $n-1$ or $n-2$ is a triangular number.*

These results are proved using a careful analysis of the optimal moves on stars together with a bound arising from a separation of the vertices into two sets. Here $G[A]$ denotes the subgraph of G induced by A, and $[A, B]$ is the set of edges with endpoints in A and B.

Lemma 1 ([7]). *If G is a graph and $\{A, B\}$ is a partition of $V(G)$, then*
$$\mathring{s}(G[A]) + \mathring{s}(G[B]) \le \mathring{s}(G) \le \mathring{s}(G[A]) + \mathring{s}(G[B]) + |[A, B]|.$$

[7] showed also that on trees the cost equals a parameter due to Bonamy and Meeks [1] called the "interactive sum choice number". The parameters differ already on even cycles.

Building on the results for trees, Gutowski, Krawczyk, Maziarz, West, Zajac, and Zhu [3] studied bounds on $\mathring{s}(G)$ for various families of sparse graphs. One general upper bound, proved by a strategy for Painter motivated by Wu's argument, uses a partition of the vertices into subsets on whose induced subgraphs Painter has a good strategy.

Theorem 3 ([3]). *Let G be an n-vertex graph. If $\mathring{s}(G[V_i]) \le c_i|V_i|$ for $1 \le i \le t$, where $V(G)$ is the disjoint union of $V_1, \dots, V_t$, then*

$$\mathring{s}(G) \le \left(\sum_i \sqrt{c_i|V_i|} \right)^2 \le \left(\sum_i c_i \right) n.$$

When $V_1, \dots, V_n$ are independent sets, this bound yields $\mathring{s}(K_{r_1,\dots,r_k}) \le n + 2\sum_{i<j} \sqrt{r_i r_j}$, which was proved for $k = 2$ in [6] and shows that Wu's lower bound on $\mathring{s}(K_{r_1,\dots,r_k})$ is almost exact. Wu improved this to $n + \sum_{i<j} \sqrt{2r_i - 1}\sqrt{2r_j - 1}$ using a more difficult argument.

Using Theorem 3 and the basic upper bound for forests, [3] proved the following upper bounds. A graph is k-*degenerate* if every subgraph has a vertex of degree at most k, and it is *acyclically k-colorable* (see [2]) if it has a proper coloring with no 2-colored cycle.

Theorem 4 ([3]). *If G is an n-vertex graph, then*

(a) $\mathring{s}(G) \leq \frac{3k+4}{4}n$ *when G is k-degenerate (improving to* $\frac{3k+3}{4}n$ *when k is odd),*

(b) $\mathring{s}(G) \leq 3n$ *when G is a planar graph whose dual has a spanning cycle, and*

(c) $\mathring{s}(G) \leq \frac{1}{k}[\sqrt{3/4}(k-1)+1]^2 n$ *when G is acyclically k-colorable and k is odd (the coefficient is 2.4881 when k = 3 and 3.9857 when k = 5).*

The most difficult results in [3] use a *potential method*. Here a potential function ϕ assigns a potential to each vertex and edge of the graph, depending on such properties as vertex degree, membership in triangles, etc. The potential $\Phi(G)$ of a graph G is the sum of the potentials of the vertices and edges.

When Lister marks a set M in a graph G in family closed under taking subgraphs, Painter will seek an independent set $X \subseteq M$ such that $|M| \leq \Phi(G) - \Phi(G - X)$. That is, the total score in the current round should be at most the loss in potential by coloring X. Since the potential is reduced to 0 when the game is over, always being able to find such a set X yields $\mathring{s}(G) \leq \Phi(G)$. This technique yields the following results.

Theorem 5 ([3]). *If G is an n-vertex graph with m edges, then*

(a) $\mathring{s}(G) \leq (8n+3m)/5$ *when G is 4-colorable (hence* $\mathring{s}(G) \leq 3.4n$ *when G is planar), and*

(b) $\mathring{s}(G) \leq (7/3)n$ *when G is outerplanar.*

Except on the families of trees and complete multipartite graphs, the bounds given above are not expected to be sharp. In particular, the worst known constructions for n-vertex outerplanar graphs have cost $2n$ and for n-vertex planar graphs have cost $2.5n$. Also, Puleo showed that the cartesian product of a 4-cycle and a path has cost $1.75n - 1$, which is the highest known for bipartite planar graphs.

The slides for the talk and the papers [3, 6, 7] are available at https://faculty.math. illinois.edu/west/pubs/publink.html.

References

1. Bonamy, M., Meeks, K.: The interactive sum choice number of graphs (2017). arXiv:1703. 05380
2. Borodin, O.V.: On acyclic colorings of planar graphs. Discrete Math. **25**, 211–236 (1979)
3. Gutowski, G., Krawczyk, T., Maziarz, K., West, D.B., Zajac, M., Zhu, X.: The slow-coloring game on sparse graphs: k-degenerate, planar, and outerplanar (submitted)
4. Kubicka, E.M.: The chromatic sum and efficient tree algorithms. ProQuest LLC, Ann Arbor, MI, Thesis (Ph.D.)–Western Michigan University (1989)
5. Kubicka, E.: The chromatic sum of a graph: history and recent developments. Int. J. Math. Math. Sci. **29–32**, 1563–1573 (2004)
6. Mahoney, T., Puleo, G.J., West, D.B.: Online sum-paintability: the slow-coloring game on graphs. Discrete Math. **341**,1084–1093 (2018)
7. Puleo, G.J., West, D.B.: Online sum-paintability: slow-coloring of trees. Discrete Appl. Math. **262**, 158–168 (2019)
8. Wu, H.: Personal Communication and Lecture at International Workshop on Graph Theory. Ewha Woman's University, Seoul, Korea, 5 January 2018

Contents

Combinatorial Optimization

Distributed Algorithms

Combinatorial Algorithms

Computational Complexity

Computational Geometry

Graph Algorithms

Complexity of Restricted Variant
of Star Colouring

M. A. Shalu$^{(\boxtimes)}$ and Cyriac Antony(iD)

IIITDM Kancheepuram, Chennai, India
{shalu,mat17d001}@iiitdm.ac.in

Abstract. Restricted star colouring is a variant of star colouring introduced to design heuristic algorithms to estimate sparse Hessian matrices.
For $k \in \mathbb{N}$, a k-restricted star (k-rs) colouring of a graph G is a function $f : V(G) \to \{0, 1, \dots, k-1\}$ such that (i) $f(x) \neq f(y)$ for every
edge xy of G, and (ii) there is no bicoloured 3-vertex path(P_3) in G
with the higher colour on its middle vertex. We show that for $k \geq 3$,
it is NP-complete to decide whether a given planar bipartite graph of
maximum degree k and girth at least six is k-rs colourable, and thereby
answer a problem posed by Shalu and Sandhya (Graphs and Combinatorics 2016). In addition, we design an $O(n^3)$ algorithm to test whether
a chordal graph is 3-rs colourable.

Keywords: Graph coloring · Star coloring · Restricted star coloring ·
Unique superior coloring · Vertex ranking · Ordered coloring ·
Complexity

1 Introduction

Many large scale optimization problems involve a multi-variable function $f :
\mathbb{R}^n \to \mathbb{R}$. The (second-order) approximation of f using Taylor series expansion
requires an estimation of the Hessian matrix of f. Vertex colouring of graphs and
its variants have been found immensely useful as models for estimation of sparse
Hessian and Jacobian matrices (see [8] for a survey). To compute a compressed
form of a given sparse matrix, Curtis et al. [5] partitioned the set of columns of
the matrix in such a way that columns that do not share non-zero entries along
the same row are grouped together. By exploiting symmetry, Powell and Toint
[13] designed a heuristic algorithm for partitioning columns of a sparse Hessian
matrix implicitly using restricted star colouring. Restricted star colouring was
also studied independently in the guise of unique superior colouring, a generalization of ordered colouring [2,9] (Ordered colouring is also known as vertex
ranking, and has applications in parallel processing and VLSI circuit design [6]).
Heuristic algorithms for restricted star colouring are given in ColPack software
suite [7] as well as [4,8]. Note that restricted star colouring appears unnamed in
[8] and under the name independent set star partition in [15].

First author is supported by SERB (DST), MATRICS scheme MTR/2018/000086.

M. Changat and S. Das (Eds.): CALDAM 2020, LNCS 12016, pp. 3–14, 2020.
https://doi.org/10.1007/978-3-030-39219-2_1

In this paper, we consider only vertex colourings of finite simple undirected graphs. A k-*star colouring* of a graph G is a function $f : V(G) \to \{0, 1, \ldots, k-1\}$ such that (i) $f(x) \neq f(y)$ whenever xy is an edge in G, and (ii) G contains no bicoloured P_4 (as subgraph). Let $V_i := \{v \in V(G) : f(v) = i\}$. Note that for $i \neq j$, every component of $G[V_i \cup V_j]$ is a star ($K_{1,p}$ where $p \geq 0$). A k-*restricted star* (k-rs) *colouring* of a graph G is a function $f : V(G) \to \{0, 1, \ldots k-1\}$ such that (i) $f(x) \neq f(y)$ whenever xy is an edge in G, and (ii) G contains no bicoloured P_3 with the higher colour on its middle vertex (i.e., no path x, y, z with $f(y) > f(x) = f(z)$). In other words, whenever $i < j$, every vertex in V_j has at most one neighbour in V_i. Note that for $i < j$, every non-trivial component of $G[V_i \cup V_j]$ is a star with its centre in V_i. Hence, every k-restricted star colouring is a k-star colouring; but the converse is not true [15].

The rs chromatic number of a graph G is defined as $\chi_{rs}(G) := \min\{k : G$ admits a k-rs colouring$\}$. We use n to denote the number of vertices unless otherwise specified. The length of a shortest cycle in a graph G is called its *girth*. A graph with maximum degree at most three is called a *subcubic* graph. The known results on restricted star colouring include the following: (i) $\chi_{rs}(G) = O(\log n)$ for a planar graph G [9], (ii) $\chi_{rs}(T) = O(\frac{\log n}{\log\log n})$ for a tree T [9], (iii) $\chi_{rs}(G) \leq 7$ for a subcubic graph G [2], (iv) $\chi_{rs}(Q_d) = d + 1$ for the hypercube Q_d [2], and (v) $\chi_{rs}(G) \leq 4\alpha(G)$ for a graph G of girth at least five where $\alpha(G)$ is the independence number of G [15].

We focus on the following decision problems in this paper.

k-RS COLOURABILITY   RS COLOURABILITY
Instance: A graph G Instance: A graph G, an integer $k \leq |V(G)|$
Question: Is $\chi_{rs}(G) \leq k$?   Question: Is $\chi_{rs}(G) \leq k$?

Note that a graph G is 2-rs colourable (or 2-star colourable) if and only if every component of G is a star. Hence, 2-RS COLOURABILITY is in P.

Question: Is k-RS COLOURABILITY NP-complete for $k \geq 3$? [15]

In this paper, we prove that for $k \geq 3$, k-RS COLOURABILITY is NP-complete for the class of planar bipartite graphs of maximum degree k and girth at least six (in Sect. 3). This answers the above question in the affirmative. In addition, we present a linear-time algorithm to test whether a tree is 3-rs colourable (in Sect. 4), and an $O(n^3)$ algorithm to test whether a chordal graph is 3-rs colourable (in Sect. 5).

The organization of the rest of this paper is as follows. In Sect. 2, we present necessary preliminaries. Sections 3, 4, 5 and 6 deal with planar bipartite graphs, trees, chordal graphs and cobipartite graphs respectively.

2 Preliminaries

We follow West [16] for graph theory terminology and notation. We denote (i) by $a_1, a_2, \ldots, a_n$, a path with vertex set $\{a_1, a_2, \ldots, a_n\}$ and edges a_1a_2, a_2a_3, $\ldots$,

$a_{n-1}a_n$, and (ii) by $(a_1, a_2, \ldots, a_n)$, a cycle with vertex set $\{a_1, a_2, \ldots, a_n\}$ and edges $a_1 a_2, \ldots, a_{n-1}a_n, a_n a_1$. We call a vertex of degree three or more as a *3-plus vertex*. The following observations are pivotal to our results on 3-RS COLOURABILITY. Proof of Observation 2 is easy, and hence omitted.

Observation 1. *Let G be a graph. If $f : V(G) \to \{0, 1, 2\}$ is a 3-rs colouring of G, then f has the following properties.*

(P1) If a is a 3-plus vertex in G, then $f(a) = 0$ or 1.
(P2) If ab is an edge joining 3-plus vertices a and b in G, then $f(b) = 1 - f(a)$.
(P3) Both end vertices of a P_3 in G cannot be coloured 0 by f. So, if a, b, c is a P_3 in G such that $f(a) = 0$ and c is a 3-plus vertex, then $f(c) = 1$.

Note: Properties above are numbered as a mnemonic. Property P2 is about path P_2, and Property P3 is about path P_3.

Proof. We prove the contrapositive of Property P1. Suppose $f(a) = 2$ for a vertex a in G. Since $f(a) = 2$, by definition of 3-rs colouring, a has at most one neighbour coloured 0 and at most one neighbour coloured 1. Hence, $\deg(a) \leq 2$. This proves Property P1. Observe that whenever $f(a), f(b) \in \{0, 1\}$, either $f(b) = f(a)$ or $f(b) = 1 - f(a)$. Therefore, Property P2 follows directly from Property P1 as $f(b) \neq f(a)$. Contrary to Property P3, assume that there is a path a, b, c in G with $f(a) = f(c) = 0$. Then, a, b, c is a bicoloured P_3 with the higher colour on its middle vertex b. This is a contradiction. $\square$

Observation 2. *Let u, v, w, x be a path in a graph G, and let f be a 3-rs colouring of G such that $f(u) = 0$ and $f(v) = 1$. Then, $f(w) = 2$, and therefore $f(x) = 0$.*

3 Planar Bipartite Graphs

In this section, we prove that for $k \geq 3$, k-RS COLOURABILITY is NP-complete for the class of planar bipartite graphs of maximum degree k and girth at least six. First, we show that 3-RS COLOURABILITY is NP-complete for subcubic planar bipartite graphs of girth at least six using a reduction from CUBIC PLANAR POSITIVE 1-IN-3 SAT. To describe the latter problem, we introduce necessary terminology assuming that the reader is familiar with satisfiability problems.

A CNF formula $\mathcal{B} = (X, C)$, where X is the set of variables and C is the set of clauses, is called a positive CNF formula if no clause contains a negated literal; in other words, the clauses are subsets of X. Let $\mathcal{B} = (X, C)$ be a positive CNF formula with $X = \{x_1, x_2, \ldots, x_n\}$ and $C = \{C_1, C_2, \ldots, C_m\}$. The *graph of formula* $\mathcal{B}$, denoted by $G_{\mathcal{B}}$, is the graph with vertex set $X \cup C$ and edges $x_i C_j$ for every variable x_i in clause C_j ($i = 1, 2, \ldots, n$, $j = 1, 2, \ldots, m$). Figure 1a shows the graph $G_{\mathcal{B}}$ for the formula $\mathcal{B} = (X, C)$ where $X = \{x_1, x_2, x_3, x_4\}$, $C = \{C_1, C_2, C_3, C_4\}$, $C_1 = \{x_1, x_2, x_3\}$, $C_2 = \{x_1, x_2, x_4\}$, $C_3 = \{x_1, x_3, x_4\}$ and $C_4 = \{x_2, x_3, x_4\}$.

CUBIC PLANAR POSITIVE 1-IN-3 SAT (CPP 1-IN-3 SAT)

Instance: A positive 3-CNF formula $\mathcal{B} = (X, C)$ such that
$G_{\mathcal{B}}$ is a cubic planar graph
Question: Is there a truth assignment for X such that
every clause in C has exactly one true variable?

This problem is proved to be NP-complete by Moore and Robson [12] (Note: in [12], the problem is called Cubic Planar Monotone 1-in-3 Sat. We use 'positive' rather than 'monotone' to be unambiguous). Observe that the graph $G_{\mathcal{B}}$ is cubic if and only if each clause contains three variables and each variable occurs in exactly three clauses. As a result, in a CPP 1-IN-3 SAT instance, the number of variables equals the number of clauses, that is $m = n$.

Theorem 1. 3-RS COLOURABILITY *is NP-complete for the class of subcubic planar bipartite graphs of girth at least six.*

Proof. 3-RS COLOURABILITY is in NP because given a 3-colouring f (certificate) of the input graph, we can verify in polynomial time that all bicoloured paths x, y, z satisfy $f(y) < f(x)$.

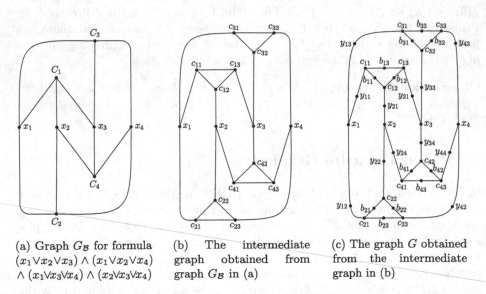

(a) Graph $G_{\mathcal{B}}$ for formula $(x_1 \vee x_2 \vee x_3) \wedge (x_1 \vee x_2 \vee x_4) \wedge (x_1 \vee x_3 \vee x_4) \wedge (x_2 \vee x_3 \vee x_4)$

(b) The intermediate graph obtained from graph $G_{\mathcal{B}}$ in (a)

(c) The graph G obtained from the intermediate graph in (b)

Fig. 1. Example for construction of graph G from $G_{\mathcal{B}}$

To prove NP-hardness, we transform CUBIC PLANAR POSITIVE 1-IN-3 SAT (CPP 1-IN-3 SAT) problem to 3-RS COLOURABILITY problem. Let $\mathcal{B} = (X, C)$ be an instance of CPP 1-IN-3 SAT where $X = \{x_1, x_2, \ldots, x_n\}$ and $C = \{C_1, C_2, \ldots, C_n\}$. We recall that $\mathcal{B}$ is a positive CNF formula and $G_{\mathcal{B}}$ is a cubic planar graph. We construct a graph G from $G_{\mathcal{B}}$ as follows. First, an intermediate graph is constructed. For each clause $C_j = \{x_{j_1}, x_{j_2}, x_{j_3}\}$, replace vertex C_j in $G_{\mathcal{B}}$ by a triangle (c_{j1}, c_{j2}, c_{j3}) and replace edges $x_{j_1}C_j$, $x_{j_2}C_j$, $x_{j_3}C_j$ in $G_{\mathcal{B}}$ by edges $x_{j_1}c_{j1}$, $x_{j_2}c_{j2}$, $x_{j_3}c_{j3}$ (see Fig. 1).

The graph G is obtained by subdividing each edge of this intermediate graph exactly once. Let us call the new vertex introduced upon subdividing the edge x_ic_{jk} as y_{ij} and the new vertex introduced upon subdividing the edge $c_{jk}c_{j\,k+1}$ as b_{jk}. We call (i) each vertex x_i in the constructed graph G as the gadget for variable x_i, (ii) each 6-vertex cycle $(c_{j1}, b_{j1}, c_{j2}, b_{j2}, c_{j3}, b_{j3})$ as the gadget for clause C_j, and (iii) each path x_i, y_{ij}, c_{jk} as the gadget that represents the occurrence of variable x_i in clause C_j.

Since the intermediate graph is a cubic planar graph of girth three (see Fig. 1b), G is a subcubic planar bipartite graph of girth six (see Fig. 1c). The graph $G_{\mathcal{B}}$ can be constructed in $O(n)$ time, and it has $2n$ vertices and $3n$ edges. Also, G can be constructed from $G_{\mathcal{B}}$ in $O(n)$ time since there are only $10n$ vertices and $12n$ edges in G. All that remains is to prove that G admits a 3-rs colouring if and only if $\mathcal{B}$ is a yes instance of CPP 1-IN-3 SAT. This is established with the help of following claims.

(CL1) If f is a 3-rs colouring of G, then for each j, exactly one vertex among c_{j1}, c_{j2}, c_{j3} is coloured 0 by f

(CL2) If f is a 3-rs colouring of G, then $f(c_{jk}) = 1 - f(x_i)$ whenever x_i, y_{ij}, c_{jk} is a path in G

Since x_i's and c_{jk}'s are 3-plus vertices in G, $f(x_i), f(c_{jk}) \in \{0,1\}$ for $1 \leq i, j \leq n$, $1 \leq k \leq 3$ by Property P1. Observe that every pair of vertices from c_{j1}, c_{j2}, c_{j3} is at a distance two. Since Property P3 forbids assigning colour 0 to both end vertices of a P_3, at most one vertex among c_{j1}, c_{j2}, c_{j3} is coloured 0 by f. Thus to prove claim CL1, it suffices to show that at least one of them is coloured 0 by f. On the contrary, assume that $f(c_{j1}) = f(c_{j2}) = f(c_{j3}) = 1$. Since c_{j1}, b_{j1}, c_{j2} is a bicoloured P_3 with colour 1 at the end vertices, its middle vertex must be coloured 0. That is, $f(b_{j1}) = 0$. Similarly, $f(b_{j2}) = 0$. This means b_{j1}, c_{j2}, b_{j2} is a P_3 with $1 = f(c_{j2}) > f(b_{j1}) = f(b_{j2}) = 0$; a contradiction. This proves claim CL1.

Next, we prove claim CL2 for $k = 1$. The proof is similar for other values of k. Let x_i, y_{ij}, c_{j1} be a path in G where $1 \leq i, j \leq n$. Recall that $f(x_i), f(c_{j1}) \in \{0,1\}$. If $f(x_i) = 0$, then $f(c_{j1}) = 1$ due to Property P3. So, it suffices to prove that $f(x_i) = 1$ implies $f(c_{j1}) = 0$. On the contrary, assume that $f(x_i) = f(c_{j1}) = 1$. Since x_i, y_{ij}, c_{j1} is a bicoloured P_3 with colour 1 at its end vertices, its middle vertex y_{ij} must be coloured 0 (by f). Thus, we have $f(y_{ij}) = 0$ and $f(c_{j1}) = 1$. Therefore, by applying Observation 2 on paths $y_{ij}, c_{j1}, b_{j1}, c_{j2}$ and $y_{ij}, c_{j1}, b_{j3}, c_{j3}$, we have $f(b_{j1}) = f(b_{j3}) = 2$ and $f(c_{j2}) = f(c_{j3}) = 0$ (see Fig. 2). The equation $f(c_{j2}) = f(c_{j3}) = 0$ contradicts Property P3. This completes the proof of claim CL2.

Now, we are ready to prove that G is 3-rs colourable if and only if $\mathcal{B}$ is a yes instance of CPP 1-IN-3 SAT. Suppose that G has a 3-rs colouring f. We define a truth assignment $\mathcal{A}$ for X by setting variable $x_i \leftarrow$ true if $f(x_i) = 1$,

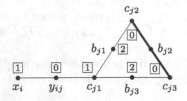

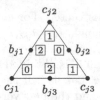

Fig. 2. $f(x_i) = f(c_{j1}) = 1$ leads to a contradiction

Fig. 3. Scheme used to assign colours to vertices b_{j1}, b_{j2}, b_{j3} (when $f(c_{j1}) = 0$)

and $x_i \leftarrow$ false if $f(x_i) = 0$. We claim that each clause C_j has exactly one true variable under $\mathcal{A}$ ($1 \leq j \leq n$). Let $C_j = \{x_p, x_q, x_r\}$. By claim CL1, exactly one vertex among c_{j1}, c_{j2}, c_{j3} is coloured 0 under f. Without loss of generality, assume that $f(c_{j1}) = 0$ and $f(c_{j2}) = f(c_{j3}) = 1$. As x_p, y_{pj}, c_{j1} is a path in G, $f(x_p) = 1 - f(c_{j1}) = 1$ by claim CL2. Similarly, $f(x_q) = 1 - f(c_{j2}) = 0$ and $f(x_r) = 1 - f(c_{j3}) = 0$ (consider path x_q, y_{qj}, c_{j2} and path x_r, y_{rj}, c_{j3}). Hence, by definition of $\mathcal{A}$, x_p is true whereas x_q and x_r are false. Therefore, $\mathcal{A}$ is a truth assignment such that every clause C_j has exactly one true variable ($1 \leq j \leq n$).

Conversely, suppose that X has a truth assignment $\mathcal{A}$ such that each clause has exactly one true variable. We produce a 3-rs colouring f of G as follows. First, colour vertices x_i by the rule $f(x_i) := 1$ if x_i is true; otherwise $f(x_i) := 0$. Next, vertices y_{ij} and c_{jk} are coloured. Assign colour 2 to all y_{ij}'s. Whenever x_i, y_{ij}, c_{jk} is a path in G, assign $f(c_{jk}) = 1 - f(x_i)$. This ensures that the path x_i, y_{ij}, c_{jk} is not bicoloured. Finally, for each $j = 1, 2, \ldots, n$, colour vertices b_{j1}, b_{j2}, b_{j3} using the scheme shown in Fig. 3 ('rotate' the colours when $f(c_{j2}) = 0$ or $f(c_{j3}) = 0$; by claim CL1, exactly one vertex among c_{j1}, c_{j2}, c_{j3} is coloured 0).

We claim that f is a 3-rs colouring of G. Since no two adjacent vertices in G have the same colour, f is indeed a 3-colouring of G. The bicoloured P_3's in G are either entirely inside a clause gadget, or of the form (i) $y_{ij}, x_i, y_{ij'}$, or (ii) $y_{ij}, c_{jk}, b_{jk'}$. Every bicoloured P_3 within clause gadgets has colour 0 at its middle vertex (see Fig. 3). Every bicoloured P_3 of the form $y_{ij}, x_i, y_{ij'}$ or $y_{ij}, c_{jk}, b_{jk'}$ has colour 2 at its end vertices because all y_{ij}'s are coloured 2. So, no bicoloured P_3 in G has the higher colour at its middle vertex, and thus f is a 3-rs colouring of G. Therefore, G is 3-rs colourable.

So, $\mathcal{B}$ is a yes instance of CPP 1-IN-3 SAT if and only if G is 3-rs colourable. This completes the proof for NP-hardness of 3-RS COLOURABILITY problem when restricted to the class of subcubic planar bipartite graphs of girth at least six. $\qquad\square$

Theorem 1 can be generalized using a simple operation. For a graph G with $\Delta(G) = k$, the graph G^+ is defined as the graph obtained from G by adding enough pendant vertices at every vertex v of G so as to ensure that $\deg_{G+}(v) = k + 1$. Hence, each vertex in G^+ has degree 1 or $k + 1$. As we are only adding pendant vertices, G^+ preserves the planarity, bipartiteness and girth of G. Moreover, we have $\Delta(G^+) = k + 1$. Further, this operation is useful in the construction of graphs of desired rs chromatic number.

Observation 3. *Let G be a graph with $\Delta(G) = k$ where $k \in \mathbb{N}$. Then, G is k-rs colourable if and only if G^+ is $(k+1)$-rs colourable.*

Proof. If G is k-rs colourable, we can colour the new pendant vertices added to G with a new colour k so that G^+ is $(k+1)$-rs colourable. Conversely, suppose that G^+ admits a $(k+1)$-rs colouring f. Recall that for every vertex u of G^+, $\deg_{G^+}(u) = 1$ or $k+1$. We claim that a vertex of degree $k+1$ cannot receive colour k under a $(k+1)$-rs colouring. This is a generalization of Property P1, and can be proved similarly. Hence, no non-pendant vertex in G^+ is coloured k by f. Observe that the set of non-pendant vertices in G^+ is precisely $V(G)$. Since only colours 0 to $k-1$ are used on non-pendant vertices of G^+ (under f), the restriction of f to $V(G)$ is a k-rs colouring of G. □

Since G^+ preserves planarity, bipartiteness and girth of G, Observation 3 helps us to generalize Theorem 1 as follows.

Theorem 2. *For $k \geq 3$, k-RS COLOURABILITY is NP-complete for the class of planar bipartite graphs of maximum degree k and girth at least six.*

4 Trees

The only known result on rs colouring of trees is that $\chi_{rs}(T) = O(\log n / \log \log n)$ and the bound is tight [9]. We design a linear-time algorithm to test 3-rs colourability of trees. We sketch the outline of the algorithm, and leave the details for the longer version of the paper.

A simple observation regarding 3-rs colouring of a tree T is that some subgraphs of T force colours on vertices of T. For instance, if u, v, w is a path in T where u, v and w are 3-plus vertices, then every 3-rs colouring of T must assign $f(v) = 0$ and $f(u) = f(w) = 1$ (by Properties P1 and P2). Colouring forced on v and w this way in turn force colours on neighbours of w and their neighbours (see Observation 2). Therefore, an algorithm to test 3-rs colourability must identify it when colours assigned to some vertices in a tree T force colours on other vertices of T.

To make things simpler, we view the input tree T (not a path) as a rooted tree with a 3-plus vertex as its root. The following definitions are helpful in presenting the key ideas at play. Let T be a partially coloured tree. If u is a vertex of T with $\deg_T(u) \neq 2$, the *rooted subtree of T at u*, denoted by T_u, is the subgraph of T induced by u and its descendants in T with parent-child relations and colours inherited from T. If we 'split' T_u at u, each resulting piece is called a *branch of T at u* (i.e., each branch of T at u is $T_u[V_i \cup \{u\}]$ with inherited parent-child relations and colours, where V_i is the vertex set for some component of $T_u - u$).

Note that if T has a branch at a vertex v, then v must be a 3-plus vertex. Consider a branch B of T (at a vertex v) comprised of a rooted subtree T_u and a path $u_1, u_2, \ldots, u_d, v$ where $u_1 = u$ and $d \in \mathbb{N}$. We claim that if no vertex in $V(T) \backslash V(T_u)$ is coloured, then the branch B can affect 3-rs colouring of the rest of the tree only in a limited number of ways as follows.

(I) B doesn't admit 3-rs colouring extension. (For all other cases, B admits 3-rs colouring extension).

(II) For every 3-rs colouring extension ϕ of B, $\phi(v) = 0$ (and no further restrictions).

(III) For every 3-rs colouring extension ϕ of B, $\phi(u_d) = 0$ and $\phi(v) = 1$.

(IV) For every 3-rs colouring extension ϕ of B, $\phi(v) = 1$ (and no further restriction).

(V)   For every 3-rs colouring extension ϕ of B, either (i) $\phi(v) = 0$, or
      (ii) $\phi(u_d) = 0$, $\phi(v) = 1$.

(VI) For every 3-rs colouring extension ϕ of B, $\phi(v) = 0$ or 1 (by Property P1, and no further restriction).

This gives a partition of branches. We call the corresponding equivalence classes as Class (I), Class (II), ... , Class (VI). Similarly, we partition rooted subtrees T_v as follows (based on how they affect 3-rs colouring on the rest of the tree).

(A) T_v doesn't admit 3-rs colouring extension. (For all other cases, T_v admits 3-rs colouring extension).

(B)   For every 3-rs colouring extension ϕ of T_v, $\phi(v) = 0$ (and no further restrictions).

(C)   For every 3-rs colouring extension ϕ of T_v, $\phi(v) = 1$ and $\phi(w) = 0$ for a child w of v.

(D)   For every 3-rs colouring extension ϕ of T_v, $\phi(v) = 1$ (and no further restrictions).

(E)   For every 3-rs colouring extension ϕ of T_v, either (i) $\phi(v) = 0$, or
      (ii) $\phi(v) = 1$, $\phi(w) = 0$ for a child w of v.

(F)   For every 3-rs colouring extension ϕ of T_v, $\phi(v) = 0$ or 1 (and no further restriction).

(G) T_v is a single vertex (that is, v is a leaf in T).

The above partition induces an equivalence relation, and we call the equivalence classes under this relation as Class (A), Class (B), ... , Class (G).

For a branch B of T at a vertex v comprised of a rooted subtree T_u and a path from u to v, the equivalence class of B can be determined from Table 1 based on the equivalence class of T_u and the distance $d = \text{dist}_T(u, v)$ (proof is omitted). The equivalence class of a rooted subtree T_v can be determined from Fig. 4. (Let p, q, r, s, t denote respectively the number of Class (II) branches, Class (III) branches, ... , Class (VI) branches of T at v where $p, q, r, s, t \geq 0$). (Proof is omitted).

The algorithm visits vertices of T in bottom-up order (by post-order traversal) and determines the equivalence classes of all branches and rooted subtrees of T until it comes across a branch/rooted subtree that doesn't admit 3-rs colouring extension (i.e., a Class (I) branch or a Class (A) rooted subtree), or all vertices of T are visited. In the former case, T is not 3-rs colourable. In the latter case, T itself is in one of the equivalence classes Class (B), Class (C), ..., Class(G), and therefore, T is 3-rs colourable.

Theorem 3. 3-RS COLOURABILITY *is in P for the class of trees.*

Table 1. Equivalence class of branch B in terms of the equivalence class of T_u (column) and value of d (row)

	B	C	D	E	F	G
1	III	I	II	III	V	VI
2	IV	II	V	VI	VI	"
3	II	III	IV	V	"	"
4	V	IV	VI	VI	"	"
5	IV	II	V	"	"	"
6	VI	V	VI	"	"	"
7	V	IV	"	"	"	"
8	VI	VI	"	"	"	"
9	"	V	"	"	"	"
≥ 10	"	VI	"	"	"	"

$p \geq 0,\ q + r \geq 0 \implies T_v \in \text{Class}(A)$
$p \geq 0,\ q = r = 0 \implies T_v \in \text{Class}(B)$
If $p = 0$ and $q + r \geq 0$,
- $q + s = 0 \implies T_v \in \text{Class}(D)$
- $q + s = 1 \implies T_v \in \text{Class}(C)$
- $q + s \geq 2 \implies T_v \in \text{Class}(A)$

If $p = q = r = 0$,
- $s = 0, t \geq 0 \implies T_v \in \text{Class}(F)$
- $s = 1 \implies T_v \in \text{Class}(E)$
- $s \geq 2 \implies T_v \in \text{Class}(B)$
- $s = t = 0 \implies T_v \in \text{Class}(G)$

Fig. 4. Rules for finding the equivalence class of a rooted subtree

5   Chordal Graphs

In this section, we show that 3-RS COLOURABILITY is polynomial-time decidable for the class of chordal graphs. This is possible because triangles in a graph G limit the number of 3-rs colourings of G. The following observation is a direct consequence of Property P1.

Observation 4. *Let G be a graph and (a, b, c) be a triangle in G. If a, b, c are all 3-plus vertices in G, then G is not 3-rs colourable.*

The following theorem shows that we can get rid of triangles in a given graph without affecting its 3-rs colourability status.

Theorem 4. *Let G be a graph and let (a, b, c) be a triangle in G. Let c be a vertex of degree two in G. Let G' be the graph obtained from $G - c$ by attaching two pendant vertices each at a and b (see Fig. 5). Then, G is 3-rs colourable if and only if G' is 3-rs colourable.*

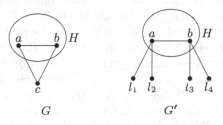

Fig. 5. Graphs G and G'

Proof. Suppose that G admits a 3-rs colouring. Then, G admits a 3-rs colouring f such that $f(c) = 2$ (for instance, if a 3-rs colouring of G assign colour 2 at b, then $\deg_G(b) = 2$ and hence swapping colours at b and c gives a 3-rs colouring of G that assign colour 2 at c). Now, the restriction of f to $V(G - c)$ can be extended into a 3-rs colouring of G' by assigning colour 2 on the newly added pendant vertices. Conversely, suppose that G' admits a 3-rs colouring f'. Since a and b are 3-plus vertices in G', $f'(a), f'(b) \in \{0,1\}$ by Property P1. Then, f' restricted to $V(G - c)$ can be extended into a 3-rs colouring of G by assigning colour 2 at c.   □

We can test whether a given connected chordal graph G is 3-rs colourable as follows. List all triangles in G in $O(n^3)$ time. If G contains a triangle (a, b, c) with all a, b, c being 3-plus vertices, then G is not 3-rs colourable by Observation 4. If the graph passes this test, we can get rid of all triangles in it by repeated application of Theorem 4 (see Fig. 6). The resultant graph will be a tree T on at most $n + 3\binom{n}{3} = O(n^3)$ vertices. Thus, we can test 3-rs colourability of G in $O(n^3)$ time by testing 3-rs colourability of T.

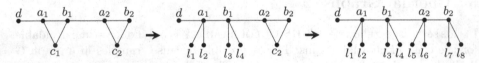

Fig. 6. Construction of a tree T from a chordal graph G such that $\chi_{rs}(T) \leq 3$ if and only if $\chi_{rs}(G) \leq 3$

Theorem 5. 3-RS Colourability *is in P for the class of chordal graphs.*

6   Cobipartite Graphs

Star colouring and ordered colouring are colouring variants closely related to restricted star colouring. A k-ordered colouring of a graph G is a function $f : V(G) \to \{0, 1, \ldots, k - 1\}$ such that (i) $f(x) \neq f(y)$ for every edge xy of G, and (ii) every non-trivial path with the same colour at its end vertices contains a vertex of higher colour [10]. Let us denote by $\chi_s(G)$ (resp. $\chi_o(G)$), the least integer k such that G admits a k-star (resp. k-ordered) colouring. Analogous to the problem k-RS Colourability (resp. RS Colourability), we can define problems k-Star Colourability and k-Ordered Colourability (resp. Star Colourability and Ordered Colourability).

For every graph G, $\chi_s(G) \leq \chi_{rs}(G) \leq \chi_o(G)$ [9]. It is known that the three parameters can be arbitrarily apart [9]. On the other hand, all three parameters are equal for cographs [11,14]. We prove that the same is true for cobipartite graphs.

Theorem 6. *For a cobipartite graph G, $\chi_s(G) = \chi_{rs}(G) = \chi_o(G)$.*

Proof. Let G be a cobipartite graph whose vertex set is partitioned into two cliques A and B. Let $k = \chi_s(G)$, and let f be a k-star colouring of G. To prove the theorem, it suffices to show that G admits a k-ordered colouring. Clearly, each colour class under f contains at most one vertex from A and at most one vertex from B. Let $\{U_0, U_1, \ldots, U_{k-1}\}$ be the set of colour classes under f. Let t be the number of colour classes with cardinality two. We assume that $t \neq 0$ (If $t = 0$, then $k = n$ and f itself is a k-ordered colouring of G). W.l.o.g., we assume that $|U_i| = 2$ for $0 \leq i \leq t-1$, and $|U_i| = 1$ for $t \leq i \leq k-1$. Observe that if $a_i \in U_i \cap A$, $b_i \in U_i \cap B$, $a_j \in U_j \cap A$ and $b_j \in U_j \cap B$, then $a_i b_j \notin E(G)$ and $a_j b_i \notin E(G)$ (if not, the path a_j, a_i, b_j, b_i or the path a_i, a_j, b_i, b_j is a P_4 in G bicoloured by f). Therefore, there are no edges between $A \cap \bigcup_{i=0}^{t-1} U_i$ and $B \cap \bigcup_{i=0}^{t-1} U_i$ in G. We claim that $h : V(G) \to \{0, 1, \ldots, k-1\}$ defined as $h(v) = i$ $\forall v \in U_i$ is a k-ordered colouring of G. Suppose that u, v are distinct vertices in G with $h(u) = h(v) = j$ (where $0 \leq j \leq k-1$). Then, $j \leq t-1$. W.l.o.g., we assume that $u \in A$ and $v \in B$. Then, $u \in A \cap \bigcup_{i=0}^{t-1} U_i$ and $v \in B \cap \bigcup_{i=0}^{t-1} U_i$. Hence, every u, v-path Q in G must contain a vertex w_Q from $U_t \cup \cdots \cup U_{k-1}$. Since $f(w_Q) > f(u)$ and u, v, Q are arbitrary, h is indeed a k-ordered colouring of G. □

Since k-ORDERED COLOURABILITY is in P for every k, and ORDERED COLOURABILITY in NP-complete for cobipartite graphs [3], we have the following corollary.

Corollary 1. *For the class of cobipartite graphs, problems k-RS COLOURA-BILITY and k-STAR COLOURABILITY are in P for all k, whereas problems RS COLOURABILITY and STAR COLOURABILITY are NP-complete.*

7 Conclusion

Deciding whether a planar bipartite graph admits a 3-star colouring is NP-complete [1]. We prove that deciding whether a subcubic planar bipartite graph of girth at least six admits a 3-restricted star colouring is NP-complete. We also present an $O(n^3)$ algorithm to test whether a chordal graph is 3-restricted star colourable. The complexity of RS COLOURABILITY in classes of trees and chordal graphs remains open.

References

1. Albertson, M.O., Chappell, G.G., Kierstead, H.A., Kündgen, A., Ramamurthi, R.: Coloring with no 2-colored P_4's. Electron. J. Comb. **11**(1), 26 (2004)
2. Almeter, J., Demircan, S., Kallmeyer, A., Milans, K.G., Winslow, R.: Graph 2-rankings. Graphs Comb. **35**(1), 91–102 (2019). https://doi.org/10.1007/s00373-018-1979-4
3. Bodlaender, H.L., et al.: Rankings of graphs. SIAM J. Discrete Math. **11**(1), 168–181 (1998). https://doi.org/10.1137/S0895480195282550

4. Bozdağ, D., Çatalyürek, Ü.V., Gebremedhin, A.H., Manne, F., Boman, E.G., Özgüner, F.: Distributed-memory parallel algorithms for distance-2 coloring and related problems in derivative computation. SIAM J. Sci. Comput. **32**(4), 2418–2446 (2010). https://doi.org/10.1137/080732158
5. Curtis, A.R., Powell, M.J., Reid, J.K.: On the estimation of sparse Jacobian matrices. J. Inst. Math. Appl. **13**(1), 117–120 (1974). https://doi.org/10.1093/imamat/13.1.117
6. Dereniowski, D.: Rank coloring of graphs. In: Kubale, M. (ed.) Graph Colorings, pp. 79–93, Chap. 6. American Mathematical Society (2004). https://doi.org/10.1090/conm/352/06
7. Gebremedhin, A., Nguyen, D., Patwary, M.M.A., Pothen, A.: ColPack: software for graph coloring and related problems in scientific computing. ACM Trans. Math. Softw. (TOMS) **40** (2013). https://doi.org/10.1145/2513109.2513110
8. Gebremedhin, A.H., Manne, F., Pothen, A.: What color is your Jacobian? Graph coloring for computing derivatives. SIAM Rev. **47**(4), 629–705 (2005). https://doi.org/10.1137/S0036144504444711
9. Karpas, I., Neiman, O., Smorodinsky, S.: On vertex rankings of graphs and its relatives. Discrete Math. **338**(8), 1460–1467 (2015). https://doi.org/10.1016/j.disc.2015.03.008
10. Katchalski, M., McCuaig, W., Seager, S.: Ordered colourings. Discrete Math. **142**(1–3), 141–154 (1995). https://doi.org/10.1016/0012-365X(93)E0216-Q
11. Lyons, A.: Acyclic and star colorings of cographs. Discrete Appl. Math. **159**(16), 1842–1850 (2011). https://doi.org/10.1016/j.dam.2011.04.011
12. Moore, C., Robson, J.M.: Hard tiling problems with simple tiles. Discrete Comput. Geom. **26**(4), 573–590 (2001). https://doi.org/10.1007/s00454-001-0047-6
13. Powell, M., Toint, P.L.: On the estimation of sparse Hessian matrices. SIAM J. Numer. Anal. **16**(6), 1060–1074 (1979). https://doi.org/10.1137/0716078
14. Scheffler, P.: Node ranking and searching on graphs. In: 3rd Twente Workshop on Graphs and Combinatorial Optimization, Memorandum No. 1132 (1993)
15. Shalu, M.A., Sandhya, T.P.: Star coloring of graphs with girth at least five. Graphs Comb. **32**(5), 2121–2134 (2016). https://doi.org/10.1007/s00373-016-1702-2
16. West, D.B.: Introduction to Graph Theory, 2nd edn. Prentice Hall, Upper Saddle River (2001)

Partitioning Cographs into Two Forests and One Independent Set

Pavol Hell[1] , César Hernández-Cruz[2] , and Anurag Sanyal[1]([✉])

[1] School of Computing Science, Simon Fraser University, University Dr 8888,
Burnaby, BC V5A 1S6, Canada
{pavol,anurag_sanyal}@sfu.ca
[2] Facultad de Ciencias, Universidad Nacional Autónoma de México,
Av. Universidad 3000, Circuito Exterior S/N,
Ciudad Universitaria, 04510 CDMX, Mexico
chc@ciencias.unam.mx

Abstract. We consider a variation of arboricity, where a graph is partitioned into p forests and q independent sets. These problems are NP-complete in general, but polynomial-time solvable in the class of cographs; in fact, for each p and q there are only finitely many minimal non-partitionable cographs. In previous investigations it was revealed that when $p = 0$ or $p = 1$, these minimal non-partitionable cographs can be uniformly described as one family of obstructions valid for all values of q. We investigate the next case, when $p = 2$; we provide the complete family of minimal obstructions for $p = 2, q = 1$, and find that they include more than just the natural extensions of the previously described obstructions for $p = 2, q = 0$. Thus a uniform description for all q seems unlikely already in the case $p = 2$.

Our result gives a concrete forbidden induced subgraph characterization of cographs that can be partitioned into two forests and one independent set. Since our proof is algorithmic, we can apply our characterization to complement the recognition algorithm for partitionable cographs by an algorithm to certify non-partitionable cographs by finding a forbidden induced subgraph.

Keywords: Vertex arboricity · Independent vertex feedback set · Cograph · Forbidden subgraph characterization · Colouring · Partition

1 Introduction and Motivation

The *vertex-arboricity* of a graph G is the minimum integer p such that the vertices of G can be partitioned into p parts each of which induces a *forest*. It is, in general, NP-complete to decide if a graph G has arboricity less than or equal to a fixed p, $p \geq 2$ [9]. This is a situation analogous to deciding if a graph G has chromatic number less than or equal a fixed q, $q \geq 3$ [6]. Both problems can be

This research was supported by the first author's NSERC Discovery Grant and the second author's SEP-CONACYT grant A1-S-8397.

M. Changat and S. Das (Eds.): CALDAM 2020, LNCS 12016, pp. 15–27, 2020.
https://doi.org/10.1007/978-3-030-39219-2_2

efficiently solved on the class of cographs, and in [8], the authors have studied, for cographs, a blended problem, whereby a graph is partitioned into p parts inducing forests and q parts that are independent sets. Each of these problems can be efficiently solved in the class of cographs, and in fact characterized by a finite number of minimal cograph obstructions. This parallels the situation for a similar blended problem studied earlier, where a cograph G is to be partitioned into k independent sets and ℓ cliques [2,10].

Cographs are one of the most popular and intensively studied classes of perfect graphs. We say that G is a *cograph* if it has no induced subgraph isomorphic to P_4, the path on 4 vertices. Equivalently [1], cographs can be recursively defined as follows: (i) The graph on single vertex is a cograph; (ii) If $G_1, G_2, ..., G_k$ are cographs then so is their union, $G_1 \cup G_2 \cup ... \cup G_k$; and (iii) If G is a cograph, then so is its complement $\overline{G}$. Since cographs are perfect, many intractable problems can be solved in polynomial time on the class of cographs [7]. Moreover, the recursive description of cographs corresponds to a natural data structure (called a *co-tree* [1]), and partition problems like the chromatic number or arboricity can be solved in linear time directly on the co-tree. This is explicitly done for the chromatic number in [1], and can be done in a very similar fashion for vertex-arboricity. In fact, in [8], the authors similarly solve, for cographs, the blended problem of partition into p forests and q independent sets (and for even more general partitions). Furthermore, it follows from [3] that each of these problems has a characterization by a finite set of minimal cograph obstructions. Here a *minimal cograph obstruction* is a cograph G that does not admit a required partition, but each proper induced subgraph of G does admit such a partition. Thus a cograph admits a required partition if and only if it does not contain an induced subgraph isomorphic to a minimal cograph obstruction.

Minimal cograph obstructions for partition into k independent sets and ℓ cliques were described in [2,4,5,10]; they have $(k+1)(\ell+1)$ vertices, and admit a partition into $k+1$ independent sets of size $\ell+1$ as well as a partition into $\ell+1$ cliques of size $k + 1$. In particular, the unique minimal cograph obstruction for partition into k independent sets is K_{k+1}, and the minimal cograph obstruction for partition into ℓ cliques is $\overline{K_{\ell+1}}$ (as is required for perfect graphs).

Minimal cograph obstructions for partition into p forests and q independent sets were investigated in [8]. Consider first the special case of $q = 0$, that is partitions into forests (arboricity). Since cographs are perfect, there are two minimal cograph obstructions for being a forest, i.e., admitting a partition with $p = 1$: these are the cycles C_3 and C_4. For partitions into $p = 2$ forests, there turn out to be exactly 7 minimal cograph obstructions, forming the family $\mathcal{A}_2$ depicted in Fig. 1.

Each of these obstructions has a natural generalization to minimal cograph obstruction for partition into p forests. For example, K_5 generalizes to K_{2p+1}, $\overline{3K_3}$ generalizes to $\overline{(p + 1)K_{p+1}}$, and so on. These 7 generalizations form a family $\mathcal{A}_p$, given by an explicit uniform description in [8]. They are all minimal cograph obstructions to partition into p forests. Nevertheless, it turns out that there are in general many additional minimal cograph obstructions, and in fact the number

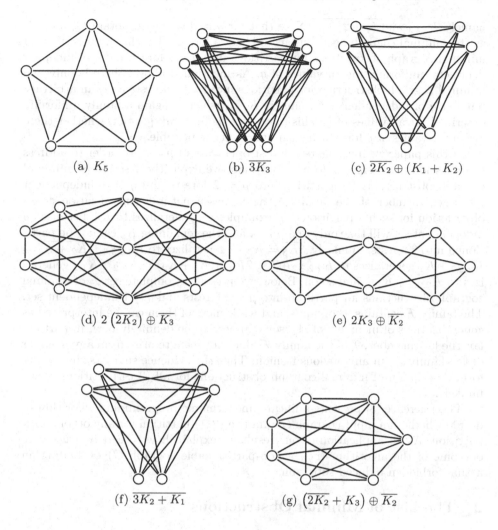

(a) K_5 (b) $\overline{3K_3}$ (c) $\overline{2K_2} \oplus (K_1 + K_2)$

(d) $2\left(\overline{2K_2}\right) \oplus \overline{K_3}$ (e) $2K_3 \oplus \overline{K_2}$

(f) $\overline{3K_2 + K_1}$ (g) $\left(\overline{2K_2} + K_3\right) \oplus \overline{K_2}$

Fig. 1. The family $\mathcal{A}_2$.

of minimal cograph obstructions for partition to p forests grows exponentially with p [8].

There is, however, a class of partition problems in which minimal cograph obstructions can be uniformly described. This is the class of problems generalizing the problem of independent vertex feedback set [8]. A *q-colourable vertex feedback set* of a graph G is a set V of vertices such that $G \backslash V$ admits a q-colouring. Thus a graph admits a q-colourable vertex-feedback set if and only if it has a partition into $p = 1$ forest and q independent sets. It is shown in [8] that there are precisely two minimal cograph obstructions for such a partition, namely K_{q+3} and $\overline{(q+2)K_2}$. (Note that for $q = 0$ we again obtain C_3 and C_4 as the minimal cograph obstructions to being a forest.) This family describes all minimal cograph obstructions for partitions into $p = 1$ forest and q of independent sets, uniformly for all values of q. As mentioned above, there is only one minimal cograph obstruction for partitions into ($p = 0$ forests and) an arbitrary number q of independent sets, namely K_{q+1}, which is again a family uniformly described for all values of q. This motivates the natural question whether there are other values of p for which such uniformity is possible.

In this paper we investigate the first open case of $p = 2$. In order to address the question of possible unform description, we explicitly describe all minimal cograph obstructions for partition into $p = 2$ forests and $q = 1$ independent set. Each member of the family $\mathcal{A}_2$ again has a natural generalization as an obstruction for such a partition. For example, K_5 generalizes to K_6, because an independent set will take only one vertex and the remaining K_5 cannot be partitioned into 2 forests. Similarly, $\overline{3K_3} = K_{3,3,3}$ generalizes to $K_{3,3,3,3}$, $\overline{3K_2 + K_1} = K_{2,2,2} \oplus K_1$ generalizes to $K_{2,2,2,2} \oplus K_1$, $\overline{2K_2} \oplus (K_1 + K_2) = K_{2,2} \oplus \overline{K_{1,2}}$ generalizes to $K_{2,2,2} \oplus \overline{K_{1,2}}$, and so on. Below, we present a complete set $\mathcal{F}$ of minimal cograph obstructions for partition into $p = 2$ forests and $q = 1$ independent set. The family $\mathcal{F}$ contains 9 cographs, and while most of them can be interpreted as generalizations of members of $\mathcal{A}_2$, some appear to be definitely new. In particular, the last member, 9, of the family $\mathcal{F}$, does not seem to arise from any member of the family $\mathcal{A}_2$ in any obvious fashion. Thus the evidence suggests that a uniform description of minimal cograph obstructions for all $(2, q)$-partitions seems unlikely.

The reference [8] presents a linear-time dynamic programming algorithm to decide whether an input cograph G admits a $(2, 1)$-partition (or any other (p, q)-partition). As an application of our result we explain how to certify a negative outcome of the algorithm (i.e., a non-partitionable cograph G) by finding an actual forbidden induced subgraph.

2   The List of Minimal Obstructions

For brevity, we call a partition of a graph G into p forests and q independent sets, a (p, q)-*partition* of G. Thus, in the remainder of the paper, we describe all minimal cograph obstructions for $(2, 1)$-partition.

We introduce the family of cographs $\mathcal{F}$. The members of the family are:

1. K_6
2. $K_{3,3,3}$
3. $K_{2,2,2} \oplus \overline{K_{1,2}}$
4. $K_{2,2,2,2} \oplus K_1$
5. $2K_3 \oplus K_{2,2}$
6. $(K_{2,2} + K_3) \oplus K_{2,2}$
7. $2K_{2,2} \oplus K_{3,3}$
8. $2K_{2,2} \oplus 2K_{2,2}$
9. $(K_4 + K_{3,3,3}) \oplus \overline{K_2}$

Lemma 1. *Each graph in $\mathcal{F}$ is a minimal cograph obstruction to $(2,1)$-partition.*

Proof. It is clear from their descriptions that each graph in the $\mathcal{F}$ family is a cograph. We claim that each of these graphs is a minimal obstruction for $(2,1)$-partition.

Consider first K_6: it does not have a $(2,1)$-partition, because any forest in K_6 can have at most two vertices, and hence two forests can have at most four vertices. This leaves at least two vertices, but no two vertices in K_6 form an independent set. Moreover, when a vertex is removed we have K_5, which has an obvious $(2,1)$-partition where each forest is one edge and the independent set is a single vertex. Therefore K_6 is a minimal obstruction.

For the graph $G = K_{3,3,3,3}$, we observe that any induced forest in G has at most four vertices, and this happens only when the forest is a tree. Thus two forests can cover at most eight vertices, and since $K_{3,3,3,3}$ has no independent set of size four, it does not have a $(2,1)$-partition. When a vertex is removed, we obtain $K_{2,3,3,3}$, where we can take one independent set consisting of a part with three vertices, and cover the vertices of the remaining two parts of size three by stars centered at the remaining two vertices. Hence $K_{3,3,3,3}$ is also a minimal obstruction.

The proof for most of the remaining obstructions follows a similar approach, and we skip the details (which are included in the last section). We do include the proof for the last two obstructions on our list, which are more interesting.

Consider the graph $2K_{2,2} \oplus 2K_{2,2}$ from 8. Any independent set must be on one side of the join, and include at most four vertices. The remaining vertices contain an induced $2\,\overline{(2K_2)} \oplus \overline{K_3}$, which is one of the obstructions for $(2,0)$-partition from Fig. 1. When a vertex is deleted, we obtain the graph $((K_{1,2} + K_{2,2}) \oplus 2K_{2,2})$, which has the following $(2,1)$-partition: one independent set of four vertices on the bigger side of the join, one forest consisting of $2K_{1,2}$ on the smaller side of the join, and one forest which is a star on five vertices. Thus $2K_{2,2} \oplus K_{2,2}$ is indeed a minimal cograph obstruction for $(2,1)$-partition.

Finally, we prove that the graph $(K_4 + K_{3,3,3}) \oplus \overline{K_2}$ is a minimal cograph obstruction for $(2,1)$-partition. We consider what an independent set S must contain in order for none of the minimal cograph obstructions for $(2,0)$-partition (from Fig. 1) to remain after S is removed. Note that our graph contains $K_{2,3,3,3}$, while in Fig. 1 there is both a $K_{3,3,3}$ and a $K_{1,2,2,2}$. Moreover, when S is removed there must not remain a copy of K_5. To satisfy just these restrictions, S must contain one vertex of the K_4, and three vertices of one entire part of the $K_{3,3,3}$. Since this is a maximal independent set, S must be this set; but then its removal results in a graph containing an induced $\overline{(2K_2 + K_3)} \oplus \overline{K_2}$ (the last graph in Fig. 1). It remains to partition the graphs resulting from deleting a vertex from $(K_4 + K_{3,3,3}) \oplus \overline{K_2}$. If a vertex in the K_4 is deleted, then we obtain a $(2,1)$-partition by taking the independent set S as above, and two stars centered at the two vertices of the $\overline{K_2}$, each involving one 3-vertex part of the $K_{3,3,3}$ and one vertex of the K_4. If a vertex of the $\overline{K_2}$ is deleted, we can take again the independent set S, one forest consisting of an edge from the K_4 and one part of the $K_{3,3,3}$, and one star centered at the other vertex of the $\overline{K_2}$. If a vertex v in the $K_{3,3,3}$ is deleted, we can take for the independent set the vertices in the $\overline{K_2}$, and partition the remaining vertices into two forests each consisting of one edge of the K_4 and one star on four vertices.

3   The Completeness of the List

We now prove that the list of minimal cograph obstructions for $(2,1)$-partition given in Lemma 1 is complete.

Theorem 1. *A cograph has a $(2,1)$-partition if and only if it is does not contain an induced subgraph from $\mathcal{F}$.*

Proof. Let G be a cograph. It is easy to see that a disconnected cograph G admits a $(2,1)$-partition if and only if each connected component of G admits a $(2,1)$-partition. Thus we may assume G is a connected cograph which does not contain an induced subgraph from $\mathcal{F}$, and proceed to prove it has a $(2,1)$-partition.

For brevity, we shall say that a graph is F-free if it does not contain F as an induced subgraph, and $\mathcal{F}$-free, if it doesn't contain any member of the family $\mathcal{F}$ as an induced subgraph.

Since G is connected, there exist cographs G_1 and G_2 such that $G = G_1 \oplus G_2$. If G_1 and G_2 are forests, then G trivially has a $(2,1)$-partition. So, at least one of G_1, G_2 must contain an induced cycle. Without loss of generality, assume that at least G_1 has an induced cycle; since G_1 is a cograph, the only cycles possible are C_3 or C_4.

1.   **Assume G_1 is C_3-free.** In this case G_1 has an induced C_4; moreover, G_1 is a bipartite graph. We will take a concrete bipartition and refer to (X, Y) as the parts. If G_2 is a forest, then we have a trivial $(2,1)$-partition with two independent sets and a forest. Thus we may assume that G_2 also has a cycle. We have the following two subcases:

(a) **Both G_1 and G_2 are C_3 free.** This implies that both cographs G_1 and G_2 are bipartite, and each has an induced C_4. Both G_1 and G_2 cannot have more than one connected component with C_4 because G is $2K_{2,2} \oplus 2K_{2,2}$-free. Hence without loss of generality we may assume that G_2 has exactly one component, say A, with a C_4, and the other components are trees. Note that A must be a complete bipartite graph since G_2 has no induced P_4. The graph G_1 must also contain at least one connected component, say B, which is a complete bipartite graph. If G_1 has other components with an induced C_4, then one of the parts of A in G_2 has exactly two vertices, because G is $2K_{2,2} \oplus K_{3,3}$-free. If the other connected components of G_1 are trees, then one of the subgraphs A or B has a bipartition with one of the parts having exactly two vertices, since G is $K_{3,3,3,3}$-free. In either case, we can obtain a $(2,1)$-partition of G as follows. Suppose the connected component A of the graph G_2 has a bipartition (X, Y), where X has exactly two vertices. The first forest is obtained by taking one vertex from X, the entire other part Y, and the remaining tree components of G_2. Since graph G_1 is also bipartite, another forest can be obtained by taking one of the parts of

G_1 and the remaining vertex in X. The remaining vertices form an independent set in G_1.

(b) G_1 **is C_3 free but G_2 contains a C_3.** Since G_1 contains a $K_{2,2}$ and since G is $2K_3 \oplus K_{2,2}$-free and $(K_{2,2} + K_3) \oplus K_{2,2}$-free, there is exactly one component of G_2 with a C_3, and the other components of G_2 are forests. Let the set (v_1, v_2, v_3) induce a C_3 in G_2, and let B be the component of G_2 containing it. Since B is a connected cograph, we have $B = B_1 \oplus B_2$ for cographs B_1, B_2. The component B cannot contain an induced K_4 and hence none of the graphs B_1 and B_2 have a C_3. So, we assume without loss of generality that $v_1, v_2 \in V(B_1)$, and $v_3 \in V(B_2)$; moreover, B_2 must be an independent set since G is K_6-free. If B_2 has at least two elements, then B_1 must be a K_2, since G is $K_{2,2,2} \oplus \overline{K_{1,2}}$-free and $K_{2,2,2,2} \oplus K_1$-free. Hence either B_1 is a K_2 or B_2 is a K_1. We construct a $(2,1)$-partition in both the cases.

When $B_1 = K_2$, then taking one of the parts of the bipartite graph G_1 along with one vertex in B_1 we obtain one first forest in our partition. To construct the second forest we include B_2 along with the remaining vertex in B_1 and the remaining tree components of G_2. The remainder in G_1 is the independent set in the partition.

When B_2 consists of a single vertex, then taking this vertex with one of the parts of the bipartite graph G_1 yields the first forest in the partition. The remaining parts of G_2 form a forest which becomes the second forest in the partition. The remaining part of G_1 is our independent set in the $(2,1)$-partition.

This concludes the first case.

2. **Assume that G_1 contains C_3.** Without loss of generality we can assume that G_2 is a forest as otherwise we have a K_6, or a situation symmetric to the case 1(b). We consider several possible cases, noting that in all the cases where G_2 has at least one edge, G_1 does not contain K_4, since G is K_6-free.

(a) **Suppose first that G_2 has at least three vertices and at least one edge.** Consider a copy of C_3 on v_1, v_2, v_3 in G_1, and the component B of G_1 containing it. Since B is a connected cograph, we have $B = B_1 \oplus B_2$ for cographs B_1, B_2. Since B does not contain a K_4, neither B_1 nor B_2 can contain a C_3. So, we assume without loss of generality that $v_1, v_2 \in V(B_1)$, and $v_3 \in V(B_2)$; moreover, B_2 is an independent set. If B_1 has an induced C_4, then B_2 will be just a single vertex because G is $K_{2,2,2} \oplus \overline{K_{1,2}}$-free and $K_{2,2,2,2} \oplus K_1$-free. (Note that G_2 contains a copy of $\overline{K_{1,2}}$ or $K_{1,2}$.) In conclusion, each component $B = B_1 \oplus B_2$ of G_1 which contains a C_3 either has a single vertex in B_2 and a bipartite B_1, or an independent set B_2 and a forest B_1. Each component of G_1 without a C_3 is bipartite.

We find a $(2,1)$-partition of the graph G as follows. One forest will be formed by the vertices in G_2. We partition G_1 into a forest and an independent set; it suffices to partition each component B of G_1 separately. A component $B = B_1 \oplus B_2$ with C_3 which has a single vertex v in B_2 yields a star centered at v and using one part of the bipartition of B_1, with the other part of the bipartition yielding an independent set. In a component $B = B_1 \oplus B_2$ with C_3 where B_2 is an independent set and B_1 is a forest, we trivially have a desired partition.

Finally, each remaining component B is bipartite and one part can be taken as a forest and the other part as an independent set.

(b) **Assume G_2 has exactly two vertices, which are adjacent.** Since G_1 does not contain K_4, it is is three colourable. One of the colour classes along with one vertex of G_2 forms one forest of the partition. Another colour class with the other vertex of G_2 yields another forest. The remainder is a single colour class which forms the independent set of the partition.

(c) **Assume G_2 has exactly two vertices, which are not adjacent.** If G_1 does not contain an induced K_4, we obtain a partition of G as in case 2(b); so we assume that G_1 has a K_4. Note that we may take G_2 for the independent set of a $(2, 1)$-partition, and it remains to find a partition of G_1 into two forests (a $(2, 0)$-partition). Clearly, it suffices to find such a partition for each component B of G_1 separately.

Note that while at least one component of G_1 has a K_4, there could be other components B of G_1 without a K_4. Such components B must have a $(2, 0)$-partition because otherwise G_1 contains a minimal cograph obstruction for $(2, 0)$-partition from the family $\mathcal{A}_2$, and adding the independent set G_2 would yield a member of $\mathcal{F}$. (This can be easily seen by comparing the two families.)

Now we consider components $B = B_1 \oplus B_2$ of G_1 which do contain a K_4.

Suppose first that both B_1, B_2 are bipartite. Note that both B_1 and B_2 cannot contain an induced a C_4 since G is $K_{2,2,2,2} \oplus K_1$-free. If both B_1 and B_2 are forests, then we have a trivial partition of B into two forests. Hence, say, B_1 has a C_4 and B_2 is a forest. In fact, B_2 must be just an edge, say uv, because G is $K_{2,2,2} \oplus K_1$-free and $K_{2,2,2} \oplus \overline{K_{1,2}}$-free. In this case a $(2, 0)$-partition of B is formed by taking one star centered at u with one part of the bipartition of B_1, and one star centered at v with the other part of B_1.

Thus we may assume that one of B_1, B_2, say B_1, contains a C_3. Since G is K_6-free, B is K_5-free, and so B_2 must an independent set. Now we further consider each component $D = D_1 \oplus D_2$ of B_1. At least one such component D' must contain a C_3, but there could also be bipartite components D; all must be K_4-free.

If B_2 has at least two vertices, then exactly one component, namely D', of B_1 has a cycle (specifically a C_3). Bipartite components D cannot have a cycle (i.e., a C_4), because G is $(K_3 + K_{2,2}) \oplus K_{2,2}$-free. Moreover no other component $D \neq D'$ can have a C_3, because G is $2K_3 \oplus K_{2,2}$-free. Hence if B_2 has at least two vertices then all the components D of B_1, other than D', are forests.

Suppose that v_1, v_2, v_3 form a C_3 in D. Since $D = D_1 \oplus D_2$ is K_4-free, neither of the graphs D_1, D_2 has a C_3. So we may assume $v_1, v_2 \in V(D_1)$ and $v_3 \in V(D_2)$; moreover we may assume D_1 is a bipartite graph and D_2 is an independent set.

If the bipartite graph D_1 contains a C_4, then both D_2 and B_2 must consist of a single vertex because G is $K_{2,2,2,2} \oplus K_1$-free.

If D_1 is a forest with more than the two vertices v_1, v_2, then it contains an induced $K_{1,2}$ or $\overline{K_{1,2}}$. Therefore, at least one of D_2, B_2 must be a single vertex, since G is $K_{2,2,2} \oplus K_1$-free and $K_{2,2,2} \oplus \overline{K_{1,2}}$-free.

Otherwise D_1 is just the edge $v_1 v_2$.

Finally, if there is no C_3 in D, i.e., D is bipartite, then D_1 is an independent set.

We now describe a $(2,0)$-partition of $B = B_1 \cup B_2$. Recall that B_2 is an independent set, and B_1 consists of components $D = D_1 \oplus D_2$ where each D_2 is an independent set and each D_1 is bipartite, with the following four possibilities: (i) D_1 contains a C_4, in which case D_2, as well as B_2, has a single vertex; (ii) D_1 is a forest of more than two vertices, in which case D_2 or B_2 has a single vertex; (iii) D_1 is an edge $v_1 v_2$; or (iv) D_1 is an independent set. Moreover, in cases (ii–iv), if B_2 has more than one vertex, then all but one component D of B_1 are forests.

We first describe a $(2,0)$-partition of $B = B_1 \cup B_2$ when B_2 has at least two vertices. In this case, there is one component $D' = D'_1 \oplus D'_2$ of B_1 with D'_1 a forest with one or more vertices (cases (ii), (iii)), and all other components D of B are forests themselves. We obtain a $(2,0)$-partition of G_1 as follows. If D'_1 is just an edge, say xy, the first forest consists of a star centred at the vertex x covering the independent set D'_2, along with the rest of the forest components of B'_1. The second forest is a star centred at the remaining vertex v covering the independent set B_2. If D'_1 has at least two vertices, then D'_2 is a single vertex u, and we can take D'_1 together with all other components D as one forest; the other forest will be the star centered at u and covering B_2.

Now consider a component $B = B_1 \cup B_2$ of G_1 when B_2 has a single vertex, say v. We put together one forest for a $(2,0)$-partition of B from the following forests in the various components $D = D_1 \oplus D_2$ of B. From components D of type (i) we take the star centered at the single vertex of D_2 and covering one part of the bipartition of D_1; from components D of type (ii-iv) we take the forests D_1. The other forest for a $(2,0)$-partition of B will be formed by a star centered at v and covering all the remaining vertices. (These are the other parts of all D_1 for components of type (i), as well as all D_2 for components of type (ii-iv); note that this is an independent set of vertices.)

(d) **Finally, we assume that G_2 is just a single vertex, say v.** The proof here is similar to the case 2(c), except that in the case (i), when D_1 contains an induced C_4, we can only claim that B_2 or D_2 is a single vertex, and in the case (ii), when D_1 is a forest with more than two vertices, we cannot claim anything about the size of B_2 or D_2 (Fig. 2).

Nevertheless, there is a $(2,1)$-partition of the entire G. (Since G_2 is a single vertex v, we may use v to form a star for the forests of the partition, and we no longer use G_2 as the independent set.) Before describing the partition, recall that G consists of a vertex v adjacent to all other vertices, and $G \backslash v$ has components $B = B_1 \oplus B_2$ of two kinds, either B_2 is a single vertex, or B_2 is an independent set with at least two vertices. For components $B' = B'_1 \oplus B'_2$ of the first kind (where B'_2 is a single vertex w), we only note that B'_1 consists of bipartite components

D. For the components B of the second kind (where B_2 is a larger independent set), we distinguish components $D' = D_1' \oplus D_2'$ in which D_2' consists of a single vertex z, and other components $D = D_1 \oplus D_2$ where D_2 is a larger independent set and D_1 is a forest. We now describe the first forest of a $(2, 1)$-partition of G. It is a star centered at v and covering the sets D_2 of all components D of B_1' for the components B' of the first kind (where B_2' is a single vertex), as well as the sets B_2 of all components B of the second kind. The second forest of the partition contains, for each component B' of the first kind (where B_2' is a single vertex w), a star centered at w and covering all first parts of the bipartitions of all D_1 of the components D of B_1'. It also contains, for each component B of the second kind, and each component D' in which D_2' consists of a single vertex z, a star centered at z and covering the first part of the bipartition of D_1', and containing D_1 for each component D in which D_1 is a forest. The remaining vertices are easily seen to form an independent set which we take for the desired $(2, 1)$-partition of G.

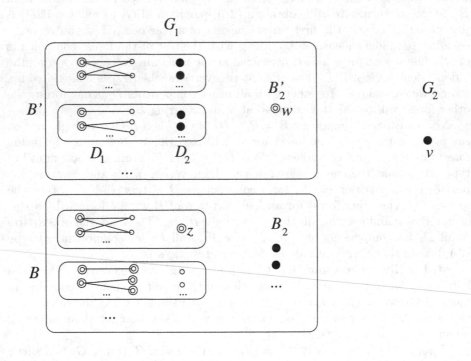

Fig. 2. An illustration of the case 2(d): one forest is indicated by large filled circles, the other forest by double circles, the remainder is independent

4   Conclusions

Theorem 1 implies that any cograph that is not $(2,1)$-partitionable must have a member of $\mathcal{F}$ as an induced subgraph. In [8] there is a linear-time dynamic programming algorithm to recognize $(2,1)$-partitionable cographs. It computes for each cograph G the set of all triples (p,q,r) such that $p + q + r \leq 3, p \leq 2, q \leq 1$, and G has a set R with r vertices such that $G \backslash R$ is (p,q)-partitionable. The cograph G is $(2,1)$-partitionable if and only if $(2,1,0)$ is included in this set of triples. To compute these triples is trivial when G has one vertex, and explicit formulas are given for computing them when $G = G_1 + G_2$ and when $G = G_1 \oplus G_2$. Thus a bottom-up calculation on the cotree of G (a data structure that captures the description of the recursive construction of G) allows us to compute the triples for G. If the triple $(2,1,0)$ is not present, we can apply a top-down process to actually recover a forbidden induced subgraph from the family $\mathcal{F}$. We can identify in the cotree a cograph G', subgraph of G, which does not have a $(2,1)$-partition, but all of whose descendants have a $(2,1)$-partition. Clearly, this means that G' was obtained by a join operation, and we may assume $G' = G_1 \oplus G_2$. Reading the proof of Theorem 1, we see that the relevant information we need includes whether or not G_1, G_2 contain an induced C_3 and where (or whether they are bipartite), and similarly for C_4, or K_4; we would also like to keep track of how many vertices they contain. Moreover, if G_1 or G_2 are themselves obtained by a disjoint union operation, we need similar information about their descendants. It is easy to see all this information can be computed during the construction of the cotree, so we assume we have it available for G' and its descendants. Then the proof of Theorem 1 directly specifies how to find a forbidden induced subgraph, since G' is known to not be $(2,1)$-partitionable. For example, if G_1 and G_2 are both bipartite and both are joins of two bipartite cographs each of which contains a C_4, then case 1(a) of the proof explains we can identify (from those C_4's) an induced copy of $2K_{2,2} \oplus 2K_{2,2}$. The remaining cases are similar, with the exception of components B without K_4 in case 2(c). In this case we rely on the corresponding results of [8], which identify forbidden induced subgraphs for a $(2,0)$-partition, to which we need to add one or both vertices of G_2. It is not difficult to implement all this in linear time. A detailed implementation will be presented in the third author's M.Sc. thesis.

5   The Remaining Proofs for Lemma 1

For the graph $G = K_{2,2,2} \oplus \overline{K_{1,2}}$, we note that any subgraph of G on at least four vertices contains an induced cycle. Hence, one forest in the partition can cover at most three vertices and two forest can cover at most six vertices, and since G has no independent set of size three, $K_{2,2,2} \oplus \overline{K_{1,2}}$ is an obstruction. For $H = K_{1,2,2} \oplus \overline{K_{1,2}}$, one of the two forests will be $\overline{K_{1,2}}$, and removing another forest on three vertices, the remainder is an independent set on two vertices, yielding a required $(2,1)$-partition. For $H = K_{2,2,2} \oplus K_2$, take one of the vertices of K_2 with one of the parts in $K_{2,2,2}$ to obtain one forest. We obtain the other forest in similar way and the remainder is just an independent set of size two. Hence, $K_{2,2,2} \oplus \overline{K_{1,2}}$ is a minimal obstruction.

To prove that $G = K_{2,2,2,2} \oplus K_1$, is an obstruction, note that any forest must be a tree and hence can have at most three vertices. Two forests can cover at most six vertices, and the remaining three vertices will contain an edge, and hence not be independent. To prove that G is indeed minimal, note that both $H_1 = K_{2,2,2,2}$ and $H_2 = K_{1,2,2,2} \oplus K_1$ have a $(2,1)$-partition in which each forest is a tree on three vertices and the independent set has two vertices.

For $G = 2K_3 \oplus K_{2,2}$, one of the two forests can cover at most four vertices and the other forest can cover at most three vertices. Hence, two forests can cover at most seven vertices and there is no independent set of size at least three in G. Hence, G does not have a $(2,1)$-partition. Now we will show that $H_1 = (K_2 + K_3) + \oplus K_{2,2}$ and $2K_3 \oplus K_{1,2}$ have a $(2,1)$-partition. For H_1, the partition consists of one forest that is $2K_2$ and has four vertices. The other forest is a tree on three vertices and the remainder is just an independent set on two vertices. In H_2, the first forest consists of the middle vertex in $K_{1,2}$ along with one vertex in the each of the K_3. The other forest has four vertices consisting of $2K_2$ each. The remainder is an independent set on two vertices. Thus $G = 2K_3 \oplus K_{2,2}$ is indeed minimal.

Similarly, for $G = (K_{2,2} + K_3) \oplus K_{2,2}$, one of the two forests in G can have at most five vertices, and then the other forest can have at most three vertices. The remainder will have at least 3 vertices. Since G does not have an independent set of size three, G does not have a $(2,1)$-partition. We will prove that all the graphs obtained from deleting one vertex from G have a $(2,1)$-partition. That is, $H_1 = (K_{1,2} + K_3) \oplus K_{2,2}$, $H_2 = (K_{2,2} + K_2) \oplus K_{2,2}$ and $H_3 = (K_{2,2} + K_3) \oplus K_{1,2}$, each have a $(2,1)$-partition. For the graph H_1, the partition has one forest on five vertices consisting of $K_{1,2}$ and $K_{2,}$, and the other forest is just a $K_{1,2}$, leaving an independent set on two vertices.

For $G = 2K_{2,2} \oplus K_{3,3}$, two forests can cover at most ten vertices. Either there is only one forest on six vertices and the other forest then can have at most four vertices, or one can obtain two forest on five vertices each. There is no independent set on four vertices., so G does not admit a $(2,1)$-partition. To see that G is indeed minimal, note that both $H_1 = (K_{2,2} + K_{1,2}) \oplus K_{3,3}$ and $H_2 = 2K_{2,2} \oplus K_{2,3}$ have a $(2,1)$-partition. For H_1 one such partition has one forest consisting of two copies of $K_{1,2}$ and other forest is a $K_{1,3}$, leaving an independent set of three vertices. For H_2, a partition can be obtained with two forests which are stars on five vertices each, and the remainder is just an independent set on three vertices.

References

1. Corneil, D.G., Lerchs, H., Burlingham, L.S.: Complement reducible graphs. Discrete Appl. Math. **3**, 163–174 (1981)
2. Demange, M., Ekim, T., de Werra, D.: Partitioning cographs into cliques and stable sets. Discrete Optim. **2**, 145–153 (2005)

3. Damaschke, P.: Induced subgraphs and well-quasi-orderings. J. Graph Theory **14**, 427–435 (1990)
4. Epple, D.D.A., Huang, J.: A note on the bichromatic numbers of graphs. J. Graph Theory **65**, 263–269 (2010)
5. Feder, T., Hell, P., Hochstättler, W.: Generalized colourings (matrix partitions) of cographs. In: Graph Theory in Paris, Trends in Mathematics, pp. 149–167. Birkhauser Verlag (2006)
6. Garey, M.R., Johnson, D.S.: Computers and Intractability: A Guide to the Theory of NP-Completeness. W. H. Freeman & Co., New York (1979)
7. Golumbic, M.C.: Algorithmic Graph Theory and Perfect Graphs. Academic Press, New York (1980)
8. González Hermosillo de la Maza, S., Hell, P., Hernández-Cruz, C., Hosseini, S.A., Valadkhan, P.: Vertex arboricity of cographs. arXiv:1907.07286 [math.CO]
9. Hakimi, S.L., Schmeichel, E.F.: A note on the vertex arboricity of a graph. SIAM J. Discrete Math. **2**, 64–67 (1989)
10. de Souza Francisco, R., Klein, S., Nogueira, L.T.: Characterizing (k, ℓ)-partitionable cographs. Electron. Notes Discrete Math. **22**, 277–280 (2005)

Monitoring the Edges of a Graph Using Distances

Florent Foucaud[1(✉)], Ralf Klasing[1], Mirka Miller[2,3], and Joe Ryan[4]

[1] Univ. Bordeaux, Bordeaux INP, CNRS, LaBRI, UMR5800, Talence, France
florent.foucaud@gmail.com, ralf.klasing@labri.fr
[2] School of Mathematical and Physical Sciences, The University of Newcastle,
Callaghan, Australia
[3] Department of Mathematics, University of West Bohemia, Pilsen, Czech Republic
[4] School of Electrical Engineering and Computing, The University of Newcastle,
Callaghan, Australia
joe.ryan@newcastle.edu.au

Abstract. We introduce a new graph-theoretic concept in the area of network monitoring. A set M of vertices of a graph G is a *distance-edge-monitoring set* if for every edge e of G, there is a vertex x of M and a vertex y of G such that e belongs to all shortest paths between x and y. We denote by $\dem(G)$ the smallest size of such a set in G. The vertices of M represent distance probes in a network modeled by G; when the edge e fails, the distance from x to y increases, and thus we are able to detect the failure. It turns out that not only we can detect it, but we can even correctly locate the failing edge.

In this paper, we initiate the study of this new concept. We show that for a nontrivial connected graph G of order n, $1 \leq \dem(G) \leq n-1$ with $\dem(G) = 1$ if and only if G is a tree, and $\dem(G) = n-1$ if and only if it is a complete graph. We compute the exact value of dem for grids, hypercubes, and complete bipartite graphs.

Then, we relate dem to other standard graph parameters. We show that $\dem(G)$ is lower-bounded by the arboricity of the graph, and upper-bounded by its vertex cover number. It is also upper-bounded by five times its feedback edge set number.

Then, we show that determining $\dem(G)$ for an input graph G is an NP-complete problem, even for apex graphs. There exists a polynomial-time logarithmic-factor approximation algorithm, however it is NP-hard to compute an asymptotically better approximation, even for bipartite graphs of small diameter and for bipartite subcubic graphs. For such instances, the problem is also unlikely to be fixed parameter tractable when parameterized by the solution size.

This paper is dedicated to the memory of Mirka Miller, who sadly passed away in January 2016. This research was started when Mirka Miller and Joe Ryan visited Ralf Klasing at the LaBRI, University of Bordeaux, in April 2015.

© Springer Nature Switzerland AG 2020
M. Changat and S. Das (Eds.): CALDAM 2020, LNCS 12016, pp. 28–40, 2020.
https://doi.org/10.1007/978-3-030-39219-2_3

1 Introduction

The aim of this paper is to introduce a new concept of network monitoring using distance probes, called *distance-edge-monitoring*. Our networks are naturally modeled by finite undirected simple connected graphs, whose vertices represent computers and whose edges represent connections between them. We wish to be able to monitor the network in the sense that when a connection (an edge) fails, we can detect this failure. We will select a (hopefully) small set of vertices of the network, that will be called *probes*. At any given moment, a probe of the network can measure its graph distance to any other vertex of the network. Our goal is that, whenever some edge of the network fails, one of the measured distances changes, and thus the probes are able to detect the failure of any edge.

Probes that measure distances in graphs are present in real-life networks, for instance this is useful in the fundamental task of *routing* [6,11]. They are also frequently used for problems concerning *network verification* [2–4].

We will now proceed with the formal definition of our main concept. In this paper, by *graphs* we refer to connected simple graphs (without multiple edges and loops). A graph with loops or multiple edges is called a *multigraph*.

We denote by $d_G(x, y)$ the distance between two vertices x and y in a graph G. When there is no path connecting x and y in G, we let $d_G(x, y) = \infty$. For an edge e of G, we denote by $G - e$ the graph obtained by deleting e from G.

Definition 1. *For a set M of vertices and an edge e of a graph G, let $P(M, e)$ be the set of pairs (x, y) with x a vertex of M and y a vertex of $V(G)$ such that $d_G(x, y) \neq d_{G-e}(x, y)$. In other words, e belongs to all shortest paths between x and y in G.*

For a vertex x, let $M(x)$ be the set of edges e such that there exists a vertex v in G with $(x, v) \in P(\{x\}, e)$. If $e \in M(x)$, we say that e is monitored by x.

A set M of vertices of a graph G is distance-edge-monitoring if every edge e of G is monitored by some vertex of M, that is, the set $P(M, e)$ is nonempty. Equivalently, $\bigcup_{x \in M} M(x) = E(G)$.

We note $\mathrm{dem}(G)$ the smallest size of a distance-edge-monitoring set of G.

Note that $V(G)$ is always a distance-edge-monitoring set of G, so $\mathrm{dem}(G)$ is always well-defined.

Consider a graph G modeling a network, and a set M of vertices of G, on which we place probes that are able to measure their distances to all the other vertices. If M is distance-edge-monitoring, if a failure occurs on any edge of the network (in the sense that the communication between its two endpoints is broken), then this failure is detected by the probes.

In fact, it turns out that not only the probes can *detect* a failing edge, but they can also precisely *locate* it (the proof of the following is delayed to Sect. 2).

Proposition 2. *Let M be a distance-edge-monitoring set of a graph G. Then, for any two distinct edges e and e' in G, we have $P(M, e) \neq P(M, e')$.*

Thus, assume that we have placed probes on a distance-edge-monitoring set M of a network G and initially computed all the sets $P(M, e)$. In the case a unique edge of the network has failed, Proposition 2 shows that by measuring the set of pairs (x, y) with $x \in M$ and $y \in V(G)$ whose distance has changed, we know exactly which is the edge that has failed.

We define the decision and optimization problem associated to distance-edge-monitoring sets.

DISTANCE-EDGE-MONITORING SET
Instance: A graph G, an integer k.
Question: Do we have $\text{dem}(G) \leq k$?

MIN DISTANCE-EDGE-MONITORING SET
Instance: A graph G.
Task: Build a smallest possible distance-edge-monitoring set of G.

Related Notions. A weaker model is studied in [2,3] as a *network discovery* problem, where we seek a set S of vertices such that for each edge e, there exists a vertex x of S and a vertex y of G such that e belongs to *some* shortest path from x to y.

Distance-edge-monitoring sets are related to *resolving sets* and *edge-resolving sets*, that also model sets of sensors that can measure the distance to all other vertices in a graph. A resolving set is a set R of vertices such that for any two distinct vertices x and y in G, there is a vertex r in R such that $d_G(r, x) \neq d_G(r, y)$. The smallest size of a resolving set in G is the *metric dimension* of G [12,20]. If instead, the set R distinguishes the edges of G (that is, for any pair e, e' of edges of G, there is a vertex $r \in R$ with $d_G(x, e) \neq d_G(x, e')$, where for $e = uv$, $d_G(x, e) = \min\{d_G(x, u), d_G(x, v)\}$), we have an edge-resolving set [15].

Another related concept is the one of *strong resolving sets* [18,19]: a set R of vertices is strongly resolving if for any pair x, y of vertices, there exists a vertex z of R such that either x is on a shortest path from z to y, or y is on a shortest path from z to x. It is related to distance-edge-monitoring sets in the following sense. Given a distance-edge-monitoring set M, for every pair x, y of *adjacent* vertices, there is a vertex z of M such that either x is on *every* shortest path from z to y, or y is on *every* shortest path from z to x.

Another related concept is the one of an *(strong) edge-geodetic set*. A set S of vertices is an edge-geodetic set if for every edge e of G, there are two vertices x, y of S such that e is on some shortest path from x to y. It is a strong edge-geodetic set if to every pair x, y of S, we can assign a shortest $x - y$ path to $\{x, y\}$ such that every edge of G belongs to one of these $\binom{|S|}{2}$ assigned shortest paths [16].

Our Results. We first derive a number of basic results about distance-edge-monitoring sets in Sect. 2 (where we also give some useful definitions).

In Sect. 3, we study dem for some basic graph families (like trees and grids) and relate this parameter to other standard graph parameters such as arboricity arb, vertex cover number vc and feedback edge set number fes. We show that

$dem(G) = 1$ if and only if G is a tree. We show that for any graph G of order n, $dem(G) \geq arb(G)$. Moreover, $dem(G) \leq vc(G) \leq n - 1$ (with equality if and only if G is complete). We show that for some families of graphs G, $dem(G) = vc(G)$, for instance this is the case for complete bipartite graphs and hypercubes. Then we show that $dem(G) \leq 5fes(G) - 5$ when $fes(G) \geq 2$ (when $fes(G) \leq 1$, $dem(G) = fes(G) + 1$).

In Sect. 4, we show that DISTANCE-EDGE-MONITORING SET is NP-complete, even for apex graphs (graphs obtained from a planar graph by adding an extra vertex). Then, we show that MIN DISTANCE-EDGE-MONITORING SET can be approximated in polynomial time within a factor of $\ln(|E(G)|+1)$ by a reduction to the set cover problem. Finally, we show that no essentially better ratio can be obtained (unless P = NP), even for graphs that are of diameter 4, bipartite and of diameter 6, or bipartite and of maximum degree 3. For the same restrictions, the problem is unlikely to be fixed parameter tractable when parameterized by the solution size. These hardness results are obtained by reductions from the set cover problem.

We conclude our paper in Sect. 5. Due to the given space constraints, some proofs are not included.

2   Preliminaries

We now give some useful lemmas about basic properties of distance-edge-monitoring sets. We start with the proof of Proposition 2.

Proof (Proof of Proposition 2). Suppose by contradiction that there are two distinct edges e and e' with $P(M, e) = P(M, e')$. Since by definition, $P(M, e) \neq \emptyset$, we have $(x, v) \in P(M, e)$ (and thus $(x, v) \in P(M, e')$), where $x \in M$ and $v \in V(G)$. Thus, all shortest paths between x and v contain both e and e'. Assume that on these shortest paths, e is closer to x than e', and let w be the endpoint of e that is closest to v. Then, we have $(x, w) \in P(M, e)$ but $(x, w) \notin P(M, e')$: a contradiction.   □

An edge e in a graph G is a *bridge* if $G - e$ has more connected components than G. We will now show that bridges are very easy to monitor.

Lemma 3. *Let G be a connected graph and let e be a bridge of G. For any vertex x of G, we have $e \in M(x)$.*

Proof. Assume that $e = uv$. Since e is a bridge, we have $d_G(x, u) \neq d_G(x, v)$. If $d_G(x, u) < d_G(x, v)$, then $(x, v) \in P(\{x\}, e)$. Otherwise, we have $(x, u) \in P(\{x\}, e)$. In both cases, $e \in M(x)$.   □

Given a vertex x of a graph G and an integer i, we let $L_i(x)$ denote the set of vertices at distance i of x in G.

Lemma 4. *Let x be a vertex of a connected graph G. Then, an edge uv belongs to $M(x)$ if and only if $u \in L_i(x)$ and v is the only neighbour of u in $L_{i-1}(x)$.*

Proof. Let $uv \in M(x)$. Then, there exists a vertex y such that all shortest paths from y to x go through uv. Thus, one of u and v (say, u) is in a set $L_i(x)$ and the other (v) is in a set $L_{i-1}(x)$, for some positive integer i. Moreover, v must be the only neighbour of u in $L_{i-1}(x)$, since otherwise there would be a shortest path from y to x going through u but avoiding uv.

Conversely, if u is a vertex in $L_i(x)$ with a unique neighbour v in $L_{i-1}(x)$, then the edge uv belongs to $M(x)$ since all shortest paths from u to x use it. □

We obtain some immediate consequences of Lemma 4.

Lemma 5. *For a vertex x of a graph G, the set of edges $M(x)$ induces a forest.*

Proof. Let G_x be the subgraph of G induced by the edges in $M(x)$. By Lemma 4, an edge e belongs to $M(x)$ if and only if $e = uv$, $u \in L_i(x)$ and v is the only neighbour of u in $L_{i-1}(x)$. In this case, we let v be the *parent* of u. Each vertex in G_x has at most one parent in G_x, and each edge of G_x is an edge between a vertex and its parent. Thus, G_x is a forest. □

Lemma 6. *Let G be a graph and x a vertex of G. Then, for any edge e incident with x, we have $e \in M(x)$.*

Proof. For every vertex y of $L_1(x)$ (that is, every neighbour of x), x is the unique neighbour of y in $L_0(x) = \{x\}$. Thus, the claim follows from Lemma 4. □

3   Basic Graph Families and Bounds

In this section, we study dem for standard graph classes, and its relation with other standard graph parameters.

Theorem 7. *Let G be a connected graph with at least one edge. We have* $\mathrm{dem}(G) = 1$ *if and only if G is a tree.*

Proof. In a tree, every edge is a bridge. Thus, by Lemma 3, if G is a tree, any vertex x of G is a distance-edge-monitoring set and we have $\mathrm{dem}(G) \leq 1$ (and of course as long as there is an edge in G, $\mathrm{dem}(G) \geq 1$).

For the converse, suppose that $\mathrm{dem}(G) = 1$. Then, clearly, G must have at least one edge. Moreover, since all edges of G must belong to $M(x)$, by Lemma 5, G must be a forest. Since G is connected, G is a tree. □

Let $G_{a,b}$ denote the grid of dimension $a \times b$. By Theorem 7, we have $\mathrm{dem}(G_{1,a}) = 1$. We can compute all other values.

Theorem 8. *For any integers $a, b \geq 2$, we have* $\mathrm{dem}(G_{a,b}) = \max\{a, b\}$.

The *arboricity* $\mathrm{arb}(G)$ of a graph G is the smallest number of sets into which $E(G)$ can be partitioned and such that each set induces a forest. The *clique number* $\omega(G)$ of G is the size of a largest clique in G.

Theorem 9. *For any graph G of order n and size m, we have* $\dem(G) \geq$ *arb(G), and thus* $\dem(G) \geq \frac{m}{n-1}$ *and* $\dem(G) \geq \frac{\omega(G)}{2}$.

Proof. By Lemma 5, for each vertex x of a distance-edge-monitoring set M, $M(x)$ induces a forest. Thus, to each edge e of G we can assign one of the forests $M(x)$ such that $e \in M(x)$. This is a partition of G into $|M|$ forests, and thus $\arb(G) \leq \dem(G)$.

Moreover, it is not difficult to see that $\arb(G) \geq \frac{m}{n-1}$ (since a forest has at most $n - 1$ edges) and $\arb(G) \geq \frac{\omega(G)}{2}$ (since a clique of size $k = \omega(G)$ has $k(k-1)/2$ edges but a forest that is a subgraph of G can contain at most $k - 1$ of these edges). □

We next see that distance-edge-monitoring sets are relaxations of vertex covers. A set C of vertices is a *vertex cover* of G if every edge of G has one of its endpoints in C. The smallest size of a vertex cover of G is denoted by $\vc(G)$.

Theorem 10. *In any graph G of order n, any vertex cover of G is a distance-edge-monitoring set, and thus* $\dem(G) \leq \vc(G) \leq n - 1$. *Moreover, we have* $\dem(G) = n - 1$ *if and only if G is the complete graph of order n.*

Proof. Let C be a vertex cover of G. By Lemma 6, for every edge e, there is a vertex in x with $e \in M(x)$, thus C is distance-edge-monitoring.

Moreover, any graph G of order n has a vertex cover of size $n - 1$: for any vertex x, the set $V(G) \setminus \{x\}$ is a vertex cover of G.

Finally, suppose that $\dem(G) = n - 1$: then also $\vc(G) = n - 1$. If G is not connected, we have $\vc(G) \leq n - 2$ (starting with $V(G)$ and removing any vertex from each connected component of G yields a vertex cover), thus G is connected. Suppose by contradiction that G is not a complete graph. Then, we have three vertices x, y and z in G such that xy and yz are edges of G, but xz is not. Then, $V(G) \setminus \{x, z\}$ is a vertex cover of G, a contradiction.

This completes the proof. □

In some graphs, any distance-edge-monitoring set is a vertex cover.

Observation 11. *If, for every vertex x of a graph G, $M(x)$ consists exactly of the sets of edges incident with x, then a set M is a distance-edge-monitoring set of G if and only if it is a vertex cover of G.*

Note that Observation 11 does not provide a characterization of graphs with $\dem(G) = \vc(G)$. For example, as seen in Theorem 8, for the grid $G_{a,2}$, we have $\dem(G_{a,2}) = \vc(G_{a,2}) = a$, but for any vertex x of $G_{a,2}$, $M(x)$ consists of the whole row and column of $G_{a,2}$.

Let $K_{a,b}$ be the complete bipartite graph with parts of sizes a and b, and let H_d denote the hypercube of dimension d. We can deduce the following from Observation 11.

Corollary 12. *We have* $\dem(K_{a,b}) = \vc(K_{a,b}) = \min\{a, b\}$, *and* $\dem(H_d) = \vc(H_d) = 2^{d-1}$.

A *feedback edge set* of a graph G is a set of edges such that removing them from G leaves a forest. The smallest size of a feedback edge set of G is denoted by $\mathrm{fes}(G)$ (it is sometimes called the *cyclomatic number* of G).

Theorem 7 states that for any tree G (that is, a graph with feedback edge set number 0), we have $\dim(G) = \mathrm{fes}(G) + 1$. We now show the same bound for graphs G with $\mathrm{fes}(G) = 1$, that is, unicyclic graphs (graphs with a unique cycle).

Proposition 13. *If G is a unicyclic graph, then* $\dim(G) = 2$.

We will now give a weaker (but similar) bound for any value of $\mathrm{fes}(G)$. But first, we need the following terminology and lemma from [9] about the structure of graphs G with given feedback edge set number.

In a graph, a vertex is a *core vertex* if it has degree at least 3. The *base graph* of a graph G is the graph obtained from G by iteratively removing vertices of degree 1 (thus, the base graph of a forest is the empty graph).

The following lemma can be found in Sect. 5.3.1 of [9].

Lemma 14 ([9]). *Let G be a graph with $\mathrm{fes}(G) = k \geq 2$. The base graph of G has at most $2k - 2$ core vertices, that are joined by at most $3k - 3$ edge-disjoint paths with internal vertices of degree 2 and whose endpoints are (not necessarily distinct) core vertices.*

In other words, Lemma 14 says that the base graph of a graph G with $\mathrm{fes}(G) = k \geq 2$ can be obtained from a multigraph H of order at most $2k - 2$ and size $3k - 3$ by subdividing its edges an arbitrary number of times.

Theorem 15. *Let G be a graph with $\mathrm{fes}(G) \geq 2$. Then, $\dim(G) \leq 5\mathrm{fes}(G) - 5$.*

Proof. Let $\mathrm{fes}(G) = k$ and G_b, the base graph of G. By Lemma 14, G_b contains at most $2k - 2$ core vertices, joined by at most $3k - 3$ edge-disjoint paths with internal vertices of degree 2 and whose endpoints are core vertices (both endpoints can be the same vertex). Let $\{P_1, \ldots, P_t\}$ be the set of these paths.

By Lemma 3, any non-empty distance-edge-monitoring set of G_b is also one of G, since all edges of G not present in G_b are bridges. Thus, it is sufficient to construct a distance-edge-monitoring set M of G_b of size at most $5k - 5$. We build M as follows: first of all, M contains all core vertices of G_b. Moreover, for each path P_i of length at least 2 connecting two core vertices x and y of G_b, we add to M an internal vertex z of P_i. If $x \neq y$ or P_i has length at most 3, we let z be a neighbour of, say, x. Otherwise, $x = y$ and P_i has length at least 4; we choose z as a vertex at distance 2 of x on P_i. Thus, $|M| \leq 5k - 5$.

Let us prove that M is distance-edge-monitoring in G_b. Let e be an edge of G_b; necessarily, e belongs to some path P_i connecting two core vertices x and y. If e is incident with x or y, by Lemma 6 e is monitored by M. Thus, suppose this is not the case. So, P_i has length at least 3 and so there is an internal vertex z of P_i in M. Again if e is incident with z we are done.

If $x = y$, P_i forms a cycle and for any vertex v, $M(v) \cap E(P_i)$ consists of the two edges (if P_i has even length) or the unique edge (if P_i has odd length) that are opposite to v on the cycle. Since x and z are non-adjacent, we have $e \in M(x) \cup M(z)$ and so z is monitored.

If $x \neq y$, assume that $e = uv$ with u the vertex of e closest to x. Let d_x be the distance from x to u on P_i, and d_y the distance from y to v on P_i. If $e \in M(x) \cup M(y)$, we are done. Otherwise, there must exist a path from x to y that is edge-disjoint with P_i (otherwise e would be a bridge and would be monitored by x or y). Let p be the length of a shortest such path. Since $e \notin M(x)$, there exists a shortest path from v to x avoiding e. This shortest path must go through y, and then use a path of length p. Thus, we have $d_y + p \leq d_x + 1$. By a similar argument with u and x, we have $d_x + p \leq d_y + 1$. We obtain from this that $d_x = d_y$, $|P_i|$ is odd, and $p = 1$. But then, the unique shortest path from v to z goes through e, and we have $e \in M(z)$. This completes the proof.   □

We do not believe that the bound of Theorem 15 is tight. For the sake of simplicity, we have not tried to optimize the bound of Theorem 15, as we do not believe that this method will provide a tight bound. There are examples of graphs G where $\mathrm{dem}(G) = \mathrm{fes}(G) + 1$, for example this is the case for the grid $G_{a,2}$: by Theorem 8, when $a \geq 2$ we have $\mathrm{dem}(G_{a,2}) = a$, and $\mathrm{fes}(G_{a,2}) = a - 1$.

Note that $G_{a,b}$ for $a, b \geq 2$ provides examples of a family of graphs where for increasing a, b the difference between $\mathrm{fes}(G_{a,b})$ and $\mathrm{dem}(G_{a,b})$ is unbounded by a constant. Indeed by Theorem 8 we have $\mathrm{dem}(G_{a,b}) = \max\{a, b\}$ but $\mathrm{fes}(G_{a,b})$ is linear in the order ab.

4   Complexity

A *c-approximation algorithm* for a given optimization problem is an algorithm that returns a solution whose size is always at most c times the optimum. We refer to the books [1,13] for more details. For a decision problem Π and for some parameter p of the instance, an algorithm for Π is said to be *fixed parameter tractable* (fpt for short) if it runs in time $f(p)n^c$, where f is a computable function, n is the input size, and c is a constant. In this paper, we will always consider the solution size k as the parameter. The class FPT contains the parameterized decisions problems solvable by an fpt algorithm. The classes W[i] (with $i \geq 1$) denote complexity classes with parameterized decision problems that are believed not to be fpt. We refer to the books [8,17] for more details.

We will now use the connection to vertex covers hinted in Sect. 3 to derive an NP-hardness result. For two graphs G, H, $G \bowtie H$ denotes the graph obtained from disjoint copies of G and H with all possible edges between $V(G)$ and $V(H)$. We denote by K_1 the graph on one single vertex.

Theorem 16. *For any graph G, we have $\mathrm{vc}(G) \leq \mathrm{dem}(G \bowtie K_1) \leq \mathrm{vc}(G) + 1$. Moreover, if G has radius at least 4, then $\mathrm{vc}(G) = \mathrm{dem}(G \bowtie K_1)$.*

We deduce the following from the second part of the statement of Theorem 16 and the NP-hardness of VERTEX COVER for planar subcubic graphs with radius at least 4 [10].

Corollary 17. DISTANCE-EDGE-MONITORING SET *is NP-complete, even for graphs obtained from a planar subcubic graph by attaching a universal vertex.*

Given a hypergraph $H = (X, S)$ with vertex set X and egde set S, a *set cover* of H is a subset $C \subseteq S$ of edges such that each vertex of X belongs to at least one edge of C. We will provide reductions for MIN DISTANCE-EDGE-MONITORING SET to and from SET COVER and MIN SET COVER.

SET COVER
Instance: A hypergraph $H = (X, S)$, an integer k.
Question: Is there a set cover $C \subseteq S$ of size at most k?

MIN SET COVER
Instance: A hypergraph $H = (X, S)$.
Task: Build a smallest possible set cover of H.

It is known that MIN SET COVER is approximable in polynomial time within a factor of $\ln(|X|+1)$ [14], but, unless P = NP, not within a factor of $(1-\epsilon)\ln|X|$ (for every positive ϵ) [7]. Moreover, SET COVER is W[2]-hard (when parameterized by k) and not solvable in time $|X|^{o(k)}|H|^{O(1)}$ unless FPT = W[1] [5].

Theorem 18. MIN DISTANCE-EDGE-MONITORING SET *is approximable within a factor of* $\ln(|E(G)| + 1)$ *in polynomial time.*

Proof. Let G be a graph and consider the following hypergraph $H = (X, S)$ with $X = E(G)$ and such that S contains, for each vertex x of G, the set $S_x = \{e \in X \mid e \in M(x)\}$. Now, it is not difficult to see that there is a one-to-one correspondence between set covers of H and distance-edge-monitoring sets of G, where we associate to each vertex x of a distance-edge-monitoring set of G, the set S_x in a set cover of H. Thus, the result follows from the $\ln(|X| + 1)$-approximation algorithm for MIN SET COVER from [14]. $\square$

Theorem 19. *Even for graphs G that are (a) of diameter 4, (b) bipartite and of diameter 6, or (c) bipartite and of maximum degree 3,* MIN DISTANCE-EDGE-MONITORING SET *is not approximable within a factor of* $(1 - \epsilon) \ln |E(G)|$ *in polynomial time, unless* $P = NP$. *Moreover, for such instances,* DISTANCE-EDGE-MONITORING SET *cannot be solved in time* $|G|^{o(k)}$, *unless* FPT = W[1], *and it is W[2]-hard for parameter k.*

Proof. For an instance (H, k) of SET COVER, we will construct in polynomial time instances $(G, k + 2)$ or $(G, k + 1)$ of DISTANCE-EDGE-MONITORING SET so that H has a set cover of size k if and only if G has a distance-edge-monitoring set of size at most $k + 2$ or $k + 1$.

In our first reduction, the obtained instance has diameter 4, while in our second reduction, the obtained instance is bipartite and has diameter 6, and in our third reduction, the graph G is bipartite and has maximum degree 3. The three constructions are similar.

The statement will follow from the hardness of approximating MIN SET COVER proved in [7], the parameterized hardness of SET COVER (parameterized by solution size), and the lower bound on its running time [5].

First of all, we point out that we may assume that in an instance $(H = (X, S), k)$ of SET COVER, there is no vertex of X that belongs to a unique set of S. Indeed, otherwise, we are forced to take S in any set cover of H; thus, by removing S and all vertices in S, we obtain an equivalent instance $(H', k - 1)$. We can iterate until the instance satisfies this property.

We now describe the first reduction, in which the obtained instance has diameter 4. Let $(H, k) = ((X, S), k)$ be an instance of SET COVER, where $X = \{x_1, x_2, \ldots, x_{|X|}\}$, $S = \{C_1, C_2, \ldots, C_{|S|}\}$ and $C_i = \{c_{i,j} \mid x_j \in C_i\}$. Construct the following instance $(G, k + 2) = ((V, E), k + 2)$ of DISTANCE-EDGE-MONITORING SET, where $V = V_1 \cup V_2 \cup \ldots \cup V_5 \cup V_1' \cup V_2' \cup V_3' \cup V_4'$, $E = E_1 \cup E_2 \cup E_3 \cup E_4 \cup E_1' \cup E_2' \cup E_3' \cup E_4' \cup E_5'$ and

$V_1 = \{u_1, u_2, u_3\}$, $V_2 = \{v_i \mid 1 \le i \le |S|\}$, $V_3 = S$, $V_4 = \{c_{i,j} \mid 1 \le i \le |S|, 1 \le j \le |X|, x_j \in C_i\}$, $V_5 = X$

$V_1' = \{u_1', u_2', u_3'\}$, $V_2' = \{v_i' \mid 1 \le i \le |S|\}$, $V_3' = \{c_{i,j}' \mid 1 \le i \le |S|, 1 \le j \le |X|, x_j \in C_i\}$, $V_4' = \{w_j' \mid 1 \le j \le |X|\}$

$E_1 = \{(u_1, u_2), (u_1, u_3), (u_2, u_1'), (u_3, u_1')\}$, $E_2 = \{(u_1, v_i), (v_i, C_i) \mid 1 \le i \le |S|\}$, $E_3 = \{(C_i, c_{i,j}) \mid 1 \le i \le |S|, 1 \le j \le |X|, x_j \in C_i\}$, $E_4 = \{(c_{i,j}, x_j) \mid 1 \le i \le |S|, 1 \le j \le |X|, x_j \in C_i\}$

$E_1' = \{(u_1', u_2'), (u_2', u_3'), (u_3', u_1')\}$, $E_2' = \{(u_1', v_i'), (v_i', C_i) \mid 1 \le i \le |S|\}$, $E_3' = \{(u_1', c_{i,j}'), (c_{i,j}', c_{i,j}) \mid 1 \le i \le |S|, 1 \le j \le |X|, x_j \in C_i\}$, $E_4' = \{(u_1', w_j'), (w_j', x_j) \mid 1 \le j \le |X|\}$, $E_5' = \{(u_1', v_i) \mid 1 \le i \le |S|\}$.

An example is given in Fig. 1.

To see that G has diameter 4, observe that every vertex of G has a path of length at most 2 to the vertex u_1'.

Let C be a set cover of H of size k. Define $M = C \cup \{u_1, u_2'\}$. Then, by Lemma 4, u_1 monitors (in particular) the edges (u_1, u_2), (u_1, u_3), (u_1', u_3'), and all the edges in $E_2 \cup E_3$. Similarly, u_2' monitors the edges (u_1', u_2'), (u_2', u_3'), (u_3, u_1'), (u_2, u_1') and all the edges in $E_2' \cup E_3' \cup E_4' \cup E_5'$. It thus remains to show that all edges of E_4 are monitored. Notice that among those edges, vertex C_i of S monitors exactly all edges c_{j_1, j_2} with $x_{j_2} \in C_i$. Thus, if $e = (c_{i,j}, x_j)$, there is i' such that $x_j \in C_{i'}$ and $C_{i'} \in C$, and e is monitored by $C_{i'}$ (either $i = i'$ and the only shortest path from x_j to C_i contains e, or $i \ne i'$ and the only shortest path from $c_{i,j}$ to $C_{i'}$ contains e). Hence, M is a distance-edge-monitoring set of G of size at most $k + 2$.

Conversely, let M be a distance-edge-monitoring set of G of size at most $k + 2$. In order to monitor the edge (u_2', u_3'), either $u_2' \in M$ or $u_3' \in M$. In order to monitor the edges (u_1, u_3) and (u_1, u_2), there must be a vertex of M in $\{u_1, u_2, u_3\}$. We may replace $M \cap \{u_1, u_2, u_3\}$ by u_1 and $M \cap \{u_2', u_3'\}$ by u_2', as u_1 monitors the same edges as $\{u_1, u_2, u_3\}$ (among those not already monitored by u_2') and u_2' monitors the same edges as $\{u_2', u_3'\}$ (among those not already monitored by u_1). As seen in the previous paragraph, all edges of E_1, E_1', E_2, E_3, E_2', E_3', E_4' and E_5' are monitored by $\{u_1, u_2'\}$. However, no edge of E_4 is

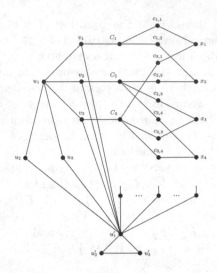

Fig. 1. First reduction from SET COVER to DISTANCE-EDGE-MONITORING SET from the proof of Theorem 19 applied to the hypergraph $(\{x_1, x_2, x_3, x_4\}, \{C_1 = \{x_1, x_2\}, C_2 = \{x_2, x_3, x_4\}, C_3 = \{x_1, x_3, x_4\}\})$. Vertices and edges of V_i' and E_i' for $i = 2, 3, 4$ are only suggested.

monitored by any vertex of $V_1 \cup V_1'$. Thus, all remaining vertices of M are needed precisely to monitor the edges of E_4.

If $v_i \in M$, let $M = M \setminus \{v_i\} \cup \{C_i\}$, and the set of monitored edges does not decrease. If $c_{i,j} \in M$, let $M = M \setminus \{c_{i,j}\} \cup \{C_i\}$, and the set of monitored edges does not decrease. If $x_j \in M$, let $M = M \setminus \{x_j\} \cup \{C_{i_0}\}$ where $i_0 = \min\{i \mid x_j \in C_i\}$, and the set of monitored edges does not decrease. If $v_i' \in M$, let $M = M \setminus \{v_i'\} \cup \{C_i\}$, and the set of monitored edges does not decrease. If $c_{i,j}' \in M$, let $M = M \setminus \{c_{i,j}'\} \cup \{C_i\}$, and the set of monitored edges does not decrease. If $w_j' \in M$, let $M = M \setminus \{w_j'\} \cup \{C_{i_0}\}$ where $i_0 = \min\{i \mid x_j \in C_i\}$, and the set of monitored edges does not decrease. Iterating this process, we finally obtain a distance-edge-monitoring set M' of G with $|M' \cap V_1| = 1$, $|M' \cap V_1'| \geq 1$, $|M' \cap V_3| \leq k$, $M' \cap (V_2 \cup V_4 \cup V_5 \cup V_2' \cup V_3' \cup V_4') = \emptyset$. Let $C = M' \cap V_3$. If C is not a set cover of H, then there is $x_j \in X$ that is not covered by C. Recall that we assumed that each vertex of X belongs to at least two edges of S. Then, according to Lemma 4, any edge $(c_{i,j}, x_j)$ is not monitored by any of the vertices in $M' = (M' \cap V_1) \cup (M' \cap V_1') \cup C$, a contradiction. Hence, C is a set cover of H of size at most k.

Due to space constraints, we omit the two other reductions. They are very similar to the above one. To obtain a bipartite graph, we remove the odd cycles of the construction at the expense of a higher diameter. To obtain maximum degree 3, we replace the set V_2 by some suitably defined binary trees. $\square$

5   Conclusion

We have introduced a new graph parameter useful in the area of network monitoring. We have related it to other standard graph parameters by the means of lower and upper bounds. It would be interesting to improve them. In particular, is it true that $\text{dem}(G) \leq \text{fes}(G) + 1$? As we have seen, this bound would be tight.

It would also be interesting to determine graph classes where DISTANCE-EDGE-MONITORING SET has a polynomial-time (or parameterized) exact or constant-factor approximation algorithm.

Acknowledgements. The authors acknowledge the financial support from the ANR project HOSIGRA (ANR-17-CE40-0022), the IFCAM project "Applications of graph homomorphisms" (MA/IFCAM/18/39), and the Programme IdEx Bordeaux – SysNum (ANR-10-IDEX-03-02).

References

1. Ausiello, G., Crescenzi, P., Gambosi, G., Kann, V., Marchetti-Spaccamela, A., Protasi, M.: Complexity and Approximation: Combinatorial Optimization Problems and Their Approximability Properties. Springer, Heidelberg (1999). https://doi.org/10.1007/978-3-642-58412-1
2. Bampas, E., Bilò, D., Drovandi, G., Gualà, L., Klasing, R., Proietti, G.: Network verification via routing table queries. J. Comput. Syst. Sci. **81**(1), 234–248 (2015)
3. Beerliova, Z., et al.: Network discovery and verification. IEEE J. Sel. Areas Commun. **24**(12), 2168–2181 (2006)
4. Bilò, D., Erlebach, T., Mihalák, M., Widmayer, P.: Discovery of network properties with all-shortest-paths queries. Theoret. Comput. Sci. **411**(14–15), 1626–1637 (2010)
5. Chen, J., et al.: Tight lower bounds for certain parameterized NP-hard problems. Inf. Comput. **201**(2), 216–231 (2005)
6. Dall'Asta, L., Alvarez-Hamelin, J.I., Barrat, A., Vázquez, A., Vespignani, A.: Exploring networks with traceroute-like probes: theory and simulations. Theoret. Comput. Sci. **355**(1), 6–24 (2006)
7. Dinur, I., Steurer, D.: Analytical approach to parallel repetition. In: Proceedings of the Forty-Sixth Annual ACM Symposium on Theory of Computing, STOC 2014, pp. 624–633 (2014)
8. Downey, R.G., Fellows, M.R.: Fundamentals of Parameterized Complexity. TCS. Springer, London (2013). https://doi.org/10.1007/978-1-4471-5559-1
9. Epstein, L., Levin, A., Woeginger, G.J.: The (weighted) metric dimension of graphs: hard and easy cases. Algorithmica **72**(4), 1130–1171 (2015)
10. Garey, M.R., Johnson, D.S.: The rectilinear Steiner tree problem is NP-complete. SIAM J. Appl. Math. **32**(4), 826–834 (1977)
11. Govindan, R., Tangmunarunkit, H.: Heuristics for Internet map discovery. In: Proceedings of the 19th IEEE International Conference on Computer Communications, INFOCOM 2000, pp. 1371–1380 (2000)
12. Harary, F., Melter, R.A.: On the metric dimension of a graph. Ars Comb. **2**, 191–195 (1976)

13. Hromkovič, J.: Algorithmics for Hard Problems: Introduction to Combinatorial Optimization, Randomization, Approximation, and Heuristics. Texts in Theoretical Computer Science. An EATCS Series, 2nd edn. Springer, Heidelberg (2004). https://doi.org/10.1007/978-3-662-05269-3

14. Johnson, D.S.: Approximation algorithms for combinatorial problems. J. Comput. Syst. Sci. **9**, 256–278 (1974)

15. Kelenc, A., Tratnik, N., Yero, I.G.: Uniquely identifying the edges of a graph: the edge metric dimension. Discrete Appl. Math. **251**, 204–220 (2018)

16. Manuel, P., Klavžar, S., Xavier, A., Arokiaraj, A., Thomas, E.: Strong edge geodetic problem in networks. Open Math. **15**, 1225–1235 (2017)

17. Niedermeier, R.: Invitation to Fixed-Parameter Algorithms. Oxford University Press, Oxford (2006)

18. Oellermann, O.R., Peters-Fransen, J.: The strong metric dimension of graphs and digraphs. Discrete Appl. Math. **155**, 356–364 (2007)

19. Sebő, A., Tannier, E.: On metric generators of graphs. Math. Oper. Res. **29**(2), 383–393 (2004)

20. Slater, P.J.: Leaves of trees. Congr. Numerantium **14**, 549–559 (1975)

The Lexicographic Method
for the Threshold Cover Problem

Mathew C. Francis$^{(\boxtimes)}$ ⓘ and Dalu Jacob

Indian Statistical Institute, Chennai Centre, Chennai, India
{mathew,dalujacob}@isichennai.res.in

Abstract. The lexicographic method is a technique that was introduced
by Hell and Huang [*Journal of Graph Theory*, 20(3): 361–374, 1995] as
a way to simplify the problems of recognizing and obtaining representations of comparability graphs, proper circular-arc graphs and proper
interval graphs. This method gives rise to conceptually simple recognition algorithms and leads to much simpler proofs for some characterization theorems for these classes. Threshold graphs are a class of graphs
that have many equivalent definitions and have applications in integer
programming and set packing problems. A graph is said to have a threshold cover of size k if its edges can be covered using k threshold graphs.
Chvátal and Hammer conjectured in 1977 that given a graph G, a suitably constructed auxiliary graph G' has chromatic number equal to the
minimum size of a threshold cover of G. Although this conjecture was
shown to be false in the general case by Cozzens and Leibowitz, it was
shown to be true for graphs having a threshold cover of size 2 by Raschle
and Simon [*Proceedings of the Twenty-seventh Annual ACM Symposium
on Theory of Computing*, STOC '95, pages 650–661, 1995]. That is, a
graph G has a threshold cover of size 2 if and only if G' is bipartite—this
is the only known forbidden structure characterization of graphs having
a threshold cover of size 2. We show how the lexicographic method can be
used to obtain a completely new and much simpler proof for this result.
This method also gives rise to a simple new LexBFS-based algorithm for
recognizing graphs having a threshold cover of size 2. Although this algorithm is not the fastest known, it is a certifying algorithm that matches
the time complexity of the fastest known certifying algorithm for this
problem. The algorithm can also be easily adapted to give a certifying
recognition algorithm for bipartite graphs that can be covered by two
chain subgraphs.

Keywords: Lexicographic method · Threshold cover · Chain graph
cover

1 Introduction

We consider only simple, undirected and finite graphs. A graph G is said to be
a *threshold graph* if it does not contain a pair of edges ab, cd such that $ad, bc \notin$

© Springer Nature Switzerland AG 2020
M. Changat and S. Das (Eds.): CALDAM 2020, LNCS 12016, pp. 41–52, 2020.
https://doi.org/10.1007/978-3-030-39219-2_4

$E(G)$; or equivalently, G is $(2K_2, P_4, C_4)$-free [1]. A graph $G = (V, E)$ is said to be *covered* by the graphs $H_1, H_2, \ldots, H_k$ if $E(G) = E(H_1) \cup E(H_2) \cup \cdots \cup E(H_k)$. A graph G is said to have a *threshold cover* of size k if it can be covered by k threshold graphs. The *threshold dimension* of a graph G is defined to be the smallest integer k such that G has a threshold cover of size k. Mahadev and Peled [12] give a comprehensive survey of threshold graphs and their applications.

Chvátal and Hammer [1] showed that the fact that a graph has a threshold cover of size k is equivalent to the following: there exist k linear inequalities on $|V(G)|$ variables such that the characteristic vector of a set $S \subseteq V(G)$ satisfies all the inequalities if and only if S is an independent set of G (see [13] for details). They further defined the auxiliary graph G' (defined in Sect. 2) corresponding to a graph G and showed that any threshold cover of G must have size at least $\chi(G')$. This gave rise to the question of whether there exist any graph G that does not have a threshold cover of size $\chi(G')$. Cozzens and Leibowitz [4] showed the existence of such graphs. In particular, they showed that for every $k \geq 4$, there exists a graph G such that $\chi(G') = k$ but G has no threshold cover of size k. The question of whether such graphs exist for $k = 2$ seems to have been intensely studied but remained open for a decade (see [11]). Ibaraki and Peled [7] showed that if G is a split graph or if G' contains at most two non-trivial components, then $\chi(G') = 2$ if and only if G has a threshold cover of size 2. They further conjectured that every graph G satisfying $\chi(G') \leq 2$ has a threshold cover of size 2. Cozzens and Halsey [3] studied some properties of graphs having a threshold cover of size 2 and show that it can be decided in polynomial time whether the complement of a bipartite graph has a threshold cover of size 2. Finally, in 1995, Raschle and Simon [13] settled the question by extending the methods of Ibaraki and Peled: they showed that every graph G whose auxiliary graph G' is bipartite has a threshold cover of size 2. This proof is very technical and involves the use of a number of complicated reductions and previously known results. In particular, they construct a set of edges that have a "threshold completion" by finding a 2-colouring of G' that is so-called "AC_{2l}-free", where $l \geq 2$ (a colouring of G' is AC_{2l}-free, if for each colour class S, there is no cyclical sequence $v_1, v_2, \ldots, v_{2l}, v_1$ of vertices in G such that $v_i v_{i+1} \in S$ if and only if i is odd). It is then shown that this reduces to finding a 2-colouring of G' which is AC_6-free. This further reduces to finding a so-called "AP_6-free" 2-colouring of G' which further reduces to finding a so-called "double AP_6-free 2-colouring" of G'. The most intricate part is the proof of correctness of an algorithm that computes this particular kind of 2-colouring of G'.

The paper of Raschle and Simon also gives an $\mathcal{O}(|E(G)|^2)$ algorithm that checks whether a graph G has a threshold cover of size 2 and outputs two threshold graphs that cover G in case it has. If the input graph G does not have a threshold cover of size 2, the algorithm detects an odd cycle in the auxiliary graph G'. This odd cycle gives edges $e_1, e_2, \ldots, e_k$ in G, where k is odd, such that the edges e_i, e_{i+1}, for $1 \leq i < k$, and the edges e_k, e_1, can never both belong to any threshold subgraph of G (because their endpoints induce a $2K_2$, P_4 or C_4 in G). In this way, the algorithm provides an easily verifiable "certificate" for the fact that there does not exist two threshold graphs that cover G. If G does have a threshold cover of size 2, then the two threshold graphs returned by the algorithm that cover G form an easily verifiable certificate for that fact. Such algorithms are called *certifying algorithms* [8].

Since as noted above, an odd cycle in the auxiliary graph G' corresponds to a structure present in G that serves as an "obstruction" to it having a threshold cover of size 2, the result of Raschle and Simon can also be seen as a "forbidden structure characterization" of graphs having a threshold cover of size 2. That is, a graph G has a threshold cover of size 2 if and only if the said obstruction is not present in G. Such characterizations are known for many different classes of graphs—for example, interval graphs [9] and circular-arc graphs [5].

In this paper, we propose a completely different and self-contained proof for the theorem of Raschle and Simon that a graph G can be covered by two threshold graphs if and only if G' is bipartite. Our proof is short and direct, and also gives rise to a simpler (although having the same asymptotic worst case running time of $\mathcal{O}(|E(G)|^2)$) certifying recognition algorithm for graphs having a threshold cover of size 2.

Note that faster algorithms for determining if a graph has a threshold cover of size 2 are known. After the algorithm of Raschle and Simon [13], Sterbini and Raschle [15] used some observations of Ma [10] to construct an $\mathcal{O}(|V(G)|^3)$ algorithm for the problem. But this algorithm is not a certifying algorithm in the sense that if the input graph G does not have a threshold cover of size 2, it does not produce an obstruction in G that prevents it from having a threshold cover of size 2. Note that there is an obvious way to make this algorithm a certifying algorithm: if the algorithm answers that the input graph G does not have a threshold cover of size 2, run a secondary algorithm that constructs G' and finds an odd cycle in it (this odd cycle can serve as a certificate). But a naive implementation of the secondary algorithm will take time $\Omega(|E(G)|^2)$, and it is not clear if there is a way to run it in time $o(|E(G)|^2)$.

In the current work, we show that a graph G has a threshold cover of size 2 if and only if its auxiliary graph G' is bipartite using a technique called the *lexicographic method* which was introduced by Hell and Huang [6]. Hell and Huang demonstrated how this method can lead to shorter proofs and simpler recognition algorithms for certain problems that can be viewed as orienting the edges of a graph satisfying certain conditions—for example, they showed how this method can lead to simpler characterization proofs and recognition algorithms for comparability graphs, proper interval graphs and proper circular-arc graphs. The method starts by taking an arbitrary ordering of the vertices of the graph. It then prescribes choosing the lexicographically smallest (with respect to the given vertex ordering) edge to orient and then orienting it in one way or the other, along with all the edges whose orientations are forced by it. Hell and Huang showed that the lexicographic approach makes it easy to ensure that the orientation so produced satisfies the necessary conditions, if such an orientation exists. We adapt this technique to the problem of generating two threshold graphs that cover a given graph, if two such graphs exist. This shows that the applicability of the lexicographic method may not be limited to only problems involving orientation of edges. However, it should be noted that in our proof, we start with a *Lex-BFS ordering* of the vertices of the graph instead of an arbitrary ordering. It is an ordering of the vertices that gives the order in which a Lex-BFS, or *Lexicographic Breadth First Search*, a graph searching algorithm that was introduced by Rose, Tarjan and Lueker [14], may visit the vertices of the graph. A Lex-BFS ordering always gives an order in which a breadth-first search can visit the vertices of the graph, but has some additional properties. Lex-BFS can be implemented to run in time linear in the size of the input graph and Rose, Tarjan and Lueker originally used this algorithm to construct a linear-time algorithm for recognizing chordal graphs. Later, Lex-BFS based algorithms were discovered for the recognition of many different graph classes (see [2] for a survey).

2   Preliminaries

Let $G = (V, E)$ be any graph. Two edges ab, cd are said to form a pair of *cross edges* in G if $ad, bc \notin E(G)$. If ab, cd form a pair of cross edges in G, we say that the set $\{a, b, c, d\}$ is a *crossing set* in G (such a set is called an AC_4 in [13]). It is easy to see that threshold graphs are exactly the graphs that contain no pairs of cross edges, or equivalently no crossing set.

For a graph G, the auxiliary graph G' is defined to be the graph with $V(G') = E(G)$ and $E(G') = \{e_1 e_2 : e_1, e_2$ form a pair of cross edges in $G\}$. We shall refer to the vertex of G' corresponding to an edge $ab \in E(G)$ alternatively as $\{a, b\}$ or ab, depending upon the context. The following lemma is just a special case of the observation of Chvátal and Hammer [1] that a graph G cannot have a threshold cover of size less than $\chi(G')$.

Lemma 1. *If a graph $G = (V, E)$ has a threshold cover of size two then G' is bipartite.*

Proof. Let G be covered by two threshold graphs H_1 and H_2. By the definition of G', if $\{ab, cd\} \in E(G')$ then $ad, bc \notin E(G)$. The fact that H_1 and H_2 are threshold subgraphs of G then implies that neither H_1 nor H_2 can contain both the edges ab and cd. We therefore conclude that the sets $E(H_1)$ and $E(H_2)$ are both independent sets in G'. Since G is covered by H_1 and H_2, we have that $V(G') = E(H_1) \cup E(H_2)$. Thus, $\{E(H_1), E(H_2) \setminus E(H_1)\}$ forms a bipartition of G' into two independent sets. This completes the proof.   □

Our goal is to provide a new proof for the following theorem of Raschle and Simon [13].

Theorem 1. *A graph G can be covered by two threshold graphs if and only if G' is bipartite.*

By Lemma 1, it is enough to prove that if G' is bipartite, then G can be covered by two threshold graphs. In order to prove this, we find a specific 2-coloring of the non-trivial components of G' using the lexicographic method of Hell and Huang [6].

Let $<$ be an ordering of the vertices of G. Given two k-element subsets $S = \{s_1, s_2, \ldots, s_k\}$ and $T = \{t_1, t_2, \ldots, t_k\}$ of $V(G)$, where $s_1 < s_2 < \cdots < s_k$ and $t_1 < t_2 < \cdots < t_k$, S is said to be *lexicographically smaller* than T, denoted by $S < T$, if $s_j < t_j$ for some $j \in \{1, 2, \ldots, k\}$, and $s_i = t_i$ for all $1 \le i < j \le k$. In the usual way, we let $S \le T$ denote the fact that either $S < T$ or $S = T$. For a set $S \subseteq V(G)$, we abbreviate $\min_< S$ to just $\min S$. Note that the relation $<$ ("is lexicographically smaller than") that we have defined on k-element subsets of $V(G)$ is a total order. Therefore, given a collection of k-element subsets of $V(G)$, the lexicographically smallest one among them is well-defined.

3 Proof of Theorem 1

Assume that G' is bipartite. Let $<$ denote a Lex-BFS ordering of the vertices of G. The following observation states a well-known property of Lex-BFS orderings [2].

Observation 1. *For $a, b, c \in V(G)$, if $a < b < c$, $ab \notin E(G)$ and $ac \in E(G)$, then there exists $x \in V(G)$ such that $x < a < b < c$, $xb \in E(G)$ and $xc \notin E(G)$.*

We shall now construct a partial 2-coloring of the vertices of G' using the colors $\{1, 2\}$. Notice that choosing a color for any vertex in a component of G' fixes the colors of all the other vertices in that component. Recall that every vertex of G' is a two-element subset of $V(G)$. For every non-trivial component C of G', perform the following operation: Choose the lexicographically smallest vertex in C (with respect to the ordering $<$) and assign the color 1 to it. This fixes the colors of all the other vertices in C. Note that after this procedure, every vertex of G' that is in a non-trivial component has been colored either 1 or 2. For $i \in \{1, 2\}$, let $F_i = \{e \in V(G') : e \text{ is colored } i\}$. Further, let F_0 denote the set of all isolated vertices in G'. Clearly, F_0 is exactly the set of uncolored vertices of G' and we have $V(G') = F_0 \cup F_1 \cup F_2$. Consider the subgraphs $H_1 = (V, F_1 \cup F_0)$ and $H_2 = (V, F_2 \cup F_0)$ of G. We claim that H_1 and H_2 are two threshold graphs that cover G. Clearly $E(G) = E(H_1) \cup E(H_2)$; so it only remains to be proven that both H_1 and H_2 are threshold graphs. Note that for any edge $ab \in E(G)$, $ab \notin E(H_1) \Rightarrow ab \in F_2$ and $ab \notin E(H_2) \Rightarrow ab \in F_1$.

Observation 2. *If ab, cd form a pair of cross edges in G, then exactly one of the following is true:*

1. *$ab \in F_1$ and $cd \in F_2$, or*
2. *$ab \in F_2$ and $cd \in F_1$.*

Therefore, ab and cd cannot be present together in either H_1 or H_2.

Proof. As ab, cd form a pair of cross edges in G, the vertices ab and cd are adjacent in G'. Therefore one of them will be colored 1 and the other 2 in the partial 2-coloring of G'. This implies that one of ab, cd belongs to F_1 and the other to F_2. Since $F_1 = E(H_1) \setminus E(H_2)$ and $F_2 = E(H_2) \setminus E(H_1)$, ab and cd cannot be both present in either $E(H_1)$ or $E(H_2)$.   □

For $i \in \{1, 2\}$, let $\mathcal{P}_i = \{\{x, y, z, w\}: xy, zw \in E(H_i), xw \notin E(G)$ and $yz \in E(G) \setminus E(H_i)\}$ and $\mathcal{C}_i = \{\{x, y, z, w\}: xy, zw \in E(H_i), xw, yz \in E(G) \setminus E(H_i)\}$. By Observation 2, it can be seen that the crossing sets in H_i are exactly the elements of $\mathcal{P}_i \cup \mathcal{C}_i$. Define $\mathcal{P} = \mathcal{P}_1 \cup \mathcal{P}_2$ and $\mathcal{C} = \mathcal{C}_1 \cup \mathcal{C}_2$. Notice that in order to show that both H_1 and H_2 are threshold graphs, we only need to prove that $\mathcal{P} \cup \mathcal{C} = \emptyset$. We shall first show that $\mathcal{P} = \emptyset$. Suppose not. Let $\{a, b, c, d\}$ be the lexicographically smallest element in $\mathcal{P}$.

Lemma 2. $\{a, b, c, d\} \notin \mathcal{P}_1$.

Proof. Suppose for the sake of contradiction that $\{a, b, c, d\} \in \mathcal{P}_1$. By definition of $\mathcal{P}_1$, we can assume without loss of generality that $ab, cd \in E(H_1)$, $ad \notin E(G)$ and $bc \in E(G) \setminus E(H_1)$. Since $E(G) = E(H_1) \cup E(H_2)$, we have that $bc \in E(H_2) \setminus E(H_1)$, which implies that $bc \in F_2$. By the definition of F_2, we have that bc belongs to a non-trivial component C of G' and has been colored 2. Therefore $\{b, c\}$ is not the lexicographically smallest vertex in C. Let $\{b_k, c_k\}$ be the lexicographically smallest vertex in C (k is defined below). Then we have $\{b_k, c_k\} < \{b, c\}$ and by our construction the vertex $\{b_k, c_k\}$ must have received color 1. Let $bc = b_0 c_0, b_1 c_1, \ldots, b_{k-1} c_{k-1}, b_k c_k$ be a path in C between $\{b, c\}$ and $\{b_k, c_k\}$, where for $0 \le i < k$, $b_i c_{i+1}, b_{i+1} c_i \notin E(G)$. Note that k is odd, $b_i c_i \in F_2$ for each even i and $b_i c_i \in F_1$ for each odd i, where $0 \le i \le k$.

We claim that $ab_i, c_i d \in E(H_1)$ for each even i and $ab_i, c_i d \in E(H_2)$ for each odd i, where $0 \le i \le k$. We prove this by induction on i. The case where $i = 0$ is trivial as $b_0 = b$ and $c_0 = c$. So let us assume that $i > 0$. Consider the case where i is odd. As $i - 1$ is even, by the induction hypothesis we have, $ab_{i-1}, c_{i-1} d \in E(H_1)$. As $ab_{i-1}, b_i c_i \in E(H_1)$ and $b_{i-1} c_i \notin E(G)$, by Observation 2, we have that $ab_i \in E(G)$. Now as $ab_i, c_{i-1} d$ form a pair of cross edges in G and $c_{i-1} d \in E(H_1)$ the same observation then implies that $ab_i \in E(H_2)$. Similarly, as $c_{i-1} d, b_i c_i \in E(H_1)$ and $b_i c_{i-1} \notin E(G)$, we have $c_i d \in E(G)$. Again, as $c_i d, ab_{i-1}$ form a pair of cross edges in G and $ab_{i-1} \in E(H_1)$ we have $c_i d \in E(H_2)$. The case where i is even is also similar and hence the claim.

By the above claim, $ab_k, c_k d \in E(H_2)$. Since $b_k c_k \in F_1$, $b_k c_k \notin E(H_2)$. Recalling that $ad \notin E(G)$, we now have that $\{a, b_k, c_k, d\} \in \mathcal{P}_2$. Since $\{b_k, c_k\} < \{b, c\}$, we have that $\{a, b_k, c_k, d\} < \{a, b, c, d\}$, which is a contradiction.   □

Lemma 3. $\{a, b, c, d\} \notin \mathcal{P}_2$.

Proof. Suppose for the sake of contradiction that $\{a, b, c, d\} \in \mathcal{P}_2$. By definition of $\mathcal{P}_2$, we can assume without loss of generality that $ab, cd \in E(H_2)$, $ad \notin E(G)$ and $bc \notin E(H_2)$. Recall that $bc \notin E(H_2) \Rightarrow bc \in F_1$. As $bc \in F_1$, the vertex bc belongs to a non-trivial component of G'. Then there exists a neighbor $b'c'$ of bc in G' such that $bc', b'c \notin E(G)$. By Observation 2, $bc \in F_1$ implies $b'c' \in F_2$. Further, $ab, b'c' \in E(H_2)$ and $bc' \notin E(G)$ implies that $ab' \in E(G)$. Now ab', cd

form a pair of cross edges in G. Since $cd \in E(H_2)$, we now have by Observation 2 that $cd \in F_2$ and $ab' \in F_1$. This implies that cd is in a non-trivial component C_1 of G'. Similarly, as $cd, b'c' \in E(H_2)$ and $b'c \notin E(G)$ we have that $c'd \in E(G)$. Now $c'd, ab$ form a pair of cross edges in G. Since $ab \in E(H_2)$, we have by Observation 2 that $ab \in F_2$ and $c'd \in F_1$. This implies that ab is in a non-trivial component C_2 of G'.

We now prove two claims using the fact that $<$ is a Lex-BFS ordering of $V(G)$.

Claim 1. $d < a < c$ is not possible.

Suppose not. Note that $da \notin E(G)$ and $dc \in E(G)$. Then by Observation 1, there exists $x \in V(G)$ such that $x < d < a < c$, $xa \in E(G)$ and $xc \notin E(G)$. Now cd, xa form a pair of cross edges in G. By Observation 2, $cd \in F_2$ implies that $xa \in F_1$. As $bc \in F_1$, $xc \notin E(G)$ and $ab \notin E(H_1)$ (recall that $ab \in F_2$) we then have that $\{x, a, b, c\} \in \mathcal{P}_1$. Further, $x < d$ implies that $\{x, a, b, c\} < \{a, b, c, d\}$ which is a contradiction.

The next claim is symmetric to the claim above, but we give a proof for the sake of completeness.

Claim 2. $a < d < b$ is not possible.

Suppose not. Note that $ad \notin E(G)$ and $ab \in E(G)$. Then by Observation 1, there exists $x \in V(G)$ such that $x < a < d < b$, $xd \in E(G)$ and $xb \notin E(G)$. Now ab, xd form a pair of cross edges in G. By Observation 2, $ab \in F_2$ implies that $xd \in F_1$. As $bc \in F_1$, $xb \notin E(G)$ and $cd \notin E(H_1)$ (recall that $cd \in F_2$) we then have that $\{x, d, c, b\} \in \mathcal{P}_1$. Further, $x < a$ implies that $\{x, d, c, b\} < \{a, b, c, d\}$ which is a contradiction.

As $cd \in F_2$, cd must have received color 2 in the partial 2-coloring of G'. This means that cd is not the lexicographically smallest vertex in the component C_1. Let $\{c_k, d_k\}$ be the lexicographically smallest vertex in C_1. Then we have $\{c_k, d_k\} < \{c, d\}$ and by our construction, the vertex $\{c_k, d_k\}$ must have received color 1. Let $cd = c_0 d_0, c_1 d_1, \ldots, c_{k-1} d_{k-1}, c_k d_k$ be a path in C_1 between cd and $c_k d_k$, where for $0 \le i < k$, $c_i d_{i+1}, c_{i+1} d_i \notin E(G)$. Note that k is odd, $c_i d_i \in F_2$ for each even i and $c_i d_i \in F_1$ for each odd i, where $0 \le i \le k$.

Claim 3. $c_i b, c' d_i \in F_1$ for each even i and $c_i b, c' d_i \in F_2$ for each odd i, where $0 \le i \le k$.

We prove this by induction on i. The case $i = 0$ is trivial as $c_0 = c$ and $d_0 = d$. So let us assume that $i > 0$. Consider the case where i is odd. As $i - 1$ is even, we have by the induction hypothesis that $c_{i-1} b, c' d_{i-1} \in F_1$. As $c_i d_i, c_{i-1} b \in F_1$ and $c_{i-1} d_i \notin E(G)$, by Observation 2, we have $c_i b \in E(G)$. Now since $c_i b, c' d_{i-1}$ form a pair of cross edges in G (recall that $bc' \notin E(G)$) and $c' d_{i-1} \in F_1$, the same observation then implies that $c_i b \in F_2$. Similarly, as $c_i d_i, c' d_{i-1} \in F_1$ and $c_i d_{i-1} \notin E(G)$, we have that $c' d_i \in E(G)$. Now since $c' d_i, c_{i-1} b$ form a pair of cross edges in G and $c_{i-1} b \in F_1$, we can deduce as before that $c' d_i \in F_2$. The case where i is even can be proved in the same way. Hence the claim.

Recall that $ab \in F_2$, ab is in a non-trivial component C_2 of G', and it has color 2 in the partial 2-coloring of G'. Therefore, there exists a lexicographically smallest vertex $\{a_k, b_k\}$ in C_2 which has been colored 1. Clearly, $\{a_k, b_k\} < \{a, b\}$. Let $ab = a_0 b_0, a_1 b_1, \ldots, a_{k-1} b_{k-1}, a_k b_k$, be a path in C_2 between $\{a_k, b_k\}$ and $\{a, b\}$, where for $0 \le i < k$, $a_i b_{i+1}, a_{i+1} b_i \notin E(G)$. Note that k is odd, $a_i b_i \in F_2$ for each even i and $a_i b_i \in F_1$ for each odd i, where $0 \le i \le k$. The following claim is symmetric to Claim 3, but we give a proof for the sake of completeness.

Claim 4. $a_i b', cb_i \in F_1$ for each even i and $a_i b', cb_i \in F_2$ for each odd i, where $0 \le i \le k$.

We prove this by induction on i. The case $i = 0$ is trivial as $a_0 = a$ and $b_0 = b$. Consider the case where i is odd. As $i - 1$ is even we have $a_{i-1} b', cb_{i-1} \in F_1$. Now as $a_i b_i, a_{i-1} b' \in F_1$ and $a_{i-1} b_i \notin E(G)$, by Observation 2, we have that $a_i b' \in E(G)$. Now $a_i b', cb_{i-1}$ form a pair of cross edges (recall that $b'c \notin E(G)$) and $cb_{i-1} \in F_1$ the same observation then implies that $a_i b' \in F_2$. Similarly, as $a_i b_i, cb_{i-1} \in F_1$ and $a_i b_{i-1} \notin E(G)$, we have that $cb_i \in E(G)$. Now $cb_i, a_{i-1} b'$ form a pair of cross edges and $a_{i-1} b' \in F_1$, implying that $cb_i \in F_2$. The case where i is even can be proved in the same way. Hence the claim.

Recall that $c_k d_k \in F_1$. By Claim 3, $c_{k-1} b \in F_1$ and $c_k b \in F_2$, implying that $c_k b \notin E(H_1)$. As $c_{k-1} d_k \notin E(G)$ we then have $\{c_{k-1}, b, c_k, d_k\} \in \mathcal{P}_1$. Similarly, as $a_k b_k \in F_1$, $a_k b_{k-1} \notin E(G)$, and by Claim 4, we have $cb_{k-1} \in F_1$ and $cb_k \notin E(H_1)$ (as $cb_k \in F_2$), we have $\{a_k, b_k, c, b_{k-1}\} \in \mathcal{P}_1$. We get the final contradiction from the following claim.

Claim 5. Either $\{a_k, b_k, c, b_{k-1}\} < \{a, b, c, d\}$ or $\{c_{k-1}, b, c_k, d_k\} < \{a, b, c, d\}$.

Suppose $d > a$. By Claim 2, we then have $d > b$. Now, since $\{a_k, b_k\} < \{a, b\}$, we have $\{a_k, b_k, c, b_{k-1}\} < \{a, b, c, d\}$, and we are done. So we shall assume that $d < a$. By Claim 1, we now have that $c < a$, implying that $a > \max\{c, d\}$. If $\min\{c_k, d_k\} < \min\{c, d\}$, then we have $\min\{c_k, d_k\} < a, c, d$, which implies that $\{c_{k-1}, b, c_k, d_k\} < \{a, b, c, d\}$, proving the claim. So we shall assume that $\min\{c_k, d_k\} \ge \min\{c, d\}$. Therefore, since $\{c_k, d_k\} < \{c, d\}$, we have $\min\{c_k, d_k\} = \min\{c, d\}$ and $\max\{c_k, d_k\} < \max\{c, d\}$. Thus we have $a > \max\{c_k, d_k\}$, implying that $\{c_{k-1}, b, c_k, d_k\} < \{a, b, c, d\}$. □

From Lemmas 2 and 3, it follows that $\mathcal{P} = \emptyset$.

Lemma 4. $\mathcal{C} = \emptyset$.

Proof. Suppose for the sake of contradiction that $\mathcal{C} \ne \emptyset$. Then there exists $i \in \{1, 2\}$ such that $\mathcal{C}_i \ne \emptyset$. Consider an element $\{a, b, c, d\} \in \mathcal{C}_i$. We can assume without loss of generality that $ab, cd \in E(H_i)$, $ad, bc \in E(G) \setminus E(H_i)$. As $ad \in E(G) \setminus E(H_i)$, it belongs to a non-trivial component of G'. Therefore there exists a neighbor $a'd'$ of ad in G' such that $ad', a'd \notin E(G)$. Therefore by Observation 2, we have that $a'd' \in E(H_i)$. As $ab, a'd' \in E(H_i)$, where $ad' \notin E(G)$, by the same observation we then have $a'b \in E(G)$. Now if $a'b \in E(H_i)$, then the fact that $cd \in E(H_i)$, $bc \in E(G) \setminus E(H_i)$ and $a'd \notin E(G)$ implies that $\{a', b, c, d\} \in \mathcal{P}_i$ which is a contradiction to our earlier observation that $\mathcal{P} = \emptyset$. Therefore

$a'b \in E(G) \setminus E(H_i)$. As $ab, a'd' \in E(H_i)$ and $ad' \notin E(G)$, it then follows that $\{a, b, a', d'\} \in \mathcal{P}_i$ which again contradicts the fact that $\mathcal{P} = \emptyset$. This completes the proof.   □

We have now shown that $\mathcal{P} \cup \mathcal{C} = \emptyset$, or in other words, there is no crossing set in either H_1 or H_2. Thus H_1 and H_2 are two threshold graphs that cover G. We have thus shown that if G' is bipartite then G has a threshold cover of size two. As we already have Lemma 1, this completes the proof of Theorem 1.

4   A Certifying Algorithm

Our proof of Theorem 1 gives an algorithm which when given a graph G as input, either constructs two threshold graphs that cover G, or produces an odd cycle in G' as a certificate that G cannot be covered by two threshold graphs.

Algorithm 2-Threshold-Cover

Input: A graph G.
Output: If G has a threshold cover of size 2, two threshold graphs H_1, H_2
       such that they cover G, otherwise the auxiliary graph G' and an
       odd cycle in it.

1. Run the Lex-BFS algorithm on G (starting from an arbitrarily chosen vertex) to produce a Lex-BFS ordering $<$ of $V(G)$.
2. Construct the auxiliary graph G'.
3. Initialize $V(H_1) = V(H_2) = V(G)$ and $E(H_1) = E(H_2) = \{e \in V(G') : e$ belongs to a trivial component of $G'\}$.
4. **While** there exist uncolored vertices in a non-trivial component C of G', **do**
   (i) Choose the lexicographically smallest vertex uv in C and assign the color 1 to it.
   (ii) Complete the 2-coloring of C by doing a BFS starting from the vertex uv. If an odd cycle is detected, return the cycle and exit. Otherwise update $E(H_1) = E(H_1) \cup \{e \in V(C) : e$ is colored 1 in $G'\}$, $E(H_2) = E(H_2) \cup \{e \in V(C) : e$ is colored 2 in $G'\}$.
5. Output H_1 and H_2.

Correctness of the algorithm follows from the proof of Theorem 1. The Lex-BFS on G can be done in $\mathcal{O}(|V(G)| + |E(G)|)$ time and the remaining steps in $\mathcal{O}(|V(G')| + |E(G')|)$ time. As G' contains at most $|E(G)|$ vertices and at most $|E(G)|^2$ edges, the running time of this algorithm is $\mathcal{O}(|E(G)|^2)$.

5   The Chain Subgraph Cover Problem

A bipartite graph $G = (A, B, E)$ is called a *chain graph* if it does not contain a pair of edges whose endpoints induce a $2K_2$ in G. A collection of chain graphs $\{H_1, H_2, \ldots, H_k\}$ is said to be a *k-chain subgraph cover* of a bipartite graph G if it is covered by $H_1, H_2, \ldots, H_k$. The problem of deciding whether a bipartite graph G can be covered by k chain graphs, i.e. whether G has a k-chain subgraph cover, is known as the k-*chain subgraph cover (k-CSC)* problem. He showed that 3-CSC is NP-complete and pointed out that using the results of Ibaraki and Peled [7], the 2-CSC problem can be solved in polynomial time as it can be reduced to the problem of determining whether a split graph can be covered by two threshold graphs. Ma and Spinrad [11] note that a direct implementation of this approach to the 2-CSC problem only gives an $\mathcal{O}(|V(G)|^4)$ algorithm and instead propose an $\mathcal{O}(|V(G)|^2)$ algorithm for the problem. This algorithm works by reducing the 2-CSC problem to the problem of deciding whether a partial order has Dushnik-Miller dimension at most 2. Note that this algorithm does not produce a directly verifiable certificate, such as a forbidden structure in the graph, in case the input graph does not have a 2-chain subgraph cover. Our algorithm can be easily modified to make it an $\mathcal{O}(|E(G)|^2)$ certifying algorithm for deciding if an input bipartite graph G has a 2-chain subgraph cover as explained below. In fact, the only modification that is needed is to change the definition of G' so that two edges of G are adjacent in G' if and only if they induce $2K_2$ in G. As shown below, we can start with an arbitrary ordering of vertices in this case, i.e. we do not need to run the Lex-BFS algorithm to produce a Lex-BFS ordering of the input graph as the first step.

Let $G = (A, B, E)$ be a bipartite graph. We now redefine the meaning of the term "cross edges". Two edges $ab, cd \in E(G)$ are now said to be cross edges if and only if $a, c \in A$, $b, d \in B$ and $ad, bc \notin E(G)$. Note that the meaning of the auxiliary graph G' now changes, but our proof that $\chi(G') \leq 2$ if and only if there exists two graphs H_1, H_2, each containing no cross edges, such that $E(G) = E(H_1) \cup E(H_2)$ still works verbatim. We could, however, let the ordering $<$ on $V(G)$ be any arbitrary ordering. In that case, we cannot use Observation 1 and any argument that uses it. Note that Observation 1 is used only in the proof of Lemma 3. We show how this proof can be modified so that Observation 1 is no longer needed. Observation 1 is used only in Claims 1 and 2, which in turn are used only in Claim 5. Remove Claims 1 and 2 and replace the proof of Claim 5 with the following proof.

Claim 5. Either $\{a_k, b_k, c, b_{k-1}\} < \{a, b, c, d\}$ or $\{c_{k-1}, b, c_k, d_k\} < \{a, b, c, d\}$.

As $c_{k-1}b, c'd \in F_1$ and $c'b \notin E(G)$, we have $c_{k-1}d \in E(G)$. From Claim 3, we have $c'd_{k-1} \in F_1$. Then, $c'd_{k-1}, cb \in F_1$ where $c'b \notin E(G)$ implies that $cd_{k-1} \in E(G)$. As $c_k d_{k-1}, c_{k-1}d_k \notin E(G)$, we can conclude that $c \neq c_k$ and $d \neq d_k$. Since $c, c_k \in A$ and $d, d_k \in B$, we further have that $c \neq d_k$ and $d \neq c_k$. Therefore we get,

$$c, d > \min\{c_k, d_k\} \qquad \text{(as } c_k d_k < cd\text{)} \tag{1}$$

From Claim 4, we have $a_{k-1}b' \in F_1$. Now $a_{k-1}b', cb \in F_1$ where $cb' \notin E(G)$ implies that $a_{k-1}b \in E(G)$. Since $cb_{k-1}, ab' \in F_1$ and $cb' \notin E(G)$, we have $ab_{k-1} \in E(G)$. As $a_k b_{k-1}, a_{k-1}b_k \notin E(G)$ we can conclude that $a \neq a_k$ and

$b \neq b_k$. Since $a, a_k \in A$ and $b, b_k \in B$, we further have that $a \neq b_k$ and $b \neq a_k$. Therefore we get,

$$a, b > \min\{a_k, b_k\} \qquad (\text{as } a_k b_k < ab) \tag{2}$$

If $a \leq \min\{c_k, d_k\}$ and $d \leq \min\{a_k, b_k\}$, we get by (1) and (2) that $a \leq \min\{c_k, d_k\} < d \leq \min\{a_k, b_k\} < a$, which is a contradiction. Therefore, either $a > \min\{c_k, d_k\}$ or $d > \min\{a_k, b_k\}$. If $a > \min\{c_k, d_k\}$, then by (1), we get $\{c_{k-1}, b, c_k, d_k\} < \{a, b, c, d\}$, and we are done. Similarly, if $d > \min\{a_k, b_k\}$, then by (2), we have $\{a_k, b_k, c, b_{k-1}\} < \{a, b, c, d\}$, again we are done. This proves the claim.

Thus Algorithm 2-Threshold-Cover can be modified into a certifying recognition algorithm for deciding if a bipartite graph has a 2-chain subgraph cover by just changing the definition of G'. Moreover, this algorithm can choose any arbitrary ordering of the vertices of the input graph to start with and hence does not require the implementation of the Lex-BFS algorithm. Note that we do not know the answer to the following question: Would Algorithm 2-Threshold-Cover correctly decide whether the input graph G has a threshold cover of size 2 even if it lets $<$ be an arbitrary ordering of $V(G)$?

6   Conclusion

Chvátal and Hammer [1] showed that the problem of deciding whether an input graph has a threshold cover of size at most k is NP-complete, when k is part of the input. Yannakakis [16] observes that a bipartite graph $G = (A, B, E)$ has a k-chain subgraph cover if and only if the split graph H obtained from G by making every pair of vertices in A adjacent to each other has a threshold cover of size k. He notes that therefore, his proof of the NP-completeness of the 3-CSC problem implies that the problem of deciding if an input graph has a threshold cover of size at most 3 is also NP-complete.

We believe that our result demonstrates once again the power of the lexicographic method in yielding short and elegant proofs for certain kinds of problems that otherwise seem to need more complicated proofs. Further research could establish the applicability of the method to a wider range of problems.

References

1. Chvátal, V., Hammer, P.L.: Aggregations of inequalities. In: Studies in Integer Programming. Annals of Discrete Mathematics, vol. 1, pp. 145–162 (1977)
2. Corneil, D.G.: Lexicographic breadth first search – a survey. In: Hromkovič, J., Nagl, M., Westfechtel, B. (eds.) WG 2004. LNCS, vol. 3353, pp. 1–19. Springer, Heidelberg (2004). https://doi.org/10.1007/978-3-540-30559-0_1
3. Cozzens, M.B., Halsey, M.D.: The relationship between the threshold dimension of split graphs and various dimensional parameters. Discrete Appl. Math. **30**(2), 125–135 (1991)

 4. Cozzens, M.B., Leibowitz, R.: Threshold dimension of graphs. SIAM J. Algebr. Discrete Methods **5**(4), 579–595 (1984)
 5. Francis, M., Hell, P., Stacho, J.: Forbidden structure characterization of circular-arc graphs and a certifying recognition algorithm. In: Proceedings of the Twenty-Sixth Annual ACM-SIAM Symposium on Discrete Algorithms, SODA 2015, pp. 1708–1727 (2015)
 6. Hell, P., Huang, J.: Lexicographic orientation and representation algorithms for comparability graphs, proper circular arc graphs, and proper interval graphs. J. Graph Theory **20**(3), 361–374 (1995)
 7. Ibaraki, T., Peled, U.N.: Sufficient conditions for graphs to have threshold number 2. In: Hansen, P. (ed.) Annals of Discrete Mathematics (11). North-Holland Mathematics Studies, vol. 59, pp. 241–268. North-Holland, Amsterdam (1981)
 8. Kratsch, D., McConnell, R., Mehlhorn, K., Spinrad, J.: Certifying algorithms for recognizing interval graphs and permutation graphs. SIAM J. Comput. **36**(2), 326–353 (2006)
 9. Lekkerkerker, C.G., Boland, J.C.: Representation of a finite graph by a set of intervals on the real line. Fundamenta Mathematicae **51**, 45–64 (1962)
10. Ma, T.H.: On the threshold dimension 2 graphs. Technical report. Institute of Information Science, Academia Sinica, Nankang, Taipei, Republic of China (1993)
11. Ma, T.H., Spinrad, J.P.: On the 2-chain subgraph cover and related problems. J. Algorithms **17**(2), 251–268 (1994)
12. Mahadev, N.V.R., Peled, U.N.: Threshold Graphs and Related Topics, vol. 56. Elsevier, Amsterdam (1995)
13. Raschle, T., Simon, K.: Recognition of graphs with threshold dimension two. In: Proceedings of the Twenty-Seventh Annual ACM Symposium on Theory of Computing, STOC 1995, pp. 650–661 (1995)
14. Rose, D.J., Tarjan, R.E., Lueker, G.S.: Algorithmic aspects of vertex elimination on graphs. SIAM J. Comput. **5**(2), 266–283 (1976)
15. Sterbini, A., Raschle, T.: An $O(n^3)$ time algorithm for recognizing threshold dimension 2 graphs. Inf. Process. Lett. **67**(5), 255–259 (1998)
16. Yannakakis, M.: The complexity of the partial order dimension problem. SIAM J. Algebr. Discrete Methods **3**(3), 351–358 (1982)

Approximating Modular Decomposition Is Hard

Michel Habib[1,3(⊠)], Lalla Mouatadid[2], and Mengchuan Zou[1,3]

[1] IRIF, UMR 8243 CNRS & Paris University, Paris, France
habib@irif.fr
[2] Department of Computer Science, University of Toronto, Toronto, ON, Canada
[3] Gang Project, Inria Paris, Paris, France

Abstract. In order to understand underlying structural regularities in a graph, a basic and useful technique, known as modular decomposition, looks for subsets of vertices that have the **exact** same neighbourhood to the outside. These are known as modules and there exist linear-time algorithms to find them. This notion however is too strict, especially when dealing with graphs that arise from real world data. This is why it is important to relax this condition by allowing some noise in the data. However, generalizing modular decomposition is far from being obvious since most of the proposals lose the algebraic properties of modules and therefore most of the nice algorithmic consequences. In this paper we introduce the notion of ϵ-**module** which seems to be a good compromise that maintains some of the algebraic structure. Among the main results in the paper, we show that minimal ϵ-modules can be computed in polynomial time, on the other hand for maximal ϵ-modules it is already NP-hard to compute if a graph admits an 1-parallel decomposition, i.e. one step of decomposition of ϵ-module with $\epsilon = 1$.

1 Introduction

Introduced by Gallai in [13] to analyze the structure of comparability graphs, modular decomposition has been used and defined in many areas of discrete mathematics, including for graphs, 2-structures, automaton, partial orders, set systems, hypergraphs, clutters, matroids, boolean, and submodular functions [8,9,11,15], see [22] for a survey on modular decomposition. Since they have been rediscovered in many fields, modules appear under various names in the literature, they have been called intervals, externally related sets, autonomous sets, partitive sets, homogeneous sets, and clans. In most of the above examples the family of modules yields a kind of partitive family [4,5], and therefore has a unique modular decomposition tree that can be computed efficiently.

Roughly speaking, elements of the module behave exactly the same with respect to the outside of the graph, and therefore a module can be contracted to a single element without losing information. This technique has been used to

This work is supported by the ANR-France Project Hosigra (ANR-17-CE40-0022).

solve many optimization problems and has led to a number of elegant graph algorithms, see for instance [21]. On the other hand, direct applications of modular decomposition in other areas include computational protein-protein interaction networks [12] and graph drawing [25], to name a few. More recently, new applications have appeared in the study of networks in social sciences [29], where a module is considered as a regularity or a community that has to be detected and understood. Although it is well-known that almost all graphs have no non-trivial modules, in some recent experiments [24] in real data, many non-trivial modules were found in these graphs. How can we explain such a phenomena? It could be that the way the data is produced can generate modules, but it could also be because we reach some known regularities as predicted by Szemerédi's regularity lemma [30]. In fact for every $\epsilon > 0$ Szemerédi's lemma asserts that $\exists n_0$ such that all undirected graphs with more than n_0 vertices admits a ϵ-regular partition of the vertices. Such a partition is a kind of approximate modular decomposition. For graphs we now have linear-time algorithms to compute a modular decomposition tree, see [16]. In this paper we study a new generalization of modular decomposition, relaxing the strict neighbourhood condition of modules with a tolerance of some errors, i.e., some missing edges. The aims of this paper are twofold: first a theoretical study of an approximation of modular decomposition, and secondly a practical application for the computation of overlapping communities in bipartite graphs.

Organization of the Paper: We begin by giving necessary notations and a background on classical modular decomposition in Sect. 2, as well as illustrating some applications of ϵ-modular decomposition on various areas, on data compression and exact encodings for instance as well as in approximation algorithms. Section 3 introduces the notion of ϵ-modules and ϵ-modular decomposition, and their first basic properties. In Sect. 4, we give algorithmic results, in particular the computation of minimal ϵ-modules, as well as testing ϵ-primality. We then focus on two classes of graphs, bipartite graphs and 1-cographs (to be defined later) and conclude our discussion in the last section. In particular for bipartite graphs we can compute in $O(n^{2 \cdot \epsilon}(n + m))$ a covering of the vertices using maximal ϵ-modules, in which two ϵ-modules can overlap on at most $2 \cdot \epsilon$ vertices. This can be of great help for community detection in bipartite graphs.

2   Approximations of Modules

Let G be a simple, loop-free, undirected graph, with vertex set $V(G)$ and edge set $E(G)$, $n = |V(G)|$ and $m = |E(G)|$ are the number of vertices and edges of G respectively. For every $X \subseteq V(G)$, we denote by $G(X)$ the induced subgraph generated by X. $N(v)$ denotes the neighbourhood of v and $\overline{N(v)}$ the non-neighbourhood, this notation could also be generalized to set of vertices, i.e. for $X \subseteq V(G)$, $N(X) = \{x \in V(G) \setminus X$ such that $\exists y \in X$ and $xy \in E(G)\}$ (resp. $\overline{N(X)} = \{x \in V(G) \setminus X$ such that $\forall y \in X$ and $xy \notin E(G)\}$). For $x, y \in V$, we call **false-twins** if $N(x) = N(y)$ and **true-twins** if $N(x) \cup \{x\} = N(y) \cup \{y\}$.

A **Moore family** on a set X is a collection of subsets $S \subset X$ closed under intersection, and the set X itself.

Formally for an undirected graph G, a module $M \subseteq V(G)$ satisfies $\forall x, y \in M$, $N(x) \setminus M = N(y) \setminus M$. In other words, $V(G) \setminus M$ is partitioned into X, Y such that there is a complete bipartite between M and X, and no edge between M and Y. For convenience let us denote X (resp. Y) by $N(M)$ (resp. $\overline{N(M)}$). It is easy to see that all vertices within a module are at least false twins.

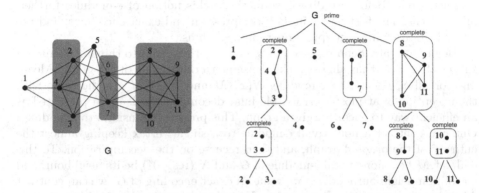

Fig. 1. Left, a graph with its maximal modules grouped. Right, its corresponding modular decomposition tree.

A single vertex $\{v\}$ and V are always modules, and called **trivial modules**. A graph that only has trivial modules is called a **prime graph**. By the Modular Decomposition Theorem [5,13], every graph admits a unique **modular decomposition tree**, in which a graph is decomposed via three types of internal nodes (operations): *parallel* (disjoint union) and *series* (connect every pair of nodes in disjoint sets X and Y), and *prime* nodes. The leaves represent the vertices of the graph, see Fig. 1.

A graph is a **complement reducible** graph if there is no prime node in its decomposition tree [7]. Complement reducible graphs are also known as **cographs** in the literature, or P_4-free graphs [28]. Cographs form a well studied graph class for which many classical NP-hard problems such as maximum clique, minimum coloring, maximum independent set, Hamiltonicity become tractable, see for instance [7].

Finding a non trivial tractable generalization of modules is not an easy task; indeed in trying to do so, we are faced with two main difficulties. The first one is to obtain a pseudo-generalization, for example if we change the definition of a module into: $\forall x, y \subset M$, $Neighbour^*(x) \setminus M = Neighbour^*(y) \setminus M$, where $Neighbour^*(x)$ means something like "vertices at distance at most k" or "joined by an odd path", etc. As it turns out, in many of these cases, the problem transforms itself into the computation of precisely the modules of some auxiliary graph built from the original one, some work in this direction avoiding this drawback can be found in [3]. The other one is NP-hardness. Consider the notion of *roles* defined in sociology, where two vertices play the same role in the social network if they have the same set of colours in the neighbourhood. If the colours of the vertices are given the problem is polynomially solvable, otherwise, this problem is a colouring one that is NP-hard to compute [10].

In this work, we will study two variations on the notion of modules both of which try to avoid these difficulties. Some aspects of them are polynomial to compute, and we believe they are worth studying further. We present the most promising one. Before we do so, we motivate this notion of ϵ-modules further, in two areas. The first one is in data compression and exact encodings, and the latter is its usefulness in approximation algorithms.

Before formally defining ϵ-modules, we want the reader to think of them as a subset of vertices that almost looks the same to the outside of the graph. Meaning, for all $x, y \in M$, an ϵ-module, $N(x) \backslash M$ and $N(y) \backslash M$ are the same with the exception of at most ϵ errors. Modular decomposition is often presented as an efficient way to encode a given graph. This property transmits to ϵ-modules. One can contract a non-trivial ϵ-module to a single vertex keeping almost the entirety of the original graph, and then recurse on the decomposition. To this end, let M be a non-trivial ϵ-module of G and X (resp. Y) be its neighbourhood (resp. non-neighbourhood). If we want an exact encoding of G, we can contract M to a unique vertex m connected to X, and not connected to Y, keep the subgraph $G(M)$ and keep tract of the errors (i.e., the edges missing in the bipartite (M, X) and the edges that appear in the bipartite (M, Y)). Thus, this new exact encoding has at least $|M| \cdot (|X| - \epsilon) - 1$ edges fewer than the original encoding.

A second application of approximate modular decomposition is in approximation algorithms. Consider the classical colouring and independent set problems on cographs. Both algorithms use modular decomposition to give optimal linear time solutions to both problems. The way the algorithms work is by computing a modular decomposition tree – known as the cotree – and keeping track of the series or parallel internal nodes by scanning the tree from the leaves to the root. We later define extensions of a cograph and a cotree to an ϵ-cograph and ϵ-cotree, and show in particular for $\epsilon = 1$, we get a simple 2-approximation for 1-cographs for these two classical problems, just by summing over all ϵ errors. In particular, when $\epsilon = 1$, this means the neighbourhood of every pair $x, y \in M$ differs by at most one neighbour/non-neighbour with respect to the outside of M.

2.1   Subset Families

Subset Families: Two sets A and B *overlap* if $A \cap B \neq \emptyset, A \backslash B \neq \emptyset$, and $B \backslash A \neq \emptyset$. Let $\mathcal{F}$ be a family of subsets of a ground set V. A set $S \in \mathcal{F}$ is called *strong* if $\forall S' \neq S \in \mathcal{F} : S$ does not overlap S'. Let Δ be the symmetric difference operation.

Definition 1 [5]. *A family of subsets $\mathcal{F}$ over a ground set V is **partitive** if it satisfies the following properties: (i) $\emptyset$, V and all singletons $\{x\}$ for $x \in V$ belong to $\mathcal{F}$. (ii) $\forall A, B \in \mathcal{F}$ that overlap, $A \cap B, A \cup B, A \backslash B$ and $A \Delta B \in \mathcal{F}$.*

Partitive families play fundamental roles in combinatorial decompositions [4,5]. Every partitive family admits a unique decomposition tree, with two types of nodes: *complete* and *prime*. It is well known that the strong elements of $\mathcal{F}$ form a tree ordered by the inclusion relation [5]. In this decomposition tree, every node corresponds to a set of the elements of the ground set V of $\mathcal{F}$, and the leaves of the tree are single elements of V.

3 ϵ-Modules and Basic Properties

One first idea to accept some errors is to say that at most k edges, for some fixed integer k, could be missing in the complete bipartite between M and $N(M)$, and symmetrically that at most k edges can exist between M and $\overline{N(M)}$. But doing so we loose most of the nice algebraic properties of modules of graphs which yield partitive families. Furthermore most algorithms for modular decomposition are based on these algebraic properties [5].

Another natural idea is to relax the condition on the complete bipartite between M and $N(M)$, for example asking for a graph that does not contain any $2K_2$. Unfortunately as shown in [27] to test whether a given graph admits such a decomposition is NP-complete. In fact they studied a generalized join decomposition solving a question asked in [18] studying perfection. This is why the following generalization of module defined for any integer ϵ, seems to be a good compromise[1].

Definition 2. *A subset $M \subseteq V(G)$ is an ϵ-module if $\forall x \in V(G) \setminus M$, either $|M \cap N(x)| \leq \epsilon$ or $|M \cap N(x)| \geq |M| - \epsilon$.*

In other words, we tolerate ϵ edges of errors per node outside the ϵ-module, and not ϵ errors per module. It should be noticed that with $\epsilon = 0$, we recover the usual definition of modules [16], i.e., $\forall x \in V(G) \setminus M$, either $M \cap N(x) = \emptyset$ or $M \cap N(x) = M$. Necessarily we will only consider $\epsilon < |V(G)| - 1$.

Let us consider the first simple properties yielded by this definition.

Proposition 1. *If M is an ϵ-module for G, then*

(a) M is an σ-module for G, for every $\epsilon \leq \sigma$.
(b) M is an ϵ-module for $\overline{G}$.
(c) M is an ϵ-module for every induced subgraph H of G such that $M \subseteq V(H)$.
(d) every ϵ-module of $G(M)$ is an ϵ-module of G.

Definition 3. *ϵ-**neighbourhoods**. For $A \subseteq V(G)$, let us denote by $N_\epsilon(A)$ (resp. $\overline{N_\epsilon(A)}$) the vertices of $V(G) \setminus A$, that are connected (resp. not connected) to M except for at most ϵ vertices. Similarly $S_\epsilon(A) = \{x \in V(G) \setminus A$ such that $\epsilon < |N(x) \cap A| < |A| - \epsilon\}$. $S_\epsilon(A)$ is called the set of ϵ-splitters of A.*

Equivalently a module can therefore be defined as a subset of vertices having no ϵ-splitter.

---

[1] We use ϵ to denote small error, despite being greater than 1.

Lemma 1. *Some easy facts, for $A \subseteq V(G)$.*

(i) *If $2 \cdot \epsilon + 1 \leq |A|$ then $N_\epsilon(A) \cap \overline{N_\epsilon(A)} = \emptyset$.*
(ii) *If $|A| \leq 2 \cdot \epsilon + 1$ then $S_\epsilon(A) = \emptyset$.*
(iii) *If $|A| = 2 \cdot \epsilon + 1$ then $N_\epsilon(A)$ and $\overline{N_\epsilon(A)}$ partition $V(G) \setminus A$.*
(iv) *If $|A| < 2 \cdot \epsilon$ then $N_\epsilon(A) = \overline{N_\epsilon(A)}$.*

So the subsets of vertices having size $2 \cdot \epsilon + 1$ seem to be crucial to study this new decomposition. If A is such a set, for every $z \notin A$ either $z \in N_\epsilon(A)$ or $z \in \overline{N_\epsilon(A)}$, but not both.

Lemma 2. *If s is a ϵ-splitter for a set A, then s is also a ϵ-splitter for every set $B \supseteq A$ such that $s \notin B$.*

Proof. Since $\epsilon < |A \cap N(x)|$ and $N(x) \cap A \subseteq N(x) \cap B$ we have: $\epsilon < |B \cap N(x)|$.
   $|A \cap N(x)| < |A| - \epsilon$ is equivalent to $|A \setminus N(x)| > \epsilon$. But $A \setminus N(x) \subseteq B \setminus N(x)$ implies $|B \setminus N(x)| > \epsilon$. So $\epsilon < |B \cap N(x)| < |B| - \epsilon$.   □

Theorem 1. *The family of ϵ-modules of a graph satisfies:*

(i) *$V(G)$ is an ϵ-module and $\forall A \subseteq V(G)$ such that $|A| \leq 2 \cdot \epsilon + 1$ are ϵ-modules.*
(ii) *$\forall A, B \subseteq V(G)$ ϵ-modules then, $A \cap B$ is an ϵ-module and for the subsets $A \setminus B$ and $B \setminus A$ their ϵ-splitters can only belong to $A \cap B$.*

Proof. (i) By definition $V(G)$ has no ϵ-splitter. Let $A \subseteq V(G)$, such that $|A| \leq 2 \cdot \epsilon + 1$ and let $x \in V(G) \setminus A$.
   Suppose $|N(x) \cap A| = k > \epsilon$ but since $|A| \leq 2 \cdot \epsilon + 1$, $\epsilon \geq |A| - \epsilon - 1$
   Therefore: $|N(x) \cap A| = k \geq |A| - \epsilon$ and A has no ϵ-splitter.
(ii) First we notice that if $A, B \subseteq V(G)$ are 2 trivial modules, obviously $A \cap B$, $A \setminus B$ and $B \setminus A$ are trivial ϵ-modules.
   Let $A, B \subseteq V(G)$ be two non trivial ϵ-modules. If $A \cap B$ has an ϵ-splitter outside of $A \cup B$ then using Lemma 2 also A, B would have an ϵ-splitter, a contradiction. Suppose now that $A \cap B$ admits an ϵ-splitter in $B \setminus A$ but then with the same Lemma we know that A would have an ϵ-splitter. Therefore $A \cap B$ is an ϵ-module. Let us now consider $A \setminus B$, if admits an ϵ-splitter in $B \setminus A$, using again Lemma 2, A would have a ϵ-splitter too. Similarly if the ϵ-splitter is outside $A \cup B$. Then the only potential ϵ-splitters for $A \setminus B$ and $B \setminus A$ are in $A \cap B$.   □

Corollary 1. *A graph G with $|V(G)| \leq 2 \cdot \epsilon + 2$ admits only trivial modules.*

By convention we will call such a graph ϵ-**degenerate** in order to distinguish with really ϵ-prime graphs.

Corollary 2. *If A, B are overlapping minimal ϵ-modules then $A \cap B$ is a trivial ϵ-module.*

We know then the ϵ-modules generate a Moore family of subsets worth studying. For usual modules as can be seen in [5,16], $A \cup B$, $B \setminus A$ and $A \setminus B$ are also modules. Unfortunately this does not always hold for ϵ-modules. Moreover we cannot bound the error as can be seen the next proposition.

Proposition 2. *Let $A, B \subseteq V(G)$ be two non trivial ϵ-modules, then*

1. *there could be $c = \Omega(\min(|A|, |B|))$, s.t. $A \cup B$ is not an ϵ-module, $\forall \epsilon \leq c$.*
2. *there could be $c = \Omega(n)$, s.t. $A\text{-}B$ is not an ϵ-module, for all $\epsilon \leq c$.*

In fact we can prove a weaker result.

Theorem 2. *Let $A, B \subseteq V(G)$ be two non trivial overlapping ϵ-modules, if $|A \cap B| \geq 2\epsilon + 1$ then $A \cup B$, $A \Delta B$ (i.e., symmetric difference) are 2ϵ-modules.*

Proof. Let $z \in V(G) \setminus B$. Since B is an ϵ-module then $S_\epsilon(B) = \emptyset$, since B is non trivial, $|B| \geq 2 \cdot \epsilon + 2$, therefore $N_\epsilon(B)$ and $\epsilon\text{-}\overline{N_\epsilon(B)}$ partition $V(G) \setminus B$, using Lemma 1. Suppose $z \in N_\epsilon(B)$, z has at most ϵ non neighbors in $A \cap B$. Therefore it has at least $\epsilon + 1$ neighbors in $A \cap B$, therefore $z \in N_\epsilon(A)$. For $A \cup B$ in the worst case z has at most ϵ non-neighbors in $A \setminus B$ and at most ϵ non-neighbors in $B \setminus A$. Therefore $A \cup B$ is a 2ϵ-module. For $A\Delta B$, the worst case is obtained when a given vertex $z \in A \cap B$ has ϵ errors in $A \setminus B$ and ϵ errors in $B \setminus A$. Therefore $A\Delta B$ is a 2ϵ-module. $\qquad\square$

Theorem 1 allows us to define a graph convexity. Since the family of ϵ-modules is closed under intersection, it yields a graph convexity and we can compute the minimal under inclusion ϵ-module $M(A)$ that contains a given set A, with strictly more that $2 \cdot \epsilon + 1$ elements, computing a **modular closure** via ϵ-splitters.

3.1   A Symmetric Variation of ϵ-Modules

One could want to restrict the definition of the ϵ-modules in a symmetric way. Here symmetric means that the condition is applied symmetrically on the vertices of the ϵ-module M and on the vertices outside, i.e., $V(G) \setminus M$.

Definition 4. *An ϵ-module M is **symmetric** if every $x \in M$ is adjacent (resp. non-adjacent) to all vertices in $N(M)$ (resp. $\overline{N(M)}$) except for at most ϵ vertices.*

In other words for $\epsilon = 1$, in the bipartite $M, N(M)$ only a matching is missing. It is a restriction of the ϵ-modules and all the previous results could be generated similarly for symmetric ϵ-modules.

Proposition 3. *If $\mathcal{P} = \{V_1, \ldots V_k\}$ is a partition of $V(G)$ into ϵ-modules, then the V_i's are necessarily symmetric ϵ-modules.*

With this definition in mind, we present extensions of the series and parallel nodes in the classical setting, as well as introduce a new graph class we call 1-*cographs*, the definition of which we present below.

Using Proposition 1(d) and mimicking the case of modular decomposition we may define an ϵ-tree decomposition as follows.

Definition 5. *An ϵ-tree decomposition is a tree whose nodes are labelled with ϵ-modules ordered by inclusion with 4 types of nodes ϵ-series, ϵ-parallel, ϵ-prime and ϵ-degenerate. Each level of the tree corresponds to a partition of $V(G)$, starting with $\{V(G)\}$ at the root and the leaves correspond to a partition of $V(G)$ into ϵ-degenerate nodes.*

For standard modular decomposition the notion of strong modules as modules that do not overlap with any other is central. For ϵ-modular decomposition we can observe that there are no strong modules other than V and $\{v\}, v \in V$ that are strong ϵ-modules. The reason is that, for $\epsilon \geq 1$, any subset of vertices of size 2 is a trivial ϵ-module, then assume there is a classical strong module $V_1 \neq V$, $|V_1| > 1$, then take any vertex $v \in V_1$ and any vertex $u \in V \setminus V_1$, then $\{u, v\}$ is a ϵ-module and overlapping with V_1.

3.2 ϵ-Series and ϵ-Parallel Operations

Definition 6. *For a graph G with $|V(G)| \geq 2\epsilon + 3$, we say that G admits an ϵ-series (resp. ϵ-parallel) decomposition if there exists a partition of $V(G)$, $\mathcal{P} = \{V_1, \ldots V_k\}$ such that: $\forall i, 1 \leq i \leq k, |V_i| \geq 2\epsilon + 1$ and $\forall x \in V_i$ and for every $j \neq i$, x is adjacent (resp. non-adjacent) to all vertices of V_j with perhaps ϵ errors.*

Using Proposition 3, all the V_i's are necessarily symmetric ϵ-modules. Furthermore in such cases every union of $V_i's$ are also symmetric ϵ-modules. Fortunately with $\epsilon = 1$ the problem of recognizing if a graph admits an 1-parallel decomposition corresponds to a nice combinatorial problem first studied in [14]. The complexity of this problem known as finding a matching cut-set is now well-known [1,6,23] and therefore we have:

Theorem 3. *Finding if a graph admits an 1-parallel decomposition is NP-hard.*

Proof. Let G be a graph with minimum degree 3, and suppose that it admits an 1-parallel decomposition into $V_1, \ldots, V_k$. Necessarily $\forall i, |V_i| > 1$, since there is no pending vertex. Therefore $\{V_1, \cup_{1 < i \leq k} V_i\}$ is a matching cut set of G. So using [6], deciding if a graph admits 1-parallel decomposition is NP-complete. □

Definition 7. *An ϵ-cograph is a graph that is decomposable with respect to ϵ-series, ϵ-parallel decompositions until we reach only degenerate subgraphs.*

Using this definition above, it is clear that cographs are precisely the 0-cographs and let us call ϵ-cotree and ϵ-modular decomposition the corresponding tree and decomposition of an ϵ-cograph.

Proposition 4. *A graph is an ϵ-cograph iff it admits a ϵ-cotree using only ϵ-series and ϵ-parallel internal nodes.*

Proof. Suppose that G admits a ϵ-series composition with a partition $\mathcal{P} = \{V_1, \ldots V_k\}$. First we must notice that these two operations are exclusive. It is the case since every part has at least $2 \cdot \epsilon + 1$ vertices, we cannot have 2 parts V_i, V_j both ϵ-connected and ϵ-disconnected. Therefore we start a ϵ-cotree starting with a node labelled ϵ-series and recurse on all the subgraphs $G(V_i)$ using proposition d.   □

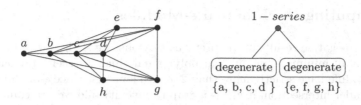

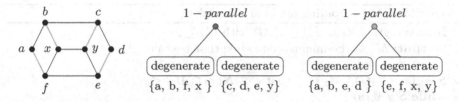

Fig. 2. 1-MD(H): A 1-cotree of a 1-cograph graph H. Notice that H is not a cograph since it contains 2 induced $P_4's$, namely $H(\{a, b, c, d\})$ and $H(\{e, f, g, h\})$.

Fig. 3. This 1-cograph G admits 2 different 1-cotrees, the internal nodes have the same label but the partitions of $V(G)$ induced by the leaves are not the same.

Let us consider now the 2 examples described in Figs. 2 and 3. The first one shows a 1-cograph H that admits a unique 1-cotree. The second one shows a 1-cograph G that admits 2 different legitimate 1-cotrees. Moreover by substituting in each vertex of G a graph isomorphic to G and if we repeat this process we can build a 1-cograph which admit **exponentially many** different legitimate 1-cotrees. At this particular time regarding for a graph the existence of ϵ-tree decomposition is not clear and as shown with ϵ-cographs we cannot ask for a unique one if it exists.

Unfortunately, it turns out, as one might expect, that finding this matching cutset is an NP-complete problem, as was shown by Chvátal in [6]. In the same work, Chvátal showed in particular that the problem is NP-hard on graphs with maximum degree four, and polynomial on graphs with maximum degree three.

Furthermore, it was shown that computing a matching cutsets in the following graph classes is polynomial: for graphs with max degree three [6], for weakly chordal graphs and line-graphs [23], for Series Parallel graphs [26], claw-free graphs and graphs with bounded clique width, as well as graphs with bounded treewidth [1], graphs with diameter 2 [2]. and for $(K_{1,4}, K_{1,4} + e)$-free graphs [20].

Therefore, to check if any of these graphs are 1-cographs, it suffices to run the corresponding matching cutset algorithms on either the graph or its complement.

But we conjecture that even 1-cographs that can be decomposed into *exactly* two cographs are hard to recognize in the general case.

4   Computing the Minimal ϵ-Modules

Despite the negative results of the previous sections, we shall now examine how to compute all minimal ϵ-modules in polynomial time. As seen previously non trivial ϵ-module have strictly more than $2 \cdot \epsilon + 1$ elements. Since ϵ-module family is closed under intersection, it yields a graph convexity and we can compute the minimal under inclusion ϵ-module $M(A)$ that contains a given set A, with strictly more that $2 \cdot \epsilon + 1$ elements, computing a **modular closure** via ϵ-splitters. In fact we built a series of subsets M_i starting with $M_0 = A$, and satisfying $M_i \subseteq M_{i+1}$.

Algorithm 1. Computing minimal ϵ-modules.

Input: a graph G and $A \subseteq V(G)$ with $|A| \geq 2 \cdot \epsilon + 2$.
Output: $M(A)$ the minimal ϵ-module that contains A
1 $M_0 \leftarrow A$, $i \leftarrow 0$;
2 $S \leftarrow \{x \in V(G) \setminus M_0$ such that $\epsilon < |N(x) \cap M_0| < |M_0| - \epsilon\}$;
3 **while** $S \neq \emptyset$ **do**
4 | $i \leftarrow i + 1$;
5 | $M_i \leftarrow M_{i-1} \cup S$;
6 | $S \leftarrow \{x \in V(G) \setminus M_i$ such that $\epsilon < |N(x) \cap M_i| < |M_i| - \epsilon\}$;
7 $M(A) \leftarrow M_i$;

Proposition 5. *Algorithm 1 computes the minimal ϵ-module that contains A.*

Proof. If A is an ϵ-module, then at line 2, $S = \emptyset$, else all the elements of S have to be added to A. In other words, using Lemma 2 there is no ϵ-module M such that: $A \subsetneq M \subsetneq A \cup S$. At the end of the While loop either $M_i = V(G)$ or we have found a non trivial ϵ-module.   □

Theorem 4. *Algorithm 1 can be implemented in $O(m + n)$.*

Proof. In fact we can implement it as a kind of graph search as follows.

At the end of this Algorithm 2 the set $M(A)$ contains a minimal ϵ-module that contains A. At first glance this algorithm requires $O(n^2)$ operations, since for each vertex we must consider all its neighbours and all its non neighbours. But if we use a partition refinement technique as defined in [17], starting with a partition of the vertices in $\{A, V(G) - A\}$, then we keep in the a same part $B(i,j)$ vertices x, y, such that $edge(x) = edge(y) = i$ and $nonedge(x) = nonedge(y) = j$. Then when visiting a vertex it suffices for each part $B(i,j)$ of the current partition to compute $B'(i+1,j) = B(i,j) \cap N(z)$ and $B''(i,j+1) = B(i,j) - N(z)$, which can be done in $O(|N(z)|)$. It should be noticed that the parts need not to be sorted in the current partition and we may have different parts with the same (edge, nonedge) values. Therefore can be implemented in $O(m+n)$.   □

Theorem 5. *Using Algorithm 1, one can compute all minimal non-trivial ϵ-modules in $O(m \cdot n^{2 \cdot \epsilon + 1})$.*

Algorithm 2. Computing minimal ϵ-modules.

Input: a graph G and $A \subseteq V(G)$ with $|A| \geq 2 \cdot \epsilon + 2$.
Output: $M(A)$ the minimal ϵ-module that contains A

```
1  OPEN ← A;
2  M(A) ← ∅;
3  ∀u ∈ V(G), CLOSED(u) ← FALSE; edge(u) ← 0; nonedge(u) ← 0;
4  while OPEN ≠ ∅ do
5  |   z ← Choice(OPEN), Delete z from OPEN ;
6  |   Add z to M(A);
7  |   CLOSED(z) ← TRUE;
8  |   ∀ neighbours u of z
9  |   if CLOSED(u) = FALSE and u ∉ M(A) then
10 |   |   edge(u) ← edge(u) + 1;
11 |   |   if ε < edge(u) and ε < nonedge(u) then
12 |   |   |   Add u to OPEN
13 |   ∀ non neighbours v of z
14 |   if CLOSED(u) = FALSE and u ∉ M(A) then
15 |   |   nonedge(v) ← nonedge(v) + 1;
16 |   |   if ε < edge(u) and ε < nonedge(u) then
17 |   |   |   Add v to OPEN
```

Proof. It suffices to use Algorithm 1 starting from every subset with $2 \cdot \epsilon + 2$ vertices. There exist $O(n^{2 \cdot \epsilon + 2})$ such subsets. And therefore this yields an algorithm in $O(m \cdot n^{2 \cdot \epsilon + 2})$. But we can do all the partition refinements in the whole, using the neighbourhood of one vertex only once. Since a vertex may belong to at most $n^{2 \cdot \epsilon + 1}$ parts, it yields an algorithm working in $O(m \cdot n^{2 \cdot \epsilon + 1})$.   □

If we consider the $\epsilon = 0$ case, this gives an implementation of the algorithm in [19] which also computes all minimal modules in $O(m \cdot n)$, to be compared to the original one in $O(n^4)$.

Corollary 3. *Using Theorem 5, one can compute a covering of $V(G)$ with an overlapping family of minimal ϵ-modules in $O(m \cdot n^{2 \cdot \epsilon + 1})$ and for any two members of the covering their overlapping is bounded by $2 \cdot \epsilon + 1$.*

Proof. Using Theorem 5, we can compute an overlapping family of minimal ϵ-modules in $O(m \cdot n^{2 \cdot \epsilon + 1})$. Perhaps it is not a covering of $V(G)$, since some vertices may not belong to any minimal non-trivial ϵ-module. To obtain a covering we simply add as singletons the remaining vertices.   □

This could be very interesting if we are looking for overlapping communities in social networks, the overlapping being bounded by $2 \cdot \epsilon + 1$.

To go a step further we can use Theorem 2 and merge every pair A, B of ϵ-modules such that $|A \cap B| \geq 2 \cdot \epsilon + 1$, either keeping $A \cup B$ as a $2 \cdot \epsilon$-module or compute $M(A \cup B)$ the minimal ϵ-module that contains $A \cup B$. But this depends on the structure of the maximal ϵ-modules, and unfortunately we do not know yet under what conditions there exists a unique partition into maximal ϵ-modules.

Corollary 4. *Checking if a graph is ϵ-prime can be done in $O(m \cdot n^{2 \cdot \epsilon + 1})$.*

Proof. It suffices to test whether for every set with $2 \cdot \epsilon + 2$ vertices if its closure is equal to $V(G)$. So either we find a non-trivial ϵ-module or the graph is ϵ-prime. Since every non-trivial ϵ-module necessarily contains one of the sets with $2 \cdot \epsilon + 2$ vertices. □

Corollary 5. *Finding for a graph G the smallest ϵ such that G has an ϵ-module can be done in $O(log n \cdot m \cdot n^{2 \cdot \epsilon + 1})$.*

Proof. To find such an ϵ we can use the above primality test in a dichotomic way, just adding a $log n$ factor to the complexity. □

5   The Bipartite Case

Let us consider now a bipartite graph $G = (X, Y, E(G))$. Unfortunately the ϵ-modules can be made up with vertices of both X, Y. But in some applications we are forced to consider X and Y separately. As for example in the case where X is a set of customers (resp. DNA sequences) and Y a set of products (resp. organisms), usually one wants to find regularities on each side of the bipartite graph. Let $\mathcal{F}_\epsilon(X) = \{M \mid \epsilon$-module of G such that $M \subseteq X\}$. It should be noticed that X is not always an ϵ-module of G.

Proposition 6. $\forall A, B \in \mathcal{F}_\epsilon(X)$, $A \cap B, A \setminus B, B \setminus A \in \mathcal{F}_\epsilon(X)$.

Proof. Using Theorem 1, the only ϵ-splitters of the sets $A \setminus B$ and $B \setminus A$ must belong to $A \cap B$. But since $A, B \subseteq X$, which is an independent set, it is impossible. □

As a consequence, using a notion of false ϵ-twins, we obtain.

Theorem 6. *For a bipartite graph $G = (X, Y, E(G))$, the maximal elements of $\mathcal{F}_\epsilon(X)$ can be computed in $O(n^{2 \cdot \epsilon}(n + m))$.*

It should be noticed that these maximal elements of $\mathcal{F}_\epsilon(X)$ may overlap, but the overlap is bounded by $2 \cdot \epsilon$. Furthermore the experimentation on real data has still to be done to evaluate the quality of the covering obtained.

5.1   Conclusions and Perspectives

The polynomial algorithms presented here have to be improved. Since it is hard to compute from the minimal ϵ-modules some hierarchy of modules – because we may have to consider an exponential number of unions of overlapping minimal ones – perhaps a good way to analyze a graph is to compute the families of minimal ones with $\epsilon = 1, 2, 3 \ldots$ and consider a hierarchy of overlapping families.

This notion of ϵ-modules yields many interesting questions both theoretical and practical. As for example for $\epsilon = 1$ to characterize 1-cographs or graphs that admits a 1-modular decomposition tree. The study of 1-primes is also worth to be done. On the other hand are there many ϵ-modules in real data? A natural consequence of this work is to extend the Courcelle's cliquewidth parameter into an ϵ-cliquewidth and similarly to define an ϵ-split operation in graphs.

Acknowledgments. The authors wish to thank anonymous reviewers for their helpful comments.

References

1. Bonsma, P.: The complexity of the matching-cut problem for planar graphs and other graph classes. JGT **62**(2), 109–126 (2009)
2. Borowiecki, M., Jesse-Józefczyk, K.: Matching cutsets in graphs of diameter 2. Theoret. Comput. Sci. **407**(1–3), 574–582 (2008)
3. Bui-Xuan, B., Habib, M., Limouzy, V., de Montgolfier, F.: Algorithmic aspects of a general modular decomposition theory. Discrete Appl. Math. **157**(9), 1993–2009 (2009)
4. Bui-Xuan, B., Habib, M., Rao, M.: Tree-representation of set families and applications to combinatorial decompositions. Eur. J. Comb. **33**(5), 688–711 (2012)
5. Chein, M., Habib, M., Maurer, M.C.: Partitive hypergraphs. Discrete Math. **37**(1), 35–50 (1981)
6. Chvátal, V.: Recognizing decomposable graphs. JGT **8**(1), 51–53 (1984)
7. Corneil, D.G., Lerchs, H., Burlingham, L.S.: Complement reducible graphs. Discrete Appl. Math. **3**(3), 163–174 (1981)
8. Ehrenfeucht, A., Harju, T., Rozenberg, G.: Theory of 2-structures. In: Fülöp, Z., Gécseg, F. (eds.) ICALP 1995. LNCS, vol. 944, pp. 1–14. Springer, Heidelberg (1995). https://doi.org/10.1007/3-540-60084-1_58
9. Ehrenfeucht, A., Rozenberg, G.: Theory of 2-structures, part II: representation through labeled tree families. Theor. Comput. Sci. **70**(3), 305–342 (1990)
10. Fiala, J., Paulusma, D.: A complete complexity classification of the role assignment problem. Theor. Comput. Sci. **349**(1), 67–81 (2005)
11. Fujishige, S.: Submodular Functions and Optimization. North-Holland, Amsterdam (1991)
12. Gagneur, J., Krause, R., Bouwmeester, T., Casari, G.: Modular decomposition of protein-protein interaction networks. Genome Biol. **5**(8), R57 (2004)
13. Gallai, T.: Transitiv orientierbare graphen. Acta Mathematica Academiae Scientiarum Hungaricae **18**, 25–66 (1967)
14. Graham, R.: On primitive graphs and optimal vertex assignments. Ann. N.Y. Acad. Sci. **175**, 170–186 (1970)
15. Habib, M., de Montgolfier, F., Mouatadid, L., Zou, M.: A general algorithmic scheme for modular decompositions of hypergraphs and applications. In: Colbourn, C.J., Grossi, R., Pisanti, N. (eds.) IWOCA 2019. LNCS, vol. 11638, pp. 251–264. Springer, Cham (2019). https://doi.org/10.1007/978-3-030-25005-8_21
16. Habib, M., Paul, C.: A survey of the algorithmic aspects of modular decomposition. Comput. Sci. Rev. **4**(1), 41–59 (2010)
17. Habib, M., Paul, C., Viennot, L.: Partition refinement techniques: an interesting algorithmic tool kit. Int. J. Found. Comput. Sci. **10**(2), 147–170 (1999)

18. Hsu, W.: Decomposition of perfect graphs. JCTB **43**(1), 70–94 (1987)
19. James, L.O., Stanton, R.G., Cowan, D.D.: Graph decomposition for undirected graphs. In: Proceedings of the 3rd Southeastern International Conference on Combinatorics, Graph Theory, and Computing, Florida Atlantic Univ., Boca Raton, Flo., pp. 281–290 (1972)
20. Kratsch, D., Le, V.B.: Algorithms solving the matching cut problem. Theor. Comput. Sci. **609**, 328–335 (2016)
21. Möhring, R.H.: Algorithmic aspects of the substitution decomposition in optimization over relations, set systems and boolean functions. Ann. Oper. Res. **6**, 195–225 (1985)
22. Möhring, R., Radermacher, F.: Substitution decomposition for discrete structures and connections with combinatorial optimization. Ann. Discret. Math. **19**, 257–356 (1984)
23. Moshi, A.M.: Matching cutsets in graphs. JGT **13**(5), 527–536 (1989)
24. Nabti, C., Seba, H.: Querying massive graph data: a compress and search approach. Future Gener. Comput. Syst. **74**, 63–75 (2017)
25. Papadopoulos, C., Voglis, C.: Drawing graphs using modular decomposition. J. Graph Algorithms Appl. **11**(2), 481–511 (2007)
26. Patrignani, M., Pizzonia, M.: The complexity of the matching-cut problem. In: Brandstädt, A., Le, V.B. (eds.) WG 2001. LNCS, vol. 2204, pp. 284–295. Springer, Heidelberg (2001). https://doi.org/10.1007/3-540-45477-2_26
27. Rusu, I., Spinrad, J.P.: Forbidden subgraph decomposition. Discrete Math. **247**(1–3), 159–168 (2002)
28. Seinsche, D.: On a property of the class of n-colorable graphs. JCTB **16**, 191–193 (1974)
29. Serafino, P.: Speeding up graph clustering via modular decomposition based compression. In: Proceedings of the 28th Annual ACM Symposium on Applied Computing, SAC 2013, Coimbra, Portugal, 18–22 March 2013, pp. 156–163 (2013)
30. Szemerédi, E.: On sets of integers containing no k elements in arithmetic progression. Acta Arithmetica **27**, 199–245 (1975)

Vertex-Edge Domination in Unit Disk Graphs

Sangram K. Jena and Gautam K. Das[✉]

Indian Institute of Technology, Guwahati, Guwahati, India
{sangram,gkd}@iitg.ac.in

Abstract. Let $G = (V, E)$ be a simple graph. A set $D \in V$ is called a vertex-edge dominating set of G if for each edge $e = (u, v) \in E$, either u or v is in D or one vertex from their neighbor is in D. Simply, a vertex $v \in V$, vertex-edge dominates every edge (u, v), as well as every edge adjacent to these edges. The vertex-edge dominating problem is to find a minimum vertex-edge dominating set of G. Herein, we study the vertex-edge dominating set problem in unit disk graphs and prove that this problem is NP-hard in that class of graphs. We also show that the problem admits a polynomial time approximation scheme (PTAS) in unit disk graphs.

Keywords: Dominating set · Vertex-edge dominating set · Unit disk graph · Approximation algorithm · Approximation scheme

1 Introduction

Let $G = (V, E)$ be a simple undirected graph. The *open neighbourhood* of a vertex $v \in V$ in G is the set $N_G(v) = \{u \in V \mid (u, v) \in E\}$ whereas the *closed neighbourhood* is the set $N_G[v] = N_G(v) \cup \{v\}$. A *dominating set* D of G is a subset of V such that every vertex in V is in D or adjacent to at least one vertex in D. A vertex $v \in D$ *dominates* all its neighbors and itself. The *dominating set problem* is to find a minimum cardinality subset $D \subseteq V$ such that D dominates all the vertices of G.

A *vertex-edge dominating set* (VEDs) of a simple undirected graph $G = (V, E)$ is a set $D \subseteq V$ of G such that every edge of G is incident with a vertex of D or a vertex adjacent to a vertex of D. The *VEDs problem* asks to find a VEDs of minimum size in a given graph. A set $D \subseteq V$ is a *double* vertex-edge dominating set if every edge $e \in E$ is vertex-edge dominated by at least two vertices in D. A set $D \subseteq V$ is called a *total* vertex-edge dominating set if every edge $e \in E$ is vertex-edge dominated by D and the graph induced by D has no isolated vertices.

2 Releated Work

The vertex-edge dominating set problem was introduced by Peters [18] and then studied further by different researchers. In particular, bounds on the vertex-edge

© Springer Nature Switzerland AG 2020
M. Changat and S. Das (Eds.): CALDAM 2020, LNCS 12016, pp. 67–78, 2020.
https://doi.org/10.1007/978-3-030-39219-2_6

domination number in several graph classes were studied in [3,14–16,20], vertex-edge degrees and vertex-edge domination polynomials of different graphs were discussed in [4,9,23,24], whereas the relations between some vertex-edge domination parameters were discussed in [3,5,12,15,16], several algorithmic aspects were discussed in [15]. Some variants of vertex-edge domination problem were studied in [2,6,11,13,19].

The minimum cardinality of a vertex-edge dominating set (double vertex-edge dominating set, respectively) of G is termed the *vertex-edge domination number* and denoted by $\gamma_{ve}(G)$ (the *double vertex-edge domination number*, $\gamma_{dve}(G)$, respectively). Krishnakumari et al. [14] proved that for every tree T of order $n \geq 3$ with ℓ leaves and s support vertices, we have $\frac{(n-\ell-s+3)}{4} \leq \gamma_{ve}(T) \leq \frac{n}{3}$. In [11], Krishnakumari et al. showed that determining $\gamma_{dve}(G)$ for bipartite graphs is NP-hard, whereas for every non-trivial connected graphs G, $\gamma_{dve}(G) \geq \gamma_{ve}(G)+1$, and for every tree T, we have $\gamma_{dve}(T) = \gamma_{ve}(T)+2$. They also provided two lower bounds on the double vertex-edge domination number of trees and unicycle graphs in terms of order n, the number of leaves and support vertices, respectively.

Boutrig et al. [3] presented a new relationship between the vertex-edge domination and some other domination parameters, answering the four open questions posed by Lewis [15]. Then, for every non-trivial connected $K_{1,k}$-free graph, with $k \geq 3$, they provided an upper bound for the independent vertex-edge domination number in terms of the vertex-edge domination number and showed that for every non-trivial tree the independent vertex-edge domination number can be bounded by the domination number. For connected C_5-free graphs, they also established an upper bound on the vertex-edge domination number. Next Boutrig and Chellali [2] studied the total vertex-edge domination. The minimum cardinality of a total vertex-edge dominating set of graph G called the *total vertex-edge domination number* and denoted by $\gamma_{ve}^t(G)$. They showed that determining $\gamma_{ve}^t(G)$ for bipartite graphs is NP-hard, and in case of tree T different from a star having order n, with ℓ leaves and s support vertices, respectively, we have $\gamma_{ve}^t(G^T) \leq \frac{(n-\ell+s)}{2}$. In the same article, they established a necessary condition for a graph G such to satisfy $\gamma_{ve}^t(G) = 2\gamma_{ve}(G)$ and for a tree T, $\gamma_{ve}^t(T) = 2\gamma_{ve}(T)$.

Later Venkatakrishnan and Kumar [22] proved that the minimum double vertex-edge dominating set problem is NP-hard for chordal graphs and APX-hard for bipartite graphs with maximum degree 5. They also proposed a linear-time algorithm for finding a minimum double vertex-edge dominating set in proper interval graphs. In addition, showed that the minimum double vertex-edge dominating set problem can not be approximated the factor $(1 - \epsilon) \ln |V|$ for any $\epsilon \geq 0$ unless $NP \subset DTIME(|V|^{O(\ln \ln |V|)})$. Finally, influence of edge removal, edge addition and edge subdivision on the double vertex-edge domination number of a graph was investigated by Krishnakumari and Venkatakrishnan [12]. Next, Horoldagva et al. [9] obtained some results on the regularity and irregularity of vertex-edge and edge-vertex degrees in graphs. Recently, Żyliński [25] proved that for any connected graph G of order $n \geq 6$, $\gamma_{ve}(G) \leq \lfloor \frac{n}{3} \rfloor$.

3   Our Contribution

We study the VEDs problem in unit disk graphs. A *unit disk graph* (UDG) is the intersection graph of equal-radii disks in the plane. Given a set $S = \{d_1, d_2, \ldots, d_n\}$ of n circular disks in the plane, each having diameter 1, the corresponding UDG $G = (V, E)$ is defined as follows: each vertex $v_i \in V$ corresponds to the disk $d_i \in S$, and there is an edge between two vertices if and only if the Euclidean distance between the relevant disk centers is at most 1.

We show that the decision version of the VEDs problem is NP-complete in unit disk graphs (Sect. 4). We also propose a polynomial-time approximation scheme for the problem (Sect. 5).

4   NP-Hardness

In this section, we show a polynomial-time reduction from the NP-hard *vertex cover* problem in planar graphs [7] to the VEDs problem to prove that the latter one is also NP-hard. The decision versions of both these problems are defined below.

The VEDs problem on UDGs (VEDS-UDG)
Instance: A unit disk graph $G = (V, E)$ and a positive integer k.
Question: Does there exist a vertex-edge dominating set D of G such that $|D| \leq k$?

The vertex cover problem on planar graphs (VC-PLA)
Instance: A planar graph $G = (V, E)$ having maximum degree 3 and a positive integer k.
Question: Does there exist a vertex cover C of G such that $|C| \leq k$?

Lemma 1 ([21]). *A planar graph $G = (V, E)$ with maximum degree 4 can be embedded in the plane using $O(|V|^2)$ area in such a way that its vertices are at integer coordinates and its edges are drawn so that they are made up of line segments of the form $x = i$ or $y = j$, for some i and j.*

This embedding is known as the *orthogonal drawing* of a graph. There is a linear-time algorithm given by Biedl and Kant [1] that gives an orthogonal drawing of a given graph with at most 2 bends along each edge (see Fig. 1).

Corollary 1. *A planar graph $G = (V, E)$ with maximum degree 3 and $|V| \geq 3$ can be embedded in the plane with its vertices are at $(4i, 4j)$ and its edges are drawn as a sequence of consecutive line segments on the lines $x = 4i$ or $y = 4j$, for some i and j.*

Lemma 2. *Let $G = (V, E)$ be an instance of VC-PLA with $|E| \geq 2$. An instance $G' = (V', E')$ of VEDS-UDG can be constructed from G in polynomial-time.*

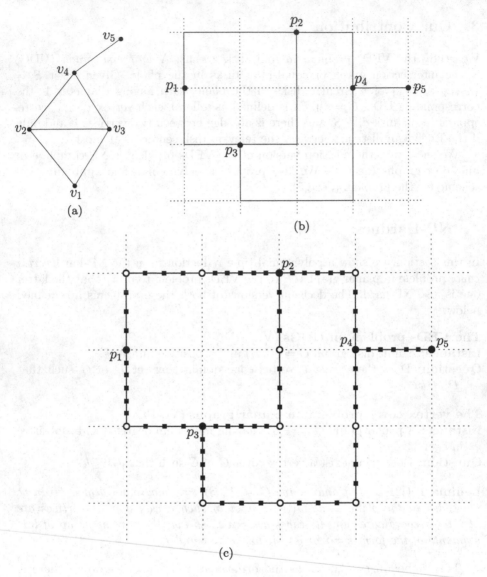

Fig. 1. (a) A planar graph G, (b) its embedding on a grid, and (c) a UDG construction from the embedding.

Proof. Our construction of G' from G is done in three steps.

Step 1: Embedding graph G into a grid of size $4n \times 4n$. G can be embedded in the plane using one of the algorithms [8,10] (see Lemma 1 and Corollary 1) with each of its edges as a sequence of connected line segment(s) of length four units. Let the total number of line segments used in the embedding is ℓ. The points $\{p_1, p_2, \ldots, p_n\}$ are termed *node points* in the embedding correspond to the vertex set $V = \{v_1, v_2, \ldots, v_n\}$ (see Fig. 1(a) and (b)).

Step 2: Adding extra points. For each edge (p_i, p_j) having length 4 units, (i) we add two points α and β on the edge (p_i, p_j) such that α is 0.8 unit apart from p_i and β is 0.8 unit appart from p_j, and (ii) add another three points between α and β with distance 0.6 unit from each other, respectively (thus adding five points in total, see edge (p_4, p_5) in Fig. 1(c)). For each edge of length greater than 4 units, we also add points as follows: (i) add a point in the joining point (grid point) of each line segments other than the *node points* and name it as a *joint point* (see empty circular points in Fig. 1(c)), (ii) for one of the two line segments whose one endpoint is associated with node point, we add five points at distances 0.8, 1.4, 2, 2.6 and 3.2 units from the node point, and for other line segments we add three points at distance 1 units from each other excluding the *joint points* (see the edge (p_3, p_4) in Fig. 1(c)). We name the points added in this step as *added points*.

Step 3: Construction step. For convenience, denote the set of node points by N and set of added points by A, respectively, that is, $N = \{p_i \mid v_i \in V\}$ and $A = \{q_1, q_2, \ldots, q_{4\ell + |E|}\}$. We construct a UDG $G' = (V', E')$, where $V' = N \cup A$ and there is an edge between two points in V' if and only if the Euclidean distance between the points is at most 1 (see Fig. 1(c)). Observe that $|N| = |V|(= n)$ and $|A| = 4\ell + |E|$, where ℓ is the total number of line segments in the embedding and $|E|$ is the total number of edges in G. Since G is planar, $|E| = O(n)$. It also follows from Lemma 1 that $\ell = O(n^2)$. Therefore both $|V'|$ and $|E'|$ are bounded by $O(n^2)$, and hence G' can be constructed in polynomial-time.   □

Theorem 1. VEDS-UDG *is NP-complete.*

Proof. For any given set $D \subseteq V$ and a positive integer k, we can verify in polynomial-time whether D is a vertex-edge dominating set of size at most k by checking whether each edge in E is vertex-edge dominated by a vertex in D or not. Hence, VEDS-UDG $\in$ NP.

Now, we need to prove VEDS-UDG $\in$ NP-hard. For the hardness proof, we show a polynomial time reduction from VC-PLA to VEDS-UDG. Let $G = (V, E)$ be an instance of VC-PLA. Construct the instance $G' = (V', E')$ of VEDS-UDG as discussed in Lemma 2. We have the following claim.

Claim. *G has a vertex cover of size at most k if and only if G' has a vertex-edge dominating set of size at most $k + \ell$.*

Necessity. Let $C \subseteq V$ be a vertex cover of G such that $|C| \leq k$. Let $N' = \{p_i \in N \mid v_i \in C\}$, i.e., N' is the set of vertices in G' that correspond to the vertices in C. The idea is to choose one vertex from each segment in the embedding such that the chosen vertex set $A'(\subseteq A)$ together with N', i.e., $N' \cup A'$ will form a VEDs of cardinality $k + \ell$ in G'. As C is a vertex cover in G, every edge in G has at least one of its endpoints in C. Let (v_i, v_j) be an edge in G and assume $v_i \in C$ (the same argument works for $v_j \in C$ or if both v_i and $v_j \in C$). It follows from the construction of G' that the edge (p_i, p_j) is represented as a sequence of line segments in the graph G', where p_i and p_j are nodes in G' corresponding to vertices v_i and v_j in G. Start traversing the segments from p_i, and add each fourth vertex to A' encountered from p_i to p_j in the traversal (see Fig. 2 for an illustration, where both big circles and squares belong to A' while traversing from p_1 to p_2, p_2 to p_3, p_1 to p_3, p_4 to p_2, p_4 to p_5 and p_4 to p_3, respectively).

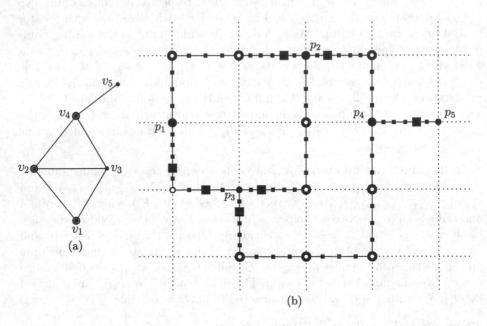

Fig. 2. (a) A vertex cover $\{v_1, v_2, v_4\}$ of G, and (b) the construction of A' in G'

Apply the same process to each chain of line segments in G' corresponding to each edge in G. Observe that the cardinality of A' is ℓ as we have chosen one vertex from each segment in the embedding. Let $D = N' \cup A'$. Now, observe that D is a vertex-edge dominating set in G' as each edge in G' is vertex-edge dominated by at least one vertex in D and $|D| = |N'| + |A'| \leq k + \ell$ as required.

Sufficiency. Let $D \subseteq V'$ be a VEDs of size at most $k + \ell$. We argue that G has a vertex cover of size at most k based upon the following claim: (i) at least one vertex on each segment in the embedding must belongs to D and hence $|A \cap D| \geq \ell$, where ℓ is the total number of segments in the embedding. We shall show that, by removing and/or replacing some vertices in D, a set of at most k vertices from N can be chosen such that the corresponding vertices in G is a vertex cover. Let $C = \{v_i \in V \mid p_i \in D \cap N\}$. If any edge (v_i, v_j) in G has none of its end vertices in C, then consider the points p_i and p_j corresponding to v_i and v_j respectively.

Case (i): If p_i is the only vertex that is connected with p_j in G', then the chain of segments (say ℓ') in the path $p_i \rightsquigarrow p_j$ in G' has at least $\ell' + 1$ vertices in D (see Fig. 3(a) for example). In this case, we delete one point from the segment containing two points in D and introduce p_i in D.

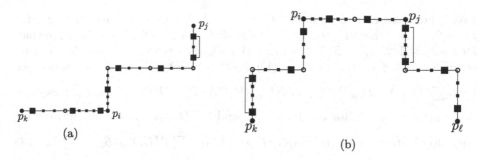

Fig. 3. (a) p_i is connected with only p_j, (b) p_i is connected with p_k and p_j is connected with p_ℓ.

Case (ii): If both p_i and p_j are connected with some points p_k and p_ℓ respectively in G', then either the chain of segments (say ℓ') in the path $p_i \leadsto p_j$ in G' has at least $\ell' + 1$ vertices in D (see Case (i)) or the chain of segments (say ℓ') in both the path $p_i \leadsto p_k$ and $(p_j \leadsto p_\ell)$ in G' has at least $\ell' + 1$ vertices in D (see Fig. 3(b) for example). In this case, we choose the segment having two points in D and remove one point of the segment from D and introduce p_j in D if $p_k \in D$ or p_ℓ is only connected with p_j, otherwise introduce p_i in D. Update C and repeat the process till every edge has at least one of its end vertices in C. Due to Claim (i), C is a vertex cover in G with $|C| \leq k$. Therefore, VEDS-UDG is NP-hard.

As VEDS-UDG is in NP as well as NP-hard, VEDS-UDG is NP-complete. □

5   Approximation Scheme

In this section, we propose a PTAS for the VEDs problem in UDGs. Let $G = (V, E)$ be a given UDG. Our PTAS is based on the concept of m-separated collection of subsets of V for some integer m. Given a graph G, let $d(u, v)$ denote the number of edges on a shortest path between u and v. For $V_1, V_2 \subseteq V$, $d(V_1, V_2)$ is defined as $d(V_1, V_2) = min_{u \in V_1, v \in V_2} \{d(u, v)\}$. We use notations $VED(A)$ and $VED_{opt}(A)$ to denote a vertex-edge dominating set of A ($\subseteq V$) in G and an optimal vertex-edge dominating set of A in G. We also define the closed neighborhood of a set $A \subseteq V$ as $N_G[A] = \bigcup_{v \in A} N_G[v]$ and the r-th neighborhood of a vertex v as $N_G^r[v] = \{u \in V \mid d(u, v) \leq r\}$ in G.

Let $\mathcal{S} = \{\mathcal{S}_1, \mathcal{S}_2, \ldots, \mathcal{S}_k\}$ be a collection of disjoint vertex subsets in G such that each $\mathcal{S}_i \subset V$ for $i = 1, 2, \ldots, k$. $\mathcal{S}$ is refered as a m-separated collection of vertices if $d(\mathcal{S}_i, \mathcal{S}_j) > m$, for $1 \leq i, j \leq k$ and $i \neq j$ (see Fig. 4 for a 4-separated collection). Nieberg and Hurink [17] considered 2-separated collection to propose a PTAS for the minimum dominating set problem on unit disk graphs.

Lemma 3. *If $\mathcal{S} = \{\mathcal{S}_1, \mathcal{S}_2, \ldots, \mathcal{S}_k\}$ is a m-separated collection in a graph $G = (V, E)$, then $\sum_{i=1}^{k} |VED_{opt}(\mathcal{S}_i)| \leq |VED_{opt}(V)|$ for each $m \geq 4$.*

Proof. For each $\mathcal{S}_i \in \mathcal{S}$, consider $P_i = \{u \in V \mid v \in S_i$ and $d(u,v) \leq 2\}$, for $i = 1, 2, \ldots, k$. Since $m \geq 4$, $P_i \cap P_j = \emptyset$ as $d(S_i, S_j) > m$ for $i \neq j$. Observe that, for each $i = 1, 2, \ldots, k$, $\mathcal{S}_i \subseteq P_i$ and $P_i \cap VED_{opt}(V)$ is a vertex-edge dominating set of $\mathcal{S}_i$. Therefore, $(P_i \cap VED_{opt}(V)) \cap (P_j \cap VED_{opt}(V)) = \emptyset$, and hence, we have $\sum_{i=1}^{k} |(P_i \cap VED_{opt}(V))| \leq |VED_{opt}(V)|$. As $P_i \cap VED_{opt}(V)$ is a vertex-edge dominating set of $\mathcal{S}_i$, for $i = 1, 2, \ldots, k$, and $VED_{opt}(V)$ is a minimum vertex-edge dominating set of the graph G, we obtain $\sum_{i=1}^{k} |VED_{opt}(\mathcal{S}_i)| \leq \sum_{i=1}^{k} |(P_i \cap VED_{opt}(V))| \leq |VED_{opt}(V)|$. $\square$

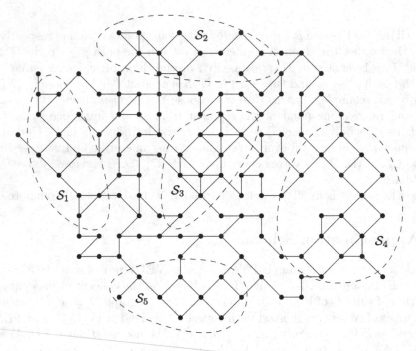

Fig. 4. A 4-separated collection $\mathcal{S} = \{\mathcal{S}_1, \mathcal{S}_2, \mathcal{S}_3, \mathcal{S}_4, \mathcal{S}_5\}$

Lemma 4. *Let* $\mathcal{S} = \{\mathcal{S}_1, \mathcal{S}_2, \ldots, \mathcal{S}_k\}$ *be an* m-*separated collection in a graph* $G = (V, E)$, $m \geq 4$, *and let* $\mathcal{R}_1, \mathcal{R}_2, \ldots, \mathcal{R}_k$ *be subsets of* V *with* $\mathcal{S}_i \subseteq \mathcal{R}_i$ *for all* $i = 1, 2, \ldots, k$. *If there exists* $\rho \geq 1$ *such that* $|VED_{opt}(\mathcal{R}_i)| \leq \rho |VED_{opt}(\mathcal{S}_i)|$ *holds for all* $i = 1, 2, \ldots, k$, *and if* $\bigcup_{i=1}^{k} VED_{opt}(\mathcal{R}_i)$ *is a vertex-edge dominating set in* G, *then* $\sum_{i=1}^{k} |VED_{opt}(\mathcal{R}_i)|$ *is at most* ρ *times the size of a minimum vertex-edge dominating set in* G.

Proof. $\sum_{i=1}^{k} |VED_{opt}(\mathcal{S}_i)| \leq |VED_{opt}(V)|$ (from Lemma 3).

Hence, $\sum_{i=1}^{k} |VED_{opt}(\mathcal{R}_i)| \leq \rho \sum_{i=1}^{k} |VED_{opt}(\mathcal{S}_i)| \leq \rho|VED_{opt}(V)|.$ □

5.1   Construction of Subsets

In this section, we discuss the process of constructing the desired 4-separated collection of subsets $\mathcal{S} = \{\mathcal{S}_1, \mathcal{S}_2, \ldots, \mathcal{S}_k\}$ and the corresponding subsets $\mathcal{R}_1, \mathcal{R}_2, \ldots, \mathcal{R}_k$ of V such that $\mathcal{S}_i \subseteq \mathcal{R}_i$ for all $i = 1, 2, \ldots, k$. The algorithm proceeds in an iterative manner. The basic idea of the algorithm is as follows: start with an arbitrary vertex $v \in V_i$, where V_i is the vertex set in the i-th iteration of the algorithm. Note that in the first iteration $V_1 = V$ and the algorithm computes $\mathcal{S}_1$ and $\mathcal{R}_1$. More specifically, for $r = 1, 2, \ldots$, we find the vertex-edge dominating set of the graphs induced by the r-th neighborhood as well as the $(r+4)$-th neighborhood of the vertex v until $|VED(N_G^{r+4}[v])| > \rho|VED(N_G^r[v])|$ holds. Here, $VED(N_G^{r+4}[v])$ and $VED(N_G^r[v])$ are vertex-edge dominating sets of $N_G^{r+4}[v]$ and $N_G^r[v]$, respectively, and $\rho = 1 + \epsilon$ ($\epsilon > 0$). Let $\hat{r}$ be the smallest r violating the above condition. Set $\mathcal{S}_i = N_G^{\hat{r}}[v]$, $\mathcal{R}_i = N_G^{\hat{r}+4}[v]$ and $V_i' = V_i \setminus N_G^{\hat{r}+3}[v]$. Note that removing $N_G^{\hat{r}+4}[v]$ from V_i implies removing the relevant edges connecting $N_G^{\hat{r}+4}[v]$ to $V_i \setminus N_G^{\hat{r}+4}[v]$ for which vertex-edge domination may not be maintained. Hence, removing $N_G^{\hat{r}+3}[v]$ from V_i removes the edges for which $VED(N_G^{\hat{r}+4}[v])$ is a vertex-edge dominating set. Let T_i be the set of vertices consisting of all singleton vertices after removing $N_G^{\hat{r}+3}[v]$ vertices from V_i in the i-th iteration of the algorithm. Set $V_{i+1} = V_i' \setminus T_i$. The process stops while $V_{i+1} = \emptyset$ and returns the sets $\mathcal{S} = \{\mathcal{S}_1, \mathcal{S}_2, \ldots, \mathcal{S}_k\}$ and $\mathcal{R} = \{\mathcal{R}_1, \mathcal{R}_2, \ldots, \mathcal{R}_k\}$. The collection of the sets $\mathcal{S}$ is a 4-separated collection.

We compute the vertex-edge dominating set of the r-th neighborhood of a vertex v, $VED(N_G^r[v])$ with respect to G as follows. Find a maximal independent set I for the graph induced by the vertices of $N_G^r[v]$. Observe that if we choose each vertex $v_i \in I$ in $VED(N_G^r[v])$, then it forms a vertex-edge dominating set for $N_G^r[v]$ (see Lemma 5).

Lemma 5. $VED(N_G^r[v])$ *is a VEDs of* $N_G^r[v]$ *in* G.

Proof. Suppose to the contrary, assume that $VED(N_G^r[v])$ is not a VEDs of the graph $G' = (V', E')$ induced by $N_G^r[v]$. That means, there exist an edge $(u, v) \in E'$ such that $N_{G'}[u] \notin VED(N_G^r[v])$ and $N_{G'}[v] \notin VED(N_G^r[v])$. It contradicts the fact that I is a maximal independent set in G'. Thus, the lemma. □

Lemma 6. *The worst case size of a vertex-edge dominating set of the r-th neighborhood of a vertex v is bounded by* $(r+2)^2$, *i.e.,* $|VED(N_G^r[v])| \leq (r+2)^2$.

Proof. We compute a maximal independent set I before computing a vertex-edge dominating set in the graph $G' = (V', E')$ induced by $N_G^r[v]$. The cardinality of a maximal independent set in the UDG G' is bounded by the number of non-intersecting unit disks packed in a disk of radius $r + 2$ centered at v. So, $|I| \le \frac{\pi(r+2)^2}{\pi(1)^2} = (r+2)^2$. From Lemma 5, in any graph, the cardinality of a minimum vertex-edge dominating set is bounded by the cardinality of maximal independent set. Therefore, $|VED(N_G^r[v])| \le (r+2)^2$.   □

Lemma 7. *For $\rho = 1 + \epsilon$, there always exists an r violating the condition $|(VED(N_G^{r+4}[v])| > \rho|VED(N_G^r[v])|$.*

Proof. Suppose to the contrary that there exists a vertex $v \in V$ such that $|(VED(N_G^{r+4}[v])| > \rho|VED(N_G^r[v])|$ for all $r = 1, 2, \ldots$.

From Lemma 6, we have $|(VED(N_G^{r+4}[v])| \le (r+6)^2$.

Therefore, if r is even,
$$(r+6)^2 \ge |(VED(N_G^{r+4}[v])| > \rho|VED(N_G^r[v])| > \cdots > \rho^{\frac{r}{2}}|VED(N_G^2[v])| \ge \rho^{\frac{r}{2}},$$
and if r is odd,
$$(r+6)^2 \ge |(VED(N_G^{r+4}[v])| > \rho|VED(N_G^r[v])| > \cdots > \rho^{\frac{r-1}{2}}|VED(N_G^1[v])| \ge \rho^{\frac{r-1}{2}}.$$

Hence,

$$(r+6) > \begin{cases} (\sqrt{\rho})^r, & \text{if r is even} \\ (\sqrt{\rho})^{r-1}, & \text{if r is odd} \end{cases} \tag{1}$$

Observe that in the inequality (1), the right side is an exponential function where as the left side is a polynomial function in r, which results in a contradiction. □

Lemma 8. *The smallest r violating inequality (1) is bounded by $O(\frac{1}{\epsilon} \log \frac{1}{\epsilon})$.*

Proof. Let $\hat{r}$ be the smallest r violating the inequalities (1). We prove $\hat{r} \le O(\frac{1}{\epsilon} \log \frac{1}{\epsilon})$ by using the inequality $\log(1+\epsilon) > \frac{\epsilon}{2}$ for $0 < \epsilon < 1$. For a fixed $\epsilon > 0$, consider the inequality $(1+\epsilon)^x < x^2$. Let $x = \frac{c}{\epsilon} \log \frac{1}{\epsilon}$, for some constant $c > 0$. By taking logarithm on the both sides of the inequality, we get $\log c + \log \log \frac{1}{\epsilon} > 0$. Note that we can always find ϵ' such that $\log c + \log \log \frac{1}{\epsilon} > 0$ for $0 < \epsilon < \epsilon'$. Therefore, $(1 + \epsilon)^x < x^2 < (x + 6)^2$ holds for sufficiently smaller ϵ values and hence, $\hat{r} \le O(\frac{1}{\epsilon} \log \frac{1}{\epsilon})$.   □

Lemma 9. *For a given $v \in V$, minimum vertex-edge dominating set $VED_{opt}(\mathcal{R}_i)$ of $\mathcal{R}_i$ can be computed in polynomial time.*

Proof. Let $G' = (V', E')$ be a graph induced by $\mathcal{R}_i \subseteq N_G^{r+4}[v]$. From Lemma 6, the size of $N_G^{r+4}[v]$ is bounded by $O(r^2)$, so we take every possible tuple of size at most $O(r^2)$ and check whether the selected tuple is a vertex-edge dominating set of the graph G'. This process takes $O(\binom{n}{r^2}) = O(n^{r^2})$ time. Since $r = O(\frac{1}{\epsilon} \log \frac{1}{\epsilon})$ by Lemma 8, $VED_{opt}(\mathcal{R}_i)$ can be computed in polynomial time.   □

Lemma 10. *For the collection of subsets $\{\mathcal{R}_1, \mathcal{R}_2, \ldots, \mathcal{R}_k\}$, $\mathcal{D} = \bigcup_{i=1}^{k} VED(\mathcal{R}_i)$ is a vertex-edge dominating set in $G = (V, E)$.*

Proof. To prove $\mathcal{D}$ is a vertex-edge dominating set of the graph G, we need to prove for every edge $(v_i, v_j) \in E$, there exists at least one vertex from $N_G[v_i]$ or $N_G[v_j]$ in $\mathcal{D}$. It follows from our construction of the subsets $\mathcal{R}_1, \mathcal{R}_2, \ldots, \mathcal{R}_k$ (Sect. 5.1) that each edge (v_i, v_j) belongs to a particular subset $\mathcal{R}_i$ and $VED(\mathcal{R}_i)$ is a vertex-edge dominating set of the graph induced by the vertices of $\mathcal{R}_i$. Thus the lemma. □

Corollary 2. $\mathcal{D}^* = \bigcup\limits_{i=1}^{k} VED_{opt}(\mathcal{R}_i)$ *is a vertex-edge dominating set in G, for the collection $\mathcal{R} = \{\mathcal{R}_1, \mathcal{R}_2, \ldots, \mathcal{R}_k\}$.*

Theorem 2. *For a given UDG, $G = (V, E)$, and an $\epsilon > 0$, we can design a $(1 + \epsilon)$-factor approximation algorithm to find a VEDs in G with running time $n^{O(c^2)}$, where $c = O(\frac{1}{\epsilon} \log \frac{1}{\epsilon})$.*

Proof. The proof of the theorem follows from Lemmas 4, 7, 9 and Corollary 2. □

6 Conclusion

In this article, we studied the minimum vertex-edge dominating set problem (VEDs) on unit disk graphs, and showed that the VEDs problem is NP-complete. We also proposed a PTAS for the VEDs problem.

References

1. Biedl, T., Kant, G.: A better heuristic for orthogonal graph drawings. Comput. Geom. **9**(3), 159–180 (1998)
2. Boutrig, R., Chellali, M.: Total vertex-edge domination. Int. J. Comput. Math. **95**(9), 1820–1828 (2018)
3. Boutrig, R., Chellali, M., Haynes, T.W., Hedetniemi, S.T.: Vertex-edge domination in graphs. Aequationes Math. **90**(2), 355–366 (2016)
4. Chellali, M., Haynes, T.W., Hedetniemi, S.T., Lewis, T.M.: On ve-degrees and ev-degrees in graphs. Discrete Math. **340**(2), 31–38 (2017)
5. Chen, X., Yin, K., Gao, T.: A note on independent vertex-edge domination in graphs. Discrete Optim. **25**, 1–5 (2017)
6. Chitra, S., Sattanathan, R.: Global vertex-edge domination sets in graph. Int. Math. Forum **7**, 233–240 (2012)
7. Garey, M.R., Johnson, D.S.: Computers and Intractability: A Guide to the Theory of NP-Completeness. Freeman, Dallas (1979)
8. Hopcroft, J., Tarjan, R.: Efficient planarity testing. J. ACM (JACM) **21**(4), 549–568 (1974)
9. Horoldagva, B., Das, K.C., Selenge, T.: On ve-degree and ev-degree of graphs. Discrete Optim. **31**, 1–7 (2019)
10. Itai, A., Papadimitriou, C.H., Szwarcfiter, J.L.: Hamilton paths in grid graphs. SIAM J. Comput. **11**(4), 676–686 (1982)
11. Krishnakumari, B., Chellali, M., Venkatakrishnan, Y.B.: Double vertex-edge domination. Discrete Math. Algorithms Appl. **9**(4), 1–11 (2017)

12. Krishnakumari, B., Venkatakrishnan, Y.B.: Influence of the edge removal, edge addition and edge subdivision on the double vertex-edge domination number of a graph. Natl. Acad. Sci. Lett. **41**(6), 391–393 (2018)
13. Krishnakumari, B., Venkatakrishnan, Y.B.: The outer-connected vertex edge domination number of a tree. Commun. Korean Math. Soc. **33**(1), 361–369 (2018)
14. Krishnakumari, B., Venkatakrishnan, Y.B., Krzywkowski, M.: Bounds on the vertex-edge domination number of a tree. C. R. Math. **352**(5), 363–366 (2014)
15. Lewis, J.: Vertex-edge and edge-vertex parameters in graphs. Dissertation presented to Graduate School of Clemson University (2007)
16. Lewis, J., Hedetniemi, S.T., Haynes, T.W., Fricke, G.H.: Vertex-edge domination. Utilitas Math. **81**, 193–213 (2010)
17. Nieberg, T., Hurink, J.: A PTAS for the minimum dominating set problem in unit disk graphs. In: Erlebach, T., Persinao, G. (eds.) WAOA 2005. LNCS, vol. 3879, pp. 296–306. Springer, Heidelberg (2006). https://doi.org/10.1007/11671411_23
18. Peters, K.: Theoretical and algorithmic results on domination and connectivity. Dissertation presented to Graduate School of Clemson University (1987)
19. Siva Rama Raju, S., Nagaraja Rao, I.: Complementary nil vertex edge dominating sets. Proyecciones (Antofagasta) **34**(1), 1–13 (2015)
20. Thakkar, D., Jamvecha, N.P.: About ve-domination in graphs. Ann. Pure Appl. Math. **14**(2), 245–250 (2017)
21. Valiant, L.G.: Universality considerations in VLSI circuits. IEEE Trans. Comput. **100**(2), 135–140 (1981)
22. Venkatakrishnan, Y.B., Naresh Kumar, H.: On the algorithmic complexity of double vertex-edge domination in graphs. In: Das, G.K., Mandal, P.S., Mukhopadhyaya, K., Nakano, S. (eds.) WALCOM 2019. LNCS, vol. 11355, pp. 188–198. Springer, Cham (2019). https://doi.org/10.1007/978-3-030-10564-8_15
23. Vijayan, A., Nagarajan, T.: Vertex-edge dominating sets and vertex-edge domination polynomials of wheels. IOSR J. Math. **10**(5), 14–21 (2014)
24. Vijayan, A., Nagarajan, T.: Vertex-edge domination polynomial of graphs. Int. J. Math. Arch. **5**(2), 281–292 (2014)
25. Żyliński, P.: Vertex-edge domination in graphs. Aequationes Math. **93**(4), 735–742 (2019)

Geometric Planar Networks
on Bichromatic Points

Sayan Bandyapadhyay[1] (ID), Aritra Banik[2] (ID), Sujoy Bhore[3($\boxtimes$)] (ID),
and Martin Nöllenburg[3] (ID)

[1] Department of Informatics, University of Bergen, Bergen, Norway
[2] School of Computer Sciences, NISER, Bhubaneswar, India
[3] Algorithms and Complexity Group, Technische Universität Wien, Vienna, Austria
sujoy.bhore@gmail.com

Abstract. We study four classical graph problems – Hamiltonian path, Traveling salesman, Minimum spanning tree, and Minimum perfect matching on geometric graphs induced by bichromatic (red and blue) points. These problems have been widely studied for points in the Euclidean plane, and many of them are NP-hard. In this work, we consider these problems in two restricted settings: (i) collinear points and (ii) equidistant points on a circle. We show that almost all of these problems can be solved in linear time in these constrained, yet non-trivial settings.

1 Introduction

In this article, we study four classical graph problems on geometric graphs induced by bichromatic (red and blue) points. Suppose, we are given a set R of n red points and a set B of m blue points in the Euclidean plane. Consider the complete bipartite graph $G(R, B, E)$ on $R \cup B$, where the set E of edges contains all bichromatic edges between the red points and the blue points. Also, suppose the graph $G(R, B, E)$ is embedded in the plane: the points are the vertices and each edge is represented by the segment between the two corresponding endpoints. We denote these edges as *bichromatic segments*, where each bichromatic segment connects a red point with a blue point. A subgraph of $G(R, B, E)$ (or equivalently a subset of edges of E) is called non-crossing (or planar) if no pair of the edges of the subgraph cross each other. Next, we discuss the four graph problems on the bipartite graph $G(R, B, E)$ induced by $R \cup B$.

In the BICHROMATIC HAMILTONIAN PATH problem, the objective is to find a path in $G(R, B, E)$ that spans all the red and blue points. Equivalently, one would like to find a polygonal chain that connects all the red points and the blue points alternately through bichromatic segments. It is not hard to see that a Hamiltonian path exists in $G(R, B, E)$ if and only if $m - 1 \leq n \leq m + 1$, and if there exists one, it can be computed efficiently, as $G(R, B, E)$ is a

Research of Sujoy Bhore and Martin Nöllenburg is supported by the Austrian Science Fund (FWF) grant P 31119.

(a)   (b)

Fig. 1. Two instances without a non-crossing Hamiltonian path. Figure (a) is not in general position, Figure (b) (borrowed from [14]) is in general position. (Color figure online)

complete bipartite graph. A more interesting problem is the NON-CROSSING BICHROMATIC HAMILTONIAN PATH problem where the objective is to find a non-crossing Hamiltonian path. Note that one can construct instances with $m - 1 \le n \le m + 1$, where it is not possible to find any non-crossing Hamiltonian path. In Fig. 1, we demonstrate two such instances.[1] Figure 1(a) has eight points where four of them lie on a horizontal line L and the remaining four lie on a line parallel to L. Notice that, there must be one red and one blue point with degree 1. One can verify by enumerating all possible paths that there is no non-crossing Hamiltonian path that spans these points. The example in Fig. 1(b) has thirteen points in general position, i.e., no pair of the points are collinear, and also does not admit a non-crossing Hamiltonian path. Indeed, if $2 \le n \le 12$, then for any given $\lfloor n/2 \rfloor$ red (resp. blue) points and $\lceil n/2 \rceil$ blue (resp. red) points in general position, there exists a non-crossing Hamiltonian path [12]. Due to the uncertainty of the existence of non-crossing Hamiltonian paths in the general case, researchers have also considered the problem of finding a non-crossing alternating path of length as large as possible [13, 16].

A related problem is the BICHROMATIC TRAVELING SALESMAN PATH (BICHROMATIC TSP) problem where one would like to find a minimum weight Hamiltonian path in $G(R, B, E)$. The weight of each edge is the (Euclidean) length of the corresponding segment. The weight of a path is the sum of the weights of the edges along the path. For simplicity, we assume $n = m$. A straightforward reduction from the (monochromatic) Euclidean TSP [4] (replace each point by a bichromatic pair that is small distance apart) shows that Bichromatic TSP is also NP-hard. One simple, but powerful fact is that an optimum Euclidean TSP is always non-crossing. This helps to obtain a PTAS [4] for the problem. However, an optimum Bichromatic TSP is not necessarily non-crossing which makes its computation much harder compared to Euclidean TSP. The best known approximation factor for the Bichromatic TSP problem is 2 due to Frank *et al.* [11] who improved the 2.5-approximation of Anily *et al.* [3]. For a

---

[1] In all the figures throughout the paper, we show red (resp. blue) points by squares (resp. disks).

set of collinear points, Evans *et al.* [10] gave a quadratic time algorithm for computing an optimum non-crossing TSP, every edge of which is a poly-line with at most two bends.

Next, we consider the BICHROMATIC SPANNING TREE problem where the objective is to compute a minimum weight spanning tree of $G(R, B, E)$. Note that this problem can be solved efficiently by any standard minimum spanning tree algorithm. A more interesting problem is the NON-CROSSING BICHROMATIC SPANNING TREE problem where additionally the computed tree must be planar (non-crossing). Borgelt *et al.* [9] showed that this problem is NP-hard. For points in general position, they gave a near-linear time $O(\sqrt{n})$-approximation. On the other hand, for points in convex position, they gave an exact cubic-time algorithm. Another line of work that received much attention is where the task is to find a degree-bounded non-crossing spanning tree [8].

Finally, we consider the BICHROMATIC MATCHING problem. Again assume that $n = m$ for simplicity. We would like to find a minimum weight perfect matching in $G(R, B, E)$. The weight of an edge is the Euclidean distance between its corresponding points. It is a well-known fact that a minimum weight bichromatic matching (for points in general position) in the plane is always non-crossing, which follows from the observation that the sum of the diagonals of a convex quadrilateral is strictly larger than the sum of any pair of opposite sides. This implies that using any standard bipartite matching algorithm one can solve Bichromatic matching exactly. But, algorithms with better running time have been designed by exploiting the underlying geometry of the plane. Recently, Kaplan *et al.* [15] designed an $O(n^2 \text{poly}(\log n))$ algorithm for the problem improving the $O(n^{2+\epsilon})$ algorithm due to Agarwal *et al.* [2], where poly(.) is a polynomial function. Abu-Affash *et al.* [1] and Biniaz *et al.* [7] studied variants of the bichromatic matching problem.

In this article, we consider the above mentioned problems in two restricted settings: (i) for collinear points and (ii) for equidistant points on a circle. For all problems, we assume that $n = m$ for simplicity. We note that the case of non-crossing graphs on collinear points is closely related to topological 1-page or 2-page book embeddings [6], which have all vertices placed on a line (called the spine) and the edges drawn without crossings in one or two of the halfplanes (but not in both) defined by the spine (called the pages). In our case we assume that the edges are drawn as (circular) arcs or 1-bend polylines either above or below the spine. We assume that their weight is given by the Euclidean distance of their endpoints. If the arcs are drawn infinitesimally close to the spine, these weights correspond to the lengths of the arcs.

Our Results. The main results obtained in this work are the following.

➡ NON-CROSSING HAMILTONIAN PATH FOR COLLINEAR POINTS – We prove that for any collinear configuration of the points, there always exists a non-crossing Hamiltonian path. We give a linear-time algorithm for computing such a path (Sect. 2).

➡ MINIMUM SPANNING TREE FOR COLLINEAR POINTS – We give a linear-time algorithm for computing a minimum weight spanning tree, and a quadratic-time algorithm for computing a minimum weight non-crossing spanning tree for collinear points and edges on a single page (Sect. 3).

➡ MINIMUM NON-CROSSING MATCHING FOR COLLINEAR POINTS – We give a linear-time algorithm that computes a minimum-weight non-crossing perfect matching for collinear points and edges on a single page (Sect. 4).

➡ TS TOUR FOR CHUNKED POINTS ON A CIRCLE – We give a linear-time algorithm for computing an optimum traveling salesman tour for equidistant points on a circle that form alternately colored and equally sized chunks (Sect. 5).

The linear-time algorithms assume that the points are given in some sorted order. We note that even in this simple one-dimensional case these problems become sufficiently challenging if one is constrained to use only linear (or near-linear) time. Throughout the paper we assume that there are no collocated points. We refer the reader to the full version [5] of our paper for missing proofs of lemmas and theorems.

2   Non-crossing Hamiltonian Path for Collinear Points

If we would require for the collinear point set that each edge of a Hamiltonian path is a straight-line segment the problem becomes trivial: an input instance can have a non-crossing Hamiltonian path if and only if the colors of the points alternate. Therefore, we consider the case where edges are represented by circular arcs drawn in the halfplane either above or below the line.

Definition 1. *Non-crossing Hamiltonian path for collinear points. Given a set of n red points and a set of n blue points on the line $H: y = 0$, find a non-crossing geometric path π in the plane such that π consists of circular arcs above or below H, each of which connects a red and a blue point and π spans all the input points.*

Note that in the above definition if the path is allowed to use arcs only from above (resp. below) H, then there might not exist such a Hamiltonian path (see Fig. 2).

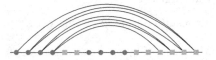

Fig. 2. Figure demonstrating a set of collinear points for which non-crossing Hamiltonian path does not exist if the arcs can be drawn only above the line. (Color figure online)

First, we give a constructive proof for the existence of such a path for any configuration of points. The construction itself takes polynomial time, hence giving a polynomial time algorithm for computation of such a path. In the following, we describe the construction.

2.1   The Construction

To construct the path, we start with any bichromatic matching (not necessarily crossing free) of the points. Note that each matched edge is a segment on H. We will connect these edges to obtain a Hamiltonian path. First, we form a hierarchical structure of these matched edges. Informally, the matched edges are hierarchical if any two edges are either disjoint or one is contained within the other.

Definition 2. *A set of matched edges M are hierarchical if for any two edges $(u, v), (w, x) \in M$ with $u < v$, $w < x$ and $u < w$, either $u < v < w < x$ ($(u, v), (w, x)$ are uncrossed) or $u < w < x < v$ ((u, v) contains (w, x)).*

Given any matching M for $R \cup B$, we can change it to a hierarchical matching in the following way. If there are two edges $(u, v), (w, x) \in M$ with $u < v$, $w < x$, $u < w$ that are not disjoint and none of them contains the other, then it must be the case that $u < w < v < x$. Now, there are two subcases, depending on the colors of u and w. If u, w are red or u, w are blue, we replace the edges $(u, v), (w, x)$ by the two bichromatic edges $(u, x), (w, v)$. Otherwise, either u is red, w is blue or u is blue, w red. In that case, we replace the edges $(u, v), (w, x)$ by the two bichromatic edges $(u, w), (v, x)$. Note that in all the cases, the new pair of edges do not violate the hierarchical structure. We repeat the process for each pair of edges that violate the condition. Newly formed edges might violate the condition with respect to other edges. However, it is easy to verify that if an edge is removed, it is never added back, and thus the process will eventually stop at some point when no pair of edges violate the condition.

Next, we associate levels with each matched edge of M in a recursive way. In the base case, for each edge that does not span any other edge, set its level to 1. Now, suppose we have defined edges of level j for each $j \leq i - 1$ for $i \geq 2$. An edge (u, v) has level i, if it contains a level $i - 1$ edge, and for any level $i - 1$ edge (w, x) that it contains, there is no other edge that is contained in (u, v) that also contains (w, x). Note that the level of each edge is unique. Let L be the maximum level.

For any edge (u, v) of M with level j, call the points that lie between u and v including u and v as a level j block. Thus, a level l block is a union of blocks of levels at most $l - 1$ and two special points which are the first and last point of the block. One can easily verify that a block contains the same number of red and blue points. We compute the Hamiltonian path for all the blocks in a bottom up manner. The path of a level 1 block is the matched edge itself which defines the block. Additionally, for each block, we compute a path for the block that satisfies the following two invariants.

- The first point of the block is an endpoint of the path.
- If an endpoint p of the path is not an endpoint of the block, then the path cannot contain two edges $(u,v),(w,x)$ with $u < v$ and $w < x$, such that (u,v) lies above H, (w,x) lies below H, $u < p < v$, and $w < p < x$.

Informally, the second condition states that, the endpoint of the path that is not an endpoint of the block should be available for connecting with an edge at least from one side. Note that the paths for level 1 blocks trivially satisfy the invariants. Now, assume that we have computed the paths for all the level j blocks for $j \leq l-1$ and $l \geq 2$ that satisfy the invariants. We show how to compute the path for a level l block S that also satisfies the invariants. Let u,v be the endpoints of the block. Also let $S_1,\ldots,S_t$ be the blocks, sorted w.r.t the index of the first point in increasing order, whose union with the set $\{u,v\}$ forms the block S. As S_i has level at most $l-1$, we have already computed the path of S_i for all i. We show, by induction, how to construct the path T' for the points in $\cup_{j=1}^{i} S_j$ for all $2 \leq i \leq t$. Then, we show how to join the edge (u,v) with T' to obtain the path for the block S. For simplicity, we also refer to the set of points $\cup_{j=1}^{i} S_j$ as a block. Now, we prove the following lemma.

Lemma 1. *A non-crossing Hamiltonian path of $\cup_{j=1}^{i} S_j$ can be computed for all $1 \leq i \leq t$ that satisfies the two invariants.*

The next lemma completes the induction step for showing the construction of the path for the level l block.

Lemma 2. *A non-crossing Hamiltonian path for the level l block S can be computed that satisfies the two invariants.*

To compute the path of all the points in $R \cup B$ one can note that $R \cup B$ is a union of a set of blocks having levels at most the maximum level L. By Lemma 2, we can compute the paths for all such blocks that satisfy the invariants. Then we can merge those paths using the construction in Lemma 1 to get the path for the points in $R \cup B$. It is easy to verify that the overall construction can be done in polynomial time. Thus, we get the following theorem.

Theorem 1. *For any set R of red points and B of blue points on $y = 0$ with $|R| = |B|$, there always exists a non-crossing Hamiltonian path whose edges are circular arcs that lie above or below $y = 0$. Moreover, such a path can be computed in polynomial time.*

2.2   A Linear Time Algorithm for Non-crossing Hamiltonian Path

Recall that all the input points lie on $H : y = 0$. We assume that the points are given in sorted order w.r.t their x coordinates. For a point p (except the last one), let $S(p)$ be the point which is the successor of p in this order. We use the following algorithm to compute a non-crossing Hamiltonian path. The algorithm processes the points from left to right and extends the Hamiltonian path constructed so far by connecting the current point with an appropriately chosen point. In particular, in every iteration, we consider a point p and connect it by adding one or more edges. Initially, p is the leftmost point, and all points are active. We store the constructed path in a set of edges Π, which is initially empty.

- Let $l(r)$ and $l(b)$ be the rightmost (or last in the order) red and blue points, respectively, which are active.
- If the color of p is different from the color of $S(p)$, we simply add a small arc $(p, S(p))$ to Π that lie above H. Make p inactive.
- Otherwise, there are two cases.
  (i). If p is red, add two edges $(p, l(b))$ and $(l(b), S(p))$ to Π. These two edges are drawn above H as circular arcs. Make p and $l(b)$ inactive.
  (ii). If p is blue, add two edges $(p, l(r))$ and $(l(r), S(p))$ to Π. These two edges are drawn below H as circular arcs. Make p and $l(r)$ inactive.
- If $S(p)$ is active, assign $S(p)$ to p (i.e., $p \leftarrow S(p)$) and repeat all the steps. Otherwise, terminate the algorithm.

We discuss the correctness of the algorithm. First, we have the following observation.

Observation 3. *Consider any iteration of the algorithm. Then, any red point on the right of $l(r)$ (if any) is inactive and has degree 2. Similarly, any blue point on the right of $l(b)$ (if any) is inactive and has degree 2. Moreover, any point on the left of p (if any) is inactive and except the first point all of them have degree 2.*

Lemma 4. *The algorithm correctly computes a bichromatic Hamiltonian path.*

Proof. Note that when the algorithm terminates, $S(p)$ is inactive. Thus, its degree must be 2. If $S(p)$ is red (resp. blue), then it had become $l(r)$ (resp. $l(b)$) at some point and its degree is 2. By Observation 3, all the points whose colors are same as the color of $S(p)$ and lie on the right of $S(p)$ have degree 2. Also, the degree of all the points on the left of p except the first point is 2. It is easy to see that the degree of p and the first point is 1. As the number of red and blue points are same, all the points that lie on the right of $S(p)$ must have degree 2. Thus, Π is a valid bichromatic Hamiltonian path. $\square$

Next, we argue that the computed Hamiltonian path is non-crossing. It is easy to see that the small arcs added in the second step do not cross any other drawn edges. Also, the edges drawn above H do not cross any edges drawn below H. Moreover, the edges $(p, l(r))$ and $(l(r), S(p))$ (or $(p, l(b))$ and $(l(b), S(p))$) drawn in the same iteration do not cross each other. The following observation completes the claim.

Observation 5. *Consider two edges (u, v) and (u', v') which are drawn as circular arcs above (resp. below) H and added to Π in different iterations. Then, either u, v lie in between u' and v', or u', v' lie in between u and v.*

The algorithm can be implemented to run in linear time. This is because, one can use three pointers to keep track of p, $l(r)$ and $l(b)$, and these pointers move in one direction – either from left to right or from right to left.

Theorem 2. *For any set R of red points and B of blue points on $y = 0$ with $|R| = |B|$, a non-crossing Hamiltonian path can be computed in linear time whose edges are circular arcs that lie above or below $y = 0$.*

3   Minimum Spanning Tree for Collinear Points

Definition 3. *Spanning tree for collinear points. Given a set of n red points and a set of n blue points all of which lie on the line $y = 0$, find a minimum weight geometric tree T in the plane such that each edge of T is represented by a circular arc that lies above $y = 0$, each arc connects a red and a blue point, and T spans all the input points. The weight of an arc is given by the Euclidean distance of its endpoints. In the non-crossing version of the problem, one would like to compute such a tree so that the corresponding circular arcs are non-crossing.*

The problem of computing an optimum, i.e., minimum-weight, spanning tree, which potentially has crossings can be solved in linear time [5]. Next, we discuss the algorithm for the non-crossing case.

Let $P_1, P_2, \ldots, P_m$ be the alternating monochromatic chunks of points ordered from left to right for $m \in [n]$. Thus, the color of the points in P_i is different from the color of the points in P_{i+1} for all $1 \leq i \leq m-1$. We start with the following observation.

Observation 6. *Consider a point $p \in P_i$. If an arc (p, q) is contained in a minimum spanning tree, then either $q \in P_{i-1}$ or $q \in P_{i+1}$.*

The observation follows from the idea that if a point is connected to a point that belongs to a non-consecutive chunk, then one can find a cheaper spanning tree by replacing the connecting arc with another arc having lower weight. As the spanning tree we want to compute is non-crossing, by the above observation, it follows that all the arcs between two consecutive chunks are nested.

Observation 7. *Consider any two arcs (p_1, q_1) and (p_2, q_2) in a minimum non-crossing spanning tree such that $p_1, p_2 \in P_i$ and $q_1, q_2 \in P_{i+1}$. Then, either $p_1 < p_2 < q_2 < q_1$ or $p_2 < p_1 < q_1 < q_2$.*

The above observation implies that the outermost arcs between consecutive chunks form a path (an umbrella) between the first and the last point and all the other arcs lie inside this umbrella (see Fig. 3). Next, we give a simple algorithm to compute an optimum spanning tree inside such an outermost arc. Suppose $p_0, p_1, \ldots, p_l, \ldots, p_{k+1}$ be points in sorted order such that $\{p_0, p_1, \ldots, p_l\} \subseteq P_i$ and $\{p_{l+1}, \ldots, p_{k+1}\} \subseteq P_{i+1}$. We would like to construct an optimum spanning tree of the points $p_0, p_1, \ldots, p_l, \ldots, p_{k+1}$ which contains the arc (p_0, p_{k+1}). Our algorithm is based on the following observation.

Fig. 3. Figure showing a spanning tree with the umbrella shown by dashed arcs. (Color figure online)

Observation 8. *Any optimum spanning tree that contains* (p_0, p_{k+1}) *must also contain either* (p_0, p_k) *or* (p_1, p_{k+1}) *whichever has lower weight.*

In our algorithm, we select the shorter arc among (p_0, p_k) and (p_1, p_{k+1}). Then, we recursively solve the problem inside the selected arc by treating it as an outermost arc. It is easy to see that this problem can be solved in linear time. Next, we give an algorithm for deciding which outermost arcs to choose.

Let $p_1, p_2, \ldots$ be the input points. Our algorithm incrementally computes a non-crossing spanning tree starting from the left and by connecting a new point in each step. Let $P_1 = \{p_1, \ldots, p_l\}$ and $P_2 = \{p_{l+1}, \ldots, p_k\}$. To initialize, for each $l+1 \leq j \leq k$, we compute the cost of optimum spanning tree of $\{p_1, \ldots, p_j\}$ that contains the outermost arc (p_1, p_j) using the above algorithm. Now, suppose we want to connect a new point $p_i \in P_{t+1}$ for $t \geq 2$. We have already computed the cost of an optimum spanning tree of $\{p_1, \ldots, p_q\}$ with any valid outermost arc (r, s), where $r \in P_{t-1}$ and $s \in P_t$. In the new spanning tree, p_i must be connected to a point s of P_t. For each such s, we compute the cost of the spanning tree that contains the arc (s, p_i). In particular, the total cost is the sum of three costs: (i) the cost of (s, p_i), (ii) the cost of connecting the points inside (s, p_i) and (iii) the cost of the optimum spanning tree of $p_1, \ldots, s$ that contains (r, s) for some $r \in P_{t-1}$. We select the arc (r, s) in our spanning tree that minimizes the total cost.

Note that each step of this dynamic programming based algorithm takes linear time. Thus, the optimum spanning tree can be computed in quadratic time.

Theorem 3. *For any set R of red points and B of blue points on $y = 0$, an optimum non-crossing spanning tree can be computed in quadratic time.*

4   Minimum Non-crossing Matching for Collinear Points

Note that the fact that a minimum weight bichromatic matching for points in general position is always non-crossing might not hold in the case of collinear points. Indeed, there are point sets for which no non-crossing matching exists if the edges are represented by segments. However, one can show that there is always a non-crossing matching of collinear points such that each matched edge is a circular arc drawn above the line. Again the weight of an arc is the Euclidean distance between its endpoints.

Definition 4. *Non-crossing matching for collinear points. Given a set of n red points and a set of n blue points all of which lie on the line $y = 0$, find* a set of n non-crossing circular arcs in the plane of minimum total weight such that the arcs lie above $y = 0$, each arc connects a red and a blue point, and the arcs span all the input points.

Using the bipartite matching algorithm due to Kaplan $et\ al.$ [15] along with a simple postprocessing (already described in the introduction), one can immediately solve this problem in $O(n^2\text{poly}(\log n))$ time. Here we design a simple algorithm with improved $O(n)$ time complexity.

Let $p_1, p_2, \ldots, p_{2n}$ be the input points sorted from left to right based on their x coordinates. We assume that the points are given in this order. For any point $p_i \in P$, let $col(p_i)$ denote the color of p_i. A subset of points $P_i \subseteq P$ is called $color\text{-}balanced$ if it contains an equal number of red and blue points. We traverse the points from left to right and seek for the first balanced subset (denoted by P_1). In order to obtain P_1 we use a simple method. We start with the leftmost point p_1 and maintain a counter C which is used to find the balanced subset and is initialized to 0 at the beginning. If $col(p_1) = $ red, we increase the value of C by 1, and decrease by 1, otherwise. Observe that we will get a balanced subset when the value of C becomes 0. Let $P_1 \subseteq P$ be the first balanced subset containing $2m$ (for some $m \in [n]$) points. The remaining points $P \setminus P_1$ also form a balanced subset since P contains exactly n red and n blue points. We prove the following lemma.

Lemma 9. Let $P_1 \subseteq P$ be the first color-balanced subset of P and $|P_1| = 2m$. Then $col(p_1) \neq col(p_{2m})$, and any minimum non-crossing perfect matching M_P of P contains the edge (p_1, p_{2m}).

Proof. The first part of the lemma is clearly true, otherwise the value of the counter would not be 0 at p_{2m}, which is the termination criteria to obtain the first balanced subset. Now, let us assume that M_P does not contain the edge (p_1, p_{2m}). Then one of the following two situations can happen: (1) p_1 and p_{2m} are matched with two intermediate points from P_1; (2) one or both of p_1 and p_{2m} are matched with points from $P \setminus P_1$.

Case 1: p_1 and p_{2m} are matched with two intermediate points p_k and p_ℓ, respectively. Note $\ell > k$, otherwise the matched edges cross each other. We know that $\{p_1, \ldots, p_k\}$ is not a balanced subset since P_1 is the first balanced subset. Therefore, there exists at least one point p_r (where $1 < r < k$) that is matched with a point p_s (where $s > k$). In that case, the edge (p_r, p_s) will intersect (p_1, p_k). Hence, we get a contradiction.

Case 2: Suppose both of p_1 and p_{2m} are matched with points from $P \setminus P_1$ and no other point from $\{p_2, \ldots, p_{2m-1}\}$ is matched with any point from $P \setminus P_1$. Then we can construct a new matching by adding the edge (p_1, p_{2m}) and by matching the two points in $P \setminus P_1$. The new matching has lesser cost and is non-crossing; see Fig. 4(a). If any other point in $\{p_2, \ldots, p_{2m-1}\}$ (say p_x) is also matched with a point in $P \setminus P_1$, then we know it must be of opposite color of either p_1 or p_{2m}, since $col(p_1) \neq col(p_{2m})$. Hence, we can either give the edge (p_1, p_x) or (p_x, p_{2m}) and this reduces the total cost; see Fig. 4(b). The new matching might not be non-crossing. But, using similar argument one can remove all the crossings without increasing the cost. Thus, at the end we get a cheaper non-crossing matching, which contradicts the optimality of M_P.

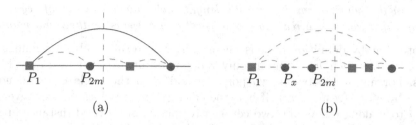

(a)   (b)

Fig. 4. Figure demonstrating the two situations in the case when both p_1 and p_{2m} are matched with points from $P \setminus P_1$. (Color figure online)

Now, if only one of p_1 or p_{2m} is matched (WLOG, assume it is p_1) with a point from $P \setminus P_1$, then we know there must be at least one other point (say $p_x \in P_1$) that is also matched with a point from $P \setminus P_1$, and $col(p_1) \neq col(p_x)$. We can apply similar arguments as above to get a contradiction, which concludes the proof of the lemma.   □

Now, we use Lemma 9 to proceed with the algorithm. First, we obtain the balanced subset P_1, and match the points p_1 and p_{2m} by an arc and include the edge (p_1, p_{2m}) in M_P. This edge partitions the point set into subsets, i.e., $P_2 = P \setminus P_1$ and $P'_1 = P_1 \setminus \{p_1, p_{2m}\}$. On each of these subsets we recursively perform the same procedure. This process is repeated until each point of P is matched.

Due to Lemma 9, we know that every edge we choose in our algorithm must be part of the optimum solution, and no two edges cross each other. It is not hard to see that all the balanced subsets can be computed in linear time in advance, as they are corresponding to matched parentheses and are at most n in number[2]. Thus, we conclude with the following theorem.

Theorem 4. *For any set R of red points and B of blue points on $y = 0$ with $|R| = |B|$, an optimum non-crossing matching can be computed in linear time.*

5   TS Tour for Chunked Points on a Circle

Finally, we study the Traveling Salesman problem on the following special point configuration on a circle.

---

[2] This algorithm can be easily implemented in the following manner. Consider the points in left to right order, and insert the leftmost point (p_1) into a stack. Now, if the next point p_2 is of same color as p_1 then insert p_2 into the stack, otherwise match p_1, p_2 and remove p_1 from the stack. Repeat this process until all points are considered.

Definition 5. *TS tour for chunked points on a circle*. *We are given a set of n red points and a set of n blue points all of which lie on a fixed circle. All points are distributed equidistantly on the circle. Further, the input points are divided into alternately-colored chunks, where each chunk contains exactly k consecutive points of the same color. The goal is to find a geometric (closed) tour π in the plane of minimum total length such that π consists of segments each of which connects a red and a blue point and π spans all the input points.*

Note that by definition, n/k is an integer. The total number of chunks is $L = 2n/k$ of which n/k contain only red points and n/k contain only blue points. For any arc between two points u and v on the circle C, we denote the arc by $c(uv)$ (resp. $a(uv)$) if it is the clock (resp. anticlock) -wise traversal from u to v along C. As any two consecutive points are a fixed distance apart we measure the length of any bichromatic edge (a straight line segment) uv by the minimum of the number of points on the arcs $c(uv)$ and $a(uv)$, respectively (including u and v). Next, we design an algorithm for computing a TS tour for the input points. We consider two cases: (i) k is even and (ii) k is odd.

*(i) k **is Even**.* The algorithm in this case is as follows.

1. Let $2p = k$. Partition each chunk into two subchunks each containing p consecutive points. Merge all consecutive subchunks of different colors to form groups. Note that each group contains p red and p blue points. We still preserve the geometry of the points of each group and identify the two peripheral red and blue points of each group as special points. We first compute a bichromatic path between the two special points for each group and later connect the special points of different groups to construct a TS tour for all the points.
2. For each group, we compute the TS tour in the following way. Consider the ordering of the groups w.r.t clockwise traversal of the points and consider the ith group in this order. WLOG, assume that the red points are visited before the blue points while traversing the points of the group in clockwise order. Let $r_1^i, r_2^i, \ldots, r_p^i, b_p^i, b_{p-1}^i, \ldots, b_1^i$ be the points in this order. Join r_p^i with b_p^i and b_{p-1}^i using two edges. For each $p-1 \geq j \geq 2$, join r_j^i with b_{j+1}^i and b_{j-1}^i. Finally, join r_1^i with b_2^i. Note that each of the points in the group except r_1^i and b_1^i is connected to two points. r_1^i and b_1^i are connected to only one point.
3. Next, we connect the special points of different groups. Recall that L is the total number of groups. Let for the first group the red points are visited before the blue points while traversing the points of the group in clockwise order. Note that the special points are $r_1^1, b_1^1, b_1^2, r_1^2, r_1^3, b_1^3, \ldots, b_1^L, r_1^L$. For $1 \leq i \leq L - 1$, we connect r_1^i to b_1^{i+1}. We also connect r_1^L to b_1^1.

It is not hard to see that the set of selected edges form a valid traveling salesman tour. This is because each of the points is connected to exactly two other points of opposite color. Now, we give a bound on the length of the tour.

Lemma 10. *The length of the computed tour is $n(k + 2 + 2/k)$.*

The case when k **is odd** is similar and the length of the computed tour is $n(k + 2 + 1/k)$.

5.1   Lower Bound

Lemma 11. *The length of any bichromatic traveling salesman tour for the configuration of the points on the circle is at least $n(k + 2 + 2/k)$ if k is even and at least $n(k + 2 + 1/k)$ if k is odd.*

Theorem 5. *For any set R of red points and B of blue points on a circle with $|R| = |B|$, an optimum non-crossing TS tour can be computed in linear time.*

References

1. Karim Abu-Affash, A., Biniaz, A., Carmi, P., Maheshwari, A., Smid, M.H.M.: Approximating the bottleneck plane perfect matching of a point set. Comput. Geom. **48**(9), 718–731 (2015)
2. Agarwal, P.K., Efrat, A., Sharir, M.: Vertical decomposition of shallow levels in 3-dimensional arrangements and its applications. SIAM J. Comput. **29**(3), 912–953 (1999)
3. Anily, S., Hassin, R.: The swapping problem. Networks **22**(4), 419–433 (1992)
4. Arora, S.: Polynomial time approximation schemes for Euclidean traveling salesman and other geometric problems. J. ACM **45**(5), 753–782 (1998)
5. Bandyapadhyay, S., Banik, A., Bhore, S., Nöllenburg, M.: Geometric planar networks on bichromatic points. CoRR, arXiv:1911.08924 (2019)
6. Bernhart, F., Kainen, P.C.: The book thickness of a graph. J. Comb. Theory Ser. B **27**(3), 320–331 (1979)
7. Biniaz, A., Bose, P., Maheshwari, A., Smid, M.H.M.: Plane geodesic spanning trees, Hamiltonian cycles, and perfect matchings in a simple polygon. Comput. Geom. **57**, 27–39 (2016)
8. Biniaz, A., Bose, P., Maheshwari, A., Smid, M.H.M.: Plane bichromatic trees of low degree. Discrete Comput. Geom. **59**(4), 864–885 (2018)
9. Borgelt, M.G., van Kreveld, M.J., Löffler, M., Luo, J., Merrick, D., Silveira, R.I., Vahedi, M.: Planar bichromatic minimum spanning trees. J. Discrete Algorithms **7**(4), 469–478 (2009)
10. Evans, W.S., Liotta, G., Meijer, H., Wismath, S.K.: Alternating paths and cycles of minimum length. Comput. Geom. **58**, 124–135 (2016)
11. Frank, A., Triesch, E., Korte, B., Vygen, J.: On the bipartite travelling salesman problem (1998)
12. Kaneko, A., Kano, M., Suzuki, K.: Balanced partitions and path covering of two sets of points in the plane, preprint
13. Kaneko, A., Kano, M.: Straight-line embeddings of two rooted trees in the plane. Discrete Comput. Geom. **21**(4), 603–613 (1999)
14. Kaneko, A., Kano, M.: Discrete geometry on red and blue points in the plane—a survey—. In: Aronov, B., Basu, S., Pach, J., Sharir, M. (eds.) Discrete and Computational Geometry, pp. 551–570. Springer, Heidelberg (2003). https://doi.org/10.1007/978-3-642-55566-4_25
15. Kaplan, H., Mulzer, W., Roditty, L., Seiferth, P., Sharir, M.: Dynamic planar Voronoi diagrams for general distance functions and their algorithmic applications. In: Symposium on Discrete Algorithms, SODA 2017, pp. 2495–2504 (2017)
16. Kyncl, J., Pach, J., Tóth, G.: Long alternating paths in bicolored point sets. Discrete Math. **308**(19), 4315–4321 (2008)

Hardness Results of Global Total k-Domination Problem in Graphs

B. S. Panda[✉] and Pooja Goyal

Computer Science and Application Group, Department of Mathematics,
Indian Institute of Technology Delhi, Hauz Khas, New Delhi 110016, India
bspanda@maths.iitd.ac.in, poojaagoyal92@gmail.com

Abstract. A set $D \subseteq V_G$ of a graph $G = (V_G, E_G)$ is called a global total k-dominating set of G if D is a total k-dominating set of both G and $\overline{G}$, the complement of G. The MINIMUM GLOBAL TOTAL k-DOMINATION problem is to find a global total k-dominating set of minimum cardinality of the input graph G and DECIDE GLOBAL TOTAL k-DOMINATION problem is the decision version of MINIMUM GLOBAL TOTAL k-DOMINATION problem. The DECIDE GLOBAL TOTAL k-DOMINATION problem is known to be NP-complete for general graphs. In this paper, we study the complexity of the MINIMUM GLOBAL TOTAL k-DOMINATION problem. We show the DECIDE GLOBAL TOTAL k-DOMINATION problem remains NP-complete for bipartite graphs and chordal graphs. Next, we show that the MINIMUM GLOBAL TOTAL k-DOMINATION problem admits a constant approximation algorithm for bounded degree graphs. Finally, we show that the MINIMUM GLOBAL TOTAL k-DOMINATION problem is APX-complete for bounded degree graphs.

1 Introduction

Let $G = (V_G, E_G)$ be a finite, simple and undirected graph with vertex set V_G and edge set E_G. A set $D \subseteq V_G$ is called a dominating set of G if every vertex $v \in V_G \setminus D$ is adjacent to at least one vertex in D. The *domination number* of G is the minimum cardinality among all dominating sets in G and it is denoted by $\gamma(G)$. The MINIMUM DOMINATION problem is to find a dominating set of minimum cardinality and DECIDE DOMINATION problem is the decision version of MINIMUM DOMINATION problem. Domination in graphs has been studied extensively and has several applications (see [4,5]).

A set $D \subseteq V_G$ is called a total dominating set of G if every vertex in V_G is adjacent to at least one vertex in D. The *total domination number* of G is the minimum cardinality among all total dominating sets in G and it is denoted by $\gamma_t(G)$. The MINIMUM TOTAL DOMINATION problem is to find a total dominating set of minimum cardinality and DECIDE TOTAL DOMINATION problem is the decision version of MINIMUM TOTAL DOMINATION problem. The MINIMUM TOTAL DOMINATION problem is also a well studied problem in graph theory (see [6,8]).

© Springer Nature Switzerland AG 2020
M. Changat and S. Das (Eds.): CALDAM 2020, LNCS 12016, pp. 92–101, 2020.
https://doi.org/10.1007/978-3-030-39219-2_8

A total k-dominating set (TkD-set) of a graph G is a set $D \subseteq V_G$ such that every vertex in V_G is adjacent to at least k vertices in D. The minimum cardinality of a TkD-set in a graph G is known as *total k-domination number* of G and is denoted by $\gamma_{kt}(G)$. The concept of total k-domination has been introduced by Henning and Kazemi in [7]. The MINIMUM TOTAL k-DOMINATION problem is to find a total k-dominating set of minimum cardinality and DECIDE TOTAL k-DOMINATION problem is the decision version of MINIMUM TOTAL k-DOMINATION problem. Pradhan [9] studied MINIMUM TOTAL k-DOMINATION problem from algorithmic point of view.

A global total k-dominating set (GTkD-set) of a graph $G = (V_G, E_G)$ is a set $D \subseteq V_G$ such that every vertex in V is adjacent as well as non-adjacent to at least k vertices in D. Equivalently, a set $D \subseteq V_G$ is a GTkD-set of a graph G if D is a total k-dominating set of both G and $\overline{G}$. The minimum cardinality of a GTkD-set in a graph G is known as *global total k-domination number* of G and is denoted by $\gamma_{kt}^g(G)$. The necessary and sufficient condition, $1 \le k \le min\{\delta(G), |V_G| - (\Delta(G) + 1)\}$, to guarantee the existence of a GTkD set in a graph G has been given by Bermudo et al. [2].

The minimum global total k-domination problem and its decision version are defined as follows:

MINIMUM GLOBAL TOTAL k-DOMINATION problem (Min GTkD)

Instance: A graph $G = (V_G, E_G)$.

Solution: A global total k-dominating set D of G.

Measure: Cardinality of the set D.

DECIDE GLOBAL TOTAL k-DOMINATION problem (Decide GTkD)

Instance: A graph $G = (V_G, E_G)$ and a positive integer r.

Question: Deciding $\gamma_{kt}^g(G) \le r$?

Bermudo et al. [2] introduced this new variant of domination, namely global total k-domination, and proved that the decision version of global total k-domination (Decide GTkD) problem is NP-complete in general graphs. In this paper, we further study the complexity of MINIMUM GLOBAL TOTAL k-DOMINATION problem.

The rest of the paper is organized as follows. In Sect. 2, we present some pertinent definitions and some preliminary results. In Sect. 3, we strengthen the NP-completeness result of the Decide GTkD problem by showing that this problem remains NP-complete for bipartite graphs and chordal graphs. In Sect. 4, we first show that the Min GTkD problem for bounded degree graphs can be approximated within a constant factor. Finally, we show that MINIMUM GLOBAL TOTAL

k-DOMINATION problem is APX-complete for bounded degree graphs. Finally, in Sect. 5, we conclude the paper.

2 Preliminaries

Let $G = (V_G, E_G)$ be a finite, simple and undirected graph with no isolated vertex. The open neighborhood of a vertex v in G is $N_G(v) = \{u \in V_G \mid uv \in E_G\}$ and the closed neighborhood is $N_G[v] = \{v\} \cup N_G(v)$. The degree of a vertex v is $|N_G(v)|$ and is denoted by $d_G(v)$. If $d_G(v) = 1$, then v is called a pendant vertex. The minimum and maximum degree of graph G will be denoted by $\delta(G)$ and $\Delta(G)$, respectively. For a non empty set $D \subseteq V_G$, and a vertex $v \in V_G$, $N_D(v)$ denotes the set of neighbors of v in D and $\overline{N}_D(v)$ denotes the set of non neighbors of v in D. The degree of v in D will be denoted by $\delta_D(v) = |N_D(v)|$ and $\overline{\delta}_D(v) = |\overline{N}_D(v)|$. For $D \subseteq V_G$, $G[D]$ denote the subgraph induced by D. For any $C \subseteq V_G$, if $G[C]$ is a complete subgraph of G then C is called a clique of G. The complete graph on n vertices is denoted by K_n. A vertex $v \in V_G$ is said to have a non-adjacent vertex in $D \subseteq V_G$ if there exists a vertex $u \in D$ such that $uv \notin E_G$. Let $[n]$ denotes $\{1, 2, \ldots, n\}$.

A *bipartite graph* is an undirected graph $G = (X, Y, E_G)$ whose vertices can be partitioned into two disjoint sets X and Y such that every edge has one end vertex in X and other in Y. A bipartite graph $G = (X, Y, E_G)$ is *complete bipartite* if for every $x \in X$ and $y \in Y$, there is an edge $xy \in E_G$. A complete bipartite graph with partitions of size $|X| = m$ and $|Y| = n$, is denoted $K_{m,n}$.

A graph $G = (V_G, E_G)$ is said to be a *chordal graph* if every cycle of length at least four has a chord i.e., an edge joining two non-consecutive vertices of the cycle. A vertex $v \in V_G$ is a *simplicial vertex* of G if $N_G[v]$ is a clique of G. An ordering $\sigma = (v_1, v_2, \ldots, v_n)$ is a *perfect elimination ordering (PEO)* of G if v_i is a simplicial vertex of $G_i = G[\{v_i, v_{i+1}, \ldots, v_n\}]$ for all i, $1 \leq i \leq n$. It is characterized that a graph G is chordal if and only if it has a PEO [3].

From approximation point of view, several variations of domination has been studied by many authors over the years. For most of the approximation related terminologies, we refer the reader to [1].

Observation 1 *(see [2]). Let $G = (V_G, E_G)$ be a graph, then $\gamma_{kt}^g(G) \geq max\{\gamma_{kt}(G), \gamma_{kt}(\overline{G})\}$.*

Observation 2 *(see [2]). Let $G = (V_G, E_G)$ be a graph and let $D \subseteq V_G$. The set D is a GTkD if and only if $\delta_D(v) \geq k$ and $\overline{\delta}_D(v) \geq k$, for every vertex $v \in V$.*

Observation 3 *(see [2]). Let $k \in \mathbb{N}$ and suppose $G = (V_G, E_G)$ be a graph of order n with minimum degree $\delta(G) \geq k$. Then $\gamma_{kt}(G) \geq \dfrac{kn}{\Delta(G)}$ and this bound is sharp.*

3   NP-completeness Results

Bermudo et al. [2] has shown that the DECIDE GLOBAL TOTAL k-DOMINATION problem is NP-complete for general graphs. In this section, we strengthen this NP-completeness result by showing that this problem remains NP-complete for bipartite graphs and chordal graphs.

3.1   NP-completeness for Bipartite Graphs

Theorem 4. *The* DECIDE GLOBAL TOTAL k-DOMINATION *problem is NP-complete for bipartite graphs.*

Proof. Given a subset $S \subseteq V_G$ of vertices of a bipartite graph $G = (V_G, E_G)$, it can be checked in polynomial time whether S is a GLOBAL TOTAL k-DOMINATING set of G. Hence, the DECIDE GLOBAL TOTAL k-DOMINATION problem is in NP for bipartite graphs. To show the hardness, we give a polynomial reduction from the DECIDE TOTAL k-DOMINATION problem for bipartite graphs, which is already known to be NP-complete (see [9]). Let $G = (V_G, E_G)$ be a given bipartite graph, where $V_G = \{v_1, v_2, \ldots, v_n\}$ and k be a fixed positive integer such that $k \leq \delta(G)$. Now, we construct a graph $H = (V_H, E_H)$ from the given graph G in polynomial time. Before the construction of the graph H, we define a subgraph $W = (V_W, E_W)$ for some fixed k, which can be obtained from k copies of $K_{k,k}$ and a vertex t, by joining the vertex t to one vertex of each copy $K_{k,k}$. Now, for the construction of graph H we take n copies of the subgraph W, say $W_1, W_2, \ldots, W_n$. Finally, the graph $H = (V_H, E_H)$ can be obtained from the given graph G in polynomial time such that $V_H = V_G \cup \left(\bigcup_{i=1}^{n} V_{W_i} \right)$ and $E_H = E_G \cup \left(\bigcup_{i=1}^{n} E_{W_i} \right) \cup \{v_i t_i \mid v_i \in V_G, t_i \in V_{W_i}, 1 \leq i \leq n\}$. Figure 1 illustrates the construction of H from a bipartite graph for $k = 2$. Clearly, the constructed graph H is a bipartite graph.

Now to complete the proof, we have to show that G has a TkD-set of cardinality at most r if and only if the graph H has a GTkD-set of cardinality at most $2nk^2 + r$. For this, we have to prove the following claims:

Claim. $\gamma_{kt}(H) = \gamma_{kt}(G) + 2nk^2$.

Proof. Let $D_G \subseteq V_G$ be a minimum total k-dominating set (TkD-set) of G. Then $D_H = D_G \cup \left(\bigcup_{i=1}^{n} \left(V_{W_i} \setminus \{t_i\} \right) \right)$ is a TkD-set of H of cardinality $2nk^2 + |D_G|$. Thus,

$$\gamma_{kt}(H) \leq 2nk^2 + \gamma_{kt}(G).$$

Now, assume that $D_H \subseteq V_H$ is a minimum TkD-set of the graph H. For every $i \in \{1, 2, \ldots, n\}$, vertices of the set $V_{W_i} \setminus \{t_i\}$ must be totally k-dominated by the set D_H, so $\bigcup_{i=1}^{n} \left(V_{W_i} \setminus \{t_i\} \right) \subset D_H$. Thus $\left| D_H \cap \left(\bigcup_{i=1}^{n} \left(V_{W_i} \setminus \{t_i\} \right) \right) \right| = 2nk^2$.

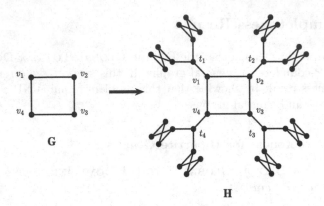

Fig. 1. An illustration of the construction of H from G in the proof of Theorem 4.

Since every $v_i \in V_G \cap V_H$ is totally k-dominated by the set D_H, so $|N_H(v_i) \cap D_H| \geq k$. Now for some fixed $j \in [n]$, if $t_j \in D_H$, then $|N_H(v_j) \cap D_H| = k$, otherwise $D_H \setminus \{t_j\}$ will also be a TkD-set of H, which is a contradiction to the assumption that D_H is the minimum cardinality TkD-set of H. Thus, $|(N_H(v_j) \setminus \{t_j\}) \cap D_H| = k - 1$. Since $\delta(G) \geq k$, there always exists a vertex $u \in N_G(v_j) \setminus D_H$ such that $(D_H \setminus \{t_j\}) \cup \{u\}$ is again a TkD-set of the graph H. Using the above process, we can construct a TkD-set of H containing no vertices of the set $\{t_i \mid 1 \leq i \leq n\}$. Hence, the set $D_H \cap V_G$ will be a TkD-set of G. Thus,

$$\gamma_{kt}(H) = |D_H| = |D_H \cap V_G| + \left| D_H \cap \left(\bigcup_{i=1}^{n} V_{W_i} \right) \right| \geq \gamma_{kt}(G) + 2nk^2.$$

This completes the proof of claim.   □

Claim. $\gamma_{kt}^g(H) = \gamma_{kt}(H)$.

Proof. It can be easily observed that every vertex v of the graph H has at least k non-adjacent vertices in the set D_H. Therefore, using Observation 2, it follows that D_H is a GTkD-set of the graph H. So, $|D_H| = \gamma_{kt}(H) \geq \gamma_{kt}^g(H)$. Thus using Observation 1, we get $\gamma_{kt}^g(H) \geq \gamma_{kt}(H)$. This completes the proof of claim.   □

Hence, from the above claims we have $\gamma_{kt}^g(H) = \gamma_{kt}(G) + 2nk^2$. Thus, we can conclude that $\gamma_{kt}(G) \leq r$ if and only if $\gamma_{kt}^g(H) \leq 2nk^2 + r$.

Therefore, the DECIDE GLOBAL TOTAL k-DOMINATION problem is NP-complete for bipartite graphs. This completes the proof of theorem.   □

3.2   NP-completeness for Chordal Graphs

Theorem 5. *The* DECIDE GLOBAL TOTAL k-DOMINATION *problem is NP-complete for chordal graphs.*

Proof. Clearly, the DECIDE GLOBAL TOTAL k-DOMINATION problem is in NP for chordal graphs. To show the hardness, we give a polynomial reduction from the DECIDE TOTAL k-DOMINATION problem for chordal graphs, which is already known to be NP-complete (see [9]). Let $G = (V_G, E_G)$ be a given chordal graph, where $V_G = \{v_1, v_2, \ldots, v_n\}$ and k be a fixed positive integer such that $k \leq \delta(G)$. Now, we construct a graph $H = (V_H, E_H)$ from the given graph G in polynomial time. Before the construction of the graph H, we define a sub-graph $W = (V_W, E_W)$ for some fixed k, which can be obtained from k copies of K_{k+1} and a vertex t, by joining the vertex t to one vertex of each copy K_{k+1}. Now, for the construction of graph H we take n copies of the subgraph W, say $W_1, W_2, \ldots, W_n$. Finally, the graph $H = (V_H, E_H)$ can be obtained from the given graph G in polynomial time such that $V_H = V_G \cup \left(\bigcup_{i=1}^{n} V_{W_i}\right)$ and $E_H = E_G \cup \left(\bigcup_{i=1}^{n} E_{W_i}\right) \cup \{v_i t_i \mid v_i \in V_G, t_i \in V_{W_i}, 1 \leq i \leq n\}$. Clearly, the obtained graph H is a chordal graph as shown in Fig. 2, where $k = 2$.

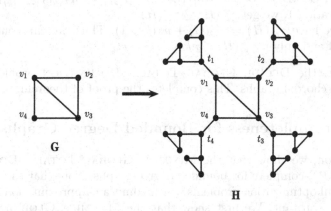

Fig. 2. An illustration of the construction of H from G in the proof of Theorem 5.

Now to complete the proof, we have to prove the following claim.

Claim. G has a TkD-set of cardinality at most r if and only if the graph H has a GTkD-set of cardinality at most $nk(k+1) + r$.

Proof. Firstly, we will show that $\gamma_{kt}(H) = \gamma_{kt}(G) + nk(k+1)$.

Let $D_G \subseteq V_G$ be a minimum total k-dominating set (TkD-set) of G. Then $D_H = D_G \cup \left(\bigcup_{i=1}^{n} (V_{W_i} \setminus \{t_i\})\right)$ is a TkD-set of H of cardinality $nk(k+1) + |D_G|$. Thus,

$$\gamma_{kt}(H) \leq nk(k+1) + \gamma_{kt}(G).$$

Next, assume that $D_H \subseteq V_H$ is a minimum TkD-set of the graph H. As we know that if degree of any vertex v in a graph is k, then every TkD-set of the graph must consists of each neighbour of v. Thus, $\bigcup_{i=1}^{n} \left(V_{W_i} \setminus \{t_i\}\right) \subset D_H$ and

$\left| D_H \cap \left(\bigcup_{i=1}^{n} \left(V_{W_i} \setminus \{t_i\} \right) \right) \right| = nk(k+1)$. Now, every $v_i \in V_G \cap V_H$ is totally k-dominated by the set D_H, so $|N_H(v_i) \cap D_H| \geq k$. Now for some fixed $j \in [n]$, if $t_j \in D_H$, then $|N_H(v_j) \cap D_H| = k$, otherwise $D_H \setminus \{t_j\}$ will also be a TkD-set of H, which is a contradiction to the assumption that D_H is the minimum cardinality TkD-set of H. Thus, $|(N_H(v_j) \setminus \{t_j\}) \cap D_H| = k-1$. Since $\delta(G) \geq k$, there always exists a vertex $u \in N_G(v_j) \setminus D_H$ such that $(D_H \setminus \{t_j\}) \cup \{u\}$ is again a TkD-set of the graph H. Using the above process, a TkD-set D_H of H can be constructed such that $D_H \cap \{t_i \mid 1 \leq i \leq n\} = \phi$. Hence, the set $D_H \cap V_G$ will be a TkD-set of G. Thus,

$$\gamma_{kt}(H) = |D_H| = |D_H \cap V_G| + \left| D_H \cap \left(\bigcup_{i=1}^{n} V_{W_i} \right) \right| \geq \gamma_{kt}(G) + nk(k+1).$$

Now, it can be easily observed that every vertex v of the graph H has at least k non-adjacent vertices in the set D_H. Therefore, using Observation 2, it follows that D_H is a GTkD-set of the graph H. So, $|D_H| = \gamma_{kt}(H) \geq \gamma_{kt}^g(H)$. Thus using Observation 1, we get $\gamma_{kt}^g(H) \geq \gamma_{kt}(H)$.

Hence, we have $\gamma_{kt}^g(H) = \gamma_{kt}(G) + nk(k+1)$. Thus, we can conclude that $\gamma_{kt}(G) \leq r$ if and only if $\gamma_{kt}^g(H) \leq nk(k+1) + r$. □

Therefore, the DECIDE GLOBAL TOTAL k-DOMINATION problem is NP-complete for chordal graphs. This completes the proof of theorem. □

4   APX-completeness for Bounded Degree Graphs

In this section, we show that the MINIMUM GLOBAL TOTAL k-DOMINATION problem is APX-complete for bounded degree graphs. Note that the class APX is the set of all optimization problems which admit a c-approximation algorithm, where c is a constant. We first show that the MINIMUM GLOBAL TOTAL k-DOMINATION problem for bounded degree graphs is in APX.

Let G be a graph of order n, maximum degree $\Delta(G)$ and k be any positive integer such that $k \leq \delta(G)$ then from Observations 1 and 3, we have $\gamma_{kt}^g(G) \geq \gamma_{kt}(G) \geq \dfrac{kn}{\Delta(G)} \geq \dfrac{n}{\Delta(G)}$. So, we immediately have the following observation.

Observation 6. *For any graph G of order n with maximum degree $\Delta(G)$,*

$$\gamma_{kt}^g(G) \geq \frac{n}{\Delta(G)}$$

Now for a given graph $G = (V_G, E_G)$, we have a global total k-dominating set of G, $D_{kt}^g = V_G$ such that $|D_{kt}^g| = |V_G| = n \leq \Delta(G) \cdot \gamma_{kt}^g(G)$. Thus we have the following theorem.

Theorem 7. MINIMUM GLOBAL TOTAL k-DOMINATION *problem in any graph G with maximum degree $\Delta(G)$ can be approximated within an approximation ratio of $\Delta(G)$.*

By Theorem 7, the MINIMUM GLOBAL TOTAL k-DOMINATION problem for bounded degree graphs can be approximated within a constant ratio. Thus the MINIMUM GLOBAL TOTAL k-DOMINATION problem for bounded degree graphs is in APX.

Next, we show that the MINIMUM GLOBAL TOTAL k-DOMINATION problem is APX-complete for graphs with maximum degree $k + 3$. For this purpose, we recall the concept of L-reduction.

Definition 1. *Given two NP optimization problems π_1 and π_2 and a polynomial time transformation f from instances of π_1 to instances of π_2, we say that f is an L-reduction if there are positive constants α and β such that for every instance x of π_1:*

1. *$opt_{\pi_2}(f(x)) \leq \alpha \cdot opt_{\pi_1}(x)$.*
2. *for every feasible solution y of $f(x)$ with objective value $m_{\pi_2}(f(x), y) = c_2$, we can find a solution y' of x in polynomial time with $m_{\pi_1}(x, y') = c_1$ such that $|opt_{\pi_1}(x) - c_1| \leq \beta \cdot |opt_{\pi_2}(f(x)) - c_2|$.*

To show the APX-completeness of a problem $\pi \in APX$, it suffices to show that there is an L-reduction from some APX-complete problem to π.

To show the APX-completeness of the MINIMUM GLOBAL TOTAL k-DOMINATION problem, we give an L-reduction from the MINIMUM TOTAL k-DOMINATION problem. For this the following theorem is required.

Theorem 8 *[9]. The* MINIMUM TOTAL k-DOMINATION *problem is APX-complete for graphs with maximum degree $k + 2$.*

Now, we are ready to prove the following theorem.

Theorem 9. *The* MINIMUM GLOBAL TOTAL k-DOMINATION *problem is APX-complete for graphs with maximum degree $k + 3$.*

Proof. Since by Theorem 8, MINIMUM TOTAL k-DOMINATION problem is APX-complete for graphs with maximum degree $k + 2$. So, it is enough to construct an L-reduction f from the instances of the MINIMUM TOTAL k-DOMINATION problem for graphs with maximum degree $k+2$ to the instances of the MINIMUM GLOBAL TOTAL k-DOMINATION problem. Given a graph $G = (V_G, E_G)$, where $V_G = \{v_1, v_2, \ldots, v_n\}$ and $k \leq \delta(G)$ be any fixed positive integer. We construct a graph $H = (V_H, E_H)$ from the graph G in the following way:

1. Construct a subgraph $W = (V_W, E_W)$ by adding k copies of K_{k+1} and another vertex, namely w, by joining the vertex w to one vertex of each copy K_{k+1}.
2. Take n copies of subgraph W, say $W_1, W_2, \ldots, W_n$.
3. Add edges between the vertex v_i of the graph G and the vertex w_i of the ith-subgraph W_i.

Note that if maximum degree of G is $k + 2$ then maximum degree of H is $k + 3$. Now, we first prove the following claim.

Claim. $\gamma_{kt}^{g}(H) = \gamma_{kt}(G) + nk(k+1)$.

Proof. Firstly, we will show that $\gamma_{kt}(H) = \gamma_{kt}(G) + nk(k+1)$ and further, we will prove that $\gamma_{kt}^{g}(H) = \gamma_{kt}(H)$.

Let $D_G \subseteq V_G$ be a minimum total k-dominating set (TkD-set) of G. Then $D_H = D_G \cup \left(\bigcup_{i=1}^{n} (V_{W_i} \setminus \{w_i\}) \right)$ is a TkD-set of H. Thus,

$$\gamma_{kt}(H) \leq nk(k+1) + \gamma_{kt}(G).$$

On the other hand, assume that $D_H \subseteq V_H$ be a minimum TkD-set of the graph H. It can be easily checked that $\bigcup_{i=1}^{n} (V_{W_i} \setminus \{w_i\}) \subset D_H$. Since every $v_i \in V_H \cap V_G$ is totally k-dominated by the set D_H, so $|N_H(v_i) \cap D_H| \geq k$. Now for some fixed $j \in [n]$, if $w_j \in D_H$, then $|N_H(v_j) \cap D_H| = k$, otherwise $D_H \setminus \{w_j\}$ will also be a TkD-set of H, which is a contradiction. Since $\delta(G) \geq k$, there always exists a vertex $u \in N_G(v_j) \setminus D_H$ such that $(D_H \setminus \{w_j\}) \cup \{u\}$ is again a TkD-set of the graph H. Using the above process, we can construct a TkD-set of H containing no vertices in the set $\{w_i \mid 1 \leq i \leq n\}$. Hence, the set $D_H \cap V_G$ will be a TkD-set of G.

$$\gamma_{kt}(H) = |D_H| = |D_H \cap V_G| + \left| D_H \cap \left(\bigcup_{i=1}^{n} V_{W_i} \right) \right| \geq \gamma_{kt}(G) + nk(k+1).$$

It can be easily observed that every vertex v of the graph H has at least k non-adjacent vertices in the set D_H. Therefore, using Observation 2, it follows that D_H is a GTkD-set of the graph H. So, $|D_H| = \gamma_{kt}(H) \geq \gamma_{kt}^{g}(H)$. Thus, using Observation 1, we get $\gamma_{kt}^{g}(H) \geq \gamma_{kt}(H) = \gamma_{kt}(G) + nk(k+1)$. This completes the proof of claim. $\square$

We now return to the proof of Theorem 9. Since the maximum degree of G is $k+2$. Therefore, from Observation 3, we note that $\gamma_{kt}(G) \geq \frac{kn}{k+2}$. Hence, we have $\gamma_{kt}^{g}(H) = \gamma_{kt}(G) + nk(k+1) \leq \gamma_{kt}(G) + (k+1)(k+2)\gamma_{kt}(G)$.

Again, $\gamma_{kt}(G) - |D_G| = \gamma_{kt}^{g}(H) - nk(k+1) - |D_H| + nk(k+1) = \gamma_{kt}^{g}(H) - |D_H|$. From these two inequalities, it is clear that the above reduction is an L-reduction with $\alpha = 1 + (k+1)(k+2)$ and $\beta = 1$. Therefore, MINIMUM GLOBAL TOTAL k-DOMINATION problem is APX-complete for graphs with maximum degree $k+3$. $\square$

5   Conclusion

In this paper, we have shown that DECIDE GLOBAL TOTAL k-DOMINATION problem is NP-complete for bipartite graphs and chordal graphs. Apart from these, we have shown that for any graph with maximum degree Δ, MINIMUM GLOBAL TOTAL k-DOMINATION problem admits a Δ approximation algorithm. We have also shown that MINIMUM GLOBAL TOTAL k-DOMINATION problem is APX-complete for graphs with maximum degree $k+3$. It would be interesting to design better approximation algorithm for MINIMUM GLOBAL TOTAL k-DOMINATION problem.

References

1. Ausiello, G., Crescenzi, P., Gambosi, G., Kann, V., Marchetti-Spaccamela, A., Protasi, M.: Complexity and Approximation: Combinatorial Optimization Problems and their Approximability Properties. Springer, Heidelberg (2012). https://doi.org/10.1007/978-3-642-58412-1
2. Bermudo, S., Martínez, A.C., Mira, F.A.H., Sigarreta, J.M.: On the global total k-domination number of graphs. Discrete Appl. Math. **263**, 42–50 (2019)
3. Fulkerson, D., Gross, O.: Incidence matrices and interval graphs. Pac. J. Math. **15**(3), 835–855 (1965)
4. Haynes, T., Hedetniemi, S., Slater, P.: Domination in Graphs: Advanced Topics. Marcel Dekker Inc., New York (1998)
5. Haynes, T., Hedetniemi, S., Slater, P.: Fundamentals of Domination in Graphs. Marcel Dekker Inc., New York (1998)
6. Henning, M.A.: A survey of selected recent results on total domination in graphs. Discrete Math. **309**(1), 32–63 (2009)
7. Henning, M.A., Kazemi, A.P.: k-tuple total domination in graphs. Discrete Appl. Math. **158**(9), 1006–1011 (2010)
8. Henning, M.A., Yeo, A.: Total Domination in Graphs. Springer, New York (2013). https://doi.org/10.1007/978-1-4614-6525-6
9. Pradhan, D.: Algorithmic aspects of k-tuple total domination in graphs. Inf. Process. Lett. **112**(21), 816–822 (2012)

Hardness and Approximation for the Geodetic Set Problem in Some Graph Classes

Dibyayan Chakraborty[1], Florent Foucaud[2], Harmender Gahlawat[1(✉)],
Subir Kumar Ghosh[3], and Bodhayan Roy[4]

[1] Indian Statistical Institute, Kolkata, India
harmendergahlawat@gmail.com
[2] Univ. Bordeaux, Bordeaux INP, CNRS, LaBRI, UMR5800, 33400 Talence, France
[3] Ramakrishna Mission Vivekananda Educational and Research Institute,
Kolkata, India
[4] Indian Institute of Technology, Kharagpur, Kharagpur, India

Abstract. In this paper, we study the computational complexity of finding the *geodetic number* of graphs. A set of vertices S of a graph G is a *geodetic set* if any vertex of G lies in some shortest path between some pair of vertices from S. The MINIMUM GEODETIC SET (MGS) problem is to find a geodetic set with minimum cardinality. In this paper, we prove that solving MGS is NP-hard on planar graphs with a maximum degree six and line graphs. We also show that unless $P = NP$, there is no polynomial time algorithm to solve MGS with sublogarithmic approximation factor (in terms of the number of vertices) even on graphs with diameter 2. On the positive side, we give an $O\left(\sqrt[3]{n}\log n\right)$-approximation algorithm for MGS on general graphs of order n. We also give a 3-approximation algorithm for MGS on solid grid graphs which are planar.

1 Introduction and Results

Suppose there is a city-road network (i.e. a graph) and a bus company wants to open bus terminals in some of the cities. The buses will go from one bus terminal to another (i.e. from one city to another) following the shortest route in the network. Finding the minimum number of bus terminals required so that any city belongs to some shortest route between some pair of bus terminals is equivalent to finding the *geodetic number* of the corresponding graph. Formally, an undirected simple graph G has vertex set $V(G)$ and edge set $E(G)$. For two vertices $u, v \in V(G)$, let $I(u, v)$ denote the set of all vertices in G that lie in some shortest path between u and v. A set of vertices S is a *geodetic set* if $\cup_{u,v \in S} I(u, v) = V(G)$. The *geodetic number*, denoted as $g(G)$, is the minimum integer k such that G has a geodetic set of cardinality k. Given a graph G, the MINIMUM GEODETIC SET (MGS) problem is to compute a geodetic set of G with minimum cardinality. In this paper, we shall study the computational complexity of MGS in various graph classes.

© Springer Nature Switzerland AG 2020
M. Changat and S. Das (Eds.): CALDAM 2020, LNCS 12016, pp. 102–115, 2020.
https://doi.org/10.1007/978-3-030-39219-2_9

The notion of geodetic sets and geodetic number was introduced by Harary et al. [18]. The notion of geodetic number is closely related to convexity and convex hulls in graphs, which have applications in game theory, facility location, information retrieval, distributed computing and communication networks [2,10,15,19,22]. In 2002, Atici [1] proved that finding the geodetic number of arbitrary graphs is NP-hard. Later, Dourado et al. [8,9] strengthened the above result to *bipartite* graphs, *chordal* graphs and *chordal bipartite* graphs. Recently, Bueno et al. [3] proved that MGS remains NP-hard even for *subcubic* graphs. On the positive side, polynomial time algorithms to solve MGS are known for *cographs* [8], *split* graphs [8], *ptolemaic* graphs [12], *outer planar* graphs [21] and *proper interval* graphs [11]. In this paper, we prove the following theorem.

Theorem 1. MGS *is NP-hard for planar graphs of maximum degree 6.*

Then we focus on *line* graphs. Given a graph G, the *line* graph of G, denoted by $L(G)$, is a graph such that each vertex of $L(G)$ represents an edge of G and two vertices of $L(G)$ are adjacent if and only if their corresponding edges share a common endpoint in G. A graph H is a *line graph* if $H \cong L(G)$ for some G. Some optimisation problems which are difficult to solve in general graphs admit polynomial time algorithms when the input is a line graph [14,17]. We prove the following theorem.

Theorem 2. MGS *is NP-hard for line graphs.*

From a result of Dourado et al. [8], it follows that solving MGS is NP-hard even for graphs with diameter at most 4. On the other hand, solving MGS on graphs with diameter 1 is trivial (since those are exactly complete graphs). In this paper, we prove that unless P = NP, there is no polynomial time algorithm with sublogarithmic approximation factor for MGS even on graphs with diameter at most 2. A *universal vertex* of a graph is adjacent to all other vertices of the graph. We shall prove the following stronger theorem.

Theorem 3. *Unless* $P = NP$, *there is no polynomial time* $o(\log n)$-*approximation algorithm for* MGS *even on graphs that have a universal vertex, where* n *is the number of vertices in the input graph.*

On the positive side, we show that a reduction to the MINIMUM RAINBOW SUBGRAPH OF MULTIGRAPH problem (defined in Sect. 3.1) gives the first sublinear approximation algorithm for MGS on general graphs.

Theorem 4. *Given a graph, there is a polynomial-time* $O(\sqrt[3]{n} \log n)$-*approximation algorithm for* MGS *where* n *is the number of vertices.*

Then we focus on *solid grid* graphs, an interesting subclass of planar graphs. A *grid embedding* of a graph is a collection of points with integer coordinates such that each point in the collection represents a vertex of the graph and two points are at a distance one if and only if the vertices they represent arc adjacent in the graph. A graph is a *grid* graph if it has a grid embedding. A graph is a *solid grid* graph if it has a grid embedding such that all interior faces have unit area. Approximation algorithms for optimisation problems like LONGEST PATH, LONGEST CYCLE, NODE-DISJOINT PATH etc. on grid graphs and solid grid graphs have been studied [4,6,20,23,25,27]. In this paper, we prove the following theorem.

Theorem 5. *Given a solid grid graph, there is an $O(n)$ time 3-approximation algorithm for* MGS, *even if the grid embedding is not given as part of the input. Here n is the number of vertices in the input graph.*

Note that recognising solid grid graphs is NP-complete [16].

Organisation of the Paper: In Sect. 2, we prove the hardness results for planar graphs, line graphs and graphs with diameter 2. In Sect. 3, we present our approximation algorithms. Finally we draw our conclusions in Sect. 4.

2   Hardness Results

In Sect. 2.1, we prove that MGS is NP-hard for planar graphs with maximum degree 6 (Theorem 1). Then in Sect. 2.2 we prove that MGS is NP-hard for line graphs (Theorem 2). In Sect. 2.3 we prove the inapproximability result (Theorem 3).

2.1   NP-hardness on Planar Graphs

Given a graph G, a subset $S \subseteq V(G)$ is a dominating set of G if any vertex in $V(G) \backslash S$ has a neighbour in S. The problem MINIMUM DOMINATING SET (MDS) consists in computing a dominating set of an input graph G with minimum cardinality. To prove Theorem 1, we reduce the NP-complete MDS on subcubic planar graphs [13] to MGS on planar graphs with maximum degree 6.

Let us describe the reduction. From a subcubic planar graph G with a given planar embedding, we construct a graph $f(G)$ as follows. Each vertex v of G will be replaced by a *vertex gadget* G_v. This vertex gadget has vertex set $\{c^v, t_0^v, t_1^v, t_2^v\} \cup \{x_{i,j}^v, y_{i,j}^v, z_{i,j}^v \mid 0 \le i < j \le 2\}$. For simplicity we will consider that $x_{i,j}^v$ and $x_{j,i}^v$ refers to the same vertex (the same holds for $y_{i,j}^v$ and $y_{j,i}^v$, and for $z_{i,j}^v$ and $z_{j,i}^v$). There are no other vertices in $f(G)$. For the edges within G_v, vertex t_i^v (for $0 \le i \le 2$) is adjacent to vertices c^v, $x_{i,i+1}^v$, $y_{i,i+1}^v$, $x_{i-1,i}^v$, $y_{i-1,i}^v$ (indices taken modulo 3). Moreover, for each pair i,j with $0 \le i < j \le 2$, $x_{i,j}^v$ is adjacent to c^v and $y_{i,j}^v$, and $y_{i,j}^v$ is adjacent to $z_{i,j}^v$. We now describe the edges outside of the vertex-gadgets. They will depend on the embedding of G. We assume that the edges incident with any vertex v are labeled e_i^v with $0 \le i < deg_G(v)$, in such a way that the numbering increases counterclockwise around v with respect to the embedding (thus the edge vw will have two labels: e_i^v and e_j^w). Consider two vertices v and w that are adjacent in G, and let e_i^v and e_j^w be the two labels of edge vw in G. Then, t_i^v is adjacent to t_j^w, $y_{i,i+1}^v$ is adjacent to $y^w_{j-1,j}$ and $y^v_{i-1,i}$ is adjacent to $y^w_{j+1,j}$ (indices are taken modulo the degree of the original vertex of G). It is clear that a planar embedding of $f(G)$ can easily be obtained from the planar embedding of G. Thus $f(G)$ is planar and has maximum degree 6. The construction is depicted in Fig. 1, where v and w are adjacent in G and the edge vw is labeled e^v_0 and e^w_0.

We will show that G has a dominating set of size k if and only if $f(G)$ has a geodetic set of size $3|V(G)| + k$.

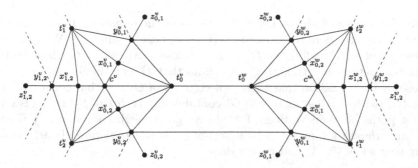

Fig. 1. Illustration of the reduction used in the proof of Theorem 1. Here, two vertex gadgets G_v, G_w are depicted, with v and w adjacent in G. Dashed lines represent potential edges to other vertex-gadgets.

Assume first that G has a dominating set D of size k. We construct a geodetic set S of $f(G)$ of size $3|V(G)| + k$ as follows. For each vertex v in G, we add the three vertices $z^v_{i,j}$ ($0 \le i < j \le 2$) of G_v to S. If v is in D, we also add vertex c^v to S.

Let us show that S is indeed a geodetic set. First, we observe that, in any vertex gadget G_v that is part of $f(G)$, the unique shortest path between two distinct vertices $z^v_{i,j}$, $z^v_{i',j'}$ has length 4 and goes through vertices $y^v_{i,j}$, t^v_k and $y^v_{i',j'}$ (where $\{k\} = \{i,j\} \cap \{i',j'\}$). Thus, it only remains to show that vertices c^v and $x^v_{i,j}$ ($0 \le i < j \le 2$) belong to some shortest path of vertices of S. Assume that v is a vertex of G in D. The shortest paths between c^v and $z^v_{i,j}$ have length 3 and one of them goes through vertex $x^v_{i,j}$. Thus, all vertices of G_v belong to some shortest path between vertices of S. Now, consider a vertex w of G adjacent to v and let $z^w_{i,j}$ be the vertex of G_w that is farthest from c_v. The shortest paths between c^v and $z^w_{i,j}$ have length 6; one of them goes through vertices c^w and $x^w_{i,j}$; two others go through the two other vertices $x^w_{i',j'}$ and $x^w_{i'',j''}$. Thus, S is a geodetic set.

For the converse, assume we have a geodetic set S' of $f(G)$ of size $3|V(G)|+k$. We will show that G has a dominating set of size k. First of all, observe that all the $3|V(G)|$ vertices of type $z^v_{i,j}$ are necessarily in S', since they have degree 1. As observed earlier, the shortest paths between those vertices already go through all vertices of type t^v_i and $y^v_{i,j}$. However, no other vertex lies on a shortest path between two such vertices: these shortest paths always go through the boundary

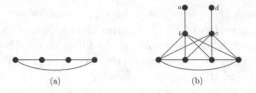

Fig. 2. (a) A triangle-free graph G and (b) the graph H_G.

6-cycle of the vertex-gadgets. Let S_0' be the set of the remaining k vertices of S'. These vertices are there to cover the vertices of type c^v and $x_{i,j}^v$. We construct a subset D' of $V(G)$ as follows: D' contains those vertices v of G whose vertex-gadget G_v contains a vertex of S_0'. We claim that D' is a dominating set of G. Suppose by contradiction that there is a vertex v of G such that neither G_v nor any of G_w (with w adjacent to v in G) contains any vertex of S_0'. Here also, the shortest paths between vertices of S always go through the boundary 6-cycle of G_v and thus, they never include vertex c_v, a contradiction. Thus, D' is a dominating set of size k, and we are done.

2.2 NP-hardness on Line Graphs

In this section, we prove that MGS remains NP-hard on line graphs. For a graph G and edges $e, e' \in E(G)$, define $d(e, e') = 1$ if e, e' shares a vertex, and $d(e, e') = i > 1$ if e, e' do not share a vertex and e' shares a vertex with an edge e'' with $d(e, e'') = i - 1$. A path between two edges e, e' is defined in the usual way.

Observation 1. *A path between two edges e, e' of a graph G corresponds to a path between the vertices e and e' in $L(G)$.*

Given a graph G, a set $S \subseteq E(G)$ is a *line geodetic set* of G if every edge $e \in E(G) \setminus S$ belongs to some shortest path between some pair of edges $\{e, e'\} \subseteq S$. Observation 1 implies the following.

Observation 2. *A graph G has a line geodetic set of cardinality k if and only if $L(G)$ has a geodetic set of size k.*

We shall show (in Lemma 8) that finding a line geodetic set of a graph with minimum cardinality is NP-hard. Then Observation 2 shall imply that solving MGS on line graphs is NP-hard. For the above purpose we need the following definition. Given a graph G, a set $S \subseteq E(G)$ is a *good edge set* if for any edge $e \in E(G) \setminus S$, there are two edges $e', e'' \in S$ such that (i) e lies in some shortest path between e' and e'', and (ii) $d(e', e'')$ is 2 or 3.

Lemma 6. *Computing a good edge set of a triangle-free graph with minimum cardinality is NP-hard.*

Proof. We shall reduce the NP-complete EDGE DOMINATING SET problem on triangle-free graphs [26] to the problem of computing a good edge set of a graph with minimum cardinality on triangle-free graphs. Given a graph G, a set $S \subseteq E(G)$ is an *edge dominating set* of G if any edge $e \in E(G) \setminus S$ shares a vertex with some edge in S. The EDGE DOMINATING SET problem is to compute an edge dominating set of G with minimum cardinality.

Let G be a triangle-free graph. For each vertex $v \in V(G)$, take a new edge $x_v y_v$. Construct a graph G^* whose vertex set is the union of $V(G)$ and the set $\{x_v, y_v\}_{v \in V(G)}$ and $E(G^*) = E(G) \cup \{vx_v\}_{v \in V(G)} \cup \{x_v y_v\}_{v \in V(G)}$. Notice that G^* is a triangle-free graph and we shall show that G has an edge dominating set of cardinality k if and only if G^* has a good edge set of cardinality $k + n$ where $n = |V(G)|$.

Let S be an edge dominating set of G. For each $v \in G$, let H_v be written as x_v, y_v, z_v. Notice that the set $S \cup \{x_v y_v\}_{v \in V(G)}$ forms a good edge set of G^* and has cardinality $k + n$. Let S' be a good edge set of G^* of size at most $k + n$. Notice that for each $v \in V(G)$, S' must contain the edge $x_v y_v$. Hence, the cardinality of the set $S' \cap E(G)$ is at most k. Moreover, for each $e \in E(G^*) \cap E(G)$, there is an edge $e' \in S'$ which is at distance 2 from e. As S' is a good edge set of G^*, any edge in $E(G) \setminus S'$ shares a vertex with some edge of S'. Hence $S' \cap E(G)$ is an edge dominating set of G of cardinality at most k.   □

For a triangle-free graph G, let H_G be the graph with $V(H_G) = V(G) \cup \{a, b, c, d\}$ and $E(H_G) = E(G) \cup \{ab, cd\} \cup E'$ where $E' = \{bv\}_{v \in V(G)} \cup \{cv\}_{v \in V(G)}$. See Fig. 2(a) and (b) for an example. We prove the following proposition.

Lemma 7. *For a triangle-free graph G, there is a line geodetic set Q of H_G with minimum cardinality such that $Q \cap E' = \emptyset$.*

Proof. For a set $S \subseteq E(H_G)$, an edge $f \in S$ *covers* an edge $e \in E(H_G)$, if there is another edge $f' \in S$ such that e lies in the shortest path between f and f'. Notice that the edges $\{ab, cd\}$ lie in any line geodetic set of H_G and all edges in E' are covered by ab and cd. First we prove the following claims.

Claim 1. *Let Q be a line geodetic set of H_G and $e \in E' \cap Q$. If e does not cover any edge of $E(G)$, then $Q \setminus \{e\}$ is a line geodetic set of H_G.*

The proof of the above claim follows from the fact that all edges in $E' \cup \{ab, cd\}$ are covered by ab and cd.

Claim 2. *Let Q be a line geodetic set of H_G and $e \in E' \cap Q$. There is another edge $e' \in E(G) \setminus Q$ such that $(Q \cup \{e'\}) \setminus \{e\}$ is a line geodetic set of H_G.*

To prove the claim above, first we define the *eccentricity* of an edge $e \in E(H_G)$ to be the maximum shortest path distance between e and any other edge in $E(H_G)$. Notice that the eccentricity of any edge in E' is two and the eccentricity of any edge of $E(G)$ in H_G is at most three. Now remove all edges from $E' \cap Q$ which do not cover any edge of $E(G)$. By Claim 1, the resulting set, say Q', is a line geodetic set of H_G. Let e be an edge $Q' \cap E'$ and let $\{f_1, f_2, \ldots, f_k\} \subseteq E(G) \setminus Q'$ be the set of edges covered by e. Since the ecentricity of e is two, there must exist $e_1, e_2, \ldots, e_k$ in Q' such that f_i has a common endpoint with both e and e_i for each $i \in \{1, 2, \ldots, k\}$. Therefore the distance between e and e_i is two for each $i \in \{1, 2, \ldots, k\}$. As G is triangle-free, $e_i \neq e_j$ for any $i, j \in \{1, 2, \ldots, k\}$. Choose any edge $f_j \in \{f_1, f_2, \ldots, f_k\}$. Observe that the distance between f_j and e_i is two when $i \neq j$. Therefore, for each $i \in \{1, 2, \ldots, j-1, j+1, \ldots k\}$, the edge f_i lies in the shortest path between f_j and e_i. Therefore, $(Q' \cup \{f_j\}) \setminus \{e\}$ is a line geodetic set of H_G.

Given any line geodetic set P of H_G, we can use the arguments used in Claims 1 and 2 repeatedly on P to construct a line geodetic set Q of H_G such that $|Q| \leq |P|$ and $Q \cap E' = \emptyset$. Thus we have the proof. $\square$

Lemma 8. *Computing a line geodetic set of a graph with minimum cardinality is NP-hard.*

Proof. We shall reduce the NP-complete problem of computing a good edge set of a triangle-free graph with minimum cardinality (Lemma 6). Let G be a triangle-free graph. Construct the graph H_G as stated above (just before Lemma 7). The set E' is also defined as before. We shall show that a triangle-free graph G has a good edge set of cardinality k if and only if H_G has a line geodetic set of cardinality $k + 2$.

Let P be a good edge set of G. Notice that, for each edge $e \in E(G)$, there are two edges $e', e'' \in P$ such that e belongs to a shortest path between e' and e'' in H_G. Also any edge of E' belongs to a shortest path between the edges ab and cd in H_G. Hence $P \cup \{ab, cd\}$ is a line geodetic set of H_G with cardinality $k + 2$.

Let Q be a line geodetic set of H_G of size $k + 2$. Notice that $\{ab, cd\} \subseteq Q$ and let $Q' = Q \setminus \{ab, cd\}$. Due to Lemma 7, we can assume that Q' does not contain any edge of E'. Let e be an edge in $E(G) \setminus Q'$ and let $e', e'' \in Q'$ such that e lies in some shortest path between e' and e'' in H_G. Since the distance between e' and e'' is at most three in H_G, it follows that Q' is a good edge set of G with cardinality k. $\square$

2.3 Inapproximability on Graphs with Diameter 2

Given a graph G, a set $S \subseteq V(G)$ is a *2-dominating set* of G if any vertex $w \in V(G) \setminus S$ has at least two neighbours in S. The 2-MDS problem is to compute a 2-dominating set of graphs with minimum cardinality. We shall use the following result.

Theorem 9 ([5,7]). *Unless $P = NP$, there is no polynomial time $o(\log n)$-approximation algorithm for the 2-MDS problem on triangle-free graphs.*

Lemma 10. *Let G be a triangle-free graph and G' be the graph obtained by adding an universal vertex v to G. A set S of vertices of G' is a geodetic set if and only if $S \setminus \{v\}$ is a 2-dominating set of G.*

Proof. Let S be a geodetic set of G'. Observe that for any vertex $u \in V(G) \setminus S$ there must exist vertices $u_1, u_2 \in S \setminus \{v\}$ such that $u \in I(u_1, u_2)$ and $u_1 u_2 \notin E(G)$. Hence, S is a 2-dominating set of G. Conversely, let S' be any 2-dominating set of G. For any vertex $u \in V(G) \setminus S'$ there exist $v, v' \in S'$ such that $uv, uv' \in E(G)$. Since G is triangle-free, v and v' are non-adjacent. Hence, $u \in I(v, v')$ and $S' \cup \{v\}$ is a geodetic set of G'. □

The proof of Theorem 3 follows due to Lemma 10 and Theorem 9.

3   Approximation Algorithms

In Sects. 3.1 and 3.2 we present approximation algorithms for MGS on general graphs and solid grid graphs, respectively.

3.1   General Graphs

We will reduce the MINIMUM GEODETIC SET problem to the MINIMUM RAINBOW SUBGRAPH OF MULTIGRAPH (MRSM) problem. A subgraph H of an edge colored multigraph G is *colorful* if H contains at least one edge of each color. Given an edge colored multigraph G, the MRSM problem is to find a colorful subgraph of G of minimum cardinality. The following is a consequence of a result due to Tirodkar and Vishwanathan [24].

Theorem 11 ([24]). *Given an edge colored multigraph G, there is a polynomial time $O(\sqrt[3]{n} \log n)$-approximation algorithm to solve the MRSM problem where $n = |V(G)|$.*

We note that Tirodkar and Vishwanathan [24] proved the above theorem for simple graphs only, but the proof works for multigraphs as well.

Given a graph G form an edge colored multigraph H_G as follows. The vertex set of H_G is the same as G. For each triplet of not necessarily distinct vertices (u, v, w) such that u lies in some shortest path between v and w, add an edge in H_G between v and w having the color u. Observe that, G has a geodetic set of cardinality k if and only if H_G has a colorful subgraph with k vertices. The proof of Theorem 4 follows from Theorem 11.

3.2   Solid Grid Graphs

In this section, we shall give a linear time 3-approximation algorithm for MGS on solid grid graphs. From now on G shall denote a solid grid graph and $\mathcal{R}$ is a grid embedding of G where every interior face has unit area.

Let G be a solid grid graph. A path P of G is a *corner path* if (i) no vertex of P is a cut vertex, (ii) both end-vertices of P have degree 2, and (iii) all vertices except the end-vertices of P have degree 3. See Fig. 3(a) for an example. Observe that for a corner path P, either the x-coordinates of all vertices of P are the same or the y-coordinates of all vertices of P are the same. Moreover, all vertices of a corner path lie in the outer face of G. The next observation follows from the definition of corner path and the fact that G is a solid grid graph.

Observation 3. *Let P be a corner path of G. Consider the set $Q = \{v \in V(G) : v \notin V(P), N(v) \cap P \neq \emptyset\}$. Then Q induces a path in G. Moreover, if the x-coordinates (resp. the y-coordinates) of all the vertices of P are the same, then the x-coordinates (resp. the y-coordinates) of all vertices in Q are the same.*

We shall use Observation 3 to prove a lower bound on the geodetic number of G in terms of the number of corner paths of G.

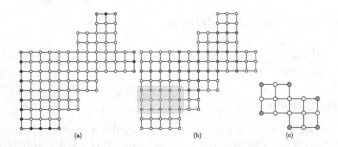

(a)   (b)   (c)

Fig. 3. (a) The black and gray vertices are the vertices of the corner paths. The gray vertices indicate the corner vertices. (b) The gray vertices are vertices of the red path. Vertices in the shaded box form a rectangular block. (c) Example of a solid grid graph whose number of corner vertices is exactly three times the geodetic number.

Lemma 12. *Any geodetic set of G contains at least one vertex from each corner path.*

Proof. Without loss of generality, we assume the x-coordinates of all vertices of P are the same. By Observation 3, the set $\{v \in V(G) : v \notin V(P), N(v) \cap P \neq \emptyset\}$ induces a path Q and the x-coordinates of all vertices in Q are the same. Now consider any two vertices $a, b \in V(G) \setminus V(P)$ and with a path P' between a and b that contains one of the end-vertices, say u, of P. Observe that P' can be expressed as $P' = a\ c_1\ c_2 \ldots c_t\ d\ f_1\ f_2 \ldots f_{t'}\ u\ g\ h_1\ h_2\ \ldots h_{t''}\ b$ such that $\{d, g\} \subseteq V(Q)$ and $\{f_1, f_2, \ldots, f_{t'}\} \subseteq V(P)$. Then there is a path $P'' = a\ c_1\ c_2 \ldots c_t\ d\ f_2' \ldots f_{t'}'\ g\ h_1\ h_2\ \ldots h_{t''}\ b$ where for $2 \leq i \leq t'$, f_i' is the vertex in Q which is adjacent to f_i in G. Observe that the length of P'' is strictly less than that of P'. Therefore $u \notin I(a, b)$ whenever $a, b \in V(G) \setminus V(P)$. Hence any geodetic set of G contains at least one vertex from P. $\qquad\square$

Any geodetic set of G contains all vertices of degree 1. Inspired by the above fact and Lemma 12, we define the term *corner vertex* as follows. A vertex v of G is a *corner vertex* if v has degree 1 or v is an end-vertex of some corner path. See Fig. 3(a) for an example. Observe that two corner paths may have at most one corner vertex in common. Moreover, a corner vertex cannot be in three corner paths. Therefore it follows that the cardinality of the set of corner vertices is at most $3 \cdot g(G)$.

Remark 13. Note that there are solid grid graphs whose number of corner vertices is exactly three times the geodetic number. See Fig. 3(c) for one such example.

Now we prove that the set of all corner vertices of G is indeed a geodetic set of G. We shall use the following proposition of Ekim and Erey [10].

Theorem 14 ([10])**.** *Let F be a graph and $F_1, \ldots, F_k$ its biconnected components. Let C be the set of cut vertices of G. If $X_i \subseteq V(F_i)$ is a minimum set such that $X_i \cup (V(F_i) \cap C)$ is a minimum geodetic set of F_i then $\cup_{i=1}^{k} X_i$ is a minimum godetic set of F.*

The next observation follows from Theorem 14.

Observation 4. *Let $C(G)$ be the set of corner vertices of G and S be the set of cut vertices of G. Let $\{H_1, H_2, \ldots, H_t\}$ be the set of biconnected components of G. The set $C(G)$ is a geodetic set of G if and only if $(C(G) \cap V(H_i)) \cup (S \cap V(H_i))$ is a geodetic set of H_i for all $1 \leq i \leq t$.*

From now on, $C(G)$ is the set of corner vertices of G and $H_1, H_2, \ldots, H_t$ are the biconnected components of G. Due to Theorem 14 and Observation 4, it is enough to show that for each $1 \leq i \leq t$, the set $(C(G) \cap V(H_i)) \cup (S \cap V(H_i))$ is a geodetic set of H_i. First, we introduce some more definitions below.

Let H be a biconnected component of G. Recall that each vertex of H is a pair of integers and each edge is a line segment with unit length. An edge $e \in E(H)$ is an *interior* edge if all interior points of e lie in an interior face of H. For a vertex $v \in V(H)$, let P_v denote the maximal path such that all edges of P_v are interior edges and each vertex in P_v has the same x-coordinate as v. Similarly, let P_v' denote the maximal path such that all edges of P_v are interior edges and each vertex in P_v' has the same y-coordinate as v. A path P of H is a *red path* if (i) there exists a $v \in V(H)$ such that $P \in \{P_v, P_v'\}$ and (ii) at least one end-vertex of P is a cut-vertex or a vertex of degree 4. A vertex v of H is *red* if v lies on some red path. See Fig. 3(b) for an example.

Definition 15. *A subgraph F of H is a rectangular block if F satisfies the following properties.*

1. *For any two vertices $(a_1, b_1), (a_2, b_2)$ of F, we have that any pair (a_3, b_3) with $a_1 \leq a_3 \leq a_2$ and $b_1 \leq b_3 \leq b_2$ is a vertex of F.*
2. *Let a, a' be the maximum and minimum x-coordinates of the vertices in F. The x-coordinate of any red vertex of F must be equal to a or a'. Similarly, let b, b' be the maximum and minimum y-coordinates of the vertices in F. The y-coordinate of any red vertex of F must be equal to b or b'.*

Observe that H can be decomposed into rectangular blocks such that each non-red vertex belongs to exactly one rectangular block. See Fig. 3(b) for an example. Let $B_1, B_2, \ldots, B_k$ be a decomposition of H into rectangular blocks. Recall that $C(G)$ is the set of corner vertices of G and S is the set of cut vertices of G. We have the following lemma.

Lemma 16. *For each $1 \leq i \leq k$, there are two vertices $x_i, y_i \in (C(G) \cap V(H)) \cup (S \cap V(H))$ such that $V(B_i) \subseteq I(x_i, y_i)$.*

Proof. Let $X \in \{B_1, B_2, \ldots, B_k\}$ be an arbitrary rectangular block. A vertex v of X is a *northern* vertex if the y-coordinate of v is maximum among all vertices of X. Analogously, *western* vertices, *eastern* vertices and *southern* vertices are defined. A vertex of X is a boundary vertex if it is either northern, western, southern or an eastern vertex of X. Let $nw(X)$ be the vertex of X which is both a northern vertex and a western vertex. Similarly, $ne(X)$ denotes the vertex which is both northern vertex and eastern vertex, $sw(X)$ denotes the vertex of X which is both southern and western vertex and $se(X)$ denotes the vertex of X which is both southern and eastern vertex.

First we prove the lemma assuming that all boundary vertices of X are red vertices. Let a (resp. b) denote the vertex with minimum y-coordinate such that P_a (resp. P_b) contains $sw(X)$ (resp. $se(X)$). Similarly, let c (resp. d) denote the vertex with maximum y-coordinate such that P_c (resp. P_d) contains $nw(X)$ (resp. $ne(X)$). Let a' (resp. c') denote the vertex with minimum x-coordinate such that $P'_{a'}$ (resp. $P'_{b'}$) contains $sw(X)$ (resp. $nw(X)$). Let b' (resp. d') denote the vertex with maximum x-coordinate such that $P'_{b'}$ (resp. $P'_{d'}$) contains $se(X)$ (resp. $ne(X)$). Observe that the vertices $a', a, b, b', d', d, c, c'$ lie on the exterior face of the embedding.

For two vertices $i, j \in \{a', a, b, b', d', d, c, c'\}$, let P_{ij} denote the path between i, j that can be obtained by traversing the exterior face of the embedding in the counter-clockwise direction starting from i. Observe that, if both $P_{a'a}$ and $P_{d'd}$ (resp. $P_{bb'}$ and $P_{cc'}$) contain a corner or cut vertex each, say f, f', then $\{sw(X), ne(X)\} \subseteq I(f, f')$ (resp. $\{nw(X), se(X)\} \subseteq I(f, f')$) and therefore $V(X) \subseteq I(f, f')$. Now consider the case when at least one of the paths in $\{P_{a'a}, P_{d'd}\}$ does not contain any corner vertex or cut vertex and when at least one of the paths in $\{P_{b'b}, P_{cc'}\}$ does not contain any corner vertex or cut vertex. Due to symmetry of rotation and reflection on grids, without loss of generality we can assume that both $P_{a'a}$ and $P_{bb'}$ have no corner vertex or cut vertex. Observe that in this case there must be a corner vertex f in P_{ab} whose x-coordinate is the same as that of b and therefore of $se(X)$. If $P_{cc'}$ contains a corner vertex f', then $\{nw(X), se(X)\} \subseteq I(f, f'))$ and therefore $V(X) \subseteq I(f, f')$. Otherwise, there must be a corner vertex f' in $P_{c'a'}$ whose y-coordinate is the same as that of c' and therefore of $nw(X)$. Hence we have $\{nw(X), se(X)\} \subseteq I(f, f')$ and therefore $V(X) \subseteq I(f, f')$ in this case also.

Now we consider the case when there are some non-red boundary vertices of X. Let v be a non-red vertex of X. Without loss of generality, we can assume that v is a western vertex of X. Now we redefine the vertices $a, a', b, b', c, c', d, d'$ as follows. Let $a' = sw(X)$, $c' = nw(X)$ and a (resp. b) be the vertex with minimum y-coordinate such that there is a path from a to $sw(X)$ (resp. from b to $se(X)$) containing vertices with the same x-coordinate as that of $sw(X)$ (resp. $se(X)$). Similarly, let c (resp. d) be the vertex with maximum y-coordinate such that there is a path from c to $nw(X)$ (resp. from d to $ne(X)$) containing vertices with the same x-coordinate as that of $nw(X)$ (resp. $ne(X)$).

Finally, let d' (resp. b') be the vertex with maximum x-coordinate such that there is a path from d' to $ne(X)$ (resp. from b' to $se(X)$) containing vertices with the same y-coordinate as that of $ne(X)$ (resp. $se(X)$). Using similar arguments on the paths P_{ij} with $i, j \in \{a', a, b, b', d', d, c, c'\}$ as before, we can show that there exists corner vertices f, f' such that $V(X) \subseteq I(f, f')$. So we have the proof.   □

By Observation 4 and Lemma 16, $C(G)$ is a geodetic set of G.

Time Complexity: If the grid embedding of G is given as part of the input, then the set of corner vertices can be computed in $O(|V(G)|)$ time by simply traversing the exterior face of the embedding. Otherwise, the set of corner vertices can be computed in $O(|V(G)|)$ time as follows (we shall only describe the procedure to find corner vertices of degree two as the other case is trivial). Let H be a biconnected component of G, v be a vertex of H having degree 2 and u_0, x_0 be its neighbours. If both u_0 and x_0 have degree 4, then v is not a corner vertex. Moreover, if at least one of u_0 and x_0 have degree 2 then v is a corner vertex. Otherwise, apply the following procedure. Assume u_0 has degree 3 and denote v as u_{-1} for technical reasons. Set $i = 0$. As H is a biconnected solid grid graph, u_i and x_i must have exactly one common neighbour which is different from u_{i-1}. Denote this vertex as x_{i+1}. Let u_{i+1} be the neighbour of u_i different from both x_{i+1} and u_{i-1}. If $deg_H(u_{i+1}) = 4$ or u_{i+1} is a cut vertex in G then terminate. If $deg_G(u_{i+1}) = 2$ then v is a corner vertex. Otherwise, set $i = i + 1$ and repeat the above steps. Observe that, when the above procedure terminates either we know that v is a corner vertex or there is no corner path that contains both u_0 and v. Now swapping roles of u_0 and x_0 in the above procedure, we can decide if v is a corner vertex. We can find all the corner vertices of H by applying the above procedure to all vertices of degree 2 of H. Similarly by applying the above procedure to all the biconnected components of G, we can find all corner vertices. Notice that, the total running time of the algorithm remains linear in the number of vertices of G.

This completes the proof of Theorem 5.

4   Conclusion

In this paper, we studied the computational complexity of MGS in various graph classes. We proved that MGS remains NP-hard on planar graphs and line graphs. We also gave an $O(\sqrt[3]{n} \log n)$-approximation algorithm for MGS on general graphs and proved that unless P = NP, there is no polynomial time $o(\log n)$-approximation algorithm for MGS even on graphs with diameter 2. This motivates the following questions.

Question 1. Are there constant factor approximation algorithms for MGS on planar graphs and line graphs?

Question 2. Is there a $O(\log n)$-approximation algorithm for MGS on general graphs ?

Acknowledgements. The authors acknowledge the financial support from the IFCAM project "Applications of graph homomorphisms" (MA/IFCAM/18/39). Florent Foucaud is supported by the ANR project HOSIGRA (ANR-17-CE40-0022). We thank Ajit Diwan for helpful discussions.

References

1. Atici, M.: Computational complexity of geodetic set. Int. J. Comput. Math. **79**(5), 587–591 (2002)
2. Buckley, F., Harary, F.: Geodetic games for graphs. Quaestiones Math. **8**(4), 321–334 (1985)
3. Bueno, L.R., Penso, L.D., Protti, F., Ramos, V.R., Rautenbach, D., Souza, U.S.: On the hardness of finding the geodetic number of a subcubic graph. Inf. Process. Lett. **135**, 22–27 (2018)
4. Călinescu, G., Dumitrescu, A., Pach, J.: Reconfigurations in graphs and grids. SIAM J. Discrete Math. **22**(1), 124–138 (2008)
5. Chlebík, M., Chlebíková, J.: Approximation hardness of dominating set problems in bounded degree graphs. Inf. Comput. **206**(11), 1264–1275 (2008)
6. Chuzhoy, J., Kim, D.H.K.: On approximating node-disjoint paths in grids. In: APPROX/RANDOM, pp. 187–211. Schloss Dagstuhl-Leibniz-Zentrum fuer Informatik (2015)
7. Dinur, I., Steurer, D.: Analytical approach to parallel repetition. In: STOC, pp. 624–633. ACM (2014)
8. Dourado, M.C., Protti, F., Rautenbach, D., Szwarcfiter, J.L.: Some remarks on the geodetic number of a graph. Discrete Math. **310**(4), 832–837 (2010)
9. Dourado, M.C., Protti, F., Szwarcfiter, J.L.: On the complexity of the geodetic and convexity numbers of a graph. In: ICDM, vol. 7, pp. 101–108. Ramanujan Mathematical Society (2008)
10. Ekim, T., Erey, A.: Block decomposition approach to compute a minimum geodetic set. RAIRO-Oper. Res. **48**(4), 497–507 (2014)
11. Ekim, T., Erey, A., Heggernes, P., van't Hof, P., Meister, D.: Computing minimum geodetic sets of proper interval graphs. In: Fernández-Baca, D. (ed.) LATIN 2012. LNCS, vol. 7256, pp. 279–290. Springer, Heidelberg (2012). https://doi.org/10.1007/978-3-642-29344-3_24
12. Farber, M., Jamison, R.E.: Convexity in graphs and hypergraphs. SIAM J. Algebraic Discrete Methods **7**(3), 433–444 (1986)
13. Garey, M.R., Johnson, D.S.: Computers and Intractability, vol. 29. W.H.Freeman, New York (2002)
14. Gerber, M.U., Lozin, V.V.: Robust algorithms for the stable set problem. Graphs Comb. **19**(3), 347–356 (2003)
15. Gerstel, O., Zaks, S.: A new characterization of tree medians with applications to distributed sorting. Networks **24**(1), 23–29 (1994)
16. Gregori, A.: Unit-length embedding of binary trees on a square grid. Inf. Process. Lett. **31**(4), 167–173 (1989)
17. Guruswami, V.: Maximum cut on line and total graphs. Discrete Appl. Math. **92**(2–3), 217–221 (1999)
18. Harary, F., Loukakis, E., Tsouros, C.: The geodetic number of a graph. Math. Comput. Modell. **17**(11), 89–95 (1993)
19. Haynes, T.W., Henning, M., Tiller, C.A.: Geodetic achievement and avoidance games for graphs. Quaestiones Math. **26**(4), 389–397 (2003)

20. Itai, A., Papadimitriou, C.H., Szwarcfiter, J.L.: Hamilton paths in grid graphs. SIAM J. Comput. **11**(4), 676–686 (1982)
21. Mezzini, M.: Polynomial time algorithm for computing a minimum geodetic set in outerplanar graphs. Theoret. Comput. Sci. **745**, 63–74 (2018)
22. Mitchell, S.L.: Another characterization of the centroid of a tree. Discrete Math. **24**(3), 277–280 (1978)
23. Sardroud, A.A., Bagheri, A.: An approximation algorithm for the longest cycle problem in solid grid graphs. Discrete Appl. Math. **204**, 6–12 (2016)
24. Tirodkar, S., Vishwanathan, S.: On the approximability of the minimum rainbow subgraph problem and other related problems. Algorithmica **79**(3), 909–924 (2017)
25. Wu, B.Y.: A 7/6-approximation algorithm for the max-min connected bipartition problem on grid graphs. In: Akiyama, J., Bo, J., Kano, M., Tan, X. (eds.) CGGA 2010. LNCS, vol. 7033, pp. 188–194. Springer, Heidelberg (2011). https://doi.org/10.1007/978-3-642-24983-9_19
26. Yannakakis, M., Gavril, F.: Edge dominating sets in graphs. SIAM J. Appl. Math. **38**(3), 364–372 (1980)
27. Zhang, W., Liu, Y.: Approximating the longest paths in grid graphs. Theoret. Comput. Sci. **412**(39), 5340–5350 (2011)

Maximum Weighted Edge Biclique Problem on Bipartite Graphs

Arti Pandey[1](✉), Gopika Sharma[1], and Nivedit Jain[2]

[1] Department of Mathematics, Indian Institute Of Technology Ropar,
Rupnagar 140001, Punjab, India
{arti,2017maz0007}@iitrpr.ac.in
[2] Department of Computer Science and Engineering,
Indian Institute of Technology Jodhpur, Karwar 342037,
Rajasthan, India
jain.22@iitj.ac.in

Abstract. For a graph G, a complete bipartite subgraph of G is called a biclique of G. For a weighted graph $G = (V, E, w)$, where each edge $e \in E$ has a weight $w(e) \in \mathbb{R}$, the MAXIMUM WEIGHTED EDGE BICLIQUE (MWEB) problem is to find a biclique H of G such that $\sum_{e \in E(H)} w(e)$ is maximum. The decision version of the MWEB problem is known to be NP-complete for bipartite graphs. In this paper, we show that the decision version of the MWEB problem remains NP-complete even if the input graph is a complete bipartite graph. On the positive side, if the weight of each edge is a positive real number in the input graph G, then we show that the MWEB problem is $O(n^2)$-time solvable for bipartite permutation graphs, and $O(m+n)$-time solvable for chain graphs, which is a subclass of bipartite permutation graphs.

Keywords: Maximum Weighted Edge Biclique · Bipartite permutation graphs · Chain graphs · NP-completeness · Graph algorithms

1 Introduction

Let $G = (V, E)$ be a graph. A *biclique* of G is a complete bipartite subgraph of G. The MAXIMUM VERTEX BICLIQUE (MVB) problem is to find a biclique of G with maximum number of vertices. The decision version of the MVB problem is NP-complete for general graphs [1], but the MVB problem is polynomial time solvable for bipartite graphs [1]. The MAXIMUM EDGE BICLIQUE (MEB) problem is to find a biclique in G with maximum number of edges. The decision version of the MEB problem is NP-complete for general graphs [1] and it also remains NP-complete for bipartite graphs [2]. Many researchers have also studied some other variations of these problems, see [1,3–5]. The MAXIMUM EDGE BICLIQUE problem was first introduced in [1] and further studied in [2,6–9]. The MEB problem has applications in biclustering analysis techniques, where one is

© Springer Nature Switzerland AG 2020
M. Changat and S. Das (Eds.): CALDAM 2020, LNCS 12016, pp. 116–128, 2020.
https://doi.org/10.1007/978-3-030-39219-2_10

interested to capture the relationship between genes and conditions. The goal of biclustering algorithms is to find a subset of genes J and a subset of conditions C such that the change in the expression level of each $j \in J$ with respect to each $c \in C$ is significant. More details about the application of the MEB problem can be found in [3,8]. Since the MEB problem is also hard to approximate in bipartite graphs within n^δ for some $\delta > 0$ [10,11] under certain assumptions such as random 4-SAT or 3-SAT hardness hypothesis, researchers have also studied the problem for subclasses of bipartite graphs. The MEB problem is polynomial time solvable for the following subclasses of bipartite graphs: chordal bipartite graphs, convex bipartite graphs and bipartite permutation graphs [8,12–16]. Some other hardness results are also available for the MEB problem based on some assumptions [6,9,17,18]. In this paper, we study the weighted version of the MEB problem.

The weighted version of the MEB problem, namely MAXIMUM WEIGHTED EDGE BICLIQUE (MWEB) problem is also studied in literature, see [3–5,7]. Given a weighted graph $G = (V, E, w)$, where each edge $e \in E$ has a weight $w(e) \in \mathbb{R}$, the MWEB problem is to find a biclique C of G such that the sum of the weights of edges of C is maximum. Note that the MAXIMUM WEIGHTED EDGE BICLIQUE problem is the generalized version of the MAXIMUM EDGE BICLIQUE problem. So, the hardness results for the MAXIMUM EDGE BICLIQUE problem are also valid for the MAXIMUM WEIGHTED EDGE BICLIQUE problem. Given a graph G and an integer $k > 0$, the WEIGHTED EDGE BICLIQUE DECISION PROBLEM (WEBDP) is to find a biclique C of G such that the sum of edge weights of C is at least k. In this paper, we show that WEBDP is NP-complete even for complete bipartite graphs.

There exists a restricted version of the MWEB problem, namely the S-MWEB problem, where S is a subset of real numbers from which edge weights are taken and the input graph is a bipartite graph. In 2008, Tan [7] proved that for a wide range of choices of S, no polynomial time algorithm can approximate the S-MWEB problem within a factor of n^ϵ for some $\epsilon > 0$ unless RP=NP. He also proved that the decision version of the S-MWEB problem is NP-complete even for $S = \{-1, 0, 1\}$. In this paper, we show that this problem remains NP-complete when $S = \{1, -M\}$ ($M > |E(G)|$). On the positive side, we show that for a set S of positive real numbers, the S-MWEB problem is quadratic time solvable for bipartite permutation graphs and linear-time solvable for chain graphs.

The rest of the paper is organized as follows. In Sect. 2, we give some pertinent definitions and notations used in the paper. In Sect. 3, we show that the WEIGHTED EDGE BICLIQUE DECISION PROBLEM is NP-complete even for complete bipartite graphs. In Sect. 4, we show that the S-MWEB is $O(n^2)$-time solvable for bipartite permutation graphs if S is as set of positive real numbers. In Sect. 5, we propose a linear-time algorithm to solve the S-MWEB problem in $O(m + n)$-time for chain graphs (under the assumption that S is a set of positive real numbers). Finally, Sect. 6 concludes the paper.

2   Preliminaries

We are considering undirected, simple and connected graphs throughout this paper. A graph G is called a *bipartite graph* if its vertex set can be partitioned into two sets, say V_1 and V_2 such that every edge of G has one end point in V_1 and other end point in V_2. The set $\{V_1, V_2\}$ is called a bipartition of G. We denote such a bipartite graph by $G = (V_1, V_2, E)$, where E is the edge set of G. A biclique of G is a complete bipartite subgraph of G. A biclique of a bipartite graph is called maximal if it is not a proper subgraph of any other biclique of G. Weight of a biclique is defined as the sum of weights of all edges belonging in it. For a complete bipartite graph with bipartition $\{X, Y\}$ such that $|X| = s$ and $|Y| = t$, we use the notation $K_{s,t}$. For a vertex v of a graph G, $d(v)$ denotes degree of v and $N(v)$ denotes the open neighborhood of v which is the set of vertices adjacent to v in G. For a set $S \subseteq V(G), N(S)$ denotes the union of open neighborhooods of all vertices in S. Throughout this paper, n denotes order (number of vertices) of the graph and m denotes the size (number of edges) of the graph under consideration.

A binary relation that is reflexive, symmetric and transitive on the same set, is called an *equivalence relation*. For an equivalence relation $\sim$ on a set S, the equivalence class of $x \in S$ is the set containing all elements which are related to x by $\sim$. We denote equivalence class of an element x by $[x]$. Equivalence classes of two elements are either disjoint or identical. Disjoint equivalence classes give a partition of the set on which the relation was defined.

3   NP-completeness

In this section, we show that the WEIGHTED EDGE BICLIQUE DECISION PROBLEM (WEBDP) is NP-complete for complete bipartite graphs which is a very restricted subclass of bipartite graphs.

Theorem 1. *WEBDP is NP-complete for complete bipartite graphs.*

Proof. Clearly, WEBDP is in NP. To prove the NP-hardness of the WEBDP for complete bipartite graph, we make a polynomial reduction from the unweighted version of the same problem for bipartite graphs. So, we prove a construction of a weighted complete bipartite graph from an unweighted bipartite graph.

Let $G = (X, Y, E)$ be an unweighted bipartite graph with $|X| = n_1$ and $|Y| = n_2$. We construct a new graph H which is nothing but K_{n_1, n_2}. Now, for an edge e in H, we define its weight to be 1 if $e \in E$ and $-M$ otherwise, where $M > m = |E|$. So, H is a weighted complete bipartite graph with weights as any real number. Figure 1 illustrates the construction of H from G. The dashed edges in Fig. 1 are the edges with weight $-M$.

Now to complete the proof of the theorem, we only need to prove the following claim.

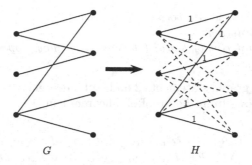

Fig. 1. An illustration to the construction of H from G.

Claim. G has a biclique of size at least $k > 0$ if and only if H has a biclique of weight at least $k > 0$.

Proof. The proof is easy and hence is omitted.   □

Hence, the theorem is proved.   □

We have observed that the WEBDP is NP-complete even for complete bipartite graphs. In next sections, we will discuss S-MWEB problem with S as the set of positive real numbers, which will be the restricted version of the MWEB problem. Throughout Sects. 4 and 5, by MWEB problem we mean S-MWEB problem, where $S = \mathbb{R}^+$.

4   Bipartite Permutation Graphs

A graph $G = (V, E)$ is called a *permutation graph* if there exists a one-to-one correspondence between its vertex set and a set of line segments between two parallel lines such that two vertices of G are adjacent if and only if their corresponding line segments intersect. A graph $G = (V, E)$ is called a *bipartite permutation graph* if it is both bipartite and permutation graph. We describe two characterizations of bipartite permutation graphs. A *strong ordering* $(<_X, <_Y)$ of a bipartite graph $G = (X, Y, E)$ consists of an ordering $<_X$ of X and an ordering $<_Y$ of Y, such that for all edges ab, $a'b'$, with $a, a' \in X$ and $b, b' \in Y$: if $a <_X a'$ and $b' <_Y b$, then ab' and $a'b$ are edges in G. An ordering $<_X$ of X has the *adjacency property* if, for every vertex in Y, its neighbors in X are consecutive in $<_X$. The ordering $<_X$ has the *enclosure property* if, for every pair of vertices y, y' of Y with $N(y) \subseteq N(y')$, the vertices of $N(y') \setminus N(y)$ appear consecutively in $<_X$. Strong ordering, adjacency property and enclosure property described above give rise to the following results which are already proven facts [19] providing two characterizations of bipartite permutation graphs.

Theorem 2. *[19] The following statements are equivalent for a graph* $G = (X, Y, E)$.

1. $G = (X, Y, E)$ is a bipartite permutation graph.
2. G has a strong ordering.
3. There exists an ordering of X which has the adjacency property and the enclosure property.

For a connected graph, statements 2 and 3 of Theorem 2 can be combined as Lemma 1 which follows from the proof of Theorem 1 in [19].

Lemma 1. *[20] Let $(<_X, <_Y)$ be a strong ordering of a connected bipartite permutation graph $G = (X, Y, E)$. Then both $<_X$ and $<_Y$ have the adjacency property and the enclosure property.*

Throughout this section, $G = (X, Y, E)$ denotes a weighted bipartite permutation graph such that $|X| = k$ and $|Y| = k'$. Weights on the edges are some positive real numbers. We assume that the strong ordering $(<_X, <_Y)$ of vertices of G are already given for the input graph. This ordering is considered as $\{x_1, x_2, \ldots, x_k\}$ and $\{y_1, y_2, \ldots, y_{k'}\}$ for X and Y respectively. We write $u <_X v$ or $u <_Y v$ for vertices u, v of G if u appears before v in the strong ordering of vertices of G. We write $u < v$ when it is clear from the context that u, v are coming from which side of the bipartition. For any edge $x_i y_j$, its weight is denoted by w_{ij}.

Now, we define first and last neighbor of a vertex in G. Since both $<_X$ and $<_Y$ satisfy adjacency property, for a vertex v of G, its neighbor set has some consecutive vertices in $<_X$ or $<_Y$. *First neighbor* of v is defined as the vertex that appears first in the strong ordering of G in its neighbor set and *last neighbor* of v is defined as the vertex that appears last in the strong ordering of G in its neighbor set. For any vertex u of G, $f(u)$ denotes the first neighbor of u and $l(u)$ denotes the last neighbor of u. For u in X, we denote $f(u)$ by y_{α_u} and $l(u)$ by y_{β_u} where $1 \leq \alpha_u \leq \beta_u \leq k'$. Combining above results, it can be observed that for a bipartite permutation graph G with its strong ordering $(<_X, <_Y)$, it has the following properties which will be used in the further discussion (See [21]):

1. Given any vertex of G, its neighbor set consists of some consecutive vertices in $<_X$ or $<_Y$.
2. For a pair of vertices u, v from X or Y, if $u < v$ then $f(u) \leq f(v)$ and $l(u) \leq l(v)$.

Now, we will discuss about the structure of a maximal biclique of G which will be used in getting a maximum biclique of G.

4.1 Maximal Bicliques

Let $G' = (X', Y', E')$ denotes a maximal biclique of G with $X' = \{x_i, x_{i+1}, \ldots, x_j\}$ and $Y' = \{y_{i'}, y_{i'+1}, \ldots, y_{j'}\}$ then edge $x_i y_{i'}$ is called the *first edge* of G'. We call an edge uv of G as a *safe edge* if it is the first edge of some maximal biclique of G. We will see that one safe edge corresponds to exactly one maximal biclique of G and vice versa.

Lemma 2. *Let $G' = (X', Y', E')$ be a biclique of G with $X' = \{x_i, x_{i+1}, \ldots, x_j\}$ and $Y' = \{y_{i'}, y_{i'+1}, \ldots, y_{j'}\}$, then G' is a maximal biclique of G if and only if the following holds for the graph G.*

(a) $l(x_i) = y_{j'}$
(b) $f(x_j) = y_{i'}$
(c) $l(y_{i'}) = x_j$
(d) $f(y_{j'}) = x_i$

Proof. First, let us assume that G' is a maximal biclique. We need to show that conditions $(a), (b), (c)$ and (d) are true. For (a), it is clear that $l(x_i) \geq y_{j'}$ since G' is a biclique. If equality holds, we are done. So, let $l(x_i) > y_{j'}$, say $l(x_i) = y_t (> y_{j'})$. Since vertices are ordered according to the strong ordering, all vertices of X' are adjacent to the vertices $y_{j'+1}, y_{j'+2}, \ldots, y_t$ in G implying that G' is not a maximal biclique of G. Now for (b), suppose that $f(x_j) < y_{i'}$, say $f(x_j) = y_p (< y_{i'})$. All vertices of X' are adjacent to $y_p, y_{p+1}, \ldots, y_{i'-1}$ in G because of the strong ordering of the vertices of G, but G' was maximal. Similarly (c) and (d) can be proven.

Conversely, we assume that the conditions $(a), (b), (c)$ and (d) are true. Let, if possible, G' is not maximal. Then there exists a vertex v in G for which one of the following conditions must be satisfied: (i) $v < x_i$ and $vy_{j'} \in E(G)$, (ii) $v < y_{i'}$ and $vx_j \in E(G)$, (iii) $v > x_j$ and $vy_{i'} \in E(G)$, and (iv) $v > y_{j'}$ and $vx_i \in E(G)$. But none of the edges $vy_{j'}, vx_j, vy_{i'}, vx_i$ can be present in G because of our assumption that $(a), (b), (c)$ and (d) are true. So, G' is a maximal biclique. $\square$

For any edge $e = uv$, the biclique corresponding to e, is the subgraph induced by the vertices $\{u, \ldots, l(v), v, \ldots, l(u)\}$. From Lemma 2, it can be observed that any maximal biclique of G can be identified from its first edge (safe edge). Given any edge uv ($u \in X$ and $v \in Y$) of G, one can easily check whether that is a safe edge or not as follows: If first neighbor of last neighbor of v is equal to v and first neighbor of last neighbor of u is equal to u, then uv qualifies as a safe edge. We observe from Lemma 2 that this condition is both necessary and sufficient for a biclique(corresponding to an edge uv) to be a maximal biclique. Hence, we can say that number of safe edges in G is equal to the number of maximal bicliques of G. We denote the maximal biclique corresponding to the safe edge e by G_e. For every vertex u of G, we define an array called *prefix sum array(psa)* of u of size $d(u)$ as an array in which each value equals the sum of weights of edges up to that position starting from $f(u)$. The *psa* of x_i(or y_j) is denoted by $A_i[\ ]$(or $B_j[\ ]$). Figure 2 represents a bipartite permutation graph. Next, we illustrate all the terminologies defined in this section using Fig. 2.

In bipartite permutation graph shown in Fig. 2, $x_2 y_2$ is a safe edge since $f(l(x_2)) = f(y_5) = x_2$ and $f(l(y_2)) = f(x_4) = y_2$ but $x_4 y_4$ is not as $f(l(x_4)) = f(y_5) = x_2 \neq x_4$. Prefix sum array of the vertex x_6 is $A_6 = \{27, 37, 48, 99\}$, where $A_6[1] = 27$, $A_6[2] = 27 + 10 = 37$, $A_6[3] = 27 + 10 + 11 = 48$ and $A_6[4] = 27 + 10 + 11 + 51 = 99$.

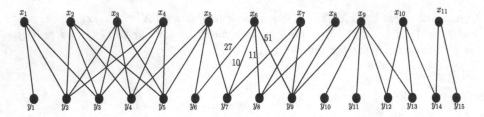

Fig. 2. An example of bipartite permutation graph.

4.2 Our Algorithm

Our idea for finding a maximum biclique is to look at all possible maximal bicliques of G and then return the one with the maximum weight. Since weights are positive real numbers any maximum biclique is some maximal biclique of G. The idea behind our algorithm is the following.

1. Find all safe edges of G.
2. Find psa of each vertex of G.
3. For each vertex $u \in X$,
      for each $v \in N(u)$ (choose vertex v in the given ordering)
          if $e = uv$ is a safe edge
              find the maximal biclique G_e
              find W_e, the weight of biclique G_e using psa of vertices.
4. Output the maximal biclique G_{e*} for which W_{e*} is maximum.

<u>Note:</u> we implement step 3 for each vertex in $O(n)$-time and hence overall complexity of step 3 is $O(n^2)$. The detailed algorithm is given in Algorithm 1.

Theorem 3. *Algorithm 1 outputs a maximum weighted edge biclique of the bipartite permutation graph G.*

Proof. The proof is omitted due to space constraints.   □

Theorem 4. *Algorithm 1 runs in $O(n^2)$-time.*

Proof. For any edge e, it will take constant time to check whether an edge qualifies as a safe edge or not by Lemma 2. So, preprocessing all safe edges take $O(m)$-time as it scans all the edges one by one. For a vertex u of G, calculating its psa will take $d(u)$ amount of time. Hence, finding psa of each vertex will take $O(m)$-time. For a vertex $u \in X$, step 3 can be implemented in $O(n)$-time. This is possible because, for all the safe edges in which one of the end point is u, we can find the weights of the corresponding bicliques in $O(n)$-time altogether. So, overall step 3 takes $O(n^2)$-time. Therefore, the algorithm returns a maximum weighted edge biclique of G in $O(m) + O(m) + O(n^2) \approx O(n^2)$-time.   □

Algorithm 1. Algorithm for finding a maximum weighted edge biclique of a bipartite permutation graph

---

Input: A bipartite permutation graph $G = (X, Y, E)$ with the strong ordering of its vertices

Output: A maximum weighted edge biclique of G

```
/* identifying safe edges                                            */
for each edge e=uv in E do
    if f(l(v))==v and f(l(u))==u then
        mark e as a safe edge
```

```
/* finding prefix sum arrays of vertices of G                        */
```
/* If $N(x_i) = \{y_{s_1}, y_{s_2}, \ldots, y_{s_{d(x_i)}}\}$, S_{x_i} denotes the set $\{1, 2, \ldots, d(x_i)\}$ */
```
for each vertex xᵢ from X do
    for every j from Sₓᵢ do
```
$A_i[j] = A_i[j - 1] + w_{is_j}$

```
/* similarly we can find psa of vertices of Y                        */
```
max:=0

for each vertex $x = x_i$ from X **do**

  sum:=0

  /* S'_x denotes the set $\{y_{a_1}, y_{a_2}, \ldots, y_{a_t}\}$ where $a_1 < a_2 < \ldots < a_t$ such that $xy_{a_1}, xy_{a_2}, \ldots, xy_{a_t}$ are safe edges   */

  for $j = 1$ to t **do**

    /* finding maximal biclique corresponding to the safe edge $e = xy_{a_j}$ */

    $X_e := \{x_i, x_{i+1}, \ldots, x_p\}$ // $x_p = l(y_{a_j})$

    $Y_e := \{y_{a_j}, y_{a_j+1}, \ldots, y_q\}$ // $y_q = l(x_i)$

    $E_e := \{uv | u \in X_e, v \in Y_e\}$, $G_e := (X_e, Y_e, E_e)$

    if $sum==0$ **then**

      for each vertex $x' = x_b$ from X_e **do**

        $sum := sum + A_b[q - \alpha_{x'} + 1] - A_b[a_j - \alpha_{x'}]$

      $W_e := sum$

    else

      $W_e := sum + W_1 - W_2$, $sum := W_e$

      /* W_1 and W_2 are the weights of the subgraphs induced by the vertices $\{x_{c+1}, \ldots l(y_{a_j}), y_{a_j}, \ldots, y_q\}$ and $\{x_i, \ldots x_c, y_{a_{j-1}}, \ldots, y_{a_j-1}\}$ respectively, where $x_c = l(y_{a_{j-1}})$. W_1 and W_2 are obtained using psa of vertices   */

    if $W_e > max$ **then**

      $max := W_e$, $e^* := e$

return G_{e^*} and max

---

5   Chain Graphs

A bipartite graph $G = (X, Y, E)$ is called a chain graph if there exists an ordering of vertices of $X = \{x_1, x_2, \ldots, x_{n_1}\}$ and an ordering of vertices of $Y = \{y_1, y_2, \ldots, y_{n_2}\}$ such that $N(x_1) \subseteq N(x_2) \subseteq \ldots \subseteq N(x_{n_1})$ and

$N(y_1) \supseteq N(y_2) \supseteq \ldots \supseteq N(y_{n_2})$. Throughout this section, $G = (X, Y, E)$ denotes a weighted chain graph with $|X| = n_1$ and $|Y| = n_2$. Weights on the edges are some positive real numbers. We assume that this ordering is given with the input graph.

For u, v in G, we define u and v to be *similar* vertices if $N(u) = N(v)$. For a set $S \subseteq V(G)$, we define S to be a *similar neighborhood set* if every two vertices from S are similar.

Now, we define a relation $\sim$ on X as for $u, v \in X$, $u \sim v$ if and only if vertices u and v are similar. One can easily observe that $\sim$ is an equivalence relation so it provides a partition P of the set X. If we define the same relation on the set Y, we will get a partition P' for the set Y. For any set $S \in P$, we keep the order of the vertices in S as it was given in the input chain graph. Order of the sets in P is also considered in such a way that taking union of all sets in that order gives the actual ordering of the vertices. We write $P = \{X_1, X_2, \ldots, X_{k_1}\}$ and $P' = \{Y_1, Y_2, \ldots, Y_{k_2}\}$, the partitions obtained for X and Y respectively from the relation $\sim$. Recall that $[x]$ denotes the equivalence class of the element x from X.

Lemma 3. *Let $\sim$ be the relation defined on X and Y as discussed above, then partitions P and P' are of same size, i.e. $|P| = |P'|$.*

Proof. We have defined the relation in such a way that vertices in one set of these partitions are similar to each other, so $N(X_1) \subset N(X_2) \subset \ldots \subset N(X_{k_1})$ and $N(Y_1) \supset N(Y_2) \supset \ldots \supset N(Y_{k_2})$ holds true. For any $i < j$, $N(X_i)$ is a proper subset of $N(X_j)$, so, say, $y \in N(X_j)$ such that $y \notin N(X_i)$. Since the graph is connected, this give rise to atleast two sets in P'. Hence, we get that $k_2 \geq k_1$. Similarly, $N(Y_i) \supset N(Y_j)$ gives $k_1 \geq k_2$ implying that $|P| = k_1 = k_2 = |P'|$. $\square$

Now, we define the representative vertex for each set of P. For a set $S \in P$, a vertex from S is called the *representative vetex* of the set S, if it is the least indexed vertex among all vertices of S. We denote representative vertex of a set S by r_S. Next we state some observations related to maximal bicliques of a chain graph which leads to a maximum weighted edge biclique of G.

5.1   Maximal Bicliques

Lemma 4. *Let $G' = (X', Y', E')$ be a maximal biclique of G, then the following holds:*

(a) If $x \in X'$, then $[x] \subseteq X'$.
(b) If, $y \in Y'$, then $[y] \subseteq Y'$.

Proof. (a) Here, we will show that $[x] \subseteq X'$ for any $x \in X'$. Let $x_0 \in [x]$, as x_0 and x are similar vertices, $N(x_0) = N(x)$. Now, $Y' \subseteq N(x) = N(x_0)$ implies that x_0 is adjacent to all vertices of Y' in G. We must have these edges in G' as it is a maximal biclique. So, $[x] \subseteq X'$ is true.

Proof of the part (b) is similar. $\square$

Below, we give a result which describes the detailed structure of a maximal biclique of a chain graph.

Lemma 5. *Let $G' = (X', Y', E')$ be a maximal biclique of G. Then there exists an index $1 \leq i \leq k$ such that $X' = X_i \cup X_{i+1} \cup \ldots \cup X_k$ and $Y' = N(r_{X_i})$.*

Proof. We know that vertices of G have an ordering as $\{x_1, x_2, \ldots, x_{n_1}\}$ and $\{y_1, y_2, \ldots, y_{n_2}\}$ for X and Y respectively. Let j be the minimum index from $\{1, 2, \ldots, n_1\}$ such that $x_j \in X'$ and there is some t such that $x_j \in X_t$. Now Lemma 4 tells that $[x_j] = X_t \subseteq X'$ implying that $x_j = r_{X_t}$. Since j is the smallest index, we get that $\{X_1 \cup X_2 \cup \ldots \cup X_{t-1}\} \cap X' = \phi$. Now, as $Y' \subseteq N(x_j)$ and G is a chain graph, $X' = X_t \cup X_{t+1} \cup \ldots \cup X_k$. Hence, for $i = t$, one part of the lemma holds. For the remaining part, it is enough to show that $N(x_j) \subseteq Y'$. So, let y be a neighbor of x_j, then y is adjacent to all vertices in the set $\{x_{j+1}, x_{j+2}, \ldots, x_{n_1}\}$ implying that $y \in Y'$. Hence, $Y' = N(r_{X_i})$ and $X' = X_i \cup X_{i+1} \cup \ldots \cup X_k$. $\square$

It can be identified from Lemma 5 that a chain graph has exactly k maximal bicliques, where k is the number of distinct equivalence classes corresponding to the relation $\sim$. Now, we define an array called *partition sum array (ptsa)* of size k for each $y \in Y$. In a partition sum array of a vertex y, each value contains the sum of weights of the edges incident on the vertex y coming from one set of P. We denote the *ptsa* of y_i by $A_i[\ ]$. Figure 3 represents a chain graph. We illustrate all the terminologies defined in this section using Fig. 3.

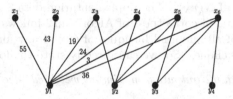

Fig. 3. An example of chain graph.

In the chain graph shown in Fig. 3, the partition $P = \{X_1, X_2, X_3, X_4\}$, where $X_1 = \{x_1, x_2\}, X_2 = \{x_3, x_4\}, X_3 = \{x_5\}$ and $X_4 = \{x_6\}$. Vertices x_3, x_4 are similar but x_1, x_3 are not and all the sets in P are similar neighborhood sets. Partition sum array of the vertex y_1 is $A_1 = \{98, 43, 3, 36\}$, where $A_1[1] = 55 + 43 = 98, A_1[2] = 19 + 24 = 43, A_1[3] = 3$ and $A_1[4] = 36$.

From, Lemma 5, we know the structure of maximal bicliques of G. One can easily see that each maximal biclique can be identified from the representative vertex of one of the X_i's from P. We use the notation G_x for the maximal biclique corresponding to the representative vertex x and, W_x for the weight of the maximal biclique G_x, where x is the representative vertex of some set in P.

5.2 Our Algorithm

Our basic idea for finding a maximum biclique in chain graphs is to find weight of each maximal biclique of G and output the one with the maximum weight. Since G has only k maximal bicliques, so, in order to get the desired biclique, we need to find out the weights of these k bicliques. Since chain graph is a subclass of bipartite permutation graph, we may also use Algorithm 1 to compute a maximum weighted edge biclique of G. The ordering of vertices of G as given in chain graph will also work for bipartite permutation graph. In this way, we will get our desired output in $O(n^2)$-time. Here, we propose an algorithm in which we use a different method to find out the sum of each maximal biclique of G which results in overall running time $O(m + n)$. The difference here is to use partition sum array instead of prefix sum array. The idea behind our algorithm is the following.

ALGORITHM 2

1. Find the partitions $P = \{X_1, X_2, \ldots, X_k\}$ and $P' = \{Y_1, Y_2, \ldots, Y_k\}$ from the equivalence relation $\sim$, say $R = < r_{X_k}, r_{X_{k-1}}, \ldots, r_{X_1} >$.
2. Calculate the ptsa for each vertex of Y.
3. For each vertex u according to the order in which it appears in R,
   find the maximal biclique G_u corresponding to the vertex u.
   find W_u, the weight of biclique G_u using ptsa of vertices from $Y \cap V(G_u)$.
4. Output the maximal biclique G_{u^*} for which W_{u^*} is maximum.

Note that we implement step 3 for each vertex $u \in R$ such that $W_{r_{X_j}}$ is calculated using ptsa of vertices of $N(r_{X_j})$. The implementation details are omitted due to space constraints. Proof of correctness of Algorithm 2 follows from the fact that it considers weights of all maximal bicliques of G and any maximum biclique is one of the maximal bicliques. So, we can directly state the following theorem.

Theorem 5. *Algorithm 2 outputs a maximum weighted edge biclique of a chain graph G.*

To analyse the running time of Algorithm 2, we need to bring some notations into consideration. We denote the cardinalities of sets in the partition P and P' by p_i, q_j for X_i, Y_j respectively, i.e. $|X_i| = p_i$ and $|Y_j| = q_j$. Now, we give a result which will be used in analyzing the running time of Algorithm 2.

Lemma 6. *Let G be a chain graph with a partition obtained from the $\sim$ relation defined on X as well as on Y. Then $m \geq kq_1 + (k-1)q_2 + \ldots + q_k$.*

Proof. We know that the relation $\sim$ made the sets from the partitions P and P' to follow the strict inclusion as $N(X_1) \subset N(X_2) \subset \ldots \subset N(X_k)$ and $N(Y_1) \supset N(Y_2) \supset \ldots \supset N(Y_k)$. Since $Y_1 \cup Y_2 \cup \ldots \cup Y_k = Y$ and for $i \neq j$, $Y_i \cap Y_j = \phi$, we can write that $m = \sum_{y \in Y_1} d(y) + \sum_{y \in Y_2} d(y) + \ldots + \sum_{y \in Y_k} d(y) = q_1 \sum_{i=1}^{k} p_i + q_2 \sum_{i=2}^{k} p_i + \ldots + q_k p_k \geq kq_1 + (k-1)q_2 + \ldots + q_k$. The last inequality follows since $|X_i| \geq 1$ for $1 \leq i \leq k$. $\square$

Theorem 6. *Algorithm 2 runs in $O(m + n)$-time.*

Proof. Step 1 will take $O(n)$-time as we have to go through all the vertices of G. To find out the time taken by step 2, we see that we are doing some number of additions during Algorithm 2. For each vertex y of Y, we are doing $d(y)$ number of additions, so overall step 2 takes $\sum_{y \in Y} d(y) = O(m)$ time. Now, to analyse step 3, we see that in our proposed algorithm, we are finding weights of maximal bicliques in the order $W_{r_{X_k}}, W_{r_{X_{k-1}}}, \ldots, W_{r_{X_1}}$. For calculating $W_{r_{X_j}}$, we are doing $\sum_{i=1}^{j} q_i + (j-1)$ number of additions, where j varies from k downto 1. Hence, step 3 performs $\sum_{i=1}^{k} q_i + \sum_{i=1}^{k-1} q_i + \ldots + q_1 + (k-1) + (k-2) + \ldots + 2 + 1 + 0 \leq kq_1 + (k-1)q_2 + \ldots + q_k + \frac{k(k+1)}{2}$ number of additions. Now we know that $m \geq \frac{k(k+1)}{2}$ since $N(X_1) \subset N(X_2) \subset \ldots \subset N(X_{k_1})$ and $N(Y_1) \supset N(Y_2) \supset \ldots \supset N(Y_{k_2})$. Now using Lemma 6, we can say that step 3 will take $O(m)$-time to execute. Clearly, choosing maximum among all the W_u's will take $O(k)$-time. Therefore, the Algorithm 2 returns a maximum weighted edge biclique of G in $O(n) + O(m) + O(m) + O(k) \approx O(m + n)$ time. $\square$

6   Conclusion

Our paper deals with the MAXIMUM WEIGHTED EDGE BICLIQUE problem. In this paper, we show that the decision version of the MAXIMUM WEIGHTED EDGE BICLIQUE problem remains NP-complete even for complete bipartite graphs, which is a subclass of bipartite graphs. On the positive side, we show that for the input graph G, if the weight of each edge is a positive real number, then the MWEB problem is $O(n^2)$-time solvable for bipartite permutation graphs and $O(m+n)$-time solvable for chain graphs. It will be interesting to try linear-time algorithm for bipartite permutation graphs, as for the unweighted graph this problem is linear-time solvable. One may also try linear-time algorithm for the MAXIMUM EDGE BICLIQUE problem in convex bipartite graphs.

References

1. Garey, M.R., Johnson, D.S.: Computers and Intractability: A Guide to the Theory of NP-Completeness, pp. 641–650 (1979)
2. Peeters, R.: The maximum edge biclique problem is NP-complete. Discrete Appl. Math. **131**(3), 651–654 (2003)
3. Dawande, M., Keskinocak, P., Tayur, S.: On the biclique problem in bipartite graphs. Technical report, Carnegie-Mellon University (1997)
4. Dawande, M., Keskinocak, P., Swaminathan, J.M., Tayur, S.: On bipartite and multipartite clique problems. J. Algorithms **41**(2), 388–403 (2001)
5. Hochbaum, D.S.: Approximating clique and biclique problems. J. Algorithms **29**(1), 174–200 (1998)

6. Manurangsi, P.: Inapproximability of maximum biclique problems, minimum k-cut and densest at-least-k-subgraph from the small set expansion hypothesis. Algorithms **11**(1), 10 (2018)
7. Tan, J.: Inapproximability of maximum weighted edge biclique and its applications. In: Agrawal, M., Du, D., Duan, Z., Li, A. (eds.) TAMC 2008. LNCS, vol. 4978, pp. 282–293. Springer, Heidelberg (2008). https://doi.org/10.1007/978-3-540-79228-4_25
8. Nussbaum, D., Pu, S., Sack, J.R., Uno, T., Zarrabi-Zadeh, H.: Finding maximum edge bicliques in convex bipartite graphs. Algorithmica **64**(2), 311–325 (2012)
9. Ambühl, C., Mastrolilli, M., Svensson, O.: Inapproximability results for maximum edge biclique, minimum linear arrangement, and sparsest cut. SIAM J. Comput. **40**(2), 567–596 (2011)
10. Feige, U.: Relations between average case complexity and approximation complexity. In: Proceedings of the Thiry-Fourth Annual ACM Symposium on Theory of Computing, pp. 534–543. ACM (2002)
11. Goerdt, A., Lanka, A.: An approximation hardness result for bipartite clique. Electronic Colloquium on Computational Complexity, Report vol. 48 (2004)
12. Alexe, G., Alexe, S., Crama, Y., Foldes, S., Hammer, P.L., Simeone, B.: Consensus algorithms for the generation of all maximal bicliques. Discrete Appl. Math. **145**(1), 11–21 (2004)
13. Dias, V.M., De Figueiredo, C.M., Szwarcfiter, J.L.: Generating bicliques of a graph in lexicographic order. Theoret. Comput. Sci. **337**(1–3), 240–248 (2005)
14. Dias, V.M., de Figueiredo, C.M., Szwarcfiter, J.L.: On the generation of bicliques of a graph. Discrete Appl. Math. **155**(14), 1826–1832 (2007)
15. Gély, A., Nourine, L., Sadi, B.: Enumeration aspects of maximal cliques and bicliques. Discrete Appl. Math. **157**(7), 1447–1459 (2009)
16. Kloks, T., Kratsch, D.: Computing a perfect edge without vertex elimination ordering of a chordal bipartite graph. Inf. Process. Lett. **55**(1), 11–16 (1995)
17. Feige, U., Kogan, S.: Hardness of approximation of the balanced complete bipartite subgraph problem. Department of Computer Science and Applied Mathematics, Weizmann Institute of Science, Rehovot, Israel, Technical report MCS04-04 (2004)
18. Ambuhl, C., Mastrolilli, M., Svensson, O.: Inapproximability results for sparsest cut, optimal linear arrangement, and precedence constrained scheduling. In: 48th Annual IEEE Symposium on Foundations of Computer Science (FOCS 2007), pp. 329–337. IEEE (2007)
19. Spinrad, J., Brandstädt, A., Stewart, L.: Bipartite permutation graphs. Discrete Appl. Math. **18**(3), 279–292 (1987)
20. Heggernes, P., van't Hof, P., Lokshtanov, D., Nederlof, J.: Computing the cutwidth of bipartite permutation graphs in linear time. In: Thilikos, D.M. (ed.) WG 2010. LNCS, vol. 6410, pp. 75–87. Springer, Heidelberg (2010). https://doi.org/10.1007/978-3-642-16926-7_9
21. Lai, T.H., Wei, S.S.: Bipartite permutation graphs with application to the minimum buffer size problem. Discrete Appl. Math. **74**(1), 33–55 (1997)

Graph Theory

Determining Number of Generalized and Double Generalized Petersen Graph

Angsuman Das$^{(\boxtimes)}$

Department of Mathematics, Presidency University, Kolkata, India
angsuman.maths@presiuniv.ac.in

Abstract. The determining number of a graph $G = (V, E)$ is the minimum cardinality of a set $S \subseteq V$ such that pointwise stabilizer of S under the action of $Aut(G)$ is trivial. In this paper, we determine the determining number of generalized Petersen graphs and double generalized Petersen graphs.

Keywords: Automorphism groups · Fixing number

1 Introduction

The *determining number* of a graph $G = (V, E)$ is the minimum cardinality of a set $S \subseteq V$ such that the automorphism group of the graph obtained from G by fixing every vertex in S is trivial. For example, $Aut(C_n) \cong D_n$ and stabilizer of any vertex is isomorphic to $\mathbb{Z}_2$. However, the pointwise stabilizer of any two non-antopodal vertices is trivial. Thus, $Det(C_n) = 2$. It was introduced independently by Boutin [2] and by Erwin and Harary (defined as *fixing number*) [6] in 2006 as a measure of destroying symmetry of a graph. Apart from proving general bounds and other results on determining number, researchers have attempted to find exact values of determining number of various families of graphs like Kneser Graphs [4], Coprime graphs [11] etc. In this paper, we find the determining numbers of generalized Petersen graphs and double generalized Petersen graphs.

For definitions and terms related to general graph theory, readers are referred to the classic book by Godsil and Royle [9]. For terms related to automorphisms of the above two families of graphs, readers are referred to [8] and [10] respectively. In Sects. 2 and 3, we study the determining sets and determining numbers of generalized Petersen graphs and double generalized Petersen graphs respectively. In particular, we prove the following theorems.

Theorem 1. *Let* $G(n, k)$ *be the generalized Petersen graph. Then*

$$Det(G(n,k)) = \begin{cases} 2, & \text{if } (n,k) \neq (4,1), (5,2), (10,3). \\ 3, & \text{if } (n,k) = (4,1), (5,2) \text{ or } (10,3). \end{cases}$$

$\square$

Theorem 2. *Let $DP(n,t)$ be the double generalized Petersen graph. Then*

$$Det(DP(n,t)) = \begin{cases} 4, & \text{if } (n,t) = (4,1). \\ 2, & \text{otherwise.} \end{cases}$$

□

2   Generalized Petersen Graphs

The generalized Petersen graph family was introduced by Coxeter [5] and was given its name by Watkins in [12].

Definition 1 (Generalized Petersen Graphs). *For integers n and k with $2 \leq 2k < n$, the Generalized Petersen graph $G(n,k)$ is defined to have vertex-set*

$$V(G(n,k)) = \{u_0, u_1, \ldots, u_{n-1}, v_0, v_1, \ldots, v_{n-1}\}$$

and edge-set $E(G(n,k))$ to consist of all edges of the form $(u_i, u_{i+1}), (u_i, v_i)$ and (v_i, v_{i+k}), where arithmetic of subscripts are to be done in modulo n.

The edges in $E(G(n,k))$ are called *outer edges, spoke edges* and *inner edges* respectively. The automorphism groups $A(n,k)$ of Generalized Petersen graphs $G(n,k)$ were studied by Frucht *et al.* [8]. Let $B(n,k)$ denote the subgroup of $A(n,k)$ which fixes the spoke edges set-wise. Define permutations ρ and δ on $V(G(n,k))$ by $\rho(u_i) = u_{i+1}, \rho(v_i) = v_{i+1}$ and $\delta(u_i) = u_{-i}, \delta(v_i) = v_{-i}$, for $i = 0, 1, \ldots, n-1$. It was proved in [5], that $\langle \rho, \delta \rangle \leq B(n,k)$. Define α on $V(G(n,k))$ by $\alpha(u_i) = v_{ki}, \alpha(v_i) = u_{ki}$, for $i = 0, 1, \ldots, n-1$. It was proved in [8], that $\alpha \in A(n,k)$ if and only if $k^2 \not\equiv \pm 1 \pmod{n}$.

In particular, they proved the following theorems:

Theorem 3 *[8].*

1. *If $k^2 \not\equiv \pm 1 \pmod{n}$, then $B(n,k) = \langle \rho, \delta : \rho^n = \delta^2 = 1; \delta\rho\delta = \rho^{-1} \rangle$.*
2. *If $k^2 \equiv 1 \pmod{n}$, then*

$$B(n,k) = \langle \rho, \delta, \alpha : \rho^n = \delta^2 = \alpha^2 = 1; \delta\rho\delta = \rho^{-1}, \alpha\delta = \delta\alpha, \alpha\rho\alpha = \rho^k \rangle.$$

3. *If $k^2 \equiv -1 \pmod{n}$, then $B(n,k) = \langle \rho, \alpha : \rho^n = \alpha^4 = 1; \alpha\rho\alpha^{-1} = \rho^k \rangle$.*

In Case 3, $\delta = \alpha^2$ and hence δ is omitted as a generator.

Theorem 4 *[8]. $B(n,k) = A(n,k)$ if and only if the ordered pair (n,k) is not one of $(4,1), (5,2), (8,3), (10,2), (10,3), (12,5), (24,5)$.*

Proposition 1. *If $k^2 \not\equiv \pm 1 \pmod{n}$ and $(n,k) \neq (10,2)$, then $Det(G(n,k)) = 2$.*

Proof : For such choice of n and k,

$$A(n, k) = \langle \rho, \delta : \rho^n = \delta^2 = 1; \delta\rho\delta = \rho^{-1} \rangle = \{\rho^i\delta^j : 0 \leq i \leq n - 1; 0 \leq j \leq 1\}.$$

We claim that $\{u_0, u_1\}$ is a determining set for $G(n, k)$. Let $\rho^i\delta^j$ be an element of $A(n, k)$ which fixes u_0 and u_1, for some $0 \leq i \leq n - 1$ and $0 \leq j \leq 1$.

If $j = 1$, then we have $\rho^i\delta(u_0) = u_0$ and $\rho^i\delta(u_1) = u_1$, i.e., $\rho^i(u_0) = u_0$ and $\rho^i(u_{-1}) = u_1$. The first equality implies $i = 0$, whereas the second one implies that $i = 2$, a contradiction. Thus $j = 0$. So, we have $\rho^i(u_0) = u_0$ and $\rho^i(u_1) = u_1$. This implies $i = 0$.

Hence, $Stab(\{u_0, u_1\})$ is trivial and $\{u_0, u_1\}$ is a determining set for $G(n, k)$. It proves that $Det(G(n, k)) \leq 2$.

If possible, let there exist a vertex whose stabilizer is trivial. Then by orbit-stabilizer theorem, the size of the orbit of that vertex is equal to $|A(n, k)| = 2n = |V(G(n, k))|$, i.e., $G(n, k)$ is vertex-transitive. However, it is shown in [8], that $G(n, k)$ is vertex-transitive if and only if $k^2 \equiv \pm 1 \ (mod \ n)$ or $n = 10$ and $k = 2$, which is a contradiction. Thus $Det(G(n, k)) = 2$. □

Proposition 2. If $k^2 \equiv 1 \ (mod \ n)$ and (n, k) is not one of $(4, 1), (8, 3)$, $(12, 5), (24, 5)$, then $Det(G(n, k)) = 2$.

Proof : For such choice of n and k,

$$A(n, k) = \langle \rho, \delta, \alpha : \rho^n = \delta^2 = \alpha^2 = 1; \delta\rho\delta = \rho^{-1}, \alpha\delta = \delta\alpha, \alpha\rho\alpha = \rho^k \rangle$$
$$= \{\rho^i\delta^j\alpha^l : 0 \leq i \leq n - 1; 0 \leq j, l \leq 1\}.$$

We claim that $\{u_0, u_1\}$ is a determining set for $G(n, k)$. Let $\rho^i\delta^j\alpha^l$ be an element of $A(n, k)$ which fixes u_0 and u_1, for some $0 \leq i \leq n - 1$ and $0 \leq j, l \leq 1$.

If possible, let $l = 1$. Then $\rho^i\delta^j\alpha(u_0) = u_0$ and $\rho^i\delta^j\alpha(u_1) = u_1$, i.e., $\rho^i\delta^j(v_0) = u_0$ and $\rho^i\delta^j(v_k) = u_1$. However, as both ρ and δ maps outer vertices to outer vertices and inner vertices to inner vertices, this leads to a contradiction. Thus, $l = 0$. So, we have $\rho^i\delta^j(u_0) = u_0$ and $\rho^i\delta^j(u_1) = u_1$.

If possible, let $j = 1$. Then $\rho^i\delta(u_0) = u_0$ and $\rho^i\delta(u_1) = u_1$, i.e., $\rho^i(u_0) = u_0$ and $\rho^i(u_{-1}) = u_1$. The first equality implies $i = 0$, whereas the second one implies that $i = 2$, a contradiction. Thus $j = 0$. So, we have $\rho^i(u_0) = u_0$ and $\rho^i(u_1) = u_1$. This implies $i = 0$.

Hence, $Stab(\{u_0, u_1\})$ is trivial and $\{u_0, u_1\}$ is a determining set for $G(n, k)$. It proves that $Det(G(n, k)) \leq 2$.

If possible, let there exist a vertex whose stabilizer is trivial. Then by orbit-stabilizer theorem, the size of the orbit of that vertex is equal to $|A(n, k)| = 4n$, which is greater than the order of $G(n, k)$, a contradiction. Thus $Det(G(n, k)) = 2$. □

Proposition 3. If $k^2 \equiv -1 \ (mod \ n)$ and $(n, k) \neq (5, 2), (10, 3)$, then $Det(G(n, k)) = 2$.

Proof : For such choice of n and k,

$$A(n,k) = \langle \rho, \alpha : \rho^n = \alpha^4 = 1; \alpha \rho \alpha^{-1} = \rho^k \rangle = \{\rho^i \alpha^j : 0 \le i \le n-1; 0 \le j \le 3\}.$$

We claim that $\{u_0, u_1\}$ is a determining set for $G(n,k)$. Let $\rho^i \alpha^j$ be an element of $A(n,k)$ which fixes u_0 and u_1, for some $0 \le i \le n-1$ and $0 \le j \le 3$.

If $j = 1$ or 3, then α^j swaps inner vertices and outer vertices and ρ^i maps outer vertices to outer vertices and inner vertices to inner vertices. Thus, $\rho^i \alpha^j$ maps u_0 to some inner vertex and hence it does not stabilize u_o. Hence, $j = 0$ or 2.

If possible, let $j = 2$. Then we have $\rho^i \alpha^2(u_0) = u_0$ and $\rho^i \alpha^2(u_1) = u_1$, i.e., $\rho^i(u_0) = u_0$ and $\rho^i(u_{-1}) = u_1$. The first equality implies $i = 0$, whereas the second one implies that $i = 2$, a contradiction. Thus $j = 0$. So, we have $\rho^i(u_0) = u_0$ and $\rho^i(u_1) = u_1$. This implies $i = 0$.

Hence, $Stab(\{u_0, u_1\})$ is trivial and $\{u_0, u_1\}$ is a determining set for $G(n,k)$. It proves that $Det(G(n,k)) \le 2$.

If possible, let there exist a vertex whose stabilizer is trivial. Then by orbit-stabilizer theorem, the size of the orbit of that vertex is equal to $|A(n,k)| = 4n$, which is greater than the order of $G(n,k)$, a contradiction. Thus $Det(G(n,k)) = 2$. □

Proposition 4. $Det(G(5,2)) = Det(G(10,3)) = Det(G(4,1)) = 3$.

Proof : $G(5,2)$ is the Petersen graph. It was shown in [2], that $Det(G(5,2)) = 3$.

It was checked using Sage that $\{u_0, u_1, v_2\}$ is a determining set of $G(10,3)$, i.e., $Stab(\{u_0, u_1, v_2\})$ is trivial. As $G(10,3)$ is vertex-transitive and $|A(10,3)| = 240$, it follows that stabilizer of any vertex is of order 12. Hence, $1 < Det(G(10,3)) \le 3$.

It is known that $G(10,3)$ is isomorphic to bipartite Kneser graph $H(5,2)$ and $Aut(H(5,2)) = S_5 \times \mathbb{Z}_2$. The vertices of $H(5,2)$ consists of all 2-subsets and 3-subsets of $\{1,2,3,4,5\}$ and two vertices are adjacent if one is a subset of the other. We prove that no two vertices form a determing set for $H(5,2)$.

If both the vertices A and B are 3-subsets, then they must have either one or two elements in their intersection. If $|A \cap B| = 1$, then they are of the form $A = \{a,b,c\}$ and $B = \{c,d,e\}$. Consider $\sigma = (a,b)(d,e) \in S_5$. σ is a non-identity element which fixes both A and B. If $|A \cap B| = 2$, then they are of the form $A = \{a,b,c\}$ and $B = \{b,c,d\}$. Then $\sigma = (b,c) \in S_5$ is a non-identity element which fixes both A and B.

If both the vertices A and B are 2-subsets, then they must have exactly one element in their intersection, i.e., they are of the form $A = \{a,b\}$ and $B = \{b,c\}$. Then $\sigma = (d,e) \in S_5$ is a non-identity element which fixes both A and B.

If A is a 3-subset and B is a 2-subset, then $|A \cap B| = 0, 1$ or 2. Then they are of the form $A = \{a,b,c\}; B = \{d,e\}$ or $A = \{a,b,c\}; B = \{c,d\}$ or $A = \{a,b,c\}; B = \{a,b\}$. In any case, $\sigma = (a,b) \in S_5$ is a non-identity element which fixes both A and B.

Thus $Det(G(10,3)) = 3$.

For $G(4,1)$, it was checked using Sage that $\{u_0, u_1, v_0\}$ is a determining set, i.e., $Det(G(4,1)) \leq 3$. Now, let us recall a result from [3].

Let H be a connected graph that is prime with respect to the Cartesian product. Then $Det(H^k) \geq \max\{Det(H), \lceil (\log k + \log |Aut(H)|)/\log |V(H)|\rceil\}$.

We note that $G(4,1) \cong C_4 \square P_2 \cong P_2 \square P_2 \square P_2 = (P_2)^3$ and P_2 is prime with respect to the Cartesian product. Thus, we have

$$Det(G(4,1)) = Det((P_2)^3) \geq \max\left\{1, \left\lceil \frac{\log 3 + \log 2}{\log 2}\right\rceil\right\} = \frac{\log 6}{\log 2} \approx 2.59.$$

Thus, we have $Det(G(4,1)) = 3$. $\hspace{3cm}\square$

Proposition 5. $Det(G(10,2)) = 2$.

Proof : $G(10,2)$ is the graph of the regular dodecahedron. Its automorphism group has already been computed in [8] to be $A(10,2) = \langle \rho, \lambda : \rho^{10} = \lambda^3 = (\lambda \rho^2)^2 = \rho^5 \lambda \rho^{-5} \lambda^{-1} = 1\rangle$, where the cycle structure of λ is given by

$$\lambda = (u_0, v_2, v_8)(u_1, v_4, u_8)(u_2, v_6, u_9)(u_3, u_6, v_9)(u_4, u_7, v_1)(u_5, v_7, v_3).$$

Observe that $\delta = (\rho\lambda)^2 \rho \lambda^{-1} \rho^{-2}$. $A(10,2)$ is isomorphic to the direct product of the alternating group A_5 with the symmetric group $\mathbb{Z}_2$. (See [7]) Thus $|A(10,2)| = 60 \times 2 = 120$.

It was checked using Sage (see Appendix) that $\{u_0, v_1\}$ is a determining set of $G(10,2)$, i.e., $Stab(\{u_0, v_1\})$ is trivial. As $G(10,2)$ is vertex-transitive and $|A(10,2)| = 120$, it follows that stabilizer of any vertex is of order 6. Hence, $Det(G(10,2)) = 2$. $\hspace{3cm}\square$

Proposition 6. $Det(G(8,3)) = Det(G(12,5)) = Det(G(24,5)) = 2$.

Proof : It was shown in [8], that for $G(n,k)$, where $(n,k) = (4,1), (8,3), (12,5)$ or $(24,5)$,

$$A(n,k) = \langle \rho, \delta, \sigma : \rho^n = \delta^2 = \sigma^3 = 1, \delta\rho\delta = \rho^{-1}, \delta\sigma\delta = \sigma^{-1}, \sigma\rho\sigma = \rho^{-1}, \sigma\rho^4 = \rho^4\sigma\rangle,$$

and $|A(n,k)| = 12n$. Note that α is superfluous and is given by $\alpha = \sigma^{-1}\rho\sigma^{-1}$ in $A(8,3)$ and $\alpha = \delta^{-1}\rho\sigma^{-1}$ in other three cases.

It was checked using Sage that $\{u_0, u_2\}$ is a determining set for each of $G(8,3), G(12,5)$ and $G(24,5)$, i.e., $Stab(\{u_0, u_2\})$ is trivial. As each of them are vertex-transitive and $|A(n,k)| = 12n$, it follows that stabilizer of any vertex is of order 6. Hence,

$$Det(G(8,3)) = Det(G(12,5)) = Det(G(24,5)) = 2.$$

$\hspace{8cm}\square$

From Propositions $1, 2, 3, 4, 5$ and 6, we have Theorem 1.

3   Double Generalized Petersen Graphs

Double Generalized Petersen Graphs $DP(n,t)$ are a natural generalization of Generalized Petersen graphs, first introduced in [13] as examples of vertex-transitive non-Cayley graphs. They are defined as follows:

Definition 2 (Double Generalized Petersen Graphs). *For integers n and t with $2 \leq 2t < n$, the Double Generalized Petersen graph $DP(n,t)$ is defined to have vertex-set*

$$V(DP(n,t)) = \{x_0, x_1, \ldots, x_{n-1}, y_0, y_1, \ldots, y_{n-1}, u_0, u_1, \ldots, u_{n-1}, v_0, v_1, \ldots, v_{n-1}\}$$

and edge-set $E(DP(n,t))$ to consist of all edges of the form: (x_i, x_{i+1}) and (y_i, y_{i+1}) (the outer edges), (x_i, u_i) and (y_i, v_i) (the spoke edges) and (u_i, v_{i+t}) and (v_i, u_{i+t}) (the inner edges), where arithmetic of subscripts are to be done in modulo n.

The automorphism groups $A(n,t)$ of Double Generalized Petersen graphs $DP(n,t)$ were studied by Kutnar and Petecki in [10]. In particular, they proved the following result.

Theorem 5 *(Corollary 3.11 [10]). The automorphism group $A(n,t)$ of the double generalized Petersen graph $DP(n,t)$ is characterized as follows:*

1. *If $n \equiv 0 \pmod 2$, $4t = n$ and $(n,t) \neq (4,1)$, then $A(n,t) = \langle \alpha, \beta, \gamma, \eta \rangle$.*
2. *If $n \equiv 0 \pmod 2$, $t^2 \equiv \pm 1 \pmod n$ and $(n,t) \neq (10,3)$, then $A(n,t) = \langle \alpha, \beta, \gamma, \delta \rangle$.*
3. *If $n \equiv 2 \pmod 4$, $t^2 \equiv k \pm 1 \pmod n$, where $n = 2k$ and $(n,t) \neq (10,2)$, then $A(n,t) = \langle \alpha, \beta, \gamma, \psi \rangle$.*
4. *If $n \equiv 0 \pmod 4$, $t^2 \equiv k \pm 1 \pmod n$, where $n = 2k$, then $A(n,t) = \langle \alpha, \beta, \gamma, \phi \rangle$.*
5. *$A(4,1) = \langle \alpha, \beta, \gamma, \delta, \eta \rangle$. $A(10,3) = \langle \alpha, \delta, \lambda \rangle$. $A(10,2) = \langle \alpha, \psi, \mu \rangle$.*
6. *$A(5,2)$ is the automorphism group of the dodecahedron.*
7. *In all cases different from the above, $A(n,t) = \langle \alpha, \beta, \gamma \rangle$,*

where $\alpha, \beta, \gamma, \delta, \eta, \psi, \phi$ are given by

$\alpha : x_i \mapsto x_{i+1}, y_i \mapsto y_{i+1}, u_i \mapsto u_{i+1}, v_i \mapsto v_{i+1}; \quad \beta : x_i \mapsto y_i, y_i \mapsto x_i, u_i \mapsto v_i, v_i \mapsto u_i$

$\gamma : x_i \mapsto x_{-i}, y_i \mapsto y_{-i}, u_i \mapsto u_{-i}, v_i \mapsto v_{-i}$

$\delta : x_{2i} \mapsto u_{2it}, x_{2i+1} \mapsto v_{(2i+1)t}, y_{2i} \mapsto v_{2it}, y_{2i+1} \mapsto u_{(2i+1)t}$
$\quad u_{2i} \mapsto x_{2it}, u_{2i+1} \mapsto y_{(2i+1)t}, v_{2i} \mapsto y_{2it}, v_{2i+1} \mapsto x_{(2i+1)t}$

$\eta : x_{2i} \mapsto x_{2i+k}, x_{2i+1} \mapsto x_{2i+1+k}, y_{2i} \mapsto y_{2i}, y_{2i+1} \mapsto y_{2i+1}$
$\quad u_{2i} \mapsto u_{2i+k}, u_{2i+1} \mapsto u_{2i+1+k}, v_{2i} \mapsto v_{2i}, v_{2i+1} \mapsto v_{2i+1}$, *where $n = 2k$.*

$\psi : x_{2i} \mapsto u_{2it}, x_{2i+1} \mapsto v_{(2i+1)t}, y_{2i} \mapsto u_{2it+k}, y_{2i+1} \mapsto v_{(2i+1)t+k}$
$\quad u_{2i} \mapsto x_{2it}, u_{2i+1} \mapsto y_{(2i+1)t}, v_{2i} \mapsto x_{2it+k}, v_{2i+1} \mapsto y_{(2i+1)t+k}$, *where $n = 2k$.*

$\phi : x_{2i} \mapsto u_{2it}, x_{2i+1} \mapsto v_{(2i+1)t}, y_{2i} \mapsto v_{2it+k}, y_{2i+1} \mapsto u_{(2i+1)t+k}$
$\quad u_{2i} \mapsto x_{2it}, u_{2i+1} \mapsto y_{(2i+1)t}, v_{2i} \mapsto y_{2it+k}, v_{2i+1} \mapsto x_{(2i+1)t+k}$, *where $n = 2k$.*

For the definition of λ and μ, please refer to [10].

Proposition 7. *If* $n \equiv 0 \pmod 2, 4t = n$ *and* $(n,t) \neq (4,1)$, *then* $Det(DP(n,t)) = 2$.

Proof : For such choice of n and t,

$$A(n,t) = \langle \alpha, \beta, \gamma, \eta \rangle = \{\alpha^i \beta^j \gamma^l \eta^s : 0 \leq i \leq n-1, 0 \leq j,l,s \leq 1\}.$$

We claim that x_0, y_1 is a determining set for $DP(n,t)$. Let $\alpha^i \beta^j \gamma^l \eta^s$ be an element of $A(n,t)$ which fixes x_0, y_1.

Since, β flips x_i's and y_i's and all others among α, γ and η maps x_i's to x_j's and y_i's to y_j's, we must have $j = 0$, i.e., it is enough to work with elements of the form $\alpha^i \gamma^l \eta^s$.

If $s = 1$, then we have $\alpha^i \gamma^l \eta(x_0) = x_0$ and $\alpha^i \gamma^l \eta(y_1) = y_1$, i.e., $\alpha^i \gamma^l(x_k) = x_0$ and $\alpha^i \gamma^l(y_1) = y_1$, where $n = 2k$. Now as α and γ has same effect on the indices of x_i's as α and γ has on the indices of y_i's, we have a contradiction. Thus, $s = 0$ and it suffices to work with $\alpha^i \gamma^l$.

If $l = 1$, we have $\alpha^i \gamma(x_0) = x_0$ and $\alpha^i \gamma(y_1) = y_1$, i.e., $\alpha^i(x_0) = x_0$ and $\alpha^i(y_{-1}) = y_1$. The first one implies $i = 0$ whereas second one implies $i = 2$, a contradiction. Thus, $l = 0$ and as a result $i = 0$.

Hence, $Stab(\{x_0, y_1\})$ is trivial and $\{x_0, y_1\}$ is a determining set for $DP(n,t)$. It proves that $Det(DP(n,t)) \leq 2$.

However, as $Stab(x_i) = Stab(u_i) = \langle \alpha^k \eta, \alpha^{2i} \gamma \rangle$ and $Stab(y_i) = Stab(v_i) = \langle \eta, \alpha^{2i} \gamma \rangle$, and each of the vertex stabilizers are isomorphic to $\mathbb{Z}_2 \times \mathbb{Z}_2$, we have $Det(DP(n,t)) = 2$. $\square$

Proposition 8. *If* $n \equiv 0 \pmod 2, t^2 \equiv \pm 1 \pmod n$ *and* $(n,t) \neq (10,3)$, *then* $Det(DP(n,t)) = 2$.

Proof : For such choice of n and t,

$$A(n,t) = \langle \alpha, \beta, \gamma, \delta \rangle = \{\alpha^i \beta^j \gamma^l \delta^s : 0 \leq i \leq n-1, 0 \leq j,l,s \leq 1\}.$$

We claim that x_0, x_1 is a determining set for $DP(n,t)$. Let $\alpha^i \beta^j \gamma^l \delta^s$ be an element of $A(n,t)$ which fixes x_0, x_1.

We claim that $s = 0$. If not, let $s = 1$ and hence $\alpha^i \beta^j \gamma^l \delta(x_0) = \alpha^i \beta^j \gamma^l(u_0) = u_p$ or v_p. Hence x_0 is not fixed. Thus $s = 0$ and it suffices to consider elements of the form $\alpha^i \beta^j \gamma^l$.

We claim that $j = 0$. Because if $j = 1$, $\alpha^i \beta \gamma^l$ maps x_0 to some y_p, a contradiction and hence we consider only elements of the form $\alpha^i \gamma^l$.

Thus $\alpha^i \gamma^l(x_0) = x_0$ and $\alpha^i \gamma^l(x_1) = x_1$. If $l = 1$, we have $\alpha^i(x_0) = x_0$ and $\alpha^i(x_{-1}) = x_1$. The first one implies $i = 0$ and the second one implies $i = 2$. Hence $l = 0$ and $i = 0$.

Hence, $Stab(\{x_0, x_1\})$ is trivial and $\{x_0, x_1\}$ is a determining set for $DP(n,t)$. It proves that $Det(DP(n,t)) \leq 2$.

If possible, let there exist a vertex whose stabilizer is trivial. Then by orbit-stabilizer theorem, the size of the orbit of that vertex is equal to $|A(n,t)| = 8n > |V(DP(n,t))|$, which is a contradiction. Thus $Det(DP(n,t)) = 2$. $\square$

Proposition 9. *If* $n \equiv 2$ *(mod* 4*)*, $t^2 \equiv k \pm 1$ *(mod* n*)*, *where* $n = 2k$ *and* $(n, t) \neq (10, 2)$, *then* $Det(DP(n, t)) = 2$.

Proof : For such choice of n and t,

$$A(n, t) = \langle \alpha, \beta, \gamma, \psi \rangle = \{\alpha^i \beta^j \gamma^l \psi^s : 0 \leq i \leq n - 1, 0 \leq j, l, s \leq 1\}.$$

We claim that x_0, x_1 is a determining set for $DP(n, t)$. Let $\alpha^i \beta^j \gamma^l \psi^s$ be an element of $A(n, t)$ which fixes x_0, x_1.

We claim that $s = 0$. If not, let $s = 1$ and hence $\alpha^i \beta^j \gamma^l \psi(x_0) = \alpha^i \beta^j \gamma^l (u_0) = u_p$ or v_p. Hence x_0 is not fixed. Thus $s = 0$ and it suffices to consider elements of the form $\alpha^i \beta^j \gamma^l$. The rest of the proof is similar to that as above.   □

Proposition 10. *If* $n \equiv 0$ *(mod* 4*)*, $t^2 \equiv k \pm 1$ *(mod* n*)*, *where* $n = 2k$, *then* $Det(DP(n, t)) = 2$.

Proof : For such choice of n and t,

$$A(n, t) = \langle \alpha, \beta, \gamma, \phi \rangle = \{\alpha^i \beta^j \gamma^l \phi^s : 0 \leq i \leq n - 1, 0 \leq j, l, s \leq 1\}.$$

We claim that x_0, x_1 is a determining set for $DP(n, t)$. Let $\alpha^i \beta^j \gamma^l \phi^s$ be an element of $A(n, t)$ which fixes x_0, x_1.

We claim that $s = 0$. If not, let $s = 1$ and hence $\alpha^i \beta^j \gamma^l \phi(x_0) = \alpha^i \beta^j \gamma^l (u_0) = u_p$ or v_p. Hence x_0 is not fixed. Thus $s = 0$ and it suffices to consider elements of the form $\alpha^i \beta^j \gamma^l$. The rest of the proof is similar to that of Proposition 8.   □

Proposition 11. $Det(DP(4, 1)) = 4$.

Proof : From Theorem 5, we get that $A(4, 1) = \langle \alpha, \beta, \gamma, \delta, \eta \rangle$. It was checked using Sage that $\{x_0, x_1, y_0, y_1\}$ is a determining set for $DP(4, 1)$. Thus $Det(DP(4, 1)) \leq 4$. We observe that

$$Stab(x_i) = Stab(u_i) = \langle \alpha^{2i} \gamma, \alpha^2 \eta, \beta \eta \beta \rangle \text{ and } Stab(y_i) = Stab(v_i) = \langle \alpha^{2i} \gamma, \eta, \alpha^2 \beta \eta \beta \rangle,$$

and each vertex stabilizer is isomorphic to $\mathbb{Z}_2 \times \mathbb{Z}_2 \times \mathbb{Z}_2$. It is clear that intersection of any two vertex stabilizers is isomorphic to $\mathbb{Z}_2 \times \mathbb{Z}_2$ and intersection of any three vertex stabilizers is isomorphic to $\mathbb{Z}_2$. Thus $Det(DP(4, 1)) = 4$.   □

Proposition 12. $Det(DP(10, 2)) = Det(DP(10, 3)) = Det(DP(5, 2)) = 2$.

Proof : It was checked using Sage that $|A(10, 2)| = 480$ and $\{x_0, v_1\}$ is a determining set for $DP(10, 2)$, i.e., $Stab(\{x_0, v_1\})$ is trivial. Hence $Det(DP(10, 2)) \leq 2$. As $DP(10, 2)$ is vertex transitive, the order of stabilizer of any vertex is $480/40 = 12$ and hence $Det(DP(10, 2)) = 2$.

As $DP(10, 2) \cong DP(10, 3)$, we have $Det(DP(10, 2)) = Det(DP(10, 3)) = 2$.

As $DP(5, 2) \cong G(10, 2)$, by Proposition 5, we have $Det(DP(5, 2)) = 2$.   □

Proposition 13. *Let* $DP(n, t)$ *be the double generalized Petersen graph, such that the parameters* n *and* t *do not satisfy any of the conditions of Propositions 7, 8, 9, 10, 11, 12. Then* $Det(DP(n, t)) = 2$.

Proof : For such choice of n and t,

$$A(n,t) = \langle \alpha, \beta, \gamma \rangle = \{\alpha^i \beta^j \gamma^l : 0 \leq i \leq n-1, 0 \leq j, l \leq 1\}.$$

We claim that x_0, x_1 is a determining set for $DP(n,t)$. Let $\alpha^i \beta^j \gamma^l$ be an element of $A(n,t)$ which fixes x_0, x_1. Mimicking the proof of Proposition 8, we can show that $Stab(\{x_0, x_1\})$ is trivial, i.e., $Det(DP(n,t)) \leq 2$.

As $|A(n,t)| = 4n$ and $DP(n,t)$ is not vertex-transitive, the order of stabilizer of any vertex should be greater than $4n/2n = 2$. Hence, there does not exist any determining set of size 1. Hence, $Det(DP(n,t)) = 2$. $\qquad\square$

From Propositions $7, 8, 9, 10, 11, 12, 13$, we have Theorem 2.

Acknowledgement. The author is thankful to the anonymous referees for their fruitful suggestions. The author also acknowledge the financial support received under the FRPDF grant of Presidency University, Kolkata and DST-SERB-SRG/2019/000475.

Appendix

In this section, we provide two Sage code for confirming the determining sets of generalized Petersen graphs and double generalized Petersen graphs. The codes are given for $G(10,2)$ and $DP(10,2)$. Readers may check for determining sets of other members of these two families by suitably editing the values of the parameters (Fig. 1).

```
n=10
k=2
u = list(var('u_%d' % i) for i in range(n))
v = list(var('v_%d' % i) for i in range(n))
V=u+v
E=[]
G=Graph()
G.add_vertices(V)
for i in range(n):
E.append((u[i],u[mod(i+1,n)]))
E.append((u[i],v[i]))
E.append((v[i],v[mod(i+k,n)]))
G.add_edges(E)
H=G.automorphism_group()
count=0
for h in H:
if h(u[0])==u[0] and h(v[1])==v[1]:
count=count+1
print count
```

```
n=10
t=2
x = list(var('x_%d' % i) for i in range(n))
y = list(var('y_%d' % i) for i in range(n))
u = list(var('u_%d' % i) for i in range(n))
v = list(var('v_%d' % i) for i in range(n))
V=x+y+u+v
E=[]
G=Graph()
G.add_vertices(V)
for i in range(n):
E.append((x[i],x[mod(i+1,n)]))
E.append((y[i],y[mod(i+1,n)]))
E.append((x[i],u[i]))
E.append((y[i],v[i]))
E.append((u[i],v[mod(i+t,n)]))
E.append((v[i],u[mod(i+t,n)]))
G.add_edges(E)
H=G.automorphism_group()
count=0
for h in H:
if h(x[0])==x[0] and h(v[1])==v[1]:
count=count+1
print count
```

Fig. 1. Sage Code for finding a determining set for $G(10,2)$ (left) and $DP(10,2)$ (right)

It is checked that $\{u_0, v_1\}$ is a determining set for $G(10,2)$ and $\{x_0, v_1\}$ is a determining set for $DP(10,2)$. The output of both the codes are 1, showing that there exists exactly one automorphism (namely, the identity automorphism) which stabilizes both u_0 and v_1, and x_0 and v_1, respectively.

References

1. Alspach, B.R.: The classification of Hamiltonian generalized Petersen graphs. J. Comb. Theory, Ser. B **34**(3), 293–312 (1983)
2. Boutin, D.L.: Identifying graph automorphisms using determining sets. Electron. J. Comb. **13**(1), 78 (2006)
3. Boutin, D.L.: The determining number of a cartesian product. J. Graph Theory **61**(2), 77–87 (2009)
4. Caceres, J., Garijo, D., Gonzalez, A., Marquez, A., Puertas, M.L.: The determining number of Kneser graphs. Discrete Math. Theor. Comput. Sci. DMTCS **15**(1), 1–14 (2013)
5. Coxeter, H.S.M.: Self-dual configurations and regular graphs. Bull. Am. Math. Soc. **56**(5), 413–455 (1950)
6. Erwin, D., Harary, F.: Destroying automorphisms by fixing nodes. Discrete Math. **306**, 3244–3252 (2006)
7. Frucht, R.: Die gruppe des Petersen'schen Graphen und der Kantensysteme der regularen Polyeder. Commentarii Mathematici Helvetici **9**(1), 217–223 (1936)
8. Frucht, R., Graver, J.E., Watkins, M.E.: The groups of the generalized Petersen graphs. Proc. Camb. Philos. Soc. **70**(2), 211–218 (1971)
9. Godsil, C., Royle, G.F.: Algebraic Graph Theory. Graduate Texts in Mathematics, vol. 207. Springer, New York (2001). https://doi.org/10.1007/978-1-4613-0163-9
10. Kutnar, K., Petecki, P.: On automorphisms and structural properties of double generalized Petersen graphs. Discrete Math. **339**, 2861–2870 (2016)
11. Pan, J., Guo, X.: The full automorphism groups determining sets and resolving sets of coprime graphs. Graphs Comb. **35**(2), 485–501 (2019)
12. Watkins, M.E.: A theorem on tait colorings with an application to the generalized petersen graphs. J. Comb. Theory **6**(2), 152–164 (1969)
13. Zhou, J.-X., Feng, Y.-Q.: Cubic vertex-transitive non-Cayley graphs of order $8p$. Electron. J. Comb. **19**, 53 (2012)

Self-centeredness of Generalized Petersen Graphs

Priyanka Singh[(✉)], Pratima Panigrahi, and Aakash Singh

Indian Institute of Technology Kharagpur, Kharagpur, India
priyankaiit22@gmail.com, pratima@maths.iitkgp.ernet.in,
aakash01iitkgp@gmail.com

Abstract. A connected graph is said to be self-centered if all its vertices have the same eccentricity. The family of generalized Petersen graphs $P(n, k)$, introduced by Coxeter [6] and named by Watkins [18], is a family of cubic graphs of order $2n$ defined by positive integral parameters n and k, $n \geq 2k$. Not all generalized Petersen graphs are self-centered. In this paper, we prove self-centeredness of $P(n, k)$ whenever k divides n and $k < \frac{n}{2}$, except the case when n is odd and k is even. We also prove non-self-centeredness of generalized Petersen graphs $P(n, k)$ when n even with $k = \frac{n}{2}$; $n = 4m + 2$ with $k = \frac{n}{2} - 1$ for some positive integer $m \geq 3$; $n \geq 9$ is odd and $k = 2$ or $k = \frac{n-1}{2}$; and $n = m(4m + 1) \pm (m + 1)$ with $k = 4m + 1$ for any positive integer $m \geq 2$. Finally, we make an exhaustive computer search and get all possible values of n and k for which $P(n, k)$ is non-self-centered.

Keywords: Eccentricity · Center of graph · Self-centered graph · Generalized Petersen graphs

1 Introduction

Graph centrality plays a significant importance in facility location problem, and has a great role in designing a communication network. In a locality, for the efficient use of resources, we place them at central nodes. Because of this, self-centered graphs are ideal as the facility can be placed (located) at any node or vertex of the locality. In the paper, by a *graph* $G = (V(G), E(G))$ (or simply G) we mean a simple finite graph with the vertex set $V(G)$ and the edge set $E(G)$. The length of a shortest u–v path in a graph G gives the *distance* between vertices u and v, which is denoted by $d_G(u, v)$ (or $d(u, v)$). The maximum of distances from a vertex v to all other vertices in a graph G is known as the *eccentricity* (denoted by $e(v)$) of the vertex v. The *radius* of G, denoted by $rad(G)$, is the minimum eccentricity of vertices in G. Similarly, the *diameter* of G, denoted by $diam(G)$, is the maximum eccentricity of vertices. Vertices with minimum eccentricity are called *central* vertices and the subgraph induced on these vertices is called the *center* $C(G)$ of the graph G. A graph G is known as a *self-centered* graph if $C(G) = G$. In other words, for a self-centered graph

© Springer Nature Switzerland AG 2020
M. Changat and S. Das (Eds.): CALDAM 2020, LNCS 12016, pp. 141–155, 2020.
https://doi.org/10.1007/978-3-030-39219-2_12

G, $rad(G)$ is equal to $diam(G)$. Further, if eccentricity of every vertex in a self-centered graph is d then the graph is known as d-*self-centered graph*.

As a generalization of the well-known Petersen graph, the generalized Petersen graph has attracted the attention of several researchers. For each positive integers n and k with $n \geq 2k$, the *generalized Petersen graph* $P(n,k)$ is a graph with vertex set $V(P(n,k)) = \{u_0, u_1, u_2, ..., u_{n-1}, v_0, v_1, v_2, ..., v_{n-1}\}$ and the edge set $E(P(n,k)) = \{u_iu_{i+1}, u_iv_i, v_iv_{i+k} : 0 \leq i \leq n-1\}$, where subscripts are addition modulo n. Throughout the paper, we refer this notation for vertex set and edge set of $P(n,k)$. For $n = 5$ and $k = 2$, $P(5,2)$ is the well known Petersen graph.

The generalized Petersen graphs, named by Watkins [18] were defined by Coxeter [6] but not with this name. The essence of the Petersen graph is a remarkable configuration that serves as a counterexample to many optimistic predictions and conjectures about what might be true for graphs in general. The generalized Petersen graphs have been studied by several authors; for instance, Tait coloring of generalized Petersen graphs have been studied and analysed in [4], generalization of generalized Petersen graphs on the basis of symmetry properties have been discussed in [13]. A result on maximum number of vertices in a generalized Petersen graph was given by authors in [1], where number of vertices is treated as a function of diameter. A formula for number of isomorphism classes of generalized Petersen graphs was presented by Steimle and Staton [17]. For works related to domination number in generalized Petersen graphs, one can refer to [5,7], and [9]. However, there were no significant work done related to self-centeredness of generalized Petersen graphs because of the complex structure of these graphs. This motivated us to work on the self-centeredness property of generalized Petersen graphs.

The theorem below gives a criteria for generalized Petersen graphs to be isomorphic.

Theorem 1. *[17] Let $n > 3$ and k, l relatively prime to n with $kl \equiv 1$ (mod n). Then $P(n,k) \cong P(n,l)$.*

The theorem stated below is useful in proving the self-centeredness of generalized Petersen graph $P(n,1)$.

Theorem 2. *[16] Let $G = G_1 \square G_2$ be the Cartesian product of graphs G_1 and G_2. If G_1 and G_2 are l- and m-self centered graphs, respectively, then G is $(l+m)$-self centered graph.*

We note that $P(n,1)$ is the Cartesian product of the cycle C_n and the complete graph K_2. Since C_n is $\lfloor \frac{n}{2} \rfloor$-self-centered graph and K_2 is 1-self-centered graph, by Theorem 2 we get the result below.

Theorem 3. *For $n \geq 3$, the generalized Petersen graph $P(n,1)$ is a d-self-centered graph, where $d = \lfloor \frac{n}{2} \rfloor + 1$.*

Vertex transitive graphs are self-centered. In [8], the authors have proved that $P(n,k)$ is vertex transitive if and only if $k^2 \equiv \pm 1 (\text{mod } n)$, or $n = 10$ and $k = 2$. So, one get the following result.

Theorem 4. *For $n \geq 3$, generalized Petersen graph $P(n,k)$ is self-centered for $k^2 \equiv \pm 1 (\mathrm{mod}\ n)$, or $n = 10$ and $k = 2$. Moreover, $P(10,2)$ is 5-self-centered and the other $P(n,k)$ are d-self-centered, where*

$$d = \begin{cases} k+1, & \text{if } n = k^2 - 1, \\ k+1, & \text{if } n = k^2 + 1 \text{ and } k \text{ is even, } k \neq 2, \\ k+2, & \text{if } n = k^2 + 1 \text{ and } k \text{ is odd.} \end{cases}$$

Self-centered graphs were studied and surveyed by many authors in the last few decades. For the same, we refer the articles [2], [3], and [10–12]. Self-centeredness of different types of graph products are studied by the authors in [14,15].

In the remaining of the paper, we assume $k \geq 2$. The main technique followed in this paper for verification of self-centeredness of $P(n,k)$ is the determination of eccentricities of u_0 and v_0, because $P(n,k)$ is symmetric on outer vertices $u_0, u_1, u_2, \ldots, u_{n-1}$ and also symmetric on inner vertices $v_0, v_1, v_2, \ldots, v_{n-1}$. If $e(u_0) = e(v_0)$ then the generalized Petersen graph is self-centered, otherwise not.

The rest of the paper is organized as follows. In Sect. 2, we prove the non-self-centeredness of generalized Petersen graphs $P(n,k)$ when n even, $k = \frac{n}{2}$ or $n = 4m + 2$ with $k = \frac{n}{2} - 1$ for some positive integer $m \geq 3$. Then we prove self-centeredness of $P(n,k)$ whenever n is even, $k < \frac{n}{2}$ and k divides n. In Sect. 3, we study self-centeredness of generalized Petersen graphs $P(n,k)$ for odd n. We prove that $P(n,k)$ is not self-centered for odd $n \geq 9$ with $k = 2$ or $k = \frac{n-1}{2}$. Then we prove the self-centeredness of $P(n,k)$ for odd n and odd k, $k < \frac{n}{2}$, and $k|n$. Also, we prove non-self-centeredness of $P(n,k)$ when $n = m(4m+1) \pm (m+1)$ with $k = 4m + 1$ for any positive integer $m \geq 2$. Finally, we make an exhaustive computer search and get all possible values of n and k for which $P(n,k)$ is non-self-centered.

2 Self-centeredness of $P(n,k)$ for an Even n

In the following result, we investigate the self-centeredness of $P(n,k)$ for an even n and $k = \frac{n}{2}$.

Theorem 5. *Let $P(n,k)$ be a generalized Petersen graph such that $n \geq 4$ is even and $k = \frac{n}{2}$. Then $P(n,k)$ is not a self-centered graph.*

Proof. To prove the result, it is sufficient to show that eccentricity of two vertices in $P(n,k)$ are not equal. We note that $C : v_0, u_0, u_1, u_2, \ldots, u_k, v_k, v_0$ induces a cycle of length $k + 3$, see Fig. 1, where the cycle C is highlighted by thick lines. We observe that $d(u_0, u_i) = d(u_0, u_{n-i})$, for $i \in \{1, 2, 3, \ldots, \frac{n}{2}\}$. Depending on the parity of k, we distinguish following two cases.

Case 1. The integer k is even.

Since $k = \frac{n}{2}$, in this case n will be a multiple of four. Consider vertices u_0 and v_0. Since u_0 lies on C and C is of length $k + 3$,

$$\max\{d(u_0, u_i) : 1 \leq i \leq \frac{n}{2}\} = \lfloor \frac{k+3}{2} \rfloor = \frac{k+2}{2} = \frac{n+4}{4} = \frac{n}{4} + 1. \quad (1)$$

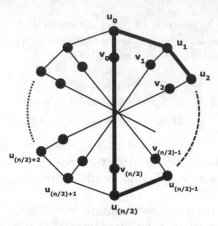

Fig. 1. Generalized Petersen graph $P(n, k)$ with n even and $k = \frac{n}{2}$

Next, we find $d(u_0, v_i)$ for $1 \leq i \leq \frac{n}{2}$. First, let $1 \leq i \leq \frac{n}{4}$. For these values of i, we get that $d(u_0, v_i) = d(u_0, u_i) + 1$. Now, $\max\{d(u_0, u_i) : 1 \leq i \leq \frac{n}{4}\} = \frac{n}{4}$ and so

$$\max\{d(u_0, v_i) : 1 \leq i \leq \frac{n}{4}\} = \max\{d(u_0, u_i) + 1 : 1 \leq i \leq \frac{n}{4}\} = \frac{n}{4} + 1. \quad (2)$$

For $\frac{n}{4} < i \leq \frac{n}{2}$, a shortest v_i–u_0 path is given by $P_i : v_i, v_{i+k}, u_{i+k}, u_{i+(k+1)}$, $u_{i+(k+2)}, \ldots, u_0$, where $l(P_i) = n - i - k + 2 = \frac{n}{2} + 2 - i$ (since $k = \frac{n}{2}$). The maximum length of P_i is for $i = \frac{n}{4} + 1$, and hence

$$\max\{l(P_i) : \frac{n}{4} < i \leq \frac{n}{2}\} = \frac{n}{4} + 1. \quad (3)$$

From Eqs. (1)–(3), we get that $e(u_0) = \frac{n}{4} + 1$.
Since v_0 lies on C and n is a multiple of four,

$$\max\{d(v_0, u_i) : 1 \leq i \leq \frac{n}{2}\} = \lfloor \frac{k+3}{2} \rfloor = \frac{n}{4} + 1. \quad (4)$$

Next for $\frac{n}{4} \leq i \leq \frac{n}{2}$, we obtain $d(v_0, v_i) = d(u_0, v_i) + 1$, and this gives

$$\max\{d(v_0, v_i) : \frac{n}{4} \leq i \leq \frac{n}{2}\} = \frac{n}{4} + 1. \quad (5)$$

Finally, for $1 \leq i \leq \frac{n}{4}$, a shortest v_i–v_0 path is given by $P_i' :$ $v_i, v_{i+k}, u_{i+k}, u_{i+(k+1)}, \ldots, u_0$, where $l(P_i') = n - i - k + 3 = \frac{n}{2} + 3 - i$, and the maximum length of P_i' is for $i = \frac{n}{4} + 1$, i.e.,

$$\max\{l(P_i') : 1 \leq i \leq \frac{n}{4}\} = \frac{n}{2} + 3 - \frac{n}{4} - 1 = \frac{n}{4} + 2. \quad (6)$$

Hence $e(v_0) = \frac{n}{4} + 2$. Thus, we get that $e(u_0) \neq e(v_0)$ and $P(n, k)$ is not a self-centered graph in this case.

Case 2. The integer k is odd.

In this case n is not a multiple of four but cycle C is of an even length. Since u_0 and v_0 lie on C, for $1 \leq i \leq \frac{n}{2}$, we have

$$\max\{d(u_0, u_i) : 1 \leq i \leq \frac{n}{2}\} = \frac{k+3}{2} \text{ and,} \tag{7}$$

$$\max\{d(v_0, u_i) : 1 \leq i \leq \frac{n}{2}\} = \frac{k+3}{2}. \tag{8}$$

For $1 \leq i \leq \lfloor \frac{n}{4} \rfloor + 1$, a shortest u_0–v_i path is given by $Q_i : u_0, u_1, u_2, \ldots, u_i, v_i$, where $l(Q_i) = i + 1$ and maximum length of Q_i is for $i = \lfloor \frac{n}{4} \rfloor + 1$, i.e.,

$$\max\{l(Q_i) : 1 \leq i \leq \lfloor \frac{n}{4} \rfloor + 1\} = \lfloor \frac{n}{4} \rfloor + 1 + 1 = \lfloor \frac{k+3}{2} \rfloor. \tag{9}$$

Again, for $\lfloor \frac{n}{4} \rfloor + 2 \leq i \leq \frac{n}{2}$, a shortest u_0–v_i path is given by $Q'_i : v_i, v_{i+k}, u_{i+k}$, $u_{i+(k+1)}, \ldots, u_0$ and the length of Q'_i is $\frac{n}{2} + 2 - i$. The path Q'_i has a maximum length for $i = \lfloor \frac{n}{4} \rfloor + 2$, i.e.,

$$\max\{l(Q'_i) : \lfloor \frac{n}{4} \rfloor + 2 \leq i \leq \frac{n}{2}\} = \frac{n}{2} - \lfloor \frac{n}{4} \rfloor = \frac{k+1}{2}. \tag{10}$$

From Eqs. (7), (9), and (10), we have $e(u_0) = \frac{k+3}{2}$.

Next, we consider the vertex v_0. For $1 \leq i \leq \lfloor \frac{n}{4} \rfloor + 1$, a shortest v_0–v_i path is given by $T_i : v_0, u_0, u_1, u_2, \ldots, u_i, v_i$. The length of the path T_i is $i + 2$. The maximum length of T_i is for $i = \lfloor \frac{n}{4} \rfloor + 1$, i.e.,

$$\max\{l(T_i) : 1 \leq i \leq \lfloor \frac{n}{4} \rfloor + 1\} = \lfloor \frac{n}{4} \rfloor + 1 + 2 = \frac{k+5}{2}. \tag{11}$$

Finally, for $\lfloor \frac{n}{4} \rfloor + 2 \leq i \leq \frac{n}{2}$, a shortest v_0–v_i path is given by $T'_i :$ $v_i, v_{i+k}, u_{i+k}, u_{i+(k+1)}, \ldots, u_n = u_0, v_0$, where $l(T'_i) = \frac{n}{2} + 3 - i$. We get the maximum length of T'_i for $\lfloor \frac{n}{4} \rfloor + 2$, i.e.,

$$\max\{l(T'_i) : \lfloor \frac{n}{4} \rfloor + 2 \leq i \leq \lfloor \frac{n}{2} \rfloor\} = \frac{k}{2} + 1. \tag{12}$$

From Eqs. (8), (11), and (12), we have $e(v_0) = \frac{k+5}{2}$. This proves that $e(u_0) \neq e(v_0)$. Hence, $P(n, k)$ is not a self-centered graph in this case also. □

In the following theorem we get some non-self-centered generalized Petersen graphs $P(n, k)$, where n is even but not divisible by k.

Theorem 6. *Let $P(n, k)$ be a generalized Petersen graph such that $n = 4m + 2$ for some positive integer $m \geq 3$ and $k = \frac{n}{2} - 1$. Then $P(n, k)$ is not a self-centered graph.*

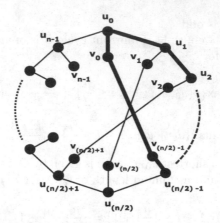

Fig. 2. Generalized Petersen graph $P(n,k)$ with $n = 4m + 2$ and $k = \frac{n}{2} - 1$

Proof. Here we obtain a cycle $C : u_0, u_1, u_2, \ldots, u_{\frac{n}{2}-1}, v_{\frac{n}{2}-1}, v_0, u_0$ of length $\frac{n+4}{2}$ i.e. of length $2m + 3$, see Fig. 2, where the cycle $\tilde{C}$ is highlighted by thick lines. Since u_0 and v_0 lie on C, we have

$$\max\{d(u_0, u_i) : 1 \leq i \leq \frac{n}{2} - 1\} = \max\{d(v_0, u_i) : 1 \leq i \leq \frac{n}{2} - 1\} = \lfloor \frac{2m+3}{2} \rfloor = m + 1. \tag{13}$$

We have $d(u_0, u_{m+1}) = d(u_0, u_{m+2}) = m + 1$ and $d(v_0, u_m) = d(v_0, u_{m+1}) = m + 1$. Next, for every $j \in \{1, 2, \ldots, m\} \cup \{m + 3, m + 4, \ldots, 2m + 1\}$, we get $d(u_0, v_j) \leq m + 1$. Similarly, for every $l \in \{1, 2, \ldots, m - 1\} \cup \{m + 2, m + 4, \ldots, 2m + 1\}$, $d(v_0, v_l) \leq m + 1$. Now for $l = m + 1$ and $m + 2$, a shortest u_0–v_i paths P_1 and P_2 are given below.

$$P_1 : u_n = u_0, u_{n-1}, u_{n-2}, \ldots, u_{n-(m-1)}, v_{n-(m-1)}, v_{m+1} = v_l, \quad \text{and}$$
$$P_2 : u_n = u_0, u_{n-1}, u_{n-2}, \ldots, u_{n-(m-2)}, v_{n-(m-2)}, v_{m+2} = v_l.$$

Also,

$$l(P_1) = m + 1 \text{ and } l(P_2) = m. \tag{14}$$

Further for $l = m$ and $m + 1$, smallest v_0–v_i paths are as given below.

$$P_3 : v_0, u_0, u_1, u_2, \ldots, u_m, v_l = v_m, \text{ and}$$

$$P_4 : v_0, u_0, u_{n-1}, u_{n-2}, \ldots, u_{n-(m-1)}, v_{n-(m-1)}, v_l = v_{m+1}. \text{ We notice that}$$

$$l(P_3) = m + 2 \text{ and } l(P_4) = m + 2. \tag{15}$$

From the above three equations, we conclude that $e(u_0) = m + 1$ and $e(v_0) = m + 2$. Thus, $e(u_0) \neq e(v_0)$, and hence $P(n, k)$ is not a self-centered graph. □

Theorem 7. *Let $P(n, k)$ be a generalized Petersen graph with n even, $k < \frac{n}{2}$ and n be divisible by k. If $n = kq$ then $P(n, k)$ is a d-self-centered graph, where*

$$d = \begin{cases} \frac{q}{2} + \lfloor \frac{k+3}{2} \rfloor, & \text{if } k \text{ divides } \frac{n}{2}, \\ \lfloor \frac{q+1}{2} \rfloor + \lfloor \frac{k}{2} \rfloor + 1, & \text{otherwise.} \end{cases}$$

Proof. To prove the result, we consider the following sets of vertices in $P(n, k)$.

- R_1 : $\{u_0, v_0, v_k, u_k, u_{k-1}, u_{k-2}, \ldots, u_3, u_2, u_1\}$ and R_1' : $\{v_1, v_2, \ldots, v_{k-1}\}$
- R_2 : $\{u_k, v_k, v_{2k}, u_{2k}, u_{2k-1}, u_{2k-2}, \ldots, u_{k+1}\}$ and R_2' : $\{v_{k+1}, v_{k+2}, \ldots, v_{2k-1}\}$
  $$\vdots$$
- R_q : $\{u_{(q-1)k}, v_{(q-1)k}, v_{qk}, u_{qk}, u_{qk-1}, u_{qk-2}, \ldots, u_{qk-k+1}\}$ and

R_q' : $\{v_{k(\frac{q}{2}-1)}, v_{k(\frac{q}{2}-1)+1}, \ldots, v_{\frac{kq}{2}-1}\}$

We see that vertices of each R_i, $i \in \{1, 2, \ldots, q\}$, induces a cycle of length $k + 3$, vertices of each R_i' induces independent set and $\bigcup_{i=1}^{q}(R_i \cup R_i') = V(P(n, k))$. Next, we find distances from u_0 and v_0 to other vertices of $P(n, k)$. It is sufficient to find the distances from u_0 and v_0 to vertices $u_1, u_2, \ldots, u_{\frac{n}{2}}$, $v_0, u_0, v_1, v_2, \ldots, v_{\frac{n}{2}}$. Here, we have following two cases.

Case 1. When k divides $\frac{n}{2}$.

Since vertices of R_1 induces a cycle $u_0, v_0, v_k, u_k, \ldots, u_2, u_1, u_0$ of length $k+3$, we have

$$\max\{d(u_0, x) : x \in R_1\} = \max\{d(v_0, x) : x \in R_1\} = \lfloor \frac{k+3}{2} \rfloor. \quad (16)$$

$$\max\{d(u_0, x) : x \in R_1'\} = \max\{d(v_0, x) : x \in R_1'\} = \lfloor \frac{k+3}{2} \rfloor + 1. \quad (17)$$

We see that for any vertex $x \in R_2$, $d(u_0, x) = d(u_0, v_0) + d(v_0, v_k) + d(v_k, x)$ and thus, we have

$$\max\{d(u_0, x) : x \in R_2\} = \lfloor \frac{k+3}{2} \rfloor + 2, \quad \text{and} \quad (18)$$

$$\max\{d(u_0, x) : x \in R_2'\} = \lfloor \frac{k+3}{2} \rfloor + 3, \quad (19)$$

and,

$$\max\{d(v_0, x) : x \in R_2\} = \lfloor \frac{k+3}{2} \rfloor + 1, \quad \text{and} \quad (20)$$

$$\max\{d(v_0, x) : x \in R_2'\} = \lfloor \frac{k+3}{2} \rfloor + 2. \quad (21)$$

Similarly, we get

$$\max\{d(u_0, x) : x \in R_q\} = \lfloor \frac{k+3}{2} \rfloor + \frac{q}{2}, \quad \text{and} \quad (22)$$

for the vertices $v_j \in R'_q$ for $j \neq \frac{k(q-1)}{2}$ and $\frac{k(q-1)}{2} + 1$ when k is even, then

$$\max\{d(u_0, x) : x \in R'_q, x \neq v_{\frac{k(q-1)}{2}}, v_{\frac{k(q-1)}{2}+1}\} = \lfloor \frac{k+3}{2} \rfloor + \frac{q}{2}. \quad (23)$$

For $j = \frac{k(q-1)}{2}$, a shortest u_0–v_j path is given by
$P : u_0, u_1, \ldots, u_{\frac{k}{2}}, v_{\frac{k}{2}}, v_{\frac{k}{2}+k}, v_{\frac{k}{2}+2k}, \ldots, v_{\frac{k}{2}+\frac{k(q-1)}{2}}$ and thus $d(u_0, u_j) = \frac{k+q}{2}$.
Due to symmetric structure of the graph, the same can be obtained for $j = \frac{k(q-1)}{2} + 1$.
Finally, consider vertices $v_j \in R'_q$ for $j \neq \frac{q(k-1)}{2}$ when k is odd. Then

$$\max\{d(u_0, x) : x \in R'_q, x \neq v_{\frac{q(k-1)}{2}}\} = \lfloor \frac{k+3}{2} \rfloor + \frac{q}{2}, \quad (24)$$

and for $j = \frac{q(k-1)}{2}$ a shortest u_0–v_j path is given by
$P' : u_0, u_1, \ldots, u_{\frac{k+1}{2}}, v_{\frac{k+1}{2}}, v_{\frac{k+1}{2}+k}, v_{\frac{k+1}{2}+2k}, \ldots, v_{\frac{k+1}{2}+k(\frac{q}{2}-1)}$ of length $\frac{k+q+1}{2}$
and thus $d(u_0, u_j) = \frac{k+q+1}{2}$.
Now, for the vertex v_0, we obtain

$$\max\{d(v_0, x) : x \in R_q\} = \lfloor \frac{k+3}{2} \rfloor + \frac{q}{2} - 1. \quad (25)$$

$$\max\{d(v_0, x) : x \in R'_q\} = \lfloor \frac{k+3}{2} \rfloor + \frac{q}{2}. \quad (26)$$

From Eqs. (16)–(26), we conclude that $e(u_0) = e(v_0) = \frac{q}{2} + \lfloor \frac{k+3}{2} \rfloor$.

Case 2. When k does not divide $\frac{n}{2}$.
Given that $n = kq$ and k does not divide $\frac{n}{2}$, so q is an odd integer. This means k must be even. In this case, the distance between the vertex u_0 and a vertex in $R_1 \cup \ldots \cup R_{\frac{q-1}{2}}$ is the same as obtained in the Case 1. That is, the maximum distance between u_0 and any vertex from $R_{\frac{q-1}{2}}$ is $\lfloor \frac{q-1}{2} \rfloor + \lfloor \frac{k+3}{2} \rfloor$. Next consider the vertices from the region $R_{\frac{q+1}{2}}$. Now, because of the symmetry of $P(n, k)$ for an even n, the vertex farthest from u_0 (v_0) lie in the region $R_{\frac{q+1}{2}}$. The vertex farthest from u_0 and v_0 are the vertices $u_{\frac{n}{2}}$ and $v_{\frac{n}{2}}$, respectively, at a distance $\lfloor \frac{q+1}{2} \rfloor + \lfloor \frac{k}{2} \rfloor + 1$, and hence the result. $\square$

Theorem 8. *The generalized Petersen graph $P(n, k)$ is not self-centered for $n = 4m(4m + 1)$ and $k = 2m(4m - 1)$ for some positive integer $m \geq 1$.*

Proof. In this case, we find that $v_{\frac{n}{2}}$ is the farthest vertex from both u_0 and v_0 and have obtained that $d(u_0, v_{\frac{n}{2}}) = 4m + 2$ and $d(v_0, v_{\frac{n}{2}}) = 4m + 1$. So, $e(u_0) = 4m + 2$ and $e(v_0) = 4m + 1$ and hence the given generalized Petersen graphs are not self-centered in this case. $\square$

3   Self-centeredness of $P(n, k)$ for Odd n

In this section, we first investigate self-centeredness of $P(n, k)$ for $k = 2$. First of all, we prove that $P(5, 2)$ and $P(7, 2)$ are self-centered graphs.

Theorem 9. *The generalized Petersen graph $P(n, 2)$ is 2- or 3-self-centered graphs for $n = 5$ or 7, respectively.*

Proof. For $n = 5$, the graph $P(n, 2)$ is the well known Petersen graph and we know that radius and diameter of $P(5, 2)$ is two. Thus, $P(5, 2)$ is 2-self-centered graph.

Let us consider the vertices u_0 and v_0 in $P(7, 2)$. The shortest path from u_0 to u_1, u_2, or u_3 is through the edges in cycle $C : u_0, u_1, u_2, u_3, u_4, u_5, u_6, u_0$. So we get that $d(u_0, u_1), d(u_0, u_2)$, and $d(u_0, u_3)$ are equal to $1, 2$, and 3, respectively. A shortest u_0-v_i path for $i = 0, 1, 2$, and 3 is (u_0, v_0), (u_0, u_1, v_1), (u_0, v_0, v_2), and (u_0, u_1, v_1, v_3) with lengths $1, 2, 2$, and 3, respectively. Similarly, a shortest $v_0 - u_i$ path for $i = 0, 1, 2$, and 3 is (v_0, u_0), (v_0, u_0, u_1), (v_0, v_2, u_2), (v_0, v_2, u_2, u_3) with lengths $1, 2, 2$, and 3, respectively. Further, a shortest v_0-v_i path for $i = 1, 2$, and 3 is (v_0, u_0, u_1, v_1), (v_0, v_2), and (v_0, v_5, v_3) with lengths $3, 1$, and 2 respectively. From this we can say that $e(u_0) = e(v_0) = 3$. Hence $P(7, 2)$ is a 3-self-centered graph.   □

Theorem 10. *The generalized Petersen graph $P(n, 2)$ is not self-centered for odd integers $n \geq 9$.*

Proof. We take $n = 4m + 1$ or $4m + 3$ for some positive integer $m \geq 2$. We shall find $d(u_0, u_i)$, $d(u_0, v_i)$, $d(v_0, u_i)$, and $d(v_0, v_i)$ for $i \in \{1, 2, \ldots, \lfloor \frac{n}{2} \rfloor\}$. First, we consider the vertex u_0. We note that $d(u_0, u_i) = i$ for $i = 1, 2, 3$, and 4.

For an even index i, $6 \leq i \leq 2m$, a shortest u_0-u_i and u_0-v_i path is given by $P_i : u_0, v_0, v_2, v_4, \ldots, v_i, u_i$ and $P_i' : u_0, v_0, v_2, \ldots, v_i$, where $l(P_i) = \frac{i+4}{2}$ and $l(P_i') = \frac{i+2}{2}$. Now,

$$\max\{l(P_i) : 6 \leq i \leq 2m, i \text{ even}\} = m + 2. \tag{27}$$

$$\max\{l(P_i') : 6 \leq i \leq 2m, i \text{ even}\} = m + 1. \tag{28}$$

For an odd index i, a shortest u_0-u_i and u_0-v_i path is given by $Q_i :$ $u_0, v_0, v_2, v_4, \ldots, v_{i-1}, u_{i-1}, u_i$ and $Q_i' : u_0, u_1, v_1, v_3, \ldots, v_i$, where $l(Q_i) = \frac{i+5}{2}$ and $l(Q_i') = \frac{i+3}{2}$. If $n = 4m + 1$ then

$$\max\{l(Q_i) : 5 \leq i \leq 2m - 1, i \text{ odd}\} = m + 2 \tag{29}$$

$$\max\{l(Q_i') : 5 \leq i \leq 2m - 1, i \text{ odd}\} = m + 1, \tag{30}$$

and for $n = 4m + 3$, we get

$$\max\{l(Q_i) : 5 \leq i \leq 2m + 1, i \text{ odd}\} = m + 3 \tag{31}$$

$$\max\{l(Q_i') : 5 \leq i \leq 2m + 1, i \text{ odd}\} = m + 2. \tag{32}$$

Next, we consider the vertex v_0. For an even index i, $6 \leq i \leq 2m$, a shortest v_0–u_i and v_0–v_i path is given by $L_i : v_0, v_2, v_4, \ldots, v_i, u_i$ and $L'_i : v_0, v_2, v_4, \ldots, v_i$, where $l(L_i) = \frac{i+2}{2}$ and $l(L'_i) = \frac{i}{2}$.

$$\max\{l(L_i) : 6 \leq i \leq 2m, i \text{ even}\} = m + 1. \tag{33}$$

$$\max\{l(L'_i) : 6 \leq i \leq 2m, i \text{ even}\} = m. \tag{34}$$

Next let i be an odd index. When $n = 4m + 1$, for $5 \leq i < 2m - 1$, a shortest v_0–v_i path is given by $M_i : v_0, v_2, v_4, \ldots, v_{i-1}, u_{i-1}, u_i, v_i$ and the length of the path M_i is $\frac{i+5}{2}$, and for $i = 2m - 1$, a shortest v_0–v_i path is $M'_i : v_0, v_{4m-1}, v_{4m-3}, \ldots, v_{4m-(2m+1)}$ with length $m + 1$. When $n = 4m + 3$, for $5 \leq i < 2m + 1$, a shortest v_0–v_i path is given by $N_i : v_0, v_2, v_4, \ldots, v_{i-1}, u_{i-1}, u_i, v_i$ and the length of the path N_i is $\frac{i+5}{2}$, and for $i = 2m + 1$, a shortest v_0–v_i path is $N'_i : v_0, v_{4m+1}, v_{4m-1}, v_{4m-3}, \ldots, v_{4m-(2m-1)}$ with length $m + 1$. Now, we get the following.

$$\max\{l(M_i) : 5 \leq i \leq 2m - 3, i \text{ odd}\} = m + 1. \tag{35}$$

$$\max\{l(N_i) : 5 \leq i \leq 2m - 1, i \text{ odd}\} = m + 2. \tag{36}$$

From the Eqs. (27)–(36), we have

$$e(u_0) = \begin{cases} m + 2, & \text{for } n = 4m + 1 \\ m + 3, & \text{for } n = 4m + 3, \end{cases}$$

and

$$e(v_0) = \begin{cases} m + 1, & \text{for } n = 4m + 1 \\ m + 2, & \text{for } n = 4m + 3 \end{cases}$$

Thus, $e(u_0) \neq e(v_0)$ and hence $P(n, 2)$ is not self-centered graph. $\qquad \square$

Corollary 1. *The generalized Petersen graph $P(n, k)$ is not self-centered for odd integers $n \geq 9$ and $k = \frac{n-1}{2}$.*

Proof. By the structure of generalized Petersen graph, for an odd integer n we get that $P(n, \frac{n+1}{2})$ and $P(n, \frac{n-1}{2})$ are isomorphic. Since $\frac{n+1}{2}$ is relatively prime with n, and n is odd, by Theorem 1 we get $P(n, 2)$ and $P(n, \frac{n+1}{2})$ are isomorphic and thus $P(n, 2)$ and $P(n, \frac{n-1}{2})$ are isomorphic. Since, $P(n, 2)$ is not a self-centered graph for $n \geq 9$ with odd n, $P(n, \frac{n-1}{2})$ is also not a self-centered graph and hence the result. $\qquad \square$

In the next theorem we prove that the generalized Petersen graph is a self-centered graph when both n and k are odd, and n is divisible by k.

Theorem 11. *Let $P(n, k)$ be a generalized Petersen graph, where n and k are both odd and k divides n. Then $P(n, k)$ is a d-self-centered graph, where $d = \frac{q+k}{2} + 1$ and $n = kq$.*

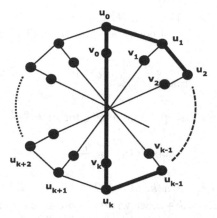

Fig. 3. Generalized Petersen graph $P(n, k)$ with both n and k odd, k divides n

Proof. Due to symmetric structure of $P(n, k)$ we have $d(u_0, u_i) = d(u_0, u_{n-i})$ for $i \in \{1, 2, \ldots, \lfloor \frac{n}{2} \rfloor\}$. We consider the cycle $C : u_0, u_1, u_2, \ldots, u_k, v_k, v_0, u_0$ of length $k + 3$, see Fig. 3, where C is highlighted by thick lines. Since $u_0, v_0 \in C$, we have

$$\max\{d(u_0, u_i) : u_i \in C\} = \max\{d(v_0, u_i) : u_i \in C\} = \frac{k+3}{2}. \tag{37}$$

Next, we determine $d(u_0, u_i)$, $d(u_0, v_i)$, $d(v_0, u_i)$, and $d(v_0, v_i)$, where u_i, $v_i \notin C$. Let $m = 1, 2, \ldots, \frac{q-1}{2} - 1, \frac{q-1}{2}$. We have following cases depending on the values of i. In the first three cases, we consider $m = 1, 2, \ldots, \frac{q-1}{2} - 1$, and in the fourth case we take $m = \frac{q-1}{2}$.

Case 1. $mk \leq i < mk + \frac{k+1}{2}$, $m = 1, 2, \ldots, \frac{q-1}{2} - 1$.

Since $mk \leq i$, we can write $i = mk + j$ for $j = 0, 1, 2, \ldots, \frac{k-1}{2}$. Now, a shortest u_0–u_i and v_0–u_i paths are given by the paths P_1 and P_2 respectively, where

$$P_1 : u_0, v_0, v_k, v_{2k}, \ldots, v_{mk}, u_{mk}, u_{mk+1}, \ldots, u_{mk+j}$$
$$P_2 : v_0, v_k, v_{2k}, \ldots, v_{mk}, u_{mk}, u_{mk+1}, \ldots, u_{mk+j}.$$

Moreover, $l(P_1) = m + j + 2$ and $l(P_2) = m + j + 1$. Both P_1 and P_2 obtain their maximum length for $m = \frac{q-1}{2} - 1$ and $j = \frac{k-1}{2}$, i.e.

$$\max\{l(P_1) : i < mk + \frac{k+1}{2}\} = \frac{q+k}{2}. \tag{38}$$

$$\max\{l(P_2) . i < mk + \frac{k+1}{2}\} = \frac{q+k}{2} - 1. \tag{39}$$

Similarly, a shortest u_0–v_i and v_0–v_i paths are given by P_3 and P_4, respectively, where

$$P_3 : v_i, v_{i-k}, v_{i-2k}, v_{i-3k}, \ldots, v_{i-mk}, u_{i-mk}, u_{i-mk-1}, \ldots, u_0$$
$$P_4 : v_0, v_k, v_{2k}, \ldots, v_{mk}, u_{mk}, u_{mk+1}, \ldots, u_{mk+j}, v_{mk+j}$$

Now, $l(P_3) = 1 + i - (k-1)m$ and $l(P_4) = m + j + 2$. Length of path P_3 is maximum for $m = \frac{q-1}{2} - 1$ and corresponding value of i i.e. $i = mk + \frac{k+1}{2} - 1$. This gives

$$\max\{l(P_3)\} = \frac{q+k}{2} - 1. \tag{40}$$

Also, the path P_4 obtains its optimum length for $m = \frac{q-1}{2} - 1$ and $j = \frac{k-1}{2}$, i.e.

$$\max\{l(P_4)\} = \frac{q+k}{2}. \tag{41}$$

Case 2. $mk + \frac{k+1}{2} < i < mk + k$, $m = 1, 2, \ldots, \frac{q-1}{2} - 1$.

Since $mk + \frac{k+1}{2} < i < mk + k$, we write $i = mk + (k-x)$ for $x = 1, 2, \ldots, \frac{k-3}{2}$. A shortest u_0-u_i and v_0-u_i paths are given below.

$P_5 : u_0, v_0, v_k, v_{2k}, \ldots, v_{mk}, v_{mk+k}, u_{mk+k}, u_{mk+(k-1)}, u_{mk+(k-2)}, \ldots, u_{mk+(k-x)}$
$P_6 : v_0, v_k, v_{2k}, \ldots, v_{mk}, v_{mk+k}, u_{mk+k}, u_{mk+(k-1)}, u_{mk+(k-2)}, \ldots, u_{mk+(k-x)}$

We see that $l(P_5) = 3 + m + x$ and $l(P_6) = 2 + m + x$. Now, maximum length of the paths P_5 and P_6 are obtained when $m = \frac{q-1}{2} - 1$ and $x = \frac{k-3}{2}$, respectively. Hence,

$$\max\{l(P_5)\} = \frac{q+k}{2} + 1, \text{and } \max\{l(P_6)\} = \frac{q+k}{2} - 1. \tag{42}$$

Further, shortest u_0-v_i and v_0-v_i paths are given by P_7 and P_8, respectively, where

$P_7 : v_i, v_{i-k}, v_{i-2k}, \ldots, v_{i-mk}, v_{n+i-mk-k}, u_{n+i-mk-k}, u_{n+i-mk-k-1}, u_{n+i-mk-k-2}, \ldots, u_0$
$P_8 : v_0, v_k, v_{2k}, v_{3k}, \ldots, v_{mk}, v_{mk+k}, u_{mk+k}, u_{mk+(k-1)}, u_{mk+(k-2)}, \ldots, u_{mk+(k-x)}, v_{mk+(k-x)}$

Moreover, $l(P_7) = 2 + m + (m+1)k - i$ and $l(P_8) = 3 + m + x$, and so

$$\max\{l(P_7)\} = \frac{q+k}{2} - 1, \text{and } \max\{l(P_8)\} = \frac{q+k}{2}. \tag{43}$$

Case 3. $i = mk + \frac{k+1}{2}$, $m = 1, 2, \ldots, \frac{q-1}{2} - 1$.

In this case, a shortest u_0-u_i, v_0-u_i, u_0-v_i, and v_0-v_i paths are given by the following paths P_9, P_{10}, P_{11}, and P_{12}, respectively, where

$P_9 : u_0, v_0, v_k, v_{2k}, \ldots, v_{mk}, u_{mk}, u_{mk+1}, u_{mk+2}, \ldots, u_{mk+\frac{k+1}{2}}$

$P_{10} : v_0, v_k, v_{2k}, \ldots, v_{mk}, u_{mk}, u_{mk+1}, u_{mk+2}, \ldots, u_{mk+\frac{k+1}{2}}$

$P_{11} : v_i, v_{i-k}, v_{i-2k}, \ldots, v_{i-mk}, u_{i-mk}, u_{i-mk-1}, u_{i-mk-2}, \ldots, u_0$

$P_{12} : v_0, v_k, v_{2k}, \ldots, v_{mk}, u_{mk}, u_{mk+1}, u_{mk+2}, \ldots, u_{mk+\frac{k+1}{2}}, v_{mk+\frac{k+1}{2}}.$

Moreover, $l(P_9) = 2 + m + \frac{k+1}{2}$, $l(P_{10}) = 1 + m + \frac{k+1}{2}$, $l(P_{11}) = 1 + i - (k-1)m$, and $l(P_{12}) = 2 + m + \frac{k+1}{2}$. These paths attain their maximum length for $m = \frac{q-1}{2} - 1$, and hence we get

$$\max\{l(P_9)\} = \frac{q+k}{2} + 1, \max\{l(P_{10})\} = \frac{q+k}{2}, \max\{l(P_{11})\} = \frac{q+k}{2},$$

$$\text{and } \max\{l(P_{12})\} = \frac{q+k}{2} + 1 \tag{44}$$

Case 4. $i = mk + l$ and $m = \frac{q-1}{2}$, where $l = 0, 1, 2, \ldots, \lfloor \frac{k}{2} \rfloor$
Shortest u_0–u_i, v_0–u_i, and v_0–v_i paths are given by:

$$P_{13} : u_0, v_0, v_k, v_{2k}, \ldots, v_{(m+1)k}, u_{(m+1)k}, u_{(m+1)k+1}, \ldots, u_{(m+1)k+l}$$

$$P_{14} : v_0, v_k, v_{2k}, \ldots, v_{(m+1)k}, u_{(m+1)k}, u_{(m+1)k+1}, \ldots, u_{(m+1)k+l},$$

$$P_{15} : v_0, v_k, v_{2k}, \ldots, v_{(m+1)k}, u_{(m+1)k}, u_{(m+1)k+1}, \ldots, u_{(m+1)k+l}, v_{(m+1)k+l}.$$

We have $l(P_{13}) = 3 + m + l$, $l(P_{14}) = 2 + m + l$, and $l(P_{15}) = 3 + m + l$, respectively. Finally, the path $P_{16} : v_i, v_{i-k}, v_{i-2k}, \ldots, v_{i-mk}, u_{i-mk}, u_{i-mk-1}, u_{i-mk-2}, \ldots, u_0$ gives a shortest u_0–v_i path and the length of the path P_{16} is $1 + m + i - mk$. The path P_{16} attains its maximum length for $m = \frac{q-1}{2}$ and $i = mk + \lfloor \frac{k}{2} \rfloor$, i.e.

$$\max\{l(P_{16})\} = \frac{q+k}{2}. \tag{45}$$

From the Eqs. (38)–(45), we conclude that $e(u_0) = e(v_0) = \frac{q+k}{2} + 1$. Thus, generalized petersen graph is a d-self-centered graph, where $d = \frac{q+k}{2} + 1$. $\quad\square$

In the theorem given below, we investigate self-centeredness of $P(n, k)$ for $n = m(4m + 1) \pm (m + 1)$ and $k = 4m + 1$.

Theorem 12. *For $n = m(4m + 1) \pm (m + 1)$, $k = 4m + 1$, and any positive integer $m \geq 2$, the generalized Petersen graph is not a self-centered graph.*

Proof. We have following two cases here.

Case 1. $k = 4m + 1$ and $n = m(4m + 1) + (m + 1) = 4m^2 + 2m + 1$ for any $m \geq 2$.
In this case, $v_{\frac{k-1}{2}}$ and $v_{n-(\frac{k-1}{2})}$ are equivalently most distant vertices from u_0 as well as v_0 and their path lengths are $\frac{k-1}{2}$ and $\frac{k-1}{2} + 1$, respectively. This implies that, $e(u_0) = \frac{k-1}{2}$ and $e(v_0) = \frac{k-1}{2} + 1$ and they differ by one.

Case 2. $k = 4m + 1$ and $n = m(4m + 1) - (m + 1) = 4m^2 - 1$ for any $m \geq 2$.
In this case, $v_{\frac{k+1}{2}}$ and $v_{n-(\frac{k+1}{2})}$ are equivalently most distant vertices from u_0 as well as v_0 and their path lengths are $\frac{k+1}{2}$ and $\frac{k+1}{2} + 1$, respectively. This implies that, $e(u_0) = \frac{k+1}{2}$ and $e(v_0) = \frac{k+1}{2} + 1$ and the eccentricities differ by one.

In both the above cases we get that $e(u_0) \neq e(v_0)$. Hence, the generalized Petersen graph is not a self-centered. $\quad\square$

4   Computer Search and Concluding Remarks

We made an exhaustive computer search and found all possible values of n and k for which $P(n, k)$ are non self-centered. In this search, we have discarded isomorphs of these generalized Petersen graphs using Theorem 1. We list all non-isomorphic generalized Petersen graphs in Tables 1 and 2 and address their theoretical proofs obtained in this paper.

Table 1. Non-self-centeredness of $P(n, k)$, n odd

n	k	Theoretical support
$n \geq 9$	$k = 2$	Theorem 9
$n = m(4m + 1) \pm (m + 1), m \geq 2$	$k = 4m + 1$	Theorem 11

Table 2. Non-self-centeredness of $P(n, k)$, n even

n	k	Theoretical support
$n \geq 4$	$k = \frac{n}{2}$	Theorem 5
$n = 4m + 2$, m is a positive integer	$k = \frac{n}{2} - 1$	Theorem 6
$n = 4m(4m + 1)$, m is a positive integer	$k = 2m(4m - 1)$	Theorem 8

For complete characterization, one has to prove theoretically that all generalized Petersen graphs $P(n, k)$ other than those in Tables 1 and 2, and their isomorphs are self-centered. Hence, we make the following conjecture.

Conjecture: The generalized Petersen graphs $P(n, k)$ other than those in Tables 1 and 2 and their isomorphs are self-centered.

Acknowledgement. We are thankful to the referees for their constructive and detail comments and suggestions which improved the paper overall.

References

1. Beenker, G.F.M., Van Lint, J.H.: Optimal generalized Petersen graphs. Philips J. Res. **43**(2), 129–136 (1988)
2. Buckley, F.: Self-centered graphs. Ann. New York Acad. Sci. **576**, 71–78 (1989)
3. Buckley, F., Miller, Z., Slater, P.J.: On graphs containing a given graph as center. J. Graph Theory **5**, 427–434 (1981)
4. Castagna, F., Prins, G.: Every generalized Petersen graph has a Tait coloring. Pac. J. Math. **40**(1), 53–58 (1972)
5. Chen, L., Ma, Y., Shi, Y., Zhao, Y.: On the [1,2]-domination number of generalized Petersen graphs. Appl. Math. Comput. **327**, 1–7 (2018)

6. Coxeter, H.S.M.: Self-dual configurations and regular graphs. Bull. Am. Math. Soc. **56**(5), 413–455 (1950)

7. Ebrahimi, B.J., Jahanbakht, N., Mahmoodian, E.S.: Vertex domination of generalized Petersen graphs. Discrete Math. **309**(13), 4355–4361 (2009)

8. Frucht, R., Graver, J.E., Watkins, M.E.: The groups of the generalized Petersen graphs. Proc. Camb. Philos. Soc. **70**(211), 211–218 (1971)

9. Fu, X., Yang, Y., Jianga, B.: On the domination number of generalized Petersen graphs $P(n, 2)$. Discrete Math. **309**(8), 2445–2451 (2009)

10. Janakiraman, T.N.: On special classes of self-centered graphs. Discrete Math. **126**, 411–414 (1994)

11. Janakiraman, T.N., Bhanumathi, M., Muthammai, S.: Self-centered super graph of a graph and center number of a graph. Ars Comb. **87**, 271–290 (2008)

12. Huilgol, M.I., Ramprakash, C.: Cyclic edge extensions- self-centered graphs. J. Math. Comput. Sci. **10**, 131–137 (2014)

13. Saražin, M.L., Pacco, W., Previtali, A.: Generalizing the generalized Petersen graphs. Discrete Math. **307**(3–5), 534–543 (2007)

14. Singh, P., Panigrahi, P.: On self-centeredness of product of graphs. Int. J. Comb. **2016**, 1–4 (2016). https://doi.org/10.1155/2016/2508156

15. Singh, P., Panigrahi, P.: On self-centeredness of tensor product of some graphs. Electron. Notes Discrete Math. **63**, 333–342 (2017)

16. Stanic, Z.: Some notes on minimal self-centered graphs. AKCE Int. J. Graphs Comb. **7**, 97–102 (2010)

17. Steimle, A., Staton, W.: The isomorphism classes of the generalized Petersen graphs. Discrete Math. **309**(1), 231–237 (2009)

18. Watkins, M.E.: A theorem on Tait colorings with an application to the generalized petersen graphs. J. Comb. Theory **6**(2), 152–164 (1969)

Weak Roman Bondage Number
of a Graph

P. Roushini Leely Pushpam and N. Srilakshmi[(✉)] [iD]

Department of Mathematics, D.B. Jain College (Affiliated to University of Madras),
Chennai 600 097, Tamil Nadu, India
roushinip@yahoo.com, srilakshmi_murali@yahoo.com

Abstract. A *Roman dominating function* (RDF) on a graph G is a labelling $f : V(G) \rightarrow \{0, 1, 2\}$ such that every vertex with label 0 has a neighbor with label 2. A vertex u with $f(u) = 0$ is said to be *undefended* with respect to f if it is not adjacent to a vertex v with the positive weight. A function $f : V(G) \rightarrow \{0, 1, 2\}$ is a *weak Roman dominating function* (WRDF) if each vertex u with $f(u) = 0$ is adjacent to a vertex v with $f(v) > 0$ such that the function $f' : V(G) \rightarrow \{0, 1, 2\}$ defined by $f'(u) = 1$, $f'(v) = f(v) - 1$ and $f'(w) = f(w)$ if $w \in V - \{u, v\}$, has no undefended vertex. The Roman bondage number $b_R(G)$ of a graph G with maximum degree at least two is the minimum cardinality of all sets $E' \subseteq E(G)$ for which $\gamma_R(G - E') > \gamma_R(G)$. We extend this concept to a weak Roman dominating function as follows: The weak Roman bondage number $b_r(G)$ of a graph G with maximum degree at least two is the minimum cardinality of all sets $E' \subseteq E(G)$ for which $\gamma_r(G - E') > \gamma_r(G)$. In this paper we determine the exact values of the weak Roman bondage number for paths, cycles and complete bipartite graphs. We obtain bounds for trees and unicyclic graphs and characterize the extremal graphs.

Keywords: Weak Roman dominating function · Weak Roman bondage number

1 Introduction

A subset S of vertices of G is a *dominating set* if $N[S] = V$. The *domination number* $\gamma(G)$ is the minimum cardinality of a dominating set of G. The *bondage number* $b(G)$ of a nonempty graph G is the minimum cardinality among all sets of edges $E' \subseteq E(G)$ for which $\gamma(G - E') > \gamma(G)$. This concept was introduced by Bauer et al. [1]. To measure the vulnerability or the stability of the domination in an interconnection network under edge failure, Fink et al. [3] also proposed the concept of the bondage number in 1990. In [21] Xu has given an elaborate survey on bondage number of graphs which includes variations and generalizations of bondage numbers.

Cockayne *et al.* [2] defined a *Roman dominating function* (RDF) in a graph G to be a function $f : V(G) \rightarrow \{0, 1, 2\}$ satisfying the condition that every vertex

© Springer Nature Switzerland AG 2020
M. Changat and S. Das (Eds.): CALDAM 2020, LNCS 12016, pp. 156–166, 2020.
https://doi.org/10.1007/978-3-030-39219-2_13

u for which $f(u) = 0$ is adjacent to at least one vertex v for which $f(v) = 2$. The weight of a Roman dominating function is the value $w(f) = \sum_{u \in V} f(u)$. The minimum weight of a Roman dominating function of a graph G is called the *Roman domination number* of G and denoted by $\gamma_R(G)$. Roman domination in graphs has been studied in [7,8,10,15,16,18–20].

Henning *et al.* [6] defined a *weak Roman dominating function* as follows: For a graph G, let $f: V(G) \rightarrow \{0,1,2\}$ be a function. A vertex u with $f(u) = 0$ is said to be *undefended* with respect to f if it is not adjacent to a vertex v with the positive weight. A function $f: V(G) \rightarrow \{0,1,2\}$ is said to be a *weak Roman dominationg function* (WRDF) if each vertex u with $f(u) = 0$ is adjacent to a vertex v with $f(v) > 0$ such that the function $f': V(G) \rightarrow \{0,1,2\}$ defined by $f'(u) = 1$, $f'(v) = f(v) - 1$ and $f'(w) = f(w)$ if $w \in V - \{u,v\}$, has no undefended vertex. We say that v *defends* u. The weight $w(f)$ of f is defined to be $\sum_{u \in V} f(u)$. The minimum weight of a weak Roman dominating function of a graph G is called the *weak Roman domination number* of G and denoted by $\gamma_r(G)$. A WRDF with weight $\gamma_r(G)$ is called a $\gamma_r(G)$-function. This concept of weak Roman domination as suggested by Henning et al. [6] is an attractive alternative as it further reduces the weight of the Roman dominating function. Weak Roman domination in graphs has been studied in [11–14,17]. A weak Roman dominating function f can also be written as $f = (V_0, V_1, V_2)$ where $V_i = \{v/f(v) = i\}$, $i = 0$, 1, 2. Notice that in a WRDF, every vertex in V_0 is dominated by a vertex in $V_1 \cup V_2$, while in an RDF every vertex in V_0 is dominated by at least one vertex in V_2. Furthermore, in a WRDF every vertex in V_0 can be defended without creating an undefended vertex. Let $v \in V_1 \cup V_2$. We say a vertex $w \in N(v) \cap V_0$ is said to be in the *dependent set* of v, denoted by $D_G(v)$ if w is defended by v alone.

Jafari Rad and Volkmann [9] defined Roman bondage number as follows: *The Roman bondage number* $b_R(G)$ of a graph G with maximum degree at least two is the minimum cardinality of all sets $E' \subseteq E(G)$ for which $\gamma_R(G - E') > \gamma_R(G)$. We extend this concept to a weak Roman dominating function as follows: The *weak Roman bondage number* $b_r(G)$ of a graph G with maximum degree at least two is the minimum cardinality of all sets $E' \subseteq E(G)$ for which $\gamma_r(G - E') > \gamma_r(G)$. In this paper we initiate a study of this parameter.

2   Notation

For notation and graph theoretic terminology, we in general follow [4,5]. Throughout this paper, we consider only simple and connected graphs. Let G be a graph with vertex set $V = V(G)$ and edge set $E = E(G)$. The order $|V|$ of G is denoted by n. For every vertex $v \in V$, the *open neighborhood* $N(v)$ is the set $\{u \in V(G) : uv \in E(G)\}$ and the *closed neighborhood* of v is the set $N[v] = N(v) \cup \{v\}$. The *degree* of a vertex v in a graph G is the number of edges that are incident to the vertex v and is denoted by $deg(v)$. The *minimum* and *maximum degree* of a graph G are denoted by $\delta = \delta(G)$ and $\Delta = \Delta(G)$. A vertex of degree zero is called an *isolated* vertex, while a vertex of degree one is called a *leaf* vertex or a *pendant* vertex of G. An edge incident to a leaf is called a *pendant edge*. A set S of vertices is called *independent* if no two vertices in S are adjacent. A simple graph in which every pair of distinct vertices are adjacent is called a complete graph. A clique of a simple graph G is a subset S of V such that $G[S]$ is complete. A connected graph having no cycles is called a *tree*. A connected graph with exactly one cycle is called an *unicyclic* graph. A *support* vertex is a vertex which is adjacent to at least one leaf vertex. A *weak support* vertex is a vertex which is adjacent to exactly one leaf vertex. A *strong support* vertex is a vertex which is adjacent to at least two leaf vertices. For two positive integers m, n, the *complete bipartite* graph $K_{m,n}$ is the graph with partition $V(G) = V_1 \cup V_2$ such that $|V_1| = m$, $|V_2| = n$ and such that $G[V_i]$ has no edges for $i = 1, 2$, and every two vertices belonging to different partition sets are adjacent to each other. A complete bipartite graph of the form $K_{1,n}$ is called a star graph. We call the vertex of $deg\ n - 1$, in a star graph as the *head* vertex.

3   Some Standard Graphs

In this section we determine the exact values for paths, cycles and complete bipartite graphs. It is clear that for complete graphs K_n, $b_r(G) = 1$.

Theorem 1. *For any graph G with $n \geq 3$, $\delta(G) = 1$ and $\Delta(G) = n - 1$, $b_r(G) \leq 2$.*

Proof. Let $f = (V_0, V_1, V_2)$ be a $\gamma_r(G)$-function. Let $v \in V(G)$ such that $deg(v) = n - 1$. If $G = P_3$, then $b_r(G) = 2$. Suppose that $G \neq P_3$ and v is adjacent to at least two leaf vertices, then $f(v) = 2$ and $\gamma_r(G - e) > \gamma_r(G)$, where e is a pendant edge of G and therefore $b_r(G) = 1$.

Suppose that v is adjacent to exactly one leaf vertex, say u and $(D_G(v) \setminus \{u\}) \cup \{v\}$ induces a clique. Consider an arbitrary edge e. If e is the pendant edge incident with v, then $G - e = K_{n-1} \cup K_1$. By assigning 1 to u and 1 to v we see that $\gamma_r(G - e) = 2$. If e is any non pendant edge incident with v, then $V(G - e)$ can be partitioned in to two sets V_1 and V_2 such that V_1 induces K_{n-2} and V_2 induces K_2 in which case $\gamma_r(G - e) = 2$. If e is a non pendant edge not incident with v, then $deg_{G-e}(v) = n - 1$ and hence $\gamma_r(G - e) = 2$. Therefore $b_r(G) \geq 2$. Now, $\gamma_r(G - \{e_1, e_2\}) > \gamma_r(G)$ where e_1 and e_2 are the pendant and a non pendant edge incident with v respectively. Hence $b_r(G) = 2$.

Suppose that $(D_G(v) \setminus \{u\}) \cup \{v\}$ does not induce a clique, then clearly for any edge e which is incident with v, $\gamma_r(G - e) > \gamma_r(G)$ and hence $b_r(G) = 1$. $\square$

Theorem 2. [6] *For $n \geq 4$, $\gamma_r(C_n) = \gamma_r(P_n) = \lceil \frac{3n}{7} \rceil$.*

Theorem 3. *For paths P_n,*

$$b_r(P_n) = \begin{cases} 2, & n = 3 \text{ and } n \equiv 5 \ (mod\ 7) \\ 1, & otherwise. \end{cases}$$

Proof. Let $P_n = (v_1, v_2, \ldots, v_n)$. One can easily verify that $b_r(P_2) = 1$ and $b_r(P_3)$ $= b_r(P_5) = 2$ and $\gamma_r(P_4 - v_1 v_2) = 3$ and therefore $b_r(P_4) = 1$. Assume that $n \geq 6$ and let $e = v_5 v_6$. Now, we consider the following cases.

Case 1. $n \not\equiv 5 \pmod 7$

$$\gamma_r(P_n - e) = \gamma_r(P_5) + \gamma_r(P_{n-5})$$
$$= 3 + \lceil \frac{3(n-5)}{7} \rceil$$
$$= 1 + \lceil \frac{3n}{7} \rceil$$
$$> \gamma_r(P_n).$$

Thus, $b_r(P_n) = 1$.

Case 2. $n \equiv 5 \pmod 7$.
We claim that $b_r(P_n) = 2$. For any arbitrary edge e, $\gamma_r(P_n - e) = \gamma_r(P_k) + \gamma_r(P_{n-k})$, where $k \equiv 0, 1, \ldots, 6 \pmod 7$ and $k = 1, 2, \ldots, \lceil \frac{n}{2} \rceil$.

$$\gamma_r(P_n - e) = \gamma_r(P_k) + \gamma_r(P_{n-k})$$
$$= \lceil \frac{3k}{7} \rceil + \lceil \frac{3(n-k)}{7} \rceil$$
$$= \lceil \frac{3n}{7} \rceil$$
$$= \gamma_r(P_n).$$

Therefore $b_r(P_n) \geq 2$. Let $e_1 = v_1 v_2$ and $e_2 = v_2 v_3$.

$$\gamma_r(P_n - \{e_1, e_2\}) = 2 + \gamma_r(P_{n-2})$$
$$= 2 + \lceil \frac{3(n-2)}{7} \rceil$$
$$= 2 + \frac{3n+5}{7}$$
$$> \gamma_r(P_n).$$

Therefore $b_r(P_n) \leq 2$. Hence, $b_r(P_n) = 2$.   □

Theorem 4. *For cycles C_n, $n \geq 4$*

$$b_r(C_n) = \begin{cases} 3, & n \equiv 5 \pmod 7 \\ 2, & \text{otherwise.} \end{cases}$$

Proof. Let $C_n = (v_1, v_2, \ldots, v_n v_1)$. If e is an arbitrary edge of C_n, then $C_n - e = P_n$. Further $\gamma_r(C_n - e) = \gamma_r(P_n) = \lceil \frac{3n}{7} \rceil$. Hence, $b_r(C_n) \geq 2$. Now, the theorem follows from Theorem 3.   □

Theorem 5. *For complete bipartite graphs* $K_{m,n}$, $1 \leq m \leq n$

$$b_r(K_{m,n}) = \begin{cases} 4, & \text{if } m = n = 3, \\ 7, & \text{if } m = n = 4, \\ m, & \text{otherwise.} \end{cases}$$

Proof. Let $X = \{x_1, x_2, \ldots, x_m\}$ and $Y = \{y_1, y_2, \ldots, y_n\}$ be the partite sets of $G = K_{m,n}$. When $m = 1$, G is a star and $b_r(G) = 1 = m$.

Next, assume that $m = 2$. If $n = 2$, then $G = C_4$ and by Theorem 4, $b_r(G) = 2$. If $n \geq 3$, for any edge e of G, $G - e$ is a graph which contains a pendant edge. Without loss of generality, let $e = x_1 y_1$. Now by assigning 1 to x_1 and x_2 and 0 elsewhere we see that x_2 defends each vertex in $Y \setminus \{y_1\}$ and x_1 defends y_1. Hence $\gamma_r(G - e) = \gamma_r(G) = 2$. Therefore $b_r(G) \geq 2$. Now $\gamma_r(G - \{e_1, e_2\}) = 3 > \gamma_r(G) = 2$, where e_1, e_2 are the two edges incident with y_1 and thus $b_r(G) \leq 2$. Therefore $b_r(G) = 2 = m$.

When $G = K_{3,3}$, $\gamma_r(G) = 3$. We claim that $b_r(G) = 4$. By assigning 1 to each member of X and 0 to each member of Y, we see that the removal of any edge will leave a graph G' in which $K_{2,3}$ is a subgraph and $\gamma_r(G') = 3$. Suppose that two edges are removed and G' is the resulting graph. If the two edges are incident. Then, clearly $C_4 \cup K_2$ is a spanning subgraph of G' for which $\gamma_r(G') = 3$. If the two edges are non incident, then clearly C_6 is a spanning subgraph of G' for which $\gamma_r(G') = 3$. Again, if three edges are removed to obtain a graph G', then the three edges form either a $K_{1,3}$ or $3K_2$ or $P_3 \cup K_2$ or P_4. The graph G' in the respective cases are $G' = K_1 \cup K_{2,3}$, C_6, P_5, $C_4 \cup K_2$. In all the cases, we see that $\gamma_r(G') = 3$. Hence, from the above arguments, $b_r(G) \geq 4$. Now, $\gamma_r(G - \{e_1, e_2, e_3, e_4\}) > \gamma_r(G)$, where e_1, e_2, e_3, e_4 form a C_4 in G. Therefore, $b_r(G) = 4$.

When $G = K_{4,4}$, let E be a subset of edges such that $\gamma_r(G - E) > \gamma_r(G) = 4$. Assume that $|E| < 7$. It is evident that $K_{4,4} - E$ contains either of the subgraphs $K_{2,4}$ or $4K_2$ which implies that $\gamma_r(G - E) = \gamma_r(G) = 4$. Hence, $|E| \geq 7$. Now let E be the set of all edges incident with both x_1 and y_1. Then, $\gamma_r(K_{4,4} - E) = 5 > \gamma_r(K_{4,4})$. Hence $|E| \leq 7$ and thus $b_r(G) = 7$.

When $m \geq 3, n \neq 3, 4$, $\gamma_r(G) = 4$. If E is the set of edges with $|E| < m$ and $G_1 = G - E$, then there are two vertices $x \in X$ and $y \in Y$ such that $N_{G_1}(x) = Y$ and $N_{G_1}(y) = X$. It follows that $\gamma_r(G_1) = 4 = \gamma_r(G)$ and $b_r(G) \geq m$. However, if we remove all the edges incident with a vertex in Y, we obtain a graph G_2 such that $\gamma_r(G_2) = 5$. This shows that $b_r(G) = m$.   □

4   Trees

In this section we prove that for trees T, $b_r(T)$ is bounded above by 3. Further we characterize trees with $b_r(T) = 3$.

Theorem 6. *For any tree T with $n \geq 3$, $b_r(T) \leq 3$.*

Proof. Let T be a tree and let $f = (V_0, V_1, V_2)$ be a $\gamma_r(T)$-function. Let v be a support vertex such that v has exactly one non leaf neighbor. If v is adjacent to at least three leaf vertices, then clearly, $\gamma_r(T - e) > \gamma_r(T)$, where e is the pendant edge incident to v. Hence, $b_r(T) = 1$. If v is adjacent to exactly two leaf vertices, then $\gamma_r(T - \{e_1, e_2, e_3\}) > \gamma_r(T)$, where e_1 and e_2 are the pendant edges incident with v and e_3 is the non-pendant edge incident with v. Therefore, $b_r(T) \leq 3$. If v is a weak support, let u, w be the leaf and non-leaf neighbors of v respectively. Let e_1 and e_2 be the pendant and non-pendant edge incident to v. If $f(v) = 2$, then $u, w \in D_T(v)$. If $f(v) = 1$, then $u \in D_T(v)$. If $f(v) = 0$, then $f(u) = 1$ and $v \in D_T(u)$. In all the cases we see that $\gamma_r(T - \{e_1, e_2\}) > \gamma_r(T)$, which implies that $b_r(T) \leq 2$ and the theorem is proved.   □

Now, we characterize trees for which $b_r(T) = 3$. To facilitate our discussion, we define the following.

A tree T of order $n \geq 3$ is said to be a *galaxy* if every non-leaf vertex is a support and is adjacent to exactly two leaf vertices.

We define two families of trees $\Im$ and $\Im_1$ as follows. A tree $T \in \Im_1$ if one of the following holds.

(i) T contains a support vertex which is adjacent to at least three leaf vertices.
(ii) T contains a weak support vertex of degree two.
(iii) T contains two non support vertices x_1 and x_2 such that $N(x_i)$, $i = 1, 2$ contains exactly $deg(x_i) - 1$ weak support vertices and a vertex $z \in N(x_1)$ $\cap N(x_2)$ (Refer Fig. 1).

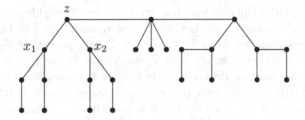

Fig. 1. A Tree $T \in \Im_1$

Now consider a tree $T = T_1 \notin \Im_1$. Identify the non-leaf, non strong support neighbors of strong support vertices and remove them from T_1 to obtain a graph T_2. Again remove such vertices from T_2 to obtain T_3.

In general, let T_i, $1 \leq i \leq k$ be the graph obtained from T_{i-1} by removing such vertices from T_{i-1}. If $T_i \in \Im_1$, then stop the process. Otherwise, repeat the process until no such vertices remains.

Now, $T \in \Im$ if the following conditions hold.

1. Each component of T_k is either a K_1 or a galaxy.
2. Each vertex x removed in the process satisfies the following conditions.
   (i) x is adjacent to at most two K_1s.
   (ii) At least three vertices in $N(x)$ are in the set $A \cup B \cup C$, where $C = \cup K_1$, A is the set of all vertices in a subgraph of T which is a galaxy, B is the set of all removed vertices in T that are adjacent to exactly two K_1s.
   (iii) If a component of T_k is a P_3, then at least one neighbor of its head vertex is adjacent to two K_1s.

Theorem 7. *For any tree $T \in \Im_1$, $b_r(T) \leq 2$.*

Proof. If either of the conditions (i) or (ii) of $\Im_1$ are satisfied, then as discussed in the proof of Theorem 6, $b_r(T) \leq 2$. Suppose that T contains two non-support vertices x_1 and x_2 such that $N(x_i)$, $i = 1, 2$ contains exactly $deg(x_i) - 1$ weak support vertices and a vertex $z \in N(x_1) \cap N(x_2)$. Then any γ_r-function f of T will assign 1 to z and $x_1, x_2 \in D_T(z)$. Hence $\gamma_r(T - zx_1) > \gamma_r(T)$ which implies $b_r(T) = 1$. Thus if $T \in \Im_1$, $b_r(T) \leq 2$. □

Theorem 8. *For any tree T with $n \geq 4$, $b_r(T) = 3$ if and only if $T \in \Im$.*

Proof. Let T be a tree and $f = (V_0, V_1, V_2)$ be a $\gamma_r(T)$-function with $b_r(T) = 3$. We claim that $T \in \Im$.

Let $T = T_1$. Suppose that $T_1 \in \Im_1$, then by Theorem 7, we get a contradiction. Thus, $T = T_1 \notin \Im_1$. If every non-leaf vertex of T is a strong support and is adjacent to exactly two leaf vertices, then T is a galaxy and $T \in \Im$.

Otherwise, identify the non-leaf, non strong support neighbors of strong support vertices and remove them. Let T_2 be the resulting graph. Let X_1 be the set of all vertices removed from T_1 to obtain T_2. Suppose that $T_2 \in \Im_1$, then there exists a support vertex say v in T_2 such that either v is adjacent to at least three leaf vertices or degree of v is two or two non support vertices x_1 and x_2 exist satisfying condition (iii) of $\Im_1$. Suppose that v is adjacent to at least three leaves. Let v_1, v_2, v_3 be the leaf neighbors of v in T_2. If $deg(v_1) = deg(v_2) = 1$ and $deg(v_3) > 1$, then clearly $N(v_3) \backslash \{v\} \subseteq X_1$. Hence, $f(v) = 2$ and $f(v_i) = 0$, $i = 1, 2, 3$. Now, $\gamma_r(T - vv_1) > \gamma_r(T)$ which implies that $b_r(T) = 1$, a contradiction. If $deg(v_1) = 1$, $deg(v_2)$, $deg(v_3) > 1$, then as before $\gamma_r(T - vv_1) > \gamma_r(T)$, a contradiction. If $deg(v_i) > 1$, $i = 1, 2, 3$, then either $f(v) = 2$ or 1. Suppose $f(v) = 2$, then $f(v_i) = 0$, $i = 1, 2, 3$ and $\gamma_r(T - vv_1) > \gamma_r(T)$, a contradiction. Suppose $f(v) = 1$. Then for each $i = 1, 2, 3$ there exists a z_i such that $z_i \in N(v_i \backslash \{v\})$, $f(z_i) = 1$ and $|D_G(z_i)| \geq 1$. Therefore, $\gamma_r(T - vv_1) > \gamma_r(T)$, a contradiction. Suppose that $deg(v) = 2$. Let v_1 be the leaf neighbor of v and v_2 be its non leaf neighbor. If $deg(v_1) > 1$ and $deg(v) > 2$, then clearly $N(v_1) \backslash \{v\} \subseteq X_1$ and $N(v) \backslash \{v_1, v_2\} \subseteq X_1$. Now, $\gamma_r(T - vv_1, vv_2) > \gamma_r(T)$. This implies $b_r(T) \leq 2$, a contradiction. If $deg(v) > 2$ and $deg(v_1) = 1$, then $N(v) \backslash \{v_1, v_2\} \subseteq X_1$. Now, $\gamma_r(T - vv_1, vv_2) > \gamma_r(T)$. This implies $b_r(T) \leq 2$, a contradiction. A similar argument holds when $deg(v) = 2$ and $deg(v_1) > 1$.

Suppose that two non support vertices x_1 and x_2 exist satisfying condition (iii) of $\Im_1$. Let $z \in N(x_1) \cap N(x_2)$, then by weighing all options we see that $\gamma_r(T - zx_1) > \gamma_r(T)$, and $b_r(T) = 1$, contradiction. Hence if $T_2 \in \Im_1$, stop the process of pruning. Otherwise we repeat the process until no such vertex remains. Let T_k be the final graph. Let $X = \{x_1, x_2, \ldots, x_r\}$ be the vertices removed from T in the process of successive pruning.

We now prove the following claims.

Claim 1. Each component of T_k is either a K_1 or a galaxy.

Suppose to the contrary that there is a component of T_k, say $T*$ which is neither a K_1 nor a galaxy. Then $T*$ is a K_2 and $\gamma_r(T-e) > \gamma_r(T)$, where e is the edge of K_2. Therefore, $b_r(T) = 1$, a contradiction. Now, consider a $K_1 = \{w\}$ in T_k. It is clear that $N(w) \subseteq X$. Further each vertex $x \in X$ is adjacent to a strong support vertex in some $T_i, 1 \leq i \leq k - 1$ and hence f can be modified in such a way that $X \subseteq V_0$. Hence, w will receive the weight 1, under f. Therefore, we conclude that each K_1 receives the weight 1.

Claim 2. Each x_i, $1 \leq i \leq r$ is adjacent to at most two K_1s.

Suppose to the contrary that some x_j is adjacent to at least three K_1s. Now, let g be a function from $V \rightarrow \{0, 1, 2\}$ such that $g(x_j) = 2$, $g(u) = 0$ where u forms a K_1 in T_k and is adjacent to x_j and $g(x) = f(x)$ otherwise. However $\gamma_r(T - x_j w) > \gamma_r(T)$ where w forms a K_1 adjacent to x_j. Hence $b_r(T) = 1$, a contradiction. Thus, every x_i is adjacent to at most two K_1s.

Claim 3. At least three vertices in $N(x_i)$ for each i are in the set $A \cup B \cup C$.

Before proving the claim, we observe that any vertex in T which is not in $A \cup B \cup C$ is a member of X, which is adjacent to at most one K_1. Hence, such vertices will receive the weight 0 under f. Now, by default, at least one member in $N(x_i)$ is a member of A, say y. Suppose to the contrary that no member other than y in $N(x_i)$ $1 \leq i \leq r$ are members of $A \cup B \cup C$. Then clearly $\gamma_r(T - e) > \gamma_r(T)$, where $e = x_i y$. Therefore, $b_r(T) = 1$, a contradiction. Also, if exactly one member in $N(x_i)$ other than y is in $A \cup B \cup C$, $\gamma_r(T - \{e_1, e_2\}) > \gamma_r(T)$ where e_1, e_2 are the edges incident with x_i and having other ends in $A \cup B \cup C$ Therefore $b_r(T) \leq 2$, a contradiction.

Claim 4. If a component of T_k is a P_3, then at least one neighbor of its head vertex is adjacent to two K_1s.

Suppose to the contrary that none of the neighbors of the head vertex a of P_3 are adjacent to two K_1s. Let e_1 and e_2 be the edges that are incident with a. Let w be a neighbor of a. If w is adjacent to exactly one $K_1 = \{b\}$, then $N(w) \setminus \{a, b\} \subset A \cup X$. Now, $\gamma_r(T - \{e_1, e_2\}) > \gamma_r(T)$. Therefore, $b_r(T) = 2$, a contradiction. Also if every member in $N(w) \setminus \{a\}$ are strong supports, then clearly $w \in X$ and $\gamma_r(T - \{e_1, e_2\}) > \gamma_r(T)$. Therefore, $b_r(T) = 2$, a contradiction.

Hence, from the above arguments $T \in \Im$.

Conversely, suppose that $T \in \mathfrak{S}$. Then, by the structure of T one can easily observe that the removal of any single edge or two edges will not alter the value of $\gamma_r(T)$. Further the removal of the three edges incident with any end support will increase the value of $\gamma_r(T)$. Hence $b_r(T) = 3$ (Refer Fig. 2).   □

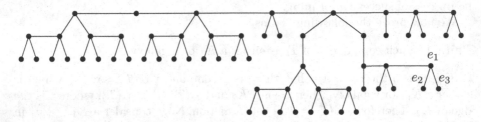

Fig. 2. A Tree $T \in \mathfrak{S}$, where $\gamma_r(T - \{e_1, e_2, e_3\}) > \gamma_r(T)$

5   Unicyclic Graphs

In this section we prove that for unicyclic graphs G, $b_r(G)$ is bounded above by 4. Further, we characterize unicyclic graphs which attain the upper bound.

Theorem 9. *For any unicyclic graph G with $\Delta(G) \geq 3$, $b_r(G) \leq 4$.*

Proof. Let G be a unicyclic graph with cycle C and $\Delta(G) \geq 3$ and let $f = (V_0, V_1, V_2)$ be a $\gamma_r(G)$-function. Suppose that G contains a vertex which is not adjacent to any vertex in C. Let v be an end support vertex in G. Then, as in Theorem 6, $b_r(G) \leq 3$. Suppose that every vertex not in C is adjacent to some vertex in C. That is every vertex not in C is of degree one. Then, there exists at least one vertex in C say z such that z is a support vertex. If z is adjacent to at least three leaf vertices, then $\gamma_r(G - e) > \gamma_r(G)$, where e is a pendant edge incident with z. Hence, $b_r(G) = 1$. If z is adjacent to exactly one leaf vertex then, $\gamma_r(G - \{e_1, e_2, e_3\}) > \gamma_r(G)$, where e_1 is the pendant edge incident with z and e_2, e_3 are the non-pendant edges incident with z. Hence, $b_r(G) \leq 3$. If z is adjacent to exactly two leaf vertices, then $\gamma_r(G - \{e_1, e_2, e_3, e_4\}) > \gamma_r(G)$, where e_1, e_2 are the pendant edges incident with z and e_3, e_4 are the non-pendant edges incident with z. Hence, $b_r(G) \leq 4$.   □

Now we characterize unicyclic graphs with $b_r(G) = 4$.

Theorem 10. *For any unicyclic graph G with cycle C and $\Delta(G) \geq 3$, $b_r(G) = 4$ if and only if every vertex on the cycle C is of degree 4 and every vertex not in the cycle is of degree 1.*

Proof. Let G be a unicyclic graph with $\Delta(G) \geq 3$ and let C be the unique cycle in G. Let $f = (V_0, V_1, V_2)$ be a $\gamma_r(G)$-function. As in the proof of Theorem 9, if G contains a vertex which is not adjacent to any vertex in C, then we see that $b_r(G) \leq 3$, a contradiction. Hence, every vertex not in C is of degree one.

Now, we claim that every vertex in C is of degree four. Since $\Delta(G) \geq 3$, choose a vertex w in C such that $deg(w) \geq 4$. If $deg(w) \geq 5$, then clearly, $\gamma_r(G-e) > \gamma_r(G)$ where e is the pendant edge incident with w. Hence, $b_r(G) = 1$, a contradiction. Suppose that $deg(w) = 3$, then $\gamma_r(G - \{e_1, e_2, e_3\}) > \gamma_r(G)$, where e_i, $i = 1, 2, 3$ are the edges incident at w. Hence, $b_r(G) \leq 3$, a contradiction. If $deg(w) = 4$ and at least one neighbor of w, say x in C is of degree two. Now, if $f(x) = 1$, then $|D_G(x)| = 1$. Therefore, $\gamma_r(G - \{e_1, e_2, e_3\}) > \gamma_r(G)$, where e_1, e_2, e_3 are the edges incident with w other than wx. Hence, $b_r(G) \leq 3$, a contradiction. A similar argument holds if $f(x) = 0$. Therefore, every vertex on the cycle C is of degree 4 and every vertex not in the cycle is of degree 1 (Refer Fig. 3).

Converse is straightforward.   □

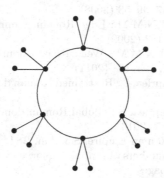

Fig. 3. A unicyclic graph G with $b_r(G) = 4$

Acknowledgement. The authors are thankful to the referees for their valuable suggestions for the improvement of the paper.

References

1. Bauer, D., Harary, F., Nieminen, J., Suffel, C.L.: Domination alteration sets in graphs. Discrete Math. **47**, 153–161 (1983)
2. Cockayne, E.J., Dreyer, P.A., Hedetniemi, S.M.: Roman domination in graphs. Discrete Math. **78**, 11–22 (2004)
3. Fink, J.F., Jacobson, M.S., Kinch, L.F., Roberts, J.: The bondage number of a graph. Discrete Math. **86**, 47–57 (1990)
4. Haynes, T.W., Hedetniemi, S.T., Slater, P.J. (eds.): Domination in Graphs. Advanced Topics. Marcel Dekker, New York (1998)
5. Haynes, T.W., Hedetniemi, S.T., Slater, P.J. (eds.): Fundamentals of Domination in Graphs. Marcel Dekker, New York (1998)
6. Hedetniemi, S.T., Henning, M.A.: Defending the Roman Empire - A new strategy. Discrete Math. **266**, 239–251 (2003)
7. Henning, M.A.: A characterization of Roman trees. Discuss. Math. Graph Theory. **22**, 325–334 (2002)

8. Henning, M.A.: Defending the Roman Empire from multiple attacks. Discrete Math. **271**, 101–115 (2003)
9. Jafari Rad, N., Volkmann, L.: Roman bondage in graphs. Discuss. Math. **31**, 763–773 (2011)
10. Revelle, C.S.: Can you protect the Roman Empire? John Hopkins Mag. **2**, 70 (1997)
11. Pushpam, P.R.L., Kamalam, M.: Efficient weak Roman domination in graphs. IJPAM **101**(5), 701–710 (2015)
12. Pushpam, P.R.L., Kamalam, M.: Efficient weak Roman domination in Myscielski graphs. Int. J. Pure Eng. Math. **3**(2), 93–100 (2015)
13. Pushpam, P.R.L., Kamalam, M.: Stability of weak Roman domination upon vertex deletion. Asian J. Math. Comput. Res. **25**(2), 97–105 (2018)
14. Pushpam, P.R.L., Kamalam, M.: Effects of vertex deletion on the weak Roman domination number of a graph. AKCE Int. J. Graphs Comb. **16**(2), 204–212 (2019)
15. Pushpam, P.R.L., Mai, T.N.M.M.: On efficient Roman dominatable graphs. J. Comb. Math. Comput. **67**, 49–58 (2008)
16. Pushpam, P.R.L., Mai, T.N.M.M.: Edge Roman domination in graphs. J. Comb. Math. Comput. **69**, 175–182 (2009)
17. Pushpam, P.R.L., Mai, T.N.M.M.: Weak Roman domination in graphs. Discuss. Math. Graph Theory. **31**, 115–128 (2011)
18. Pushpam, P.R.L., Padmapriea, S.: Restrained Roman domination in graphs. Trans. Comb. **4**(1), 1–17 (2015)
19. Pushpam, P.R.L., Padmapriea, S.: Global Roman domination in graphs. Discrete Appl. Math. **200**, 176–185 (2016)
20. Stewart, I.: Defend the Roman Empire. Sci. Am. **281**, 136–139 (1991)
21. Xu, J.-M.: On bondage numbers of graphs - a survey with some comments. Int. J. Comb. **2013** (2013). 34 pages

On the Geodetic and Hull Numbers of Shadow Graphs

S. V. Ullas Chandran[1]([⊠]), Mitre C. Dourado[2][ORCID], and Maya G. S. Thankachy[3]

[1] Department of Mathematics, Mahatma Gandhi College,
Kesavadasapuram, Thiruvananthapuram 695004, India
`svuc.math@gmail.com`
[2] Instituto de Matemtica, Universidade Federal do Rio de Janeiro,
Rio de Janeiro, Brazil
`mitre@dcc.ufrj.br`
[3] Department of Mathematics, Mar Ivanios College, University of Kerala,
Thiruvananthapuram 695015, India
`mayagsthankachy@gmail.com`

Abstract. Given two vertices u, v in a graph G, a shortest (u, v)-path in G is called an (u, v)-geodesic. Let $I_G[u, v]$ denote the set of all vertices in G lying on some (u, v)-geodesic. Given a set $T \subseteq V(G)$, let $I_G[T] = \cup_{u,v \in T} I_G[u, v]$. If $I_G[T] = T$, we call T a convex set. The convex hull, denoted by $\langle T \rangle_G$, is the smallest convex set containing T. A subset T of vertices of a graph G is a hull set if $\langle T \rangle_G = V(G)$. Moreover, T is a geodetic if $I_G[T] = V(G)$. The hull number $h(G)$ of a graph G is the minimum size of a hull set. The geodetic number $g(G)$ of G is the minimum size of a geodetic set. The shadow graph, denoted by $S(G)$, of a graph G is the graph obtained from G by adding a new vertex v' for each vertex v of G and joining v' to the neighbors of v in G. In this paper, we study the geodetic and hull numbers of shadow graphs. Bounds for the geodetic and hull numbers of shadow graphs are obtained and for several classes exact values are determined. Graphs G for which $g(S(G)) \in \{2, 3\}$ are characterized.

Keywords: Convex set · Geodetic number · Hull number · Simplicial vertex · Shadow graphs

AMS Subject Classification: 05C12 · 05C76

1 Introduction

Convexities in graphs are extensively studied due to their prominent role in graph theory as well as their contributions to axiomatic convexity theory. Given

M. C. Dourado—Partially supported by Conselho Nacional de Desenvolvimento Científico e Tecnológico, Brazil.
M. G. S. Thankachy—Supported by the University of Kerala for providing University JRF.

M. Changat and S. Das (Eds.): CALDAM 2020, LNCS 12016, pp. 167–177, 2020.
https://doi.org/10.1007/978-3-030-39219-2_14

a finite set X, a family $\mathcal{C}$ of subsets of X is a *convexity* on X if $\varnothing \in \mathcal{C}, X \in \mathcal{C}$, and $\mathcal{C}$ is closed under intersections [12,13,21]. A set $T \subseteq X$ is said to be $\mathcal{C}$-*convex* if $T \in \mathcal{C}$. The $\mathcal{C}$-*convex hull* of $T \subseteq X$, $\langle T \rangle_{\mathcal{C}}$, is the minimum $\mathcal{C}$-convex set containing T. The cardinality of minimum set whose convex hull is X is the *hull number of* $\mathcal{C}$.

The most studied graph convexities are convexities defined by a family of paths $\mathcal{P}$, in a way that a set T of vertices of G is convex if and only if each vertex that lies on an (u, v)-path of $\mathcal{P}$ belongs to T. In this paper, we consider the geodetic convexity in graphs. In this convexity, $\mathcal{P}$ is the family of geodesics (shortest paths) of the graph.

One of the most studied numbers associated with graphs is the chromatic number. The chromatic number $\chi(G)$ is the minimum number of colors that can be assigned to the vertices of G so that adjacent vertices are colored differently. It is clear that $\chi(G) \geq \omega(G)$, where $\omega(G)$ is the size of a largest clique in G. However, a graph G may have arbitrarily large chromatic number without triangles ($\omega(G) = 2$). In 1955 Jan Mycielski used a fascinating construction called the Mycielskian or Mycielski graph [9,16]. His construction preserves the property of being triangle-free but increases the chromatic number. Applying the construction repeatedly to a triangle-free starting graph, we obtain a triangle-free graph with arbitrarily large chromatic number. A graph closely related to this construction is called the shadow graph. The *shadow graph* $S(G)$ of a graph G is the graph obtained from G by adding a new vertex v' for each vertex v of G and joining v' to the neighbors of v in G (the vertex v' is called the *shadow vertex* of v). The *star shadow graph* of a graph G is the graph obtained from the shadow graph $S(G)$ of G by adding a new vertex s^* (*star vertex*) and joining s^* to all shadow vertices. The Mycielski's construction consists of repeatedly finding the star shadow of the previous one beginning with the cycle C_5. Particularly, $s^*(C_5)$ is called the *Grötzsch graph* (a triangle-free graph) with chromatic number four. The term shadow graph was coined in [9,11].

In this paper, we continue our investigation of hull and geodetic numbers on shadow graphs. In Sect. 2, we fix the notation, terminologies and discuss some preliminary results of the geodetic and hull numbers already available in the literature.

2 Preliminaries

Let G be a connected graph and $u, v \in V(G)$. The *distance* $d_G(u, v)$ between u and v is the minimum number of edges on a (u, v)-path. The maximum distance between all pairs of vertices of G is the *diameter* $diam(G)$ of G. A (u, v)-path of length $d_G(u, v)$ is called an (u, v)-*geodesic*. Then, the *geodetic interval* $I_G[u, v]$ between vertices u and v of a graph G is the set of vertices x such that there exists a (u, v)-geodesic which contains x. For $T \subseteq V(G)$ we set $I_G[T] = \bigcup_{u,v \in T} I_G[u, v]$. The set T is a geodetic set if $I_G[T] = V(G)$. The geodetic number, denoted by $g(G)$, is the size of a minimum geodetic set. To simplify the writing, we may omit the index G in the above notation provided that G is clear from the context. The geodetic number of a graph was introduced in [14] and [1,2,6–8,17,18] contain numerous results and references concerning geodetic sets and the geodetic number.

The set T is *convex* in G if $I_G[T] = T$. The *convex hull* $\langle T \rangle_G$ of T is the smallest convex set that contains T, and T is a *hull set* of G if $\langle T \rangle_G$ is the whole vertex set of G. A smallest hull set is a *minimum hull set* of G, its cardinality is the *hull number* $h(G)$ of G. The convex hull $\langle T \rangle_G$ can also be formed from the sequence $\{I_G^k[T]\}$, $k \geq 0$, where $I_G^0[T] = T$, $I_G^1[T] = I_G[T]$ and $I_G^k[T] = I_G[I_G^{k-1}[T]]$ for $k \geq 2$. From some term on, this sequence must be constant. Let p be the smallest number such that $I_G^p[T] = I_G^{p+1}[T]$. Then $I_G^p[T]$ is the convex hull $\langle T \rangle_G$. The hull number of a graph was introduced by Everett and Seidman in [15]. See [2,5,10,17] for recent developments on the hull sets and the hull number of a graph. The hull number of composition, cartesian product, and strong product of graphs were studied in [3,4] and [19], respectively. A vertex v is called a simplicial vertex G if the subgraph induced by the neighbors of v is complete. The set of all simplicial vertices in a graph G is denoted by $simp(G)$ and $sp(G) = |simp(G)|$. A graph is *chordal* if it contains no induced cycle of length greater than three. A graph G is an *extreme hull graph* if the set of all simplicial vertices forms a hull set. In this paper, we make use of the following result.

Lemma 1. [2,10] *In a connected graph G, each simplicial vertex belongs to every hull set of G.*

3   Hull Number of Shadow Graphs

In this section, we estimate the upper and lower bounds of the hull number of shadow graphs and simplify the exact values for the shadows of complete graphs, hyper-cubes, grids, cycles and complete bipartite graphs. We prove a formula for the hull number of the shadow graph of a tree.

Lemma 2. *For any non-trivial connected graph G, a vertex v in $S(G)$ is a simplicial vertex of $S(G)$ if and only if v is a shadow of a simplicial vertex in G.*

Proof. Since $N_{S(G)}(v') = N_G(v)$, it follows that v' is a simplicial vertex in $S(G)$ for any simplicial vertex v in G. On the otherhand, let x be any simplicial vertex of $S(G)$. Then observe that x must be a shadow vertex, say $x = v'$, shadow of v. Now, if v is non-simplicial, then there exist non-adjacent neighbors, say s and t of v in G. This shows that s and t are also non-adjacent neighbors of v' in $S(G)$, impossible. Thus v must be simplicial in G.   ∎

Lemma 3. (i) *Let x and y be non-adjacent vertices in G. Then*
    (1) $d_{S(G)}(x,y) = d_G(x,y)$.
    (2) $d_{S(G)}(x,y') = d_G(x,y)$.
    (3) $d_{S(G)}(x',y') = d_G(x,y)$.
(ii) *Let x and y be adjacent vertices in G. Then*
    (1) $d_{S(G)}(x,y) = d_G(x,y) = 1$.

(2) $d_{S(G)}(x, y') = d_{S(G)}(x, y) = 1$.

(3) $d_{S(G)}(x', y') = \begin{cases} 2 & \text{if } xy \text{ lies on an induced } K_3 \\ 3 & \text{otherwise} \end{cases}$

Proof. (i) First consider the case x and y are non-adjacent in G. Let $d_G(x, y) = d \geq 2$ and let $P : x = x_0, x_1, \ldots, x_d = y$ be a (x, y)-geodesic in G. This shows that x' is adjacent with x_1 and y' is adjacent with x_{d-1} in $S(G)$. Hence $x', x_1, x_2, \ldots, x_{d-1}, y'$ is an (x', y')-path of length d in $S(G)$ and so $d_{S(G)}(x', y') \leq d$. Now, suppose that $d_{S(G)}(x', y') = k < d$. Let $Q : x' = y_0, y_1, \ldots, y_k = y'$ be an (x', y')-geodesic of length k. Then it follows from the definition of $S(G)$ that $y_1, y_{k-1} \in V(G)$ and $xy_1, yy_{k-1} \in E(G)$. Now, suppose that the (y_1, y_{k-1})-subpath Q_1 of Q in $S(G)$ contains a shadow vertex u' of u, say $y_i = u'$. Then $2 \leq i \leq k - 2$ and $y_{i-1}, y_{i+1} \in V(G)$. Then it follows from the definition of $S(G)$, the vertex u is adjacent to both y_{i-1} and y_{i+1} in G. This shows that (y_1, y_{i-1})-subpath of Q_1 together with the path y_{i-1}, u, y_{i+1} and the (y_{i+1}, y_{k-1})-subpath of Q_1 is a (y_1, y_{k-1})-path which has the same length of Q_1. Hence for each shadow vertex in Q_1, can be replaced with the corresponding vertex without changing the length of Q_1. Hence without loss of generality, we may assume that Q_1 has no shadow vertices and so Q_1 is a path in G. Then Q_1 together with the edges xy_1 and $y_{k-1}y$ is an (x, y)-walk of length k in G. Then $d_G(x, y) \leq k < d$, a contradiction. Thus $d_{S(G)}(x', y') = d_G(x, y)$. Proof for the remaining cases are similar. ∎

Lemma 4. *For every graph G, it holds $h(S(G)) \leq h(G) + sp(G)$.*

Proof. Let T be a hull set of G and T' be the set of all vertices of $S(G)$ formed by the shadow vertices of T. We claim that $T' \cup simp(G)$ is a hull set of $S(G)$.

Observe from Lemma 3 that $V(G) \backslash T$ is contained in the convex hull of T'. Next, let $v \in T \backslash simp(G)$ and let $u, w \in N_G(v)$ such that $uw \notin E(G)$. If $u \in T$, then $u' \in T'$, otherwise, u belongs to the convex hull of T'. Therefore, $V(G)$ is contained in the convex hull of $T' \cup simp(G)$.

Now, by Lemma 2, every shadow vertex not in T' is not a simplicial vertex of $S(G)$, and then it has two non-adjacent neighbors in $V(G)$. ∎

Theorem 1. *For any non-trivial connected graph G of order n,*

$$\max\{2, sp(G)\} \leq h(S(G)) \leq \min\{n, h(G) + sp(G)\}.$$

Proof. The left inequality is an immediate consequence of Lemmas 1 and 2. By Lemma 4, it remains to prove that $h(S(G)) \leq n$. Now, let V' be the set of shadow vertices of $V(G)$ in $S(G)$. We claim that V' is a hull set in $S(G)$. For, let v be any vertex in G. First suppose that $deg_G(v) \geq 2$. Let u and w be two distinct neighbors of v in G. Then the shadow vertices u' and w' of u and w, respectively are adjacent to v in $S(G)$. This shows that $v \in I_{S(G)}[u', w'] \subseteq \langle V' \rangle_{S(G)}$. So, assume that v is a pendent vertex in G. Let u be the unique neighbor of v in G and let v' and u' be the corresponding shadow vertices of v and u respectively. Then it follows from Lemma 3 that $d_{S(G)}(u', v') = 3$ and $v \in I_{S(G)}[u', v'] \subseteq \langle V' \rangle_{S(G)}$.

This shows that $\langle V' \rangle_{S(G)} = V(S(G))$ and so V' is a hull set of $S(G)$. Hence $h(S(G)) \leq |V'| = n$. ∎

Now, the following formulas that can be easily deduced from the Theorem 1. The k-cube Q_k has the vertex set $\{0,1\}^k$, two vertices being adjacent if they differ in precisely one coordinate.

- $h(S(K_n)) = h(K_n) = n$, where $n \geq 2$.
- $h(S(K_{m,n})) = h(K_{m,n}) = 2$, where $m, n \geq 2$.
- $h(S(Q_n)) = h(Q_n) = 2$, where $n \geq 2$.
- $h(S(C_n)) = h(C_n) = \begin{cases} 2 \text{ if } n \text{ is even} \\ 3 \text{ if } n \text{ is odd} \end{cases}$
- $h(S(G_{n,m})) = h(G_{n,m}) = 2$, where $G_{n,m}(n, m \geq 2)$ is the 2 dimensional grid of order nm.

In view of Theorem 1, we also have the following result.

Theorem 2. Let G be a connected graph of order n. If $h(S(G)) = n$, then G is a chordal graph of diameter at most three.

Proof. First suppose that $h(S(G)) = n$. Let V' denotes the set of all shadow vertices of $V(G)$ in $S(G)$. We first claim that $diam(G) \leq 3$. Assume the contrary that there is a shortest path, say $u = u_0, u_1, u_2, u_3, u_4 = v$ of length four in G. For each i in the interval $0 \leq i \leq 4$, let u_i' denote the shadow vertices of u_i. Then by Lemma 3, $d_{S(G)}(u_0', u_4') = d_G(u_0, u_4) = 4$. We prove that $V' \backslash u_2'$ is a hull set of $S(G)$. Since V' is a hull set of $S(G)$, it is enough to prove that $u_2' \in I_{S(G)}[V' \backslash u_2']$. Now, since $P : u_0', u_1, u_2', u_3, u_4'$ is a path of length four and $d_{S(G)}(u_0', u_4') = 4$, we know that P is a (u_0', u_4')-geodesic containing the vertex u_2'. Hence $u_2' \in I_{S(G)}[u_0', u_4'] \subseteq I[V' \backslash u_2']$ and so $V' \backslash u_2'$ is a hull set of $S(G)$. This shows that $h(S(G)) \leq |V' \backslash u_2'| = n - 1$, a contradiction. Thus $diam(G) \leq 3$. Now, suppose that G contains an induced cycle, say $C : u_1, u_2, \ldots, u_n, u_1$ of length $n \geq 4$. As above, we claim that $V' \backslash u_1'$ is a hull set of $S(G)$. Since $\langle V' \rangle_{S(G)} = V(S(G))$, it is enough to prove that $u_1' \in I_{S(G)}[V' \backslash u_1']$. Now, since C is chordless, it follows from Lemma 3 that $u_1 \in I_{S(G)}[u_2', u_n'], u_3 \in I_{S(G)}[u_2', u_4']$ and $u_{n-1} \in I_{S(G)}[u_{n-2}', u_n']$. Hence $u_2, u_n \in I_{S(G)}^2[V' \backslash u_1']$. Thus $u_1' \in I_{S(G)}[u_n, u_2] \subseteq I_{S(G)}^3[V' \backslash u_1']$. This shows that $u_1' \in \langle V' \backslash u_1' \rangle_{S(G)}$ and so $V' \subseteq \langle V' \backslash u_1' \rangle_{S(G)}$. Hence $\langle V' \backslash u_1' \rangle_{S(G)} = \langle V' \rangle_{S(G)} = V(S(G))$. This leads to the fact that $h(S(G)) \leq |V' \backslash u_1'| = n - 1$, a contradiction. This proves that G is a chordal graph. ∎

The converse of Theorem 2 need not be true. Consider the chordal graph of diameter 2 shown in the Fig. 1. The set $T' = \{v_1', v_4', v_5'\}$ is the set of all simplicial vertices of the shadow graph of G. Now, $I_{S(G)}[T'] = T' \cup \{v_2, v_3\}$ and $I_{S(G)}^2[T'] = I_{S(G)}[T']$ and so T' is not a hull set of $S(G)$.

On the otherhand, since the set $T' = \{v_1', v_4', v_5', v_3'\}$ is a hull set of $S(G)$. It follows that $h(S(G)) = 4 < 5 = n$. This example also shows that the inequalities in Theorem 1 can be strict.

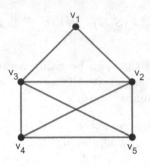

Fig. 1. G

In the following, we determine a formula for the hull number of shadow graph of a tree. A vertex v in a tree T is a *support vertex* if it is adjacent to a pendant vertex in T. A support vertex v is said to be a *first order support vertex* of T, if $T - v$ has atmost one non-trivial component.

Theorem 3. *Let T be a tree with k end vertices and l first order support vertices. Then $h(S(T)) = k + l$.*

Proof. Let $v_1, v_2, \ldots, v_k$ be the end vertices of T. Since each end vertex is a simplicial vertex, it follows from Lemma 1 that $h(S(T)) \geq k$. Let R be a hull set of $S(T)$. Then $v'_1, v'_2, \ldots, v'_k \in R$. Let v be a first order support vertex of T. Then $T - v$ has at most one non-trivial component. This shows that if v lies on a (u, w)-geodesic in T, then atleast one end, say u, must be an end vertex of T. Now, suppose that the shadow vertex v' of v lies only on a (x, y)-geodesic in $S(T)$, say $P : x = x_0, x_1, \ldots, x_i = v', \ldots, x_n = y$. Then $Q : x = x_0, x_1, \ldots, x_i = v, \ldots, x_n = y$ is also an (x, y)-geodesic in $S(T)$ containing the vertex v. Now, if Q contains any shadow vertex, then we can replace each shadow vertex by the corresponding vertex in T. Hence we can assume without loss of generality that Q is (x, y)-geodesic in T containing v. Thus $x = u$ or $y = u$, say $x = u$. Also note that the vertex u lies internally only on (v, v')-geodesic in $S(T)$. This shows that either $u \in R$ or $v' \in R$ and so $|R| \geq k + l$. Now, on the otherhand consider the set $R' = \{v'_1, v'_2, \ldots, v'_k, w'_1, w'_2, \ldots, w'_l\}$, where $w_1, w_2, \ldots, w_l$ are the first order support vertices of T. Then $I_{S(G)}[R'] = V(S(T)) - \{v_i\}$. Then $I^2_{S(G)}[R'] = V(S(T))$, implies that R' is a hull set of $S(T)$. Therefore, $h(S(T)) = |R'| = k + l$. ∎

The hull number of the shadow graph of a graph can be significantly small. For instance, in the class of wheels $W_{1,n} (n \geq 5)$, one can observe that $h(W_{1,n}) = \lfloor \frac{n}{2} \rfloor$. Whereas, one can easily check that the set $T = \{u, u'\}$ (u is the vertex with largest degree in $W_{1,n}$) is a hull set of the shadow graph of $W_{1,n}$. Thus $h(S(W_{1,n})) = 2$. In general, if a graph G of order n has a vertex of degree $n - 1$, then $h(S(G)) = 2$. In view of this observation, we leave the following problems as open.

Problem 1. Characterize graphs G for which $h(S(G)) = 2$.

Problem 2. Characterize graphs G for which $h(S(G)) = 3$.

4   Geodetic Number of Shadow Graphs

In this section, we estimate the upper and lower bounds of the geodetic number of any shadow graph and for several classes of shadow graphs the exact values are determined. We prove that K_2 is the only connected graph in which $g(S(G)) = 2$; and the graphs K_3 and P_3 are the only connected graphs in which $g(S(G)) = 3$.

Definition 1. *For any set $T' \subseteq V(S(G))$, the set $\pi(T')$ is defined as $\pi(T') = \{v \in V(G) : v \in T' \text{ or } v' \in T'\}$.*

Lemma 5. *If T' is a geodetic set of $S(G)$, then $\pi(T')$ is a geodetic set of G.*

Proof. Let x be any vertex in G such that $x \notin \pi(T')$. Then $x' \notin T'$. Since T' is a geodetic set in $S(G)$, there exist vertices $u, v \in T'$ such that $x' \in I_{S(G)}[u, v]$. Let $P : u = u_0, u_1, \ldots, u_i = x', \ldots, u_n = v$ be a (u, v)-geodesic in $S(G)$ containing the vertex x'. Then $u_{i-1}, u_{i+1} \in V(G)$.
Case 1: Both $u, v \in V(G)$. Since $d_G(u, v) \geq 2$, by Lemma 3, $d_G(u, v) = d_{S(G)}(u, v)$. Now, the path $Q : u = u_0, \ldots, u_{i-1}, x, u_{i+1}, \ldots, u_n = v$ is a (u, v)-geodesic in $S(G)$ containing the vertex x. Now, if Q contains any shadow vertex y' then we can replace y' by the corresponding vertex y. Hence without loss of generality, we may assume that Q has no shadow vertices and so Q is a (u, v)-geodesic in G containing the vertex x. Thus $x \in I_G[u, v] \subseteq I_G[\pi(T')]$.
Case 2: Both u and v are shadow vertices, say $u = r'$ and $v = t'$. Since $x' \in I_{S(G)}[r', t']$, it is clear that $d_{S(G)}(r', t') \geq 4$ and so $d_G(r, t) = d_{S(G)}(r', t')$. This shows that the path $Q : r, u_1, \ldots, u_{i-1}, x, u_{i+1}, \ldots, u_{n-1}, t$ is an (r, t)-geodesic in $S(G)$ containing the vertex x. Again without loss of generality, we may assume that Q has no shadow vertices and so Q is a (r, t)-geodesic in G containing the vertex x. Then $x \in I_G[r, t] \subseteq I_G[\pi(T')]$.
Case 3: $u = r'$ and $v \in V(G)$. Again, since $x' \in I_{S(G)}[r', v]$, we have that $v \neq r$ and $d_{S(G)}(r', v) \geq 3$. Hence similar to the above cases, we have that the path $Q : r, u_1, \ldots, u_{i-1}, x, u_{i+1}, \ldots, u_n = v$ is an (r, v)-geodesic in G containing the vertex x. Again, we may assume that Q has no shadow vertex and so Q is a (r, v)-geodesic in G containing the vertex x. Hence $x \in I_G[r, v] \subseteq I_G[\pi(T')]$. Thus in all cases $\pi(T')$ is a geodetic set in G. ∎

A set T of vertices in a graph G is an *open geodetic set* if for each vertex v in G, either (1) v is a simplicial vertex of G and $v \in T$ or (2) v is an internal vertex of an (x, y)-geodesic for some $x, y \in T$. The minimum size of an open geodetic set is the open geodetic number $og(G)$ of G. In the following, we obtain an upper bound for the geodetic number of $S(G)$ in terms of the open geodetic number of G. The open geodetic number of a graph was introduced and studied in [20].

Theorem 4. *Let G be any connected graph of order n. Then*

$$g(G) \leq g(S(G)) \leq \min\{n, og(G) + sp(G)\}.$$

Proof. Let T' be a minimum geodetic set of $S(G)$. Then by Lemma 5, $\pi(T')$ is a geodetic set of G. Thus $g(G) \leq |\pi(T')| \leq |T'| = g(S(G))$. On the other hand, let R be a minimum open geodetic set in G. We claim that $T = R \cup simp(S(G))$ is a geodetic set of $S(G)$. Observe that $V(G) \subseteq I_G[R] \subseteq I_{S(G)}[T]$. Now let v' be any shadow vertex in $S(G)$ corresponding to the vertex v of G. If v is simplicial, then $v' \in T$. So, assume that v is non-simplicial. Since R is an open geodetic set, there exist $x, y \in R$ such that $v \in I_G[x, y]$ with $v \neq x$ and $v \neq y$. Let $P : x = u_0, u_1, \ldots, u_i = v, u_{i+1}, \ldots, v_n = y$ with $1 \leq i \leq n-1$ be a (x, y)-geodesic in G containing the vertex v internally. Then by Lemma 3, the path $P : x = u_0, u_1, \ldots, v_{i-1}, v', v_{i+1}, \ldots, v_n = y$ is an (x, y)-geodesic in $S(G)$ containing the vertex v'. Hence $v' \in I_{S(G)}[T]$. This shows that T is a geodetic set of $S(G)$ and so $g(S(G)) \leq |T| = og(G) + sp(G)$. Now, as in the case of Theorem 1, one can easily verify that the set of all shadow vertices of $V(G)$ is a geodetic set of $S(G)$. Thus $g(S(G)) \leq n$. Hence the result follows. ∎

The following two observations are used to characterize graphs G in which $g(S(G)) \in \{2, 3\}$.

Observation 5. *Let u and v be two distinct vertices in G. Then $u \in I_{S(G)}[u', v']$ if and only if u and v are adjacent vertices in G having no common neighbors.*

Observation 6. *Let u be any vertex in G. Then $u' \notin I_{S(G)}[u, x]$ for any $x \in V(S(G))$ distinct from u'.*

Theorem 7. *Let G be any connected graph. Then $g(S(G)) = 2$ if and only if $G = P_2$.*

Proof. Let $T = \{x, y\}$ be a geodetic set in $S(G)$. If $x \in V(G)$, then by Observation 6, $x' \notin I_{S(G)}[x, y]$. Hence x must be a shadow vertex, say $x = v'$ for some $v \in V(G)$. Now, since T is a geodetic set of $S(G)$, we have that $v \in I_{S(G)}[x, y]$. But, in this case, one can easily observe that $v \in I_{S(G)}[x, y]$ if and only if $y = u'$ where u is adjacent to v in G. This is possible only when $G = P_2$. Hence the result follows. ∎

Theorem 8. *Let G be any connected graph. Then $g(S(G)) = 3$ if and only if G is either K_3 or P_3.*

Proof. First, suppose that $G = K_3$ or $G = P_3$, then one can easily verify that $g(S(G)) = 3$. Conversely, assume that $g(S(G)) = 3$. If G has only three vertices, then $G = K_3$ or $G = P_3$. So, assume that G contains at least four vertices. Let T be a geodetic set in $S(G)$ of size three. We consider the following cases.
Case 1: $T = \{x', y', z\}$, where x' and y' are shadow vertices of x and y in G and $z \in V(G)$. First suppose that $z = x$ or $z = y$, say $z = x$. Choose $w \in V(G)$ be such that $w \notin \{x, y, z\}$ (This is possible because G has at least

four vertices). Now, since $d_{S(G)}(x', x) = 2$, it follows that $w' \in I_{S(G)}[x', y']$. This shows that $d_{S(G)}(x', y') \geq 4$ and hence by Lemma 3, x and y are non-adjacent in G. This shows that $y \notin I_{S(G)}[x, x']$ and so $y \in I_{S(G)}[y', x']$, a contradiction to Observation 5. Hence $z \neq x$ and $z \neq y$. Now, it follows from Observation 6 that $z' \in I_{S(G)}[x', y']$. Thus $d_{S(G)}(x', y') \geq 4$. Let $x' = x_0, x_1, \ldots, x_{i-1}, x_i = z', x_{i+1}, \ldots, x_n = y'$ be an (x', y')-geodesic in $S(G)$ containing the vertex z'. Then $x_{i-1}, x_{i+1} \in V(G)$ and so the path $P : x' = x_0, x_1, x_{i-1}, z, x_{i+1}, \ldots, x_n = y'$ is also an (x', y')-geodesic containing the vertex z. This shows that $z \in I_{S(G)}[x', y']$ and so $I_{S(G)}[x', z] \subseteq I_{S(G)}[x', y']$ and $I_{S(G)}[z, y'] \subseteq I_{S(G)}[x', y']$. This shows that the set $U = \{x', y'\}$ is a geodetic set in $S(G)$, a contradiction to the fact that $g(S(G)) = 3$.

Case 2: $T = \{x', y, z\}$, where x' is a shadow vertex of x in G and $y, z \in V(G)$. Now, if $x = y$ or $x = z$, say $x = y$, then it follows from Observation 6 that $z' \in I_{S(G)}[y, y']$. This leads to the fact that $d_{S(G)}(y, y') \geq 3$, a contradiction. Hence $x \neq y$ and $x \neq z$. Again by Observation 6, $z' \in I_{S(G)}[x', y]$. Hence as in the previous case, we can show that $z \in I_{S(G)}[x', y]$. Thus $I_{S(G)}[x', z] \subseteq I_{S(G)}[x', y]$ and $I_{S(G)}[z, y] \subseteq I_{S(G)}[x', y]$. This leads to the fact that the set $U = \{x', y\}$ is a geodetic set in $S(G)$, a contradiction.

Case 3: $T = \{x, y, z\} \subseteq V(G)$. By Observation 6, we have that $y' \in I_{S(G)}[x, z]$. Hence as in the previous cases, $y \in I_{S(G)}[x, z]$ and hence $U = \{x, z\}$ is a geodetic set in $S(G)$, a contradiction.

Case 4: All the three vertices of T are shadow vertices, say $T = \{x', y', z'\}$. Choose $w \in V(G)$ be such that $w \notin \{x, y, z\}$. Since T is a geodetic set, we may assume that $w' \in I_{S(G)}[x', y']$. This shows that $d_{S(G)}(x', y') \geq 4$. Hence by Lemma 3, the vertices x and y are non-adjacent in G. Moreover, $d_G(x, y) = d_{S(G)}(x', y') \geq 4$. Now, by Observation 5, it follows that both $x, y \notin I_{S(G)}[x', y']$. Now, suppose that $x \in I_{S(G)}[x', z']$ and $y \in I_{S(G)}[y', z']$, it follows from Observation 5 that both x and y must be adjacent with z in G. Then $d_{S(G)}(x', y') = d_G(x, y) = 2$, a contradiction. Hence either $x \notin I_{S(G)}[x', z']$ or $y \notin I_{S(G)}[y', z']$, say $x \notin I_{S(G)}[x', z']$. Now, since T is a geodetic set of $S(G)$, we have that $x \in I_{S(G)}[y', z']$. Moreover, recall that $d_G(x, y) \geq 4$. This shows that $d_{S(G)}(y', z') \geq 4$ and so by Lemma 3, $d_G(y, z) = d_{S(G)}(y', z') \geq 4$. Now, let $P : y' = u_0, u_1, \ldots, u_{i-1}, u_i = x, u_{i+1}, \ldots, u_n = z'$ be a (y', z')-geodesic in $S(G)$ containing the vertex x. Note that $u_{i-1} \neq y'$ and $u_{i+1} \neq z'$. Now without loss of generality, we may assume that both u_{i-1} and u_{i+1} are vertices in G. Otherwise, if they are shadow vertices, we can replace these vertices by the corresponding vertices in G. This shows that the path $P' : y' = u_0, u_1, \ldots, u_{i-1}, x', u_{i+1}, \ldots, u_n = z'$ is a (y', z')-geodesic containing x' and so $x' \in I_{S(G)}[y', z']$. Hence as the previous cases, the set $U = \{y', z'\}$ must be a geodetic set in $S(G)$, a contradiction. Hence the result follows. ∎

The following result is an immediate consequence of Theorems 4, 7 and 8.

- $g(S(K_n)) = g(K_n) = n$, where $n \geq 2$.
- $g(S(K_{m,n})) = og(K_{m,n}) = 4$, where $m, n \geq 2$.
- $g(S(Q_n)) = og(Q_n) = 4$, where $n \geq 2$.
- $g(S(G_{n,m})) = og(G_{n,m}) = 4$, where $n, m \geq 2$.

Theorem 9. *Let T be a tree with k end vertices and m support vertices. Then $g(S(T)) = k + m$.*

Proof. Let $v_1, v_2, \ldots, v_k$ be end vertices of T, let $u_1, u_2, \ldots, u_m$ be the corresponding support vertices and let M' be a geodetic set of $S(T)$. Then by Lemma 2, $v_i' \in M'$ for all $i = 1, 2, \ldots, k$. For each $i = 1, 2, \ldots, k$, if the vertex v_i in $S(G)$ lies internally on an (x, y)-geodesic in $S(T)$, then $x = u_i'$ or $y = u_i'$, where u_i is the corresponding support vertices of v_i. This shows that either $u_i \in M'$ or $v_i \in M'$. Hence $|M'| \geq k + m$. On the other hand, consider the set $R' = \{v_1, v_2, \ldots, v_k\} \cup \{u_1', u_2', \ldots, u_m'\}$. Let x be any vertex of $S(T)$ such that $x \notin R'$.

Case 1: $x \in V(G)$. Then x lies on a (v_i, v_j)-geodesic in T, say $v_i = y_0, y_1, \ldots, y_r = x, y_{r+1}, \ldots, y_k = v_j$. Then by Lemma 3, the path $y_0', y_1, \ldots, y_r, x, y_{r+1}, \ldots, y_{k-1}, y_k'$ is an (y_0', y_k')-geodesic in $S(T)$ containing the vertex x and so $x \in I_{S(G)}[R']$.

Case 2: $x = u'$, a shadow vertex of u in T. If u is a support vertex, then $u' \in R'$. So, assume that u is not a support vertex. This shows that u lies on a (v_i, v_j)-geodesic in R', say $v_i = y_0, y_1, \ldots, y_r = u, y_{r+1}, \ldots, y_l = v_j$. Since u is not a support vertex of T, we have that $2 \leq r \leq l - 2$.

This shows that by Lemma 3, the path $v_i' = y_0, y_1, \ldots, y_{r-1}, u', y_{r+1}, \ldots, y_l = v_j'$ is a (v_i', v_j')- geodesic in $S(T)$ containing the vertex u'. Hence R' is a geodetic set of $S(T)$ and so $g(S(T)) = k + m$. ∎

References

1. Bresar, B., Klavžar, S., Horvat, A.T.: On the geodetic number and related metric sets in Cartesian product graphs. Discrete Math. **308**, 5555–5561 (2008)
2. Buckley, F., Harary, F.: Distance in Graphs. Addison-Wesley, Redwood City (1990)
3. Cagaanan, G.B., Canoy Jr., S.R.: On the hull sets and hull number of the Composition graphs. Ars Combinatoria **75**, 113–119 (2005)
4. Cagaanan, G.B., Canoy Jr., S.R.: On the hull sets and hull number of the Cartesian product of graphs. Discrete Math. **287**, 141–144 (2004)
5. Chartrand, G., Harary, F., Zhang, P.: On the hull number of a graph. Ars Combinatoria **57**, 129–138 (2000)
6. Chartrand, G., Harary, F., Swart, H.C., Zhang, P.: Geodomination in graphs. Bull. ICA **31**, 51–59 (2001)
7. Chartrand, G., Zhang, P.: Extreme geodesic graphs. Czechoslovak Math. J. **52**(127), 771–780 (2002)
8. Chartrand, G., Harary, F., Zhang, P.: On the geodetic number of a graph. Networks **39**(1), 1–6 (2002)
9. Chartrand, G., Zhang, P.: Introduction to Graph Theory. Tata McGraw-Hill Edition, New Delhi (2006)
10. Dourado, M.C., Protti, F., Rautenbach, D., Szwarcfiter, J.L.: On the hull number of triangle-free graphs. SIAM J. Discrete Math. **23**, 2163–2172 (2010)
11. Garza, G., Shinkel, N.: Which graphs have planar shadow graphs? Pi Mu Epsilon J. **11**(1), 11–20 (1999). Fall
12. Edelman, P.H., Jamison, R.E.: The theory of convex geometries. Geometriae Dedicata **19**, 247–270 (1985)

13. Farber, M., Jamison, R.E.: Convexity in graphs and hypergraphs. SIAM J. Algebraic Discrete Methods **7**, 433–444 (1986)
14. Harary, F., Loukakis, E., Tsouros, C.: The geodetic number of a graph. Math. Comput. Model. **17**(11), 89–95 (1993)
15. Everett, M.G., Seidman, S.B.: The hull number of a graph. Discrete Math. **57**, 217–223 (1985)
16. Mycielski, J.: Sur le coloriage des graphes. Colloq. Math. **3**, 161–162 (1955)
17. Pelayo, I.M.: Geodesic Convexity in Graphs. Springer Briefs in Mathematics. Springer, New York (2013). https://doi.org/10.1007/978-1-4614-8699-2
18. Santhakumaran, A.P., Ullas Chandran, S.V.: The geodetic number of strong product graphs. Discuss. Math. Graph Theory **30**(4), 687–700 (2010)
19. Santhakumaran, A.P., Ullas Chandran, S.V.: The hull number of strong product graphs. Discuss. Math. Graph Theory **31**(3), 493–507 (2011)
20. Santhakumaran, A.P., Kumari Latha, T.: On the open geodetic number of a graph. Scientia Ser. A Math. Sci. **19**, 131–142 (2010)
21. Van de Vel, M.: Theory of Convex Structures. North-Holland, Amsterdam (1993)

Indicated Coloring of Complete Expansion and Lexicographic Product of Graphs

P. Francis[1], S. Francis Raj[2]([✉]), and M. Gokulnath[2]

[1] Department of Computer Science, Indian Institute of Technology,
Palakkad 678557, India
pfrancis@iitpkd.ac.in
[2] Department of Mathematics, Pondicherry University, Puducherry 605014, India
francisraj_s@yahoo.com, gokulnath.math@gmail.com

Abstract. Indicated coloring is a slight variant of the game coloring which was introduced by Grzesik [6]. In this paper, we show that for any graphs G and H, $G[H]$ is k-indicated colorable for all $k \geq \mathrm{col}(G)\mathrm{col}(H)$. Also, we show that for any graph G and for some classes of graphs H with $\chi(H) = \chi_i(H) = \ell$, $G[H]$ is k-indicated colorable if and only if $G[K_\ell]$ is k-indicated colorable. As a consequence of this result we show that if $G \in \mathcal{G} = \Big\{$Chordal graphs, Cographs, $\{P_5, C_4\}$-free graphs, Complete multipartite graphs$\Big\}$ and $H \in \mathcal{F} = \Big\{$Bipartite graphs, Chordal graphs, Cographs, $\{P_5, K_3\}$-free graphs, $\{P_5, Paw\}$-free graphs, Complement of bipartite graphs, $\{P_5, K_4, Kite, Bull\}$-free graphs, connected $\{P_6, C_5, \overline{P_5}, K_{1,3}\}$-free graphs which contain an induced C_6, $\mathbb{K}[C_5](m_1, m_2, \ldots, m_5)$, $\{P_5, C_4\}$-free graphs, connected $\{P_5, \overline{P_2 \cup P_3}, \overline{P_5}, Dart\}$-free graphs which contain an induced $C_5\Big\}$, then $G[H]$ is k-indicated colorable for every $k \geq \chi(G[H])$. This serves as a partial answer to one of the questions raised by Grzesik in [6].

Keywords: Game chromatic number · Indicated chromatic number · Lexicographic product of graphs

2000 AMS Subject Classification: 05C15

1 Introduction

All graphs considered in this paper are simple, finite and undirected. For a family $\mathcal{F}$ of graphs, we say that a graph G is $\mathcal{F}$-free if it contains no induced subgraph which is isomorphic to a graph in $\mathcal{F}$. The coloring number of a graph G (see [7]), denoted by $\mathrm{col}(G)$, is defined as $\mathrm{col}(G) = 1 + \max_{H \subseteq G} \delta(H)$, where $H \subseteq G$ means H is a subgraph of G.

The lexicographic product of two graphs G and H, denoted by $G[H]$, is a graph whose vertex set $V(G) \times V(H) = \{(x, y) : x \in V(G) \text{ and } y \in V(H)\}$

© Springer Nature Switzerland AG 2020
M. Changat and S. Das (Eds.): CALDAM 2020, LNCS 12016, pp. 178–183, 2020.
https://doi.org/10.1007/978-3-030-39219-2_15

and two vertices (x_1, y_1) and (x_2, y_2) of $G[H]$ are adjacent if and only if either $x_1 = x_2$ and $y_1 y_2 \in E(H)$, or $x_1 x_2 \in E(G)$. For each $u \in V(G)$, $\langle u \times V(H) \rangle$ is isomorphic to H and it is denoted by H_u and for each $v \in V(H)$, $\langle V(G) \times v \rangle$ is isomorphic to G and it is denoted by G_v.

Let G be a graph on n vertices $v_1, v_2, \ldots, v_n$, and let $H_1, H_2, \ldots, H_n$ be n vertex-disjoint graphs. An expansion $G(H_1, H_2, \ldots, H_n)$ of G (see [1]) is the graph obtained from G by
(i) replacing each v_i of G by H_i, $i = 1, 2, \ldots, n$, and
(ii) by joining every vertex in H_i with every vertex in H_j whenever v_i and v_j are adjacent in G.

For $i \in \{1, 2, \ldots, n\}$, if $H_i \cong K_{m_i}$, then $G(H_1, H_2, \ldots, H_n)$ is said to be a complete expansion of G and is denoted by $\mathbb{K}[G](m_1, m_2, \ldots, m_n)$ or $\mathbb{K}[G]$. For $i \in \{1, 2, \ldots, n\}$, if $H_i \cong \overline{K_{m_i}}$, then $G(H_1, H_2, \ldots, H_n)$ is said to be an independent expansion of G and is denoted by $\mathbb{I}[G](m_1, m_2, \ldots, m_n)$ or $\mathbb{I}[G]$. It can be noted that, if $m_1 = m_2 = \ldots = m_n = m$, then $\mathbb{K}[G](m_1, m_2, \ldots, m_n) \cong G[K_m]$ and $\mathbb{I}[G](m_1, m_2, \ldots, m_n) \cong G[\overline{K_m}]$.

The idea of indicated coloring was introduced by Grzesik in [6] as a slight variant of the game coloring in the following way: in each round the first player Ann selects a vertex and then the second player Ben colors it properly, using a fixed set of colors 'C'. The aim of Ann is to achieve a proper coloring of the whole graph G, while Ben tries to "block" some vertex. A *block* vertex means an uncolored vertex which has all colors from C on its neighbors. The smallest number of colors required for Ann to win the game on a graph G is known as the indicated chromatic number of G and is denoted by $\chi_i(G)$. Clearly from the definition we see that $\omega(G) \leq \chi(G) \leq \chi_i(G) \leq \Delta(G) + 1$. For a graph G, if Ann has a winning strategy using k colors, then we say that G is k-indicated colorable.

Grzesik in [6] has raised the following question: For a graph G, if G is k-indicated colorable, will it imply that G is also $(k + 1)$-indicated colorable? One can equivalently characterize all graphs G which are k-indicated colorable for all $k \geq \chi_i(G)$. There has been already some partial answers to this question. See for instance, [2–4,8]. In this paper, we show that for any graphs G and H, $G[H]$ is k-indicated colorable for all $k \geq \mathrm{col}(G)\mathrm{col}(H)$. Also, we show that for any graph G and for some special families of graphs H with $\chi(H) = \chi_i(H) = \ell$, we show that $G[H]$ is k-indicated colorable if and only if $G[K_\ell]$ is k-indicated colorable. In addition, we prove that if the graph $G \cong \mathbb{K}[H]$, where $H \in \mathcal{G} = \Big\{$Chordal graphs, Cographs, $\{P_5, C_4\}$-free graphs, Complete multipartite graphs$\Big\}$, then G is k-indicated colorable for all $k \geq \chi(G)$. As a consequence of these results, we show that if $G \in \mathcal{G}$ and $H \in \mathcal{F} = \Big\{$Bipartite graphs, Chordal graphs, Cographs, $\{P_5, K_3\}$-free graphs, $\{P_5, Paw\}$-free graphs, Complement of bipartite graphs, $\{P_5, K_4, Kite, Bull\}$-free graphs, connected $\{P_6, C_5, \overline{P_5}, K_{1,3}\}$-free graphs which contain an induced C_6, $\mathbb{K}[C_5](m_1, m_2, \ldots, m_5)$, $\{P_5, C_4\}$-free graphs, connected $\{P_5, \overline{P_2 \cup P_3}, \overline{P_5}, Dart\}$-free graphs which contain an induced $C_5\Big\}$, then $G[H]$ is k-indicated colorable for every $k \geq \chi(G[H])$.

Notations and terminologies not mentioned here are as in [9].

2   Indicated Coloring of Lexicographic Product of Graphs

Let us start Sect. 2 by recalling a result proved in [8].

Theorem 1   ([8]). *Any graph G is k-indicated colorable for all $k \geq \mathrm{col}(G)$.*

Theorem 2, gives a relation between the indicated coloring of $G[H]$ and the coloring number of G and H.

Theorem 2. *For any graphs G and H, $G[H]$ is k-indicated colorable for all $k \geq \mathrm{col}(G)\mathrm{col}(H)$.*

By Theorem 1, we know that $\mathrm{col}(G[H])$ is an upper bound for $\chi_i(G[H])$. Without much difficulty, one can show that $\mathrm{col}(G)\mathrm{col}(H)$ is a better upper bound for $\chi_i(G[H])$, that is, $\mathrm{col}(G[H]) - \mathrm{col}(G)\mathrm{col}(H)$ can be arbitrarily large.

We now define a family $\mathcal{H}$ of graphs. A graph G belongs to $\mathcal{H}$ if Ann has a winning strategy using $\chi_i(G)$ colors which she can follow until Ben uses $\chi_i(G)$ colors for the vertices of G and for the remaining vertices she has a way of extending this to a winning strategy using k colors, for any $k \geq \chi_i(G)$.

Let us now consider the indicated coloring of the lexicographic product of any graph G with a graph $H \in \mathcal{H}$ with $\chi(H) = \chi_i(H)$.

Theorem 3. *For any graph G and for any graph $H \in \mathcal{H}$ with $\chi(H) = \chi_i(H) = \ell$, $G[H]$ is k-indicated colorable if and only if $G[K_\ell]$ is k-indicated colorable. In particular, $\chi_i(G[H]) = \chi_i(G[K_\ell])$.*

Proof. Let G be any graph and $H \in \mathcal{H}$ be a graph with $\chi(H) = \chi_i(H) = \ell$ whose vertices are $u_1, u_2, \ldots, u_n$ and $v_1, v_2, \ldots, v_{n'}$ respectively. Let us first assume that $G[K_\ell]$ is k-indicated colorable and let $st_{G[K_\ell]}$ denote a winning strategy of Ann for $G[K_\ell]$ using k colors. Also, let st_H be a winning strategy of Ann for H using ℓ colors. Corresponding to the strategy st_H of H, for $1 \leq i \leq n$, Ann can get a winning strategy for H_{u_i}, by presenting the vertex (u_i, v) whenever v is presented in the strategy st_H. Let us call this winning strategy of H_{u_i} as st_{H_i}. Using the strategies $st_{G[K_\ell]}$ and st_{H_i}, for $1 \leq i \leq n$, we shall construct a winning strategy for Ann for the graph $G[H]$ using k colors as follows.

In $st_{G[K_\ell]}$, if the first vertex presented by Ann belongs to $K_{\ell_{u_i}}$, for some i, $1 \leq i \leq n$, then let Ann present the first vertex from H_{u_i} by following the strategy st_{H_i} of H_{u_i}. If Ben colors it with a color, say c_1, then we continue with the strategy $st_{G[K_\ell]}$ by assuming that the color c_1 is given to the vertex which was presented in $K_{\ell_{u_i}}$. If the second vertex presented by Ann in the strategy $st_{G[K_\ell]}$ belongs to $K_{\ell_{u_j}}$, for some j (not necessarily distinct from i), $1 \leq j \leq n$, then as per the strategy of st_{H_j}, let Ann present the vertices of H_{u_j} until a new color is given by Ben to a vertex in H_{u_j}, say c_2. That is, if Ann presents the vertices from the same H_{u_i}, then Ann will continue presenting the vertices until a vertex from a new color class in H_{u_i} is presented. Instead, if Ann presents a vertex from H_{u_j}, $i \neq j$ and $1 \leq j \leq n$, then that vertex will be a vertex from a new color class in H_{u_j}. This is because this is the first vertex presented from H_{u_j}. Then we continue with the strategy $st_{G[K_\ell]}$ by assuming that the color c_2 is given to the vertex presented by Ann in $K_{\ell_{u_j}}$. In general, if the vertex presented by Ann in $st_{G[K_\ell]}$ belongs to $K_{\ell_{u_r}}$, for some r, $1 \leq r \leq n$, then in $G[H]$, let Ann present the vertices in H_{u_r} by continuing with the strategy st_{H_r}, until a new color is given by Ben to a vertex in H_{u_r}, say c_r. Now we shall continue with the strategy $st_{G[K_\ell]}$ by assuming that the color c_r is given to the vertex presented by Ann in $K_{\ell_{u_r}}$. Repeat this process until all the vertices in $G[K_\ell]$ have been presented using the strategy $st_{G[K_\ell]}$. While following this strategy in $G[H]$, suppose for some p, q, $1 \leq p \leq n$, $1 \leq q \leq n'$, Ben creates a block vertex (u_p, v_q) in $G[H]$. Then (u_p, v_q) must be adjacent to all the k colors. According to the Ann strategy for $G[H]$, if a vertex of $G[H]$ in H_{u_p} is adjacent with a color, then there exists a vertex of $G[K_\ell]$ in $K_{\ell_{u_p}}$ which is adjacent with the same color. Thereby, there exists an uncolored vertex of $G[K_\ell]$ in $K_{\ell_{u_p}}$ which is a block vertex, a contradiction to $st_{G[K_\ell]}$ being a winning strategy of $G[K_\ell]$. So, Ben cannot create a block vertex in $G[H]$ when Ann follows this strategy.

At this stage, that is, when all the vertices in $G[K_\ell]$ have been presented using the strategy $st_{G[K_\ell]}$ as shown above, we see that the number of colors used in H_{u_i}, for $1 \leq i \leq n$, will be exactly ℓ. Also there maybe some uncolored vertices left in $G[H]$. For $1 \leq i, j \leq n$, the colors given to the vertices of H_{u_i} cannot be given to the vertices of H_{u_j}, for any u_j such that $u_i u_j \in E(G)$. Thus Ben has at least ℓ colors available to color the remaining uncolored vertices of H_{u_i}. Also by our assumption that $H \in \mathcal{H}$, even if the number of colors available for Ben is $\ell' \geq \ell$, Ann will still have a winning strategy to present the remaining uncolored vertices of H_{u_i} using ℓ' colors. Hence $G[H]$ is k-indicated colorable.

Next, if $G[H]$ is k-indicated colorable, by using similar technique one can show that $G[K_\ell]$ is k-indicated colorable. $\quad\blacksquare$

3 Consequences of Theorem 3

Let us recall some of the results shown in [3,6] and [8].

Theorem 4 ([8]). *Let $G = G_1 \cup G_2$. If G_1 is k_1-indicated colorable for every $k_1 \geq \chi_i(G_1)$ and G_2 is k_2-indicated colorable for every $k_2 \geq \chi_i(G_2)$, then $\chi_i(G) = \max\{\chi_i(G_1), \chi_i(G_2)\}$ and G is k-indicated colorable for every $k \geq \chi_i(G)$.*

Theorem 5 ([3,6,8]). *Let $\mathcal{F} = \Big\{$ Bipartite graphs, Chordal graphs, Cographs, $\{P_5, K_3\}$-free graphs, $\{P_5, Paw\}$-free graphs, Complement of bipartite graphs, $\{P_5, K_4, Kite, Bull\}$-free graphs, connected $\{P_6, C_5, \overline{P_5}, K_{1,3}\}$-free graphs which contain an induced C_6, $\mathbb{K}[C_5](m_1, m_2, \ldots, m_5)$, $\{P_5, C_4\}$-free graphs, connected $\{P_5, \overline{P_2 \cup P_3}, \overline{P_5}, Dart\}$-free graphs which contain an induced $C_5 \Big\}$. If $G \in \mathcal{F}$, then G is k-indicated colorable for all $k \geq \chi(G)$.*

In [3,6,8], if one closely observes the proofs of the families of graphs in $\mathcal{F}$ while showing that they are k-indicated colorable for every $k \geq \chi(G)$, we can see that the winning strategy of Ann will be independent of the choice of k. Hence any graph in $\mathcal{F}$ is also a graph in $\mathcal{H}$.

Proposition 1 shows that complete expansion of tree-free graphs or cycle-free graphs is again tree-free or cycle-free respectively.

Proposition 1. *Let T be a tree on at least 3 vertices and let G be a T-free graph or a C_ℓ-free graph, $\ell \geq 4$. Then the graph $\mathbb{K}[G]$ is also either T-free or C_ℓ-free respectively.*

As a consequence of Theorem 5 and Proposition 1, we obtain Corollary 1.

Corollary 1. *If G is a chordal graph or a cograph (or) a $\{P_5, C_4\}$-free graph, then $\mathbb{K}[G]$ is k-indicated colorable for all $k \geq \chi(\mathbb{K}[G])$.*

Without much difficulty, one can observe that the complete expansion of the independent expansion of any graph G is k indicated colorable whenever $\mathbb{K}[G]$ is k-indicated colorable. As a consequence, we have Theorem 6.

Theorem 6. *Let G be a complete multipartite graph. Then $\mathbb{K}[G]$ is k-indicated colorable for all $k \geq \chi(\mathbb{K}[G])$.*

4   Conclusion

For $G \in \{$chordal graph, cograph, $\{P_5, C_4\}$-free graph, complete multipartite graph$\}$ and $H \in \mathcal{F}$, $G[H]$ is k-indicated colorable for all $k \geq \chi(G[H])$.

Acknowledgment. For the first author, this research was supported by Post Doctoral Fellowship, Indian Institute of Technology, Palakkad. And for the second author, this research was supported by SERB DST, Government of India, File no: EMR/2016/007339. Also, for the third author, this research was supported by the UGC-Basic Scientific Research, Government of India, Student id: gokulnath.res@pondiuni.edu.in.

References

1. Choudum, S.A., Karthick, T.: Maximal cliques in $\{P_2 \cup P_3, C_4\}$-free graphs. Discrete Math. **310**, 3398–3403 (2010)
2. Francis, P., Francis Raj, S.: Indicated coloring of Cartesian product of some families of graphs. (to appear in Ars Combinatoria)
3. Francis, P., Francis Raj, S., Gokulnath, M.: On indicated coloring of some classes of graphs. Graphs and Combinatorics **35**(5), 1105–1127 (2019)
4. Francis Raj, S., Pandiya Raj, R., Patil, H.P.: On indicated chromatic number of graphs. Graphs and Combinatorics **33**, 203–219 (2017)
5. Geller, D.P., Stahl, S.: The chromatic number and other functions of the lexicographic product. J. Comb. Theor. Ser. B **19**, 87–95 (1975)
6. Grzesik, A.: Indicated coloring of graphs. Discrete Math. **312**, 3467–3472 (2012)

7. Jensen, T.R., Toft, B.: Graph Coloring Problems. Wiley, New York (1995)
8. Pandiya Raj, R., Francis Raj, S., Patil, H.P.: On indicated coloring of graphs. Graphs and Combinatorics **31**, 2357–2367 (2015)
9. West, D.B.: Introduction to Graph Theory. Prentice-Hall of India Private Limited, Upper Saddle River (2005)

Smallest C_{2l+1}-Critical Graphs
of Odd-Girth $2k + 1$

Laurent Beaudou[1], Florent Foucaud[2], and Reza Naserasr[3(✉)]

[1] National Research University, Higher School of Economics,
3 Kochnovsky Proezd, Moscow, Russia
lbeaudou@hse.ru

[2] Univ. Bordeaux, Bordeaux INP, CNRS, LaBRI, UMR5800, 33400 Talence, France
florent.foucaud@gmail.com

[3] Université de Paris, IRIF, CNRS, 75013 Paris, France
reza@irif.fr

Abstract. Given a graph H, a graph G is called H-critical if G does not admit a homomorphism to H, but any proper subgraph of G does. Observe that K_{k-1}-critical graphs are the classic k-(colour)-critical graphs. This work is a first step towards extending questions of extremal nature from k-critical graphs to H-critical graphs. Besides complete graphs, the next classic case is odd cycles. Thus, given integers $l \geq k$ we ask: what is the smallest order $\eta(k, l)$ of a C_{2l+1}-critical graph of odd-girth at least $2k + 1$? Denoting this value by $\eta(k, l)$, we show that $\eta(k, l) = 4k$ for $l \leq k \leq \frac{3l+i-3}{2}$ ($2k = i \bmod 3$) and that $\eta(3, 2) = 15$. The latter is to say that a smallest graph of odd-girth 7 not admitting a homomorphism to the 5-cycle is of order 15 (there are at least 10 such graphs on 15 vertices).

1 Introduction

A k-critical graph is a graph which is k-chromatic but any proper subgraph of it is $(k - 1)$-colourable. Extremal questions on critical graphs are a rich source of research in graph theory. Many well-known results and conjectures are about this subject, see for example [4,6–8,15]. Typical questions are for example:

Problem 1. What is the smallest possible order of a k-critical graph having a certain property, such as low clique number, high girth or high odd-girth?

For example, Erdős' proof of existence of graphs of high girth and high chromatic number [5] is a starting point for Problem 1. This fact implies that each of the above questions have a finite answer. The specific question of the smallest 4-critical graph without a triangle has received considerable attention: Grötzsch built a graph on 11 vertices which is triangle-free and not 3-colourable.

This work is supported by the IFCAM project Applications of graph homomorphisms (MA/IFCAM/18/39) and by the ANR project HOSIGRA (ANR-17-CE40-0022).

M. Changat and S. Das (Eds.): CALDAM 2020, LNCS 12016, pp. 184–196, 2020.
https://doi.org/10.1007/978-3-030-39219-2_16

Harary [11] showed that any such graph must have at least 11 vertices, and Chvátal [3] showed that the Grötzsch graph is the only one with 11 vertices.

Every graph with no odd cycle being 2-colourable, in the context of colouring, it is of interest to consider the *odd-girth*, the size of a smallest odd-cycle of a graph (rather than the girth). Extending construction of Grötzsch's graph, Mycielski [14] introduced the construction, now knows as the Mycielski construction, to increase the chromatic number without increasing the clique number. A generalization of this construction is used to build 4-critical graphs of high odd-girth, more precisely every *generalized Mycielski construction on* C_{2k+1}, denoted $M_k(C_{2k+1})$ is a 4-critical graph. The graph $M_2(C_5)$ is simply the classic Mycielski construction for C_5, that is, the Grötzsch graph. $M_k(C_{2k+1})$ has odd-girth $2k+1$, and several authors (starting with Payan [18]) showed that $M_k(C_{2k+1})$ is 4-chromatic for any $k \geq 1$ and in fact 4-critical, thus providing an upper bound of $2k^2 + k + 1$ for the minimum order of a 4-critical graph of odd-girth at least $2k + 1$. We refer to [10,16,19,20] for several other proofs. Among these authors, Ngoc and Tuza [16] asked whether this upper bound of $2k^2 + k + 1$ is essentially optimal. The best known lower bound is due to Jiang [13], who proved the bound of $(k-1)^2 + 2$, which establishes the correct order of magnitude at $\Theta(k^2)$ (see [17] for an earlier but weaker lower bound).

The current work is a first step towards generalizing these extremal questions for k-critical graphs to H-critical graphs, defined using the terminology of homomorphisms. A *homomorphism* of a graph G to a graph H is a vertex-mapping that preserves adjacency, i.e., a mapping $\psi : V(G) \to V(H)$ such that if x and y are adjacent in G, then $\psi(x)$ and $\psi(y)$ are adjacent in H. If there exists a homomorphism of G to H, we may write $G \to H$ and we may say that G is H-colourable. In the study of homomorphisms, it is usual to work with the *core* of a graph, that is, a minimal subgraph which admits a homomorphism from the graph itself. It is not difficult to show that a core of any graph is unique up to isomorphism. A graph is said to be *a core* if it admits no homomorphism to a proper subgraph. We refer to the book [12] for a reference on these notions.

It is a classic fact that homomorphisms generalize proper vertex-colourings. Indeed a homomorphism of G to K_k is equivalent to a k-colouring of G. However, the extension of the notion of colour-criticality to a homomorphism-based one has been almost forgotten. As defined by Catlin [2], for a graph H, (we may assume H is a core), a graph G is said to be H-*critical* if G does not have a homomorphism to H but any proper subgraph of G does. Thus:

Observation 2. *A graph G is k-critical if and only if it is K_{k-1}-critical.*

This gives a large number of interesting extremal questions. By Observation 2, these questions are well-studied when H is a complete graph. The next most important family to be considered is when H is an odd cycle. This is the goal of this work. More precisely we ask:

Problem 3. Given positive integers k, l, what is the smallest order $\eta(k, l)$ of a C_{2l+1}-critical graph of odd-girth at least $2k + 1$?

In this work, we study Problem 3 when $l \geq 2$. As we discuss in Sect. 2, it follows from a theorem of Gerards [9] that $\eta(k,l) \geq 4k$ whenever $l \leq k$, and $\eta(k,k) = 4k$. We prove (in Sect. 3) that, surprisingly, $\eta(k,l) = 4k$ whenever $l \leq k \leq \frac{3l+i-3}{2}$ (with $2k = i \mod 3$). We then prove (in Sect. 4) that $\eta(3,2) = 15$. We conclude with further research questions in the last section. Table 1 summarizes the known bounds for Problem 3 and small values of k and l.

Note that the value of $\eta(3,2)$ indeed was the initial motivation of this work. In [1], we use the fact that $\eta(3,2) = 15$ to prove that if a graph B of odd-girth 7 has the property that any series-parallel graph of odd-girth 7 admits a homomorphism to B, then B has at least 15 vertices.

Table 1. Known values/bounds on the smallest order of a not C_{2l+1}-colourable graph of odd-girth $2k + 1$. Bold values are proved in this paper.

	$k = 1$	$k = 2$	$k = 3$	$k = 4$	$k = 5$	$k = 6$	$k = 7$	$k = 8$
$l = 1$	4	11[11]	**15**–22 [Th. 12]–[18]	17–37 [Co. 13]–[18]	20–56 [Co. 8]–[18]	27–79 [13]–[18]	38–106 [13]–[18]	51–137 [13]–[18]
$l = 2$	3	8 [Co. 6]	**15** [Th. 12]	17–37 [Co. 13]–[18]	20–56 [Co. 8]–[18]	24–79 [Co. 8]–[18]	28–106 [Co. 8]–[18]	32–137 [Co. 8]–[18]
$l = 3$	3	5	**12** [Co. 6]	**16** [Th. 10]	20–56 [Co. 8]–[18]	24–79 [Co. 8]–[18]	28–106 [Co. 8]–[18]	32–137 [Co. 8]–[18]
$l = 4$	3	5	7	**16** [Co. 6]	20 [Th. 10]	24–79 [Co. 8]–[18]	28–106 [Co. 8]–[18]	32–137 [Co. 8]–[18]
$l = 5$	3	5	7	9	20 [Co. 6]	24 [Th. 10]	28 [Th. 10]	32–137 [Co. 8]–[18]
$l = 6$	3	5	7	9	11	24 [Co. 6]	28 [Th. 10]	32 [Th. 10]
$l = 7$	3	5	7	9	11	13	28 [Co. 6]	32 [Th. 10]
$l = 8$	3	5	7	9	11	13	15	32 [Co. 6]

2   Preliminaries

This section is devoted to introduce useful preliminary notions and results.

Circular chromatic number. We recall some basic notions related to circular colourings. For a survey on the matter, consult [21]. Given two integers p and q with $\gcd(p,q) = 1$, the *circular clique* $C(p,q)$ is the graph on vertex set $\{0, \ldots, p-1\}$ with i adjacent to j if and only if $q \leq |i - j| \leq p - q$. A homomorphism of a graph G to $C(p,q)$ is called a (p,q)-colouring, and the *circular chromatic number* of G, denoted $\chi_c(G)$, is the smallest rational p/q such that G has a (p,q)-colouring. Since $C(p,1)$ is the complete graph K_p, we have $\chi_c(G) \leq \chi(G)$. On the other hand $C(2l + 1, l)$ is the cycle C_{2l+1}. Thus C_{2l+1}-colourability is about deciding whether $\chi_c(G) \leq 2 + \frac{1}{l}$. It is a well-known fact that $C(p,q) \to C(r,s)$ if and only if $\frac{p}{q} \leq \frac{r}{s}$ (e.g. see [21]), in particular we will use the fact that $C(12,5) \to C_5$.

Odd-K_4's and a theorem of Gerards. The following notion will be central in our proofs. An *odd-K_4* is a subdivision of the complete graph K_4 where each of the four triangles of K_4 has become an odd-cycle [9]. Furthermore, we call it a $(2k+1)$-*odd-K_4* if each such cycle has length exactly $2k+1$. Since subdivided triangles are the only odd-cycles of an odd-K_4, the odd-girth of a $(2k+1)$-odd-K_4 is $2k+1$. The following is an easy fact about odd-K_4's whose proof we leave as an exercise.

Proposition 4. *Let K be an odd-K_4 of odd-girth at least $2k+1$. Then, K has order at least $4k$, with equality if and only if K is a $(2k+1)$-odd-K_4. Furthermore, in the latter case any two disjoint edges of K_4 are subdivided the same number of times when constructing K.*

A $(2k+1)$-odd-K_4 is, more precisely, referred to as an (a, b, c)-*odd-K_4* if three edges of a triangle of K_4 are subdivided into paths of length a, b and c respectively (by Proposition 4 this is true for all four triangles). Note that while the terms "odd-K_4" or "$(2k+1)$-odd-K_4" refer to many non-isomorphic graphs, an (a, b, c)-odd-K_4 ($a + b + c = 2k + 1$) is unique up to a relabeling of vertices.

An *odd-K_3^2* is a graph obtained from three disjoint odd-cycles and three disjoint paths (possibly of length 0) joining each pair of cycles [9]. Thus, in such graph, any two of the three cycles have at most one vertex in common (if the path joining them has length 0). Hence, an odd-K_3^2 of odd-girth at least $2k+1$ has order at least $6k$.

Theorem 5. (Gerards [9]). *If G has neither an odd-K_4 nor an odd-K_3^2 as a subgraph, then it admits a homomorphism to its shortest odd-cycle.*

Corollary 6. *For any positive integer k, we have $\eta(k, k) = 4k$.*

Proof. Consider a C_{2k+1}-critical graph G of odd-girth $2k + 1$. It follows from Theorem 5 that G contains either an odd-K_4, or an odd-K_3^2. If it contains the latter, then G has at least $6k$ vertices. Otherwise, G must contain an odd-K_4 of odd-girth at least $2k+1$, and then by Proposition 4, G has at least $4k$ vertices. This shows that $\eta(k, k) \geq 4k$.

Moreover, any $(2k+1)$-odd-K_4 has order $4k$ (by Proposition 4) and admits no homomorphism to C_{2k+1}, showing that $\eta(k, k) \leq 4k$.   □

Since C_{2l+3} maps to C_{2l+1} and by transitivity of homomorphisms, a graph with no homomorphism to C_{2l+1} also has no homomorphism to C_{2l+3}. Thus:

Observation 7. *Let k, l be two positive integers. We have $\eta(k, l) \geq \eta(k, l + 1)$.*

We obtain this immediate consequence of Corollary 6 and Observation 7:

Corollary 8. *For any two integers k, l with $k \geq l \geq 1$, we have $\eta(k, l) \geq 4k$.*

3 Rows of Table 1

In this section, we study the behavior of $\eta(k,l)$ when l is a fixed value, that is, the behavior of each row of Table 1. Note again that whenever $l \geq k+1$, $\eta(k,l) = 2k+1$. As mentioned before, the first row (i.e. $l = 1$) is about the smallest order of a 4-critical graph of odd-girth $2k+1$ and we know $\eta(k,1) = \Theta(k^2)$ [17].

It is not difficult to observe that for $k \geq l$, the function $\eta(k,l)$ is strictly increasing, in fact with a little bit of effort we can even show the following.

Proposition 9. *For $k \geq l$, we have $\eta(k+1,l) \geq \eta(k,l) + 2$.*

Proof. Let G be a C_{2l+1}-critical graph of odd-girth $2k+3$ and order $\eta(k+1,l)$. Consider any $(2k+3)$-cycle $v_0 \cdots v_{2k+2}$ of G, and build a smaller graph by identifying v_0 with v_2 and v_1 with v_3. It is not difficult to check that the resulting graph has odd-girth exactly $2k+1$ and does not map to C_{2l+1} (otherwise, G would), proving the claim.   □

While we expect that for a fixed l, $\eta(k,l)$ grows quadratically in terms of k, we show, somewhat surprisingly, that at least just after the threshold of $k = l$, the function $\eta(k,l)$ only increases by 4 when l increases by 1, implying that Proposition 9 cannot be improved much in this formulation. More precisely, we have the following theorem.

Theorem 10. *For any $k,l \geq 3$ and $l \leq k \leq \frac{3l+i-3}{2}$ (where $2k = i$ mod 3), we have $\eta(k,l) = 4k$.*

To prove this theorem, we give a family of C_{2l+1}-critical odd-$K4$'s which are of odd-girth $2k+1$. This is done in the next theorem, after which we give a proof of Theorem 10.

Given a graph G, a *thread* of G is a path in G where the internal vertices have degree 2 in G. When G is clear from the context, we simply use the term thread.

Theorem 11. *Let $p \geq 3$ be an integer. If p is odd, any (a,b,c)-odd-K_4 with $(a,b,c) \in \{(p-1,p-1,p),(p,p,p)\}$ has no homomorphism to C_{2p+1}. If p is even, any $(p-1,p,p)$-odd-K_4 has no homomorphism to C_{2p+1}.*

Proof. Let $(a,b,c) \in \{(p-1,p-1,p),(p,p,p),(p-1,p,p)\}$ and let K be an (a,b,c)-odd-K_4. Let t, u, v, w be the vertices of degree 3 in K with the tu-thread of length a, the uv-thread of length b and the tv-thread of length c. We now distinguish two cases depending on the parity of p and the values of (a,b,c).

Case 1. Assume that p is odd and $(a,b,c) \in \{(p-1,p-1,p),(p,p,p)\}$. By contradiction, we assume that there is a homomorphism h of K to C_{2p+1}. Then, the cycle C_{tvw} formed by the union of the tv-thread, the vw-thread and the tw-thread is an odd-cycle of length $a + b + c$. Therefore, its mapping by h to C_{2p+1} must be *onto*. Thus, u has the same image by h as some vertex u' of C_{tvw}. Note that u' is not one of t, v or w, indeed by identifying u with any of these vertices we obtain a graph containing an odd-cycle of length p or $2p-1$; thus, this identification cannot be extended to a homomorphism to C_{2p+1}. Therefore, u' is an internal vertex of one of the three maximal threads in C_{tvw}. Let C_u be the odd-cycle of length $a+b+c$ containing u and u'. After identifying u and u', C_u is transformed into two cycles, one of them being odd. If $(a,b,c) = (p,p,p)$, then C_u has length $3p$. Then, the two newly created cycles have length at least $p+1$, and thus at most $2p-1$. If $(a,b,c) = (p-1,p-1,p)$, then C_u has length $3p-2$, and the two cycles have length at least p and at most $2p-2$. In both cases, we have created an odd-cycle of length at most $2p-1$. Hence, this identification cannot be extended to a homomorphism to C_{2p+1}, a contradiction.

Case 2. Assume that p is even and $(a,b,c) = (p-1,p,p)$, and that h is a homomorphism of K to C_{2p+1}. Again, the image of C_{tvw} by h is onto, and u has the same image as some vertex u' of C_{tvw}. If $u' = t$, identifying u and u' produces an odd $(p-1)$-cycle, a contradiction. If $u' \in \{v,w\}$, then we get a $(2p-1)$-cycle, a contradiction. Thus, u' is an internal vertex of one of the three maximal threads in C_{tvw}. Let C_u be the odd-cycle of length $3p-1$ containing u and u'. As in Case 1, after identifying u and u', C_u is transformed into two cycles, each of length at least p and at most $2p-1$; one of them is odd, a contradiction. $\qquad\square$

We note that Theorem 11 is tight, in the sense that if p is odd and $(a,b,c) \in \{(p-1,p-1,p),(p,p,p)\}$ or if p is even and $(a,b,c) = (p-1,p,p)$, any (a,b,c)-odd-K_4 has a homomorphism to C_{2p-1}.

We can now prove Theorem 10.

Proof (Proof of Theorem 10). By Corollary 8, we know that $\eta(k,l) \geq 4k$. We now prove the upper bound. Recall that $\eta(k,l) \leq \eta(k,l-1)$. If $2k = 0 \bmod 3$, then $p = \frac{2k+3}{3}$ is an odd integer, and $p \leq l$. By Theorem 11, a $(p-1,p-1,p)$-odd-K_4, which has order $6p-6 = 4k$, has no homomorphism to C_{2p+1}, and thus $\eta(k,l) \leq \eta(k,p) \leq 4k$. Similarly, if $2k = 1 \bmod 3$, then $p = \frac{2k+2}{3}$ is an even integer, and $p \leq l$. By Theorem 11, a $(p-1,p,p)$-odd-K_4, which has order $6p-4 = 4k$, has no homomorphism to C_{2p+1}, and thus $\eta(k,l) \leq \eta(k,p) \leq 4k$. Finally, if $2k = 2 \bmod 3$, then $p = \frac{2k+1}{3}$ is an odd integer, and $p \leq l$. By Theorem 11, a (p,p,p)-odd-K_4, which has order $6p-2 = 4k$, has no homomorphism to C_{2p+1}, and thus $\eta(k,l) \leq \eta(k,p) \leq 4k$. $\qquad\square$

4   The Value of $\eta(3,2)$

We now determine $\eta(3,2)$, which is not covered by Theorem 11. By Corollary 8, we know that $\eta(3,2) \geq 12$. In fact, we will show that $\eta(3,2) = 15$. Using a computer search, Gordon Royle (private communication, 2016) has found that there are at least ten graphs of order 15 and odd-girth 7 that do not admit a homomorphism to C_5. For example, see the three graphs of Fig. 1. Thus, $\eta(3,2) \leq 15$. Next, we prove that this upper bound is tight.

Theorem 12. *Any graph G of order at most 14 and odd-girth at least 7 admits a homomorphism to C_5, and thus $\eta(3,2) = 15$.*

Proof. We consider a C_5 on the vertex set $\{0,1,2,3,4\}$ where vertex i is adjacent to vertices $i+1$ and $i-1$ (modulo 5). Thus, in the following, to give a C_5-colouring we will give a colouring using elements of $\{0,1,2,3,4\}$ where adjacent pairs are mapped into (cyclically) consecutive elements of this set.

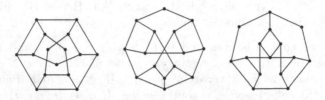

Fig. 1. Three C_5-critical graphs of order 15 and odd-girth 7.

Given a graph G and a vertex v of it, we partition $V(G)$ into four sets $\{v\}, N_1(v), N_2(v)$ and $N_{3+}(v)$ where $N_1(v)$ (respectively $N_2(v)$) designates the set of vertices at distance exactly 1 (respectively 2) of v, and $N_{3+}(v)$ the vertices at distance 3 or more of v. A proper 3-colouring of $G[N_{3+}(v)]$ using colours c_1, c_2 and c_3 is said to be v-*special* if:

(i) each vertex with colour c_3 is an isolated vertex of $G[N_{3+}(v)]$,
(ii) no vertex from $N_2(v)$ sees both colours c_1 and c_2.

A key observation is the following: given any graph G, if for some vertex v of G, there exists a v-special colouring of $G[N_{3+}(v)]$, then G maps to C_5. Such a homomorphism is given by mapping c_1-vertices to 0, c_2-vertices to 1 and c_3-vertices to 3, and then extending as follows:

- for any vertex u in $N_2(v)$, if u has a c_1-neighbour, map it to 4; otherwise, map it to 2,
- all vertices of $N_1(v)$ are mapped to 3,
- vertex v is mapped to 2 or 4.

Now, let G be a minimal counterexample to Theorem 12. We first collect a few properties of G. The previous paragraph allows us to state our first claim.

Claim 12.A. *For no vertex v of G there is a v-special colouring of $G[N_{3+}(v)]$.*

Since G is a minimal counterexample, it cannot map to a subgraph of itself (which would be a smaller counterexample):

Claim 12.B. *Graph G is a core. In particular, for any two vertices u and v of G, $N(u) \nsubseteq N(v)$.*

Recall that a *walk* between two vertices u and v is a sequence of (not necessarily distinct) vertices starting with u and ending with v, where two consecutive vertices in the sequence are adjacent. A walk between u and v is an *uv-walk*, and a *k-walk* is a walk with $k + 1$ vertices.

Claim 12.C. *For any two distinct vertices u and v of G, there is a uv-walk of length 5.*

Proof of claim. If not, identifying u and v would result in a smaller graph of yet odd-girth 7 which does not map to C_5, contradicting the minimality of G. $\square$

Claim 12.D. *Graph G has no thread of length 4 or more.*

Proof of claim. Once again, by minimality of G, if we remove a thread of length 4, the resulting graph maps to C_5. But since there is a walk of length 4 between any two vertices of C_5, this mapping could easily be extended to G. $\square$

Now we can state a more difficult claim.

Claim 12.E. *There is no vertex of G of degree 4 or more, nor a vertex of degree exactly 3 with a second neighbourhood of size 5 or more.*

Proof of claim. For a contradiction, suppose that a vertex v has degree 4 or more, or has degree 3 and a second neighbourhood of size 5 or more.

By Claim 12.B, the neighbours of v should have pairwise distinct neighbourhoods, so that even if v has degree 4 or more, we must have $|N_2(v)| \geq 4$. Thus, by a counting argument (recall that G has at most 14 vertices), $N_{3+}(v)$ has size at most 5. Since G has odd-girth 7, this means $G[N_{3+}(v)]$ is bipartite.

Suppose $G[N_{3+}(v)]$ has only one non-trivial connected component. Consider any proper 3-colouring of $G[N_{3+}(v)]$ such that colours c_1 and c_2 are used for the non-trivial connected component, and colour c_3 is used for isolated vertices. Then, no vertex of $N_2(v)$ can see both colours c_1 and c_2, as this would result in a short odd-cycle in G. Thus, any such colouring is v-special. Such colouring clearly exists, hence by Claim 12.A, we derive that $G[N_{3+}(v)]$ has at least two non-trivial components. Since it has order at most 5, it must have exactly two.

Assume now that both non-trivial connected components are isomorphic to K_2. Consider all proper 3-colourings of $G[N_{3+}(v)]$ such that colours c_1 and c_2 are used for the copies of K_2 (the potentially remaining vertex being coloured with c_3). One may check that since by Claim 12.A, none of them is v-special, and hence there is a short odd-cycle in G, which is a contradiction.

Hence, $G[N_{3+}(v)]$ is isomorphic to the disjoint union of K_2 and $K_{2,1}$. Let u_1 and u_2 be the vertices of K_2, and u_3, u_4 and u_5 be the vertices of $K_{2,1}$ such that u_4 is the central vertex.

Let φ_1 and φ_2 be two proper 2-colourings of $G[N_{3+}(v)]$ as follows: $\varphi_1(u_1) = \varphi_1(u_3) = \varphi_1(u_5) = c_1$ and $\varphi_1(u_2) = \varphi_1(u_4) = c_2$, $\varphi_2(u_2) = \varphi_2(u_3) = \varphi_2(u_5) = c_1$ and $\varphi_2(u_1) = \varphi_2(u_4) = c_2$. Since φ_1 is not v-special, there is either a vertex t_1 adjacent to u_2 and u_3 (by considering the symmetry of u_3 and u_5), or a vertex t'_1 adjacent to u_1 and u_4. Similarly, since φ_2 is not v-special, either there is a vertex t_2 adjacent to u_1 and one to u_3 and u_5, or there is a vertex t_2' adjacent to u_2 and u_4. The existence of some t_i' (for $i = 1$ or 2), together with any of these three remaining vertices would result in a short odd-cycle in G. Thus, t_1 and t_2 must exist and more precisely, t_2 has to be a neighbour of u_5.

Next, we show that u_4 has degree at least 3. Suppose not, then it has degree exactly 2. Let φ_3 be a partial C_5-colouring of G defined as follows: $\varphi_3(u_1) = \varphi_3(u_5) = 0, \varphi_3(u_2) = 1, \varphi_3(u_3) = 3$ and $\varphi_3(u_4) = 4$. Then, no vertex in $N_2(v)$ sees both 0 and 1 (by odd-girth arguments). Thus, we can extend φ_3 to $N_2(v)$ using only colours 2 and 4 on these vertices. Then, all vertices of $N_1(v)$ can be mapped to 3 and v can be mapped to 2 and $G \to C_5$, a contradiction.

Hence, there exists a vertex t_3 in $N_2(v)$ which is adjacent to u_4. Note that, by the odd-girth condition, t_3 has no other neighbour in $G[N_{3+}(v)]$ and, in particular, it must be distinct from t_1 and t_2. Vertices t_1, t_2 and t_3 are in $N_2(v)$, so there are vertices s_1, s_2 and s_3 in $N_1(v)$ such that s_i is adjacent to t_i for i between 1 and 3. Moreover, vertices t_1, t_2 and t_3 are pairwise connected by a path of length 3. Therefore, their neighbourhoods cannot intersect, so that vertices s_1, s_2 and s_3 are distinct.

Now, consider the partial C_5-colouring φ_3 again. We may extend φ_3 to $N_2(v)$ by assigning colour 0 to neighbours of u_4, colour 4 to neighbours of u_1 and u_5, and colour 2 to the rest. If no vertex of $N_1(v)$ sees both colours 0 and 4 in $N_2(v)$, we may colour $N_1(v)$ with 1 and 3 and colour v with 2, which is a contradiction. Thus, there exists some vertex x in $N_1(v)$ seeing both colours 0 and 4 in $N_2(v)$. The only vertices with colour 0 in $N_2(v)$ are neighbours of u_4 so that x must be at distance 2 from u_4. Since there is no short odd-cycle in G, vertex x cannot be at distance 2 from u_5. Thus, it is at distance 2 from u_1. Let t_4 be the middle vertex of this path from x to u_1. Now t_4 is a neighbour of u_1 which is distinct from t_1, t_2 and t_3. By the symmetry between u_1 and u_2, there must be a fifth vertex t_5 in $N_2(v)$ which is a neighbour of u_2 and distinct from t_1, t_2, t_3 and t_4. Moreover, t_5 has a neighbour y in $N_1(v)$ that is at distance 2 from u_4. We can readily check that y is distinct from x, s_1 and s_2. Thus, $N_1(v)$ has size at least 4 and $N_2(v)$ has size at least 5, which is a contradiction with the order of G (which should be at most 14). This concludes the proof of Claim 12.E. $\square$

Claim 12.F. *G contains no 6-cycle.*

Proof of claim. Suppose, by contradiction, that G contains a 6-cycle C : $v_0, \ldots, v_5$. For a pair v_i and v_{i+2} (addition in indices are done modulo 6) of vertices of C, the 5-walk connecting them (see Claim 12.C) is necessarily a 5-path, which we denote by P^i. Furthermore, at most one inner-vertex of P^i may belong to C, and if it does, it must be a neighbour of v_i or v_{i+2} (one can check that otherwise, there is a short odd-cycle in G).

Assume first that none of the six paths P^i ($0 \leq i \leq 5$) has any inner-vertex on C. In this case and by Claim 12.E, we observe that the neighbours of v_i in P^i and P^{i+4} (additions in superscript are modulo 6) are the same. Let v_i' be this neighbour of v_i.

Vertices v_i', $i = 0, 1, \ldots, 5$ are all distinct, as otherwise we have a short odd-cycle in G. Let x and y by two internal vertices of P^1 distinct from v_0' and v_2'.

By our assumption, x and y are distinct from vertices of C. We claim that they are also distinct from v_i', $i = 0, \ldots, 5$. Vertex x is indeed distinct from v_0' and v_2' by the choice of P^0. It is distinct from v_1', v_3' and v_4' as otherwise there will be a short odd-cycle. Similarly, y is distinct from $v_0', v_1', v_2', v_4', v_5'$. For the same reason, we cannot simultaneously have $x = v_5'$ and $y = v_3'$. Finally, if we have $x = v_5'$ then $\{v_0', v_1, y, v_3, v_4'\} \subseteq N_2(x)$ and $d(x) \geq 3$, contradicting Claim 12.E. As a result, since $|V(G)| \leq 14$, x and y are internal vertices of all P^i's. But then it easy to find a short odd-cycle.

Hence, we may assume, without loss of generality, that P^1 has one inner-vertex on C, say $v_1' = v_0$. Let $v_0 x_1 x_2 x_3 v_3$ be the 4-path connecting v_0 and v_3 (recall that $x_i \notin C$ for $i = 1, 2, 3$). We next assume that P^5 does not have any inner-vertex in C. Then, no vertex of P^5 is a vertex from $\{x_1, x_2, x_3\}$, for otherwise we have a short odd-cycle in G. But then, v_0 violates Claim 12.E. Therefore, an inner-vertex of P^5 lies on C, say it is v_2 and we have $P^5 : v_5 y_1 y_2 y_3 v_2 v_1$. Then, P^1 and P^5 are vertex-disjoint, for otherwise we have a short odd-cycle in G. Remark that C together with the union of the paths P^i, $i = 0, \ldots, 5$, forms a $(2k+1)$-odd K_4, in fact it is a $(1, 2, 4)$-odd-K_4 (a subgraph of $C(12, 5)$). Now, because of the odd-girth of G, and by Claim 12.E, the only third neighbour of v_1, if any, is v_4 (and vice-versa). Furthermore, the set $\{x_1, x_2, x_3, y_1, y_2, y_3\}$ of vertices induces a subgraph matching the $x_i y_i$ with $i = 1, 2, 3$. If there is no additional vertex in G, then G has order 12 and it is a subgraph of $C(12, 5)$. But then, the circular chromatic number of G is at most $12/5$, implying that G has a homomorphism to C_5, a contradiction. Thus, G has order at least 13. Again by Claim 12.E, any of the two last potentially existing vertices of G can be adjacent only to x_2 or y_2. Without loss of generality (considering the symmetries of the graph), assume that x_2 has an additional neighbour, v. Then, either v is also adjacent to y_2 (then G has order 13), or v and y_2 have a common neighbour, say w. If v is adjacent to both x_2 and y_2, then there is no edge in the set $\{x_1, x_2, x_3, y_1, y_2, y_3\}$ (otherwise we have a short odd-cycle). But then, we exhibit a homomorphism of G to C_5: map x_3, y_1, v to 0; x_2, y_2 to 1; v_1, x_1, y_3 to 2; v_0, v_2, v_4 to 3; v_3 and v_5 to 4. This is a contradiction. Thus, v and y_2 have a common neighbour w, and both v and w are of degree 2. We now create a homomorphic image of G by identifying v with y_2 and w with x_2. Then, this image of G is a subgraph of $C(12, 5)$, and thus the circular chromatic number of G is at most $12/5$, implying that G has a homomorphism to C_5, a contradiction. This completes the proof of Claim 12.F. $(\square)$

Claim 12.G. G *contains no 4-cycle.*

Proof of claim. Assume by contradiction that G contains a 4-cycle $C : tuvw$. As in the proof of Claim 12.F, there must be two 5-paths $P_{tv} : t a_1 a_2 a_3 a_4 v$ and $P_{uw} : u b_1 b_2 b_3 b_4 w$ connecting t with v and u with w, respectively. Moreover, these two paths must be vertex-disjoint because of the odd-girth of G. Thus, the union of C, P_{tv} and P_{uw} forms a $(1, 1, 5)$-odd-K_4, K. By assumption on the odd-girth of G, any additional edge inside $V(K)$ must connect an internal vertex of Ptv to an internal vertex of P_{uw}. But any such edge would either create a short odd-cycle or a 6-cycle in G, the latter contradicting Claim 12.F. Thus, the only edges in $V(K)$ are those of K. If there is no additional vertex in G, we have two 5-threads in G, contradicting Claim 12.D; thus there is at least one additional vertex in G, say x, and perhaps a last vertex, y. Note that t, u, v and w are already, in K, degree 3-vertices with a second neighbourhood of size 4, thus by Claim 12.E none of t, u, v, w, a_1, a_4, b_1 and b_4 are adjacent to any vertex not in K. Thus $N(x), N(y) \subseteq \{a_2, a_3, b_2, b_3, x, y\}$.

Assume that some vertex not in K (say x) is adjacent to two vertices of K. Then, these two vertices must be a vertex on the tu-path of K (a_2 or a_3, without loss of generality it is a_2) and a vertex on the vw-path of K (b_2 or b_3). By the automorphism of K that swaps v and w and reverses the vw-path, without loss of generality we can assume that the second neighbour of x in K is b_3. Then, we claim that G has no further vertex. Indeed, if there is a last vertex y, since a_2 and b_3 have already three neighbours, y must be adjacent to at least two vertices among $\{a_3, b_2, x\}$. If it is adjacent to both a_3 and b_2, we have a 6-cycle, contradicting Claim 12.F; otherwise, y is of degree 2 but part of a 4-cycle, implying that G is not a core, a contradiction. Thus, G has order 13 and no further edge. We create a homomorphic image of G by identifying x and a_3. The obtained graph is a subgraph of $C(12, 5)$. Hence, G has circular chromatic number at most $12/5$ and a homomorphism to C_5, a contradiction.

Thus, any vertex not in K has at most one neighbour in K. Since G has no 4-thread, one vertex not in K (say x) is adjacent to one of a_2 or a_3 (without loss of generality, say a_2), and the last vertex, y, is adjacent to one of b_2 and b_3 (as before, by the symmetries of K we can assume it is b_2). Moreover, x and y must be adjacent, otherwise they both have degree 1. Also there is no further edge in G. But then, as before, we create a homomorphic image of G by identifying x with b_2 and y with a_2. The resulting graph is a subgraph of $C(12, 5)$, which again gives a contradiction. This completes the proof of Claim 12.G. $\square$

To complete the proof, we note that since G has no homomorphism to C_5, it also has no homomorphism to C_7. Thus, by Theorem 5, G must contain either an odd-K_3^2 or an odd-K_4 of odd-girth at least 7. Since such an odd-K_3^2 must have at least 18 vertices, G contains an odd-K_4. Let H be such an odd-K_4 of G. Since its girth is at least 7, by Proposition 4 it has at least 12 vertices. We consider three cases, depending on the order of H. Due to the space limit, the proofs of these three cases are omitted but can be found in the full version of the paper. $\square$

We now deduce the following consequence of Theorem 12 and Proposition 9, that improves the known lower bounds on $\eta(4, 2)$ and $\eta(4, 1)$ (noting that for larger values of k, the bound of Corollary 8 is already stronger).

Corollary 13. *We have $\eta(4, 1) \geq \eta(4, 2) \geq 17$.*

5   Concluding Remarks

In this work, we have started investigating the smallest order of a C_{2l+1}-critical graph of odd-girth $2k + 1$. We have determined a number of previously unknown values, in particular we showed that a smallest C_5-critical graph of odd-girth 7 is of order 15. In contrast to the result of Chvátal on the uniqueness of the smallest triangle-free 4-chromatic graph [3], we have found more than one such graph: Gordon Royle showed computationally, that there are at least 10 such graphs (private communication, 2016).

Regarding Table 1, we do not know the growth rate in each row of the table. Perhaps it is quadratic; that would be to say that for a fixed l, $\eta(k, l) = \Theta(k^2)$. This is indeed true for $l = 1$, as proved by Nilli [17].

Our last remark is about Theorem 10. We think that for any given k, Theorem 10 covers all values of k for which $\eta(k, l) = 4k$.

References

1. Beaudou, L., Foucaud, F., Naserasr, R.: Homomorphism bounds and edge-colourings of K_4-minor-free graphs. J. Comb. Theor. Ser. B **124**, 128–164 (2017)
2. Catlin, P.A.: Graph homomorphisms into the five-cycle. J. Comb. Theor. Ser. B **45**, 199–211 (1988)
3. Chvátal, V.: The minimality of the mycielski graph. In: Bari, R.A., Harary, F. (eds.) Graphs and Combinatorics. LNM, vol. 406, pp. 243–246. Springer, Heidelberg (1974). https://doi.org/10.1007/BFb0066446
4. Dirac, G.A.: A property of 4-chromatic graphs and remarks on critical graphs. J. London Math. Soc. **27**, 85–92 (1952)
5. Erdős, P.: Graph theory and probability. Can. J. Math. **11**, 34–38 (1959)
6. Exoo, G., Goedgebeur, J.: Bounds for the smallest k-chromatic graphs of given girth. Discrete Math. Theor. Comput. Sci. **21**(3), 9 (2019)
7. Gallai, T.: Kritische Graphen I. Magyar Tud. Akad. Mat. Kutató Int. Közl. **8**, 165–192 (1963)
8. Gallai, T.: Kritische Graphen II. Magyar Tud. Akad. Mat. Kutató Int. Közl. **8**, 373–395 (1963)
9. Gerards, A.M.H.: Homomorphisms of graphs into odd cycles. J. Graph Theor. **12**(1), 73–83 (1988)
10. Gyárfás, A., Jensen, T., Stiebitz, M.: On graphs with strongly independent colour classes. J. Graph Theor. **46**(1), 1–14 (2004)
11. Harary, F.: Graph Theory, p. 149. Addison-Wesley, Reading (1969). Exercise 12.19
12. Hell, P., Nešetřil, J.: Graphs and Homomorphisms. Oxford Lecture Series in Mathematics and Its Applications. Oxford University Press, Oxford (2004)
13. Jiang, T.: Small odd cycles in 4-chromatic graphs. J. Graph Theor. **37**(2), 115–117 (2001)
14. Mycielski, J.: Sur le coloriage des graphes. Colloq. Math. **3**, 161–162 (1955)
15. Năstase, E., Rödl, V., Siggers, M.: Note on robust critical graphs with large odd girth. Discrete Math. **310**(3), 499–504 (2010)
16. Ngoc, N.V., Tuza, Z.: 4-chromatic graphs with large odd girth. Discrete Math. **138**(1–3), 387–392 (1995)
17. Nilli, A.: Short odd cycles in 4-chromatic graphs. J. Graph Theor. **31**(2), 145–147 (1999)
18. Payan, C.: On the chromatic number of cube-like graphs. Discrete Math. **103**(3), 271–277 (1992)
19. Tardif, C.: The fractional chromatic numbers of cones over graphs. J. Graph Theor. **38**(2), 87–94 (2001)

20. Youngs, D.A.: 4-chromatic projective graphs. J. Graph Theor. **21**(2), 219–227 (1996)
21. Zhu, X.: Circular chromatic number, a survey. Discrete Math. **229**(1–3), 371–410 (2001)

Ramsey Numbers for Line Graphs

Huzaifa Abbasi, Manu Basavaraju$^{(\boxtimes)}$, Eeshwar Gurushankar, Yash Jivani,
and Deepak Srikanth

National Institute of Technology Karnataka Surathkal, Mangalore, India
huzaifabbasi@hotmail.com, {manub,yashbalvantbhai.183cs001}@nitk.edu.in,
eeshwarg13@gmail.com, deepak.s@gatech.edu

Abstract. Given a graph, the classical Ramsey number $R(k, l)$ is the least number of vertices that need to be in the graph for the existence of a clique of size k or an independent set of size l. Finding $R(k, l)$ exactly has been a notoriously hard problem. Even $R(k, 3)$ has not been determined for all values of k. Hence finding the Ramsey number for subclasses of graphs is an interesting question. It is known that even for claw-free graphs, finding Ramsey number is as hard as for general graphs for infinite number of cases. Line graphs are an important subclass of claw-free graphs. The question with respect to line graph $L(G)$ is equivalent to the minimum number of edges the underlying graph G needs to have for the existence of a vertex with degree k or a matching of size l. Chvátal and Hanson determined this exactly for line graphs of simple graphs. Later Balachandran and Khare gave the same bounds with a different proof. In this paper we find Ramsey numbers for line graph of multi graphs thereby extending the results of Chvátal and Hanson. Here we determine the maximum number of edges that a multigraph can have, when its matching number, multiplicity, and maximum degree are bounded, and characterize such graphs.

Keywords: Ramsey numbers · Extremal graph theory · Edge extremal graphs · Line graphs

1 Introduction

Extremal graph theory is a branch of graph theory which deals with maximization or minimization of some graph parameters like cardinality of the edge set, the cardinality of the vertex set, the girth of the graph, subject to some constraints on the properties of the graph. Here, we look at the specific problem in extremal graph theory where the number of edges in the graph is the parameter to be maximized with some constraints on other parameters of the graph. Chvátal and Hanson [4] gave the maximum number of edges possible when the maximum degree and matching number are bounded. Subsequently Balachandran and Khare [2] obtained the same values with a different proof. In particular, they proved that the maximum number of edges is $(d-1)(m-1)+\lfloor\frac{d-1}{2}\rfloor\lfloor\frac{m-1}{\lceil\frac{d-1}{2}\rceil}\rfloor$, where the degree of any vertex is less than d and the matching size is less than m.

© Springer Nature Switzerland AG 2020
M. Changat and S. Das (Eds.): CALDAM 2020, LNCS 12016, pp. 197–208, 2020.
https://doi.org/10.1007/978-3-030-39219-2_17

The problem of edge extremality with these constraints has been solved for specific cases such as for claw-free graphs [5], split graphs [6], unit interval graphs [6], fuzzy circular interval graphs [3].

This problem is related to Ramsey numbers of graphs. Given a graph, the classical Ramsey number on graphs, $R(k, l)$ is the least number of vertices that need to be in the graph for the existence of a clique of size k or an independent set of size l. It is known that even for claw-free graphs, finding Ramsey number is as hard as for general graphs for infinite number of cases [7]. Line graphs are an important subclass of claw-free graphs. The question with respect to line graph $L(G)$ is equivalent to the minimum number of edges the underlying graph G needs to have for the existence of a vertex with degree k or a matching of size l.

Let $e(d-1, m-1)$ denote the maximum number of edges in a graph with maximum degree at most $d-1$ and matching number at most $m-1$. It is easy to see that $R(d, m) = e(d-1, m-1) + 1$ for line graphs. Chvátal and Hanson determined this exactly for line graphs of simple graphs. Later Balachadran and Khare gave the same bounds with a different proof. In this paper we find Ramsey numbers for line graph of multigraphs thereby extending the results of Chvátal and Hanson. Here we determine the maximum number of edges that a multigraph can have, when its matching number, multiplicity, and maximum degree are bounded, and characterize such graphs.

2   Preliminaries

The graphs that we deal with in this paper are undirected loop-less multigraphs. The maximum number of edges that can be incident on any vertex in a graph G is called the maximum degree - denoted by $\Delta(G)$ - of the graph. The size of the maximum matching of a graph G is its matching number - denoted by $\nu(G)$. Since we have multigraphs, the maximum number of edges that can exist between a pair of vertices is called the multiplicity - denoted by $\mu(G)$ - of a graph G.

We define a family of graphs $\mathcal{F}(d, m, l)$ to be the set of all graphs G which satisfy the properties that $\Delta(G) < d$, $\nu(G) < m$ and $\mu(G) < l$. Given any family of graphs, an edge extremal graph G belonging to that family is a graph such that any graph H with more edges than G violates at least one of the three bounds. We define $\mathcal{F}_e(d, m, l)$ to be the set of edge extremal graphs in $\mathcal{F}(d, m, l)$. Note that by definition, the number of edges in all the graphs in $\mathcal{F}_e(d, m, l)$ is the same.

A k-star is defined as the bipartite graph $K_{1,k}$. A graph is said to be *factor critical* if the removal of any vertex v yields a perfect matching i.e. $(G - v)$ has a perfect matching for all vertices $v \in V(G)$. A Shannon multigraph $Sh(k)$ is a multigraph of three vertices with the three possible pairs of vertices having $\lfloor \frac{k}{2} \rfloor$, $\lfloor \frac{k}{2} \rfloor$ and $\lfloor \frac{k+1}{2} \rfloor$ edges between them. Note that the number of edges in the graph $Sh(k)$ is $\lfloor \frac{3k}{2} \rfloor$.

Let a factor critical component C have r matching edges. We know from the definition of a factor critical graph that the number of vertices in such a component is $2r+1$. Let us denote a factor critical component with matching size r containing the maximum possible number of edges subject to the constraints of the problem as $FCC(r)$.

In this paper, we prove the following theorem:

Theorem 1. *Let* $s = \left\lceil \frac{d-1}{2(l-1)} \right\rceil$ *and* $k = \left\lfloor \frac{1}{2}\left(\frac{d-1}{l-1} - 1\right)\right\rfloor + 1$. *Then* $G \in \mathcal{F}_e(d, m, l)$ *if* G *is one of the following:*

(1) $l - 1 \geq 5(d-1)/12$:

$$G = (m-1) \cdot Sh(\max\{d-1, 2(l-1)\}).$$

(2) $(d-1)/3 < l - 1 < 5(d-1)/12$:

$$G = \left\lfloor \frac{m-1}{2} \right\rfloor \cdot FCC(2) + \left(\left\lceil \frac{m-1}{2} \right\rceil - \left\lfloor \frac{m-1}{2} \right\rfloor\right) \cdot Sh(2(l-1)).$$

(3) $l - 1 \leq (d-1)/3$:

(3.1) If $s = k + 1$, $m - 1 \geq s(s-1)$ *and* $(d-1) < \frac{2s(2s-1)(l-1)}{2s+1}$ *and* $(d-1) < \frac{2s(2s-1)(l-1)-1}{2s+1}$ *for even and odd value of* $d-1$ *respectively:*
$G = a \cdot FCC(s-1) + b \cdot FCC(s)$, *where* a *and* b *are integers such that* $m - 1 = a(s-1) + bs$ *and* $b < s - 1$.

(3.2) If $s = k + 1$, $m - 1 = a(s-1) + bs$ *with* $a, b \geq 0$ *and* $a < s$ *and* $(d-1) < \frac{2a(s-1)(2s-1)(l-1)}{a(2s-1)-1}$ *and* $(d-1) < \frac{2a(s-1)(2s-1)(l-1)+(a-1)}{a(2s-1)-1}$ *for even and odd value of* $d-1$ *respectively:*

$$G = a \cdot FCC(s-1) + b \cdot FCC(s).$$

(3.3) Otherwise:
$G = b \cdot FCC(s) + a \cdot K_{1,d-1}$ *such that* $m - 1 = a + bs$ *with* $a, b \geq 0$ *and* $a < s$.

3   Proof of Theorem

In this section, we find the number of edges that a graph in $\mathcal{F}_e(d, m, l)$ has. We also provide the structure of one of the graphs having this maximum number of edges for given d, m and l.

We call a component of a graph *trivial* if the matching number of that component is 1. Any component with greater matching number is called *non-trivial*.

The number of edges in a factor critical component can be restricted by either the maximum degree or by the multiplicity of edges. We have the following lemma:

Lemma 1. *Let* C *be a factor critical component having* $\nu(C) = r$, $\Delta \leq d - 1$ *and* $\mu \leq l - 1$. *Then, the maximum number of edges in* C *is equal to*

1. $(2r + 1)r(l - 1)$, if $l - 1 < \frac{d-1}{2r}$
2. $\left\lfloor \frac{(2r+1)(d-1)}{2} \right\rfloor$, if $l - 1 \geq \frac{d-1}{2r}$

Proof. 1. In this case, the maximum degree possible for any vertex is $2r(l - 1)$. Thus component C can have at most $(2r + 1)r(l - 1)$ edges. We can realize this by taking a complete multigraph on $2r+1$ vertices with each edge having multiplicity $(l - 1)$.

2. The maximum degree possible is $(d - 1)$. Now we realize maximum edges in such component C as follows: We start with an independent set of $2r + 1$ vertices. As shown by Walecki [1], a complete graph on an odd number $2r+1$ of vertices can be decomposed into r Hamiltonian cycles. We arbitrarily order these Hamiltonian cycles from 1 to r. Since each vertex in a Hamiltonian cycle has, degree 2, adding a Hamiltonian cycle increases the degree of each vertex by 2. We keep doing this till the degree condition is reached. If the Hamiltonian cycles get exhausted, we take another set of same Hamiltonian cycles and keep adding them by adding multiedges between vertices. If $d - 1$ is even, then we add $(d - 1)/2$ Hamiltonian cycles to get the required degree for each vertex, which implies that the number of edges in the component is $\frac{(2r+1)(d-1)}{2}$.

Otherwise, once we reach the degree of $\left\lfloor \frac{d-1}{2} \right\rfloor$, every vertex has degree $d - 2$. Then we take the next Hamiltonian cycle and add only the alternate edges to the graph. That is we are taking a matching of r edges and adding them to the graph. Thus, all except one vertex will have degree $d - 1$, and the remaining vertex will have degree $d - 2$. The number of edges in the component now is

$$|E(C)| = \frac{(2r)(d - 1) + (d - 2)}{2}$$
$$= \frac{(2r + 1)(d - 1) - 1}{2}.$$

As $|E(C)|$ must be a whole number, we have $|E(C)| = \left\lfloor \frac{(2r+1)(d-1)}{2} \right\rfloor$. Thus we get that the factor critical component we created always has $\left\lfloor \frac{(2r+1)(d-1)}{2} \right\rfloor$ edges.

Next we give some structural properties about $\mathcal{F}_e(d, m, l)$ which aid us in the proof of the theorem. The first part of the lemma was proved in [2].

Lemma 2. *There exists a graph in $\mathcal{F}_e(d, m, l)$ in which:*

1. *all the non-trivial components are factor critical.*
2. *the matching number of no factor critical component exceeds $s = \left\lceil \frac{d-1}{2(l-1)} \right\rceil$.*
3. *the matching number of no factor critical component is below $k = \left\lfloor \frac{1}{2} \left(\frac{d-1}{l-1} - 1 \right) \right\rfloor + 1$.*

Proof. 1. Let $G \in \mathcal{F}_e(d, m, l)$ be a graph which contains a non-trivial component which is not factor critical. Gallai's lemma [8] states that a graph is factor critical if and only if removing any vertex does not reduce the matching number of the graph. As G is not factor critical, we know that there exists a vertex v where $\nu(G \backslash v) < \nu(G)$. Let us define G' as $G' = G \backslash v \cup K_{1,d-1}$. Then $\nu(G') \leq \nu(G)$ and since removing v can remove at most $d - 1$ edges $|E(G')| \geq |E(G)|$. Thus $G' \in \mathcal{F}(d, m, l)$ contains at least as many edges as G. Thus, as long as there exists a non-trivial, non-factor critical component, it can be replaced with the union of a $K_{1,d-1}$ and a component of smaller matching number without reducing the number of edges in the graph. This process can be repeated until either there are no remaining non-trivial components, or all non-trivial components are factor critical components.

2. Let $\mathcal{F}_{fc}(d, m, l) = \{G : G \in \mathcal{F}_e(d, m, l)$ *and all the non-trivial components of* G *are factor critical* $\}$. For any graph $G \in \mathcal{F}_{fc}(d, m, l)$, let $\nu_{max}(G)$ indicate the maximum size of matching in any component of G. Suppose there exists a $G \in \mathcal{F}_{fc}(d, m, l)$ such that $\nu_{max}(G) \leq \left\lceil \frac{d-1}{2(l-1)} \right\rceil$, then we are done. Otherwise, let $G \in \mathcal{F}_{fc}(d, m, l)$ be a graph such that $\nu_{max}(G)$ is as small as possible. Let $\mathcal{C}$ be the set of components in G that have a matching of size $r = \nu_{max}(G)$. Let C' be an edge extremal factor critical graph such that $\nu(C') = \nu_{max}(G) - 1$. Let $\mathcal{C}'$ be a graph that is $|\mathcal{C}|$ copies of C' and let $G' = G \backslash \mathcal{C} \cup \mathcal{C}' \cup |\mathcal{C}| \cdot K_{1,d-1}$. That is G' is a graph obtained by removing all the components in $\mathcal{C}$ from G and adding $|\mathcal{C}|$ copies of C' as well and $|\mathcal{C}|$ copies of $K_{1,d-1}$ into it. We claim that $G' \in F_{fc}(d, m, l)$. It is easy to note that all the non-trivial components of G' are factor critical. We know that $l - 1 \geq \frac{d-1}{2r}$, where the components in $\mathcal{C}$ have a matching of size r. By using Lemma 1, we have

$$|E(G')| - |E(G)| = |\mathcal{C}| \left((d-1) + \left\lfloor \frac{(2r-1)(d-1)}{2} \right\rfloor - \left\lfloor \frac{(2r+1)(d-1)}{2} \right\rfloor \right)$$

or

$$|E(G')| - |E(G)| = |\mathcal{C}|((d-1) - (d-1)) = 0.$$

This implies that G' is edge extremal with matching size $\nu(G') = \nu(G)$. We infer that $G' \in F_{fc}(d, m, l)$. But this is a contradiction to our choice of G. Thus our assumption that there exists no graph $G \in \mathcal{F}_{fc}(d, m, l)$ such that $\nu_{max}(G) \leq \left\lceil \frac{d-1}{2(l-1)} \right\rceil$ is not correct. Hence we have proved part (2) of the Lemma.

3. Let $\mathcal{F}'_{fc}(d, m, l) = \{G : G \in \mathcal{F}_{fc}(d, m, l)$ *and the matching size of any factor critical component in* G *is at most* $\left\lceil \frac{d-1}{2(l-1)} \right\rceil$, *non-trivial components of* G *are factor critical* $\}$. Let $\nu_{min}(G)$ denote the minimum size of a matching of any non-trivial component in $G \in \mathcal{F}'_{fc}(d, m, l)$. If there exists a G such that $\nu_{min}(G) \geq k$, then we are done. Otherwise, G has a factor critical component C of matching size $r < k$. Since $r \leq \frac{d-1}{2(l-1)} - 1$, the number of edges in

C is $(2r + 1)r(l - 1)$, but r stars will have $r(d - 1)$ edges. Since $d - 1 \geq 2(r + 1)(l - 1)$, we have $r(d - 1) > (2r + 1)(l - 1)r$. Thus G is not edge extremal, a contradiction. Hence we have proved part (3) of the Lemma.

When $\nu(C) = 1$, then the component C is either a triangle or a star. In the case of a star, adding a parallel edge is equivalent to adding an edge between a new vertex and the central vertex. In the case of a triangle, parallel edges are added till either the multiplicity bound or the degree bound is reached. This splits into 3 cases and enumerating each case gives us the following observation:

Observation 1. *When $\nu = 1$, the edge extremal multigraphs are,*

$$
\begin{cases}
K_{1,d-1} & when \ (l-1) \leq (d-1)/3 \\
Sh(2(l-1)) & when \ (d-1)/3 < (l-1) < (d-1)/2 \\
Sh(d-1) & when \ (l-1) \geq (d-1)/2
\end{cases}
$$

We now come to our main result which enumerates the number of edges in a graph $G \in \mathcal{F}_e(d, m, l)$. We consider different cases depending on the relation between d and l.

Case 1: $(l - 1) \geq \frac{5(d-1)}{12}$.

Claim 1. *If $G \in \mathcal{F}_e(d, m, l)$ and $(l-1) \geq \frac{5(d-1)}{12}$ then G contains only Shannon multigraphs $Sh(\max\{d - 1, 2(l - 1)\})$.*

Proof. From Observation 1 and Lemma 2, we have that the graph G can contain only factor critical components of matching size at most 2 and Shannon multigraphs $Sh(\max\{d - 1, 2(l - 1)\})$ in this range of $(l - 1)$. Let $G' \in \mathcal{F}_e(d, m, l)$ and it has factor critical component C of matching size 2. Let $G = G' \setminus C \cup 2 \cdot Sh(\max\{d - 1, 2(l - 1)\})$. The graph G is obtained by removing factor critical component C and adding 2 Shannon multigraphs $Sh(\max\{d - 1, 2(l - 1)\})$ in graph G'. By using Lemma 1, we have

$$
|E(G)| - |E(G')| = 2 \cdot \max\left(\left\lfloor \frac{3(d - 1)}{2} \right\rfloor, 3(l - 1)\right) - \left\lfloor \frac{5(d - 1)}{2} \right\rfloor. \tag{1}
$$

Here $|E(G)| - |E(G')| > 0$ for all possible value of $(l - 1)$ in this range, which means that $G' \notin \mathcal{F}_e(d, m, l)$, a contradiction. Therefore we infer that G does not contain any non-trivial factor critical components. This proves the claim.

Case 2: $(d - 1)/3 < (l - 1) < 5(d - 1)/12$.

Claim 2. *If $G \in \mathcal{F}_e(d, m, l)$ and $(d-1)/3 < (l-1) < 5(d-1)/12$ then graph G contains at most one Shannon multigraph $Sh(2(l - 1))$ and all the other components are factor critical components of matching size 2.*

Proof. From Observation 1 and Lemma 2, we have that the graph G can contain only factor critical components of matching size at most 2 and Shannon multigraphs $Sh(2(l - 1))$ in this range of $(l - 1)$. Let $G' \in \mathcal{F}_e(d, m, l)$

be a graph with more than 1 Shannon multigraphs $Sh(2(l-1))$. Let graph $G = G' \setminus 2 \cdot Sh(2(l-1)) \cup \mathcal{C}$. That is, the graph G which is obtained by removing 2 Shannon multigraphs $Sh(2(l-1))$ and adding 1 factor critical component C of matching size 2 in G'. By using Lemma 1, we have

$$|E(G)| - |E(G')| = \left\lfloor \frac{5(d-1)}{2} \right\rfloor - 2(3(l-1)). \tag{2}$$

Here $|E(G)| - |E(G')| > 0$ for all possible value of $(l-1)$ in this range, which means $G' \notin \mathcal{F}_e(d, m, l)$, a contradiction. Hence we infer that G contains at most one Shannon multigraph $Sh(2(l-1))$ and all the other components are factor critical components of matching size 2.

Case 3: $(l-1) \le (d-1)/3$.

Note that in this case $s \ge 3$ and $k \ge 2$, where k and s are same as in statement of Lemma 2. We stat with an interesting observation.

Proposition 1. *Let $s = k+1$ and let $G \in \mathcal{F}_e(d, m, l)$. Then G does not contain both a factor critical component of matching size k and a component $K_{1,d-1}$ in it.*

Proof. Suppose not. Then G contains a factor critical component of matching size k and a component $K_{1,d-1}$ in it. We obtain a graph G' from G by deleting a component of matching size k and a component $K_{1,d-1}$ and adding a component of matching size $k + 1$. Notice that $\nu(G) = \nu(G')$. Now we compare edges of G and G': If $(d-1)$ is even:

$$|E(G')| - |E(G)| = \frac{(2s+1)(d-1)}{2} - (2s-1)(s-1)(l-1) - (d-1).$$

Since $d - 1 > 2(s-1)(l-1)$ and $s \ge 3$, we have that,

$$|E(G')| - |E(G)| = \frac{(2s-1)}{2}((d-1) - 2(s-1)(l-1)) \ge 1.$$

If $(d-1)$ is odd:

$$|E(G')| - |E(G)| = \frac{(2s+1)(d-1) - 1}{2} - (2s-1)(s-1)(l-1) - (d-1).$$

Since $d - 1 > 2(s-1)(l-1)$ and $s \ge 3$, we have that,

$$|E(G')| - |E(G)| = \frac{(2s-1)}{2}((d-1) - 2(s-1)(l-1)) - \frac{1}{2} \ge 1.$$

This is a contradiction to our assumption that G is edge extremal. Therefore we infer that G does not contain both a factor critical component of matching size k and a component $K_{1,d-1}$ in it.

From Lemma 2, Observation 1 and Proposition 1, we can infer that graph G will either contain factor critical components of matching size s and stars or factor critical components of matching size s and $s - 1$. Let $\mathcal{N}_s(G)$ denote the number of factor critical components of matching size s in G and $\mathcal{N}_{s-1}(G)$ denote the number of factor critical components of matching size $s-1$ in G. We leave out G, when the graph considered is clear from the context. Let $\mathcal{F}'_e(d, m, l)$ be a set of all possible edge extremal graphs which contains only factor critical components of size s or $s - 1$. We have the following lemma:

Lemma 3. *If $\mathcal{F}'_e(d, m, l) \neq \emptyset$, then there exist a graph $G \in \mathcal{F}'_e(d, m, l)$ such that either $\mathcal{N}_s(G) < s - 1$ or $\mathcal{N}_{s-1}(G) < s$.*

Proof. If there exist $G \in \mathcal{F}'_e(d, m, l)$ which satisfies either $\mathcal{N}_s(G) < s - 1$ or $\mathcal{N}_{s-1}(G) < s$, then we are done. If not, then all the graphs $G \in \mathcal{F}'_e(d, m, l)$ will have $\mathcal{N}_s(G) \geq s - 1$ and $\mathcal{N}_{s-1}(G) \geq s$. Suppose $(d - 1)$ is even, then let

$$(d - 1) < \frac{2s(2s - 1)(l - 1)}{2s + 1}, \tag{3}$$

and if $(d - 1)$ is odd, then let

$$(d - 1) < \frac{2s(2s - 1)(l - 1) - 1}{2s + 1}. \tag{4}$$

Let G' be a graph obtained by removing $s - 1$ factor critical components of matching size s from graph G and adding s factor critical components of matching size $s - 1$. We claim that $G' \in \mathcal{F}'_e(d, m, l)$. To prove this claim we will compare the number of edges in G and G'.
If $(d - 1)$ is even:

$$|E(G')| - |E(G)| = \frac{s(2s - 1)(2s - 1)(l - 1)}{2} - \frac{(s - 1)(2s + 1)(d - 1)}{2}$$

$$= \frac{s - 1}{2}(2s(2s - 1)(l - 1) - (2s + 1)(d - 1)). \tag{5}$$

If $(d - 1)$ is odd:

$$|E(G')| - |E(G)| = \frac{s(2s - 1)(2s - 1)(l - 1)}{2} - \frac{(s - 1)(2s + 1)(d - 1) - 1}{2}$$

$$= \frac{s - 1}{2}(2s(2s - 1)(l - 1) - (2s + 1)(d - 1) - 1). \tag{6}$$

For the given values of $d - 1$, we have that $|E(G')| - |E(G)| > 0$, a contradiction since G is edge extremal. This implies that for the given degree conditions, there exists a graph $G \in \mathcal{F}'_e(d, m, l)$ such that $\mathcal{N}_s(G) < s - 1$ or $\mathcal{N}_{s-1}(G) < s$.

On the other hand, if Eqs. 3 and 4 does not satisfy, then for even value of $(d - 1)$ we have:

$$(d - 1) \geq \frac{2s(2s - 1)(l - 1)}{2s + 1}, \tag{7}$$

and for odd value of $(d-1)$ we have:

$$(d-1) \geq \frac{2s(2s-1)(l-1)-1}{2s+1}. \tag{8}$$

Let G be an edge extremal graph that has $\mathcal{N}_s(G) \geq s-1$ and $\mathcal{N}_{s-1}(G) \geq s$ with $\mathcal{N}_{s-1}(G)$ as small as possible. Let G' be a graph obtained by removing s factor critical components of matching size $s-1$ from graph G and adding $s-1$ factor critical components of matching size s. We claim that $G' \in \mathcal{F}'_e(d,m,l)$. To prove this claim we will compare the number of edges in G and G'. if $(d-1)$ is even:

$$|E(G')| - |E(G)| = \frac{(s-1)(2s+1)(d-1)}{2} - \frac{s(2s-1)(2s-1)(l-1)}{2}$$

$$= \frac{s-1}{2}((2s+1)(d-1) - 2s(2s-1)(l-1)). \tag{9}$$

If $(d-1)$ is odd:

$$|E(G')| - |E(G)| = \frac{(s-1)(2s+1)(d-1)-1}{2} - \frac{s(2s-1)(2s-1)(l-1)}{2}$$

$$= \frac{s-1}{2}((2s+1)(d-1) + 2s(2s-1)(l-1) + 1). \tag{10}$$

For the given values of $d-1$, we have that $|E(G')| - |E(G)| \geq 0$, which implies that G' is edge extremal. Also, we have that $\mathcal{N}_{s-1}(G') < \mathcal{N}_{s-1}(G)$, a contradiction to the choice of G. This implies that for the given degree conditions, there exists a graph $G \in \mathcal{F}'_e(d,m,l)$ such that $\mathcal{N}_s(G) < s-1$ or $\mathcal{N}_{s-1}(G) < s$.
This proves the lemma.

Claim 3. *Let* $m-1 \geq s(s-1)$, $k = s-1$ *and* $(d-1)$ *obeys Eqs. 3 or 4 of Lemma 3. Then* $\mathcal{F}'_e(d,m,l) \neq \emptyset$.

Proof. Suppose $\mathcal{F}'_e(d,m,l) = \emptyset$. Then by Lemma 1, we have that there exists a graph $G' \in F_e(d,m,l)$ which is a disjoint union of stars and factor critical components of matching size s. From Lemma 2 graph G' contains at most $(s-1)$ stars and $\mathcal{N}_s(G') \geq (s-1)$. Let G be a graph obtained by removing $s-1$ factor critical components of matching size s from G' and adding s factor critical component of matching size $s-1$. Here G has more edges then G' according to Lemma 3 and Eqs. 3 or 4. This implies that the graph G' is not edge extremal, a contradiction. Hence we have that $\mathcal{F}'_e(d,m,l) \neq \emptyset$.

Claim 4. *Let* $k = s-1$ *and* $(m-1)$ *be an integer that can be written as* $m-1 = a(s-1) + bs$ *with* $a,b \geq 0$ *and* $a < s$. *Also if* $(d-1)$ *is even, let*

$$(d-1) < \frac{2a(s-1)(2s-1)(l-1)}{a(2s-1)-1}. \tag{11}$$

If $(d-1)$ *is odd, let*

$$(d-1) < \frac{2a(s-1)(2s-1)(l-1)+(a-1)}{a(2s-1)-1}. \tag{12}$$

Then $\mathcal{F}'_e(d, m, l) \neq \emptyset$.

Proof. Suppose $\mathcal{F}'_e(d, m, l) = \emptyset$. Then by Lemma 2, we have that there exists a graph $G' \in F_e(d, m, l)$ which is a disjoint union of stars and factor critical components of matching size s. Let t denote the number of stars in graph G' and $r = s - t$. Let G be a graph obtained by removing all stars and $r - 1$ factor critical components of matching size s from G' and adding r factor critical components of matching size $s - 1$. Here we can see that $r = \mathcal{N}_{s-1}(G)$. Now we try to compare the number of edges in graph G and G'. If $(d - 1)$ is even

$$
\begin{aligned}
|E(G)| - |E(G')| &= \frac{r(2s - 1)(s2 - 2)(l - 1)}{2} - (s - r)(d - 1) \\
&\quad - \frac{(r - 1)(2s + 1)(d - 1)}{2} \\
&= \frac{r(2s - 1)(2s - 2)(l - 1)}{2} - \frac{(d - 1)(r(2s - 1) - 1)}{2}.
\end{aligned}
\tag{13}
$$

If $(d - 1)$ is odd

$$
\begin{aligned}
|E(G)| - |E(G')| &= \frac{r(2s - 1)(s2 - 2)(l - 1)}{2} - (s - r)(d - 1) \\
&\quad - \frac{(r - 1)((2s + 1)(d - 1) - 1)}{2} \\
&= \frac{r(2s - 1)(2s - 2)(l - 1)}{2} - \frac{(d - 1)(r(2s - 1) - 1)}{2} - (r - 1).
\end{aligned}
\tag{14}
$$

Equations. 13 and 14 will have positive values for any possible value of $(d - 1)$ as in Eqs. 11 and 12. This implies that the graph G' is not edge extremal, a contradiction. Hence we have that $\mathcal{F}'_e(d, m, l) \neq \emptyset$.

Notice that if a graph $G \in \mathcal{F}'_e(d, m, l)$, then the conditions given in Claim 3 and 4 have to be true. Otherwise, the edge extremal graph contains only stars and factor critical components of size s. Hence we get G with a stars and b factor critical components of matching size s, where $a, b \geq 0$ and $a < s$, such that $m - 1 = a + bs$. This along with Claim 1, Claim 2, Claim 3 and Claim 4 proves Theorem 1.

We also give the number of edges corresponding to the graphs we have described in Theorem 1, below:

Theorem 2. *Let $\mathcal{F}(d, m, l)$ be the set of all edge extremal graphs G which satisfy the properties that $\Delta(G) < d$, $\nu(G) < m$ and $\mu(G) < l$. Also, let $s = \left\lceil \frac{d-1}{2(l-1)} \right\rceil$ and $k = \left\lfloor \frac{1}{2} \left(\frac{d-1}{l-1} - 1 \right) \right\rfloor + 1$. Then the number of edges in a graph $G \in \mathcal{F}(d, m, l)$ is given below:*

(1) $l - 1 \geq 5(d - 1)/12$:

$$
|E| = (m - 1) \cdot \max\left(\left\lfloor \frac{3(d - 1)}{2} \right\rfloor, 3(l - 1) \right).
$$

(2) $(d-1)/3 < l - 1 < 5(d-1)/12$:

$$|E| = \left\lfloor \frac{m-1}{2} \right\rfloor \cdot \left\lfloor \frac{5(d-1)}{2} \right\rfloor + \left(\left\lceil \frac{m-1}{2} \right\rceil - \left\lfloor \frac{m-1}{2} \right\rfloor \right) \cdot 3(l-1).$$

(3) $l - 1 \le (d-1)/3$:

(3.1) If $s = k+1$, $m - 1 \ge s(s-1)$ *and* $(d-1) < \frac{2s(2s-1)(l-1)}{2s+1}$ *and* $(d-1) < \frac{2s(2s-1)(l-1)-1}{2s+1}$ *for even and odd value of* $d-1$ *respectively:*

$$|E| = \frac{\left(s \left\lfloor \frac{m-1}{s-1} \right\rfloor - (m-1) \right)(2s-1)(2s-2)(l-1)}{2}$$
$$+ \left(m - 1 - (s-1) \left\lfloor \frac{m-1}{s-1} \right\rfloor \right) \left\lfloor \frac{(2s+1)(d-1)}{2} \right\rfloor.$$

(3.2) If $s = k+1$, $m - 1 = a(s-1) + bs$ *with* $a, b \ge 0$ *and* $a < s$ *and* $(d-1) < \frac{2a(s-1)(2s-1)(l-1)}{a(2s-1)-1}$ *and* $(d-1) < \frac{2a(s-1)(2s-1)(l-1)+(a-1)}{a(2s-1)-1}$ *for even and odd value of* $d-1$ *respectively:*

$$|E| = \frac{\left(s - m + 1 + s \left\lfloor \frac{m-1}{s} \right\rfloor \right)(2s-1)(2s-2)(l-1)}{2}$$
$$+ \left(\left\lfloor \frac{m-1}{s} \right\rfloor (1-s) - s + m \right) \left\lfloor \frac{(2s+1)(d-1)}{2} \right\rfloor.$$

(3.3) Otherwise:

$$|E| = \left(m - 1 - s \left\lfloor \frac{m-1}{s} \right\rfloor \right)(d-1) + \left(\left\lfloor \frac{m-1}{s} \right\rfloor \right) \left\lfloor \frac{(2s+1)(d-1)}{2} \right\rfloor.$$

4   Conclusion

The main result of this paper has been summarized by Theorem 2. It gives the maximum number of edges in a graph bounded by degree, matching and multiplicity. This result is a generalization of the result obtained by [2,4]. Setting $l = 2$ matches the result given by [2,4]. Here we have $k = s$ when setting $l = 2$ and also the graph does not contain Shannon multigraphs. Hence this comes in case 3.3 of our result and can be derived as below:

1. **If** $(d-1)$ **is even**

$$|E| = \left\lfloor \frac{m-1}{s} \right\rfloor \left(\frac{(2s+1)(d-1)}{2} \right) + \left((m-1) - s \left\lfloor \frac{m-1}{s} \right\rfloor \right)(d-1). \quad (15)$$

After putting value of s,

$$|E| = (m-1)(d-1) + \left(\frac{d-1}{2} \right) \left\lfloor \frac{2(m-1)}{d-1} \right\rfloor. \quad (16)$$

2. **If $(d-1)$ is odd**

$$|E| = \left\lfloor \frac{m-1}{s} \right\rfloor \left(\frac{(2s+1)(d-1)-1}{2} \right) + \left((m-1) - s \left\lfloor \frac{m-1}{s} \right\rfloor \right) (d-1).$$
(17)

After putting value of s,

$$|E| = (d-1)(m-1) + \left\lfloor \frac{d-1}{2} \right\rfloor \left\lfloor \frac{m-1}{\lceil \frac{d-1}{2} \rceil} \right\rfloor.$$
(18)

References

1. Alspach, B.: The wonderful walecki construction. Bull. Inst. Comb. Appl. **52**, 7–20 (2008)
2. Balachandran, N., Khare, N.: Graphs with restricted valency and matching number. Discrete Math. **309**(12), 4176–4180 (2009)
3. Belmonte, R., Heggernes, P., van't Hof, P., Rafiey, A., Saei, R.: Graph classes and ramsey numbers. Discrete Appl. Math. **173**, 16–27 (2014)
4. Chvátal, V., Hanson, D.: Degrees and matchings. J. Comb. Theor. Ser. B **20**(2), 128–138 (1976)
5. Dibek, C., Ekim, T., Heggernes, P.: Maximum number of edges in claw-free graphs whose maximum degree and matching number are bounded. Discrete Math. **340**(5), 927–934 (2017)
6. Maland, E.K.: Maximum number of edges in graph classes under degree and matching constraints. Master's thesis, University of Bergen, Norway (2015)
7. Matthews, M.M., Sumner, D.P.: Longest paths and cycles in $k_{1,3}$-free graphs. J. Graph Theor. **9**(2), 269–277 (1985)
8. West, D.B.: Introduction to Graph Theory, vol. 2. Prentice hall, Upper Saddle River (2001)

Δ-Convexity Number and Δ-Number
of Graphs and Graph Products

Bijo S. Anand[1]([✉]), Prasanth G. Narasimha-Shenoi[2][iD], and Sabeer Sain Ramla[3]

[1] Department of Mathematics, Sree Narayana College, Punalur,
Kollam 691305, Kerala, India
bijos_anand@yahoo.com
[2] Department of Mathematics, Government College, Chittur,
Palakkad 678104, Kerala, India
prasanthgns@gmail.com
[3] Department of Mathematics, Mar Ivanios College, University of Kerala,
Thiruvananthapuram 695015, India
sabeersainr@gmail.com

Abstract. The Δ-interval of $u, v \in V(G)$, $I_\Delta(u,v)$, is the set formed by u, v and every w in $V(G)$ such that $\{u, v, w\}$ is a triangle (K_3) of G. A set S of vertices such that $I_\Delta(S) = V(G)$ is called a Δ-set. Δ-number is the minimum cardinality of a Δ-set. Δ-graph is a graph with all the vertices lie on some triangles. If a block graph is a Δ-graph, then we say that it is a block Δ-graph. A set $S \subseteq V(G)$ is Δ-convex if there is no vertex $u \in V(G) \setminus S$ forming a triangle with two vertices of S. The convexity number of a graph G with respect to the Δ-convexity is the maximum cardinality of a proper convex subset of G. We have given an exact value for the convexity number of block Δ-graphs with diameter ≤ 3, block Δ-graphs with diameter >3 and the two standard graph products (Strong, Lexicographic products), a bound for Cartesian product. Also discussed some bounds for Δ-number and a realization is done for the Δ-number and the hull number.

Keywords: Δ-convexity · Δ-convexity number · Δ-number · Graph products

AMS Subject Classification: 05C38 · 05C76 · 05C99 · 52A01

1 Introduction

Axiomatic convexity and convexity spaces are studied in different branches of mathematics. The graph convexity has also been studied since 50 years.

P. G. Narasimha-Shenoi—Supported by Science and Engineering Research Board, a statutory body of Government of India under their MATRICS Scheme No. MTR/2018/000012.
R. Sabeer Sain—Supported by the University of Kerala for providing University JRF.

M. Changat and S. Das (Eds.): CALDAM 2020, LNCS 12016, pp. 209–218, 2020.
https://doi.org/10.1007/978-3-030-39219-2_18

Various types of graph convexities are studied, see van de Vel [8]. One of the important problems in graph convexities is related to determining convexity parameters in different families of graphs. Some of the convexity parameters that are studied in the literature are the Carathéodory, Helly and Radon number and its relationships. For more see [21].

In graphs, a family of paths are said to be a feasible path family if the graph contains at least one path between any pair of vertices. Some of the important feasible paths are "shortest paths", "the induced paths", "the all paths", and "triangle paths" where each of these concepts give rise to a convexity on the vertices of the graph G. To mention a few see [9,13,19] for geodesic convexity, [6,12,16] for induced path convexity and [5,7,10] for triangle path convexity.

Another interesting notions are some invariant properties of convexities defined on graphs. Duchet in his paper [11] computed Carathéodory, Helly and Radon type numbers for the minimal path convexity. In [5] Changat et al. considered the invariants when triangle paths are under consideration. For geodesic convexity we refer [14]. For more on these invariants and related structures see [2–4]. The concepts which have been recently studied with respect to the feasible path families are the exchange number, hull number, pre-hull number and the convexity number, for details see [3,9,15,20].

We consider finite, simple, connected and undirected graphs. Given a graph G, the Δ-interval of $u, v \in V(G)$, $I_\Delta(u,v)$, is the set formed by u, v and every w in $V(G)$ such that $\{u,v,w\}$ is a triangle (K_3) of G. For $U \subseteq V(G)$, $I_\Delta(U)$ denote the set of all vertices that belong to some $I_\Delta(u,v)$, $u, v \in U$. A set S of vertices such that $I_\Delta(S) = V(G)$ is called a Δ-set. The Δ-number of a graph G is the minimum cardinality of a Δ-set and is denoted by $\Delta(G)$. A set $S \subseteq V(G)$ is Δ-convex if there is no vertex $u \in V(G) \setminus S$ forming a triangle with two vertices of S, or equivalently, if $N(u) \cap S$ is an independent set for every $u \in V(G) \setminus S$. The Δ-convex hull $\langle S \rangle$ of a set S is the minimum Δ-convex set containing S. Convexity number of a graph G, denoted by $C_\Delta(G)$ is the maximum cardinality of a proper convex subset of G. The hull number of G in Δ-convexity is the minimum cardinality of a set S such that $\langle S \rangle = V(G)$ and is denoted by $h_n(G)$. The Δ-interval function and the associated Δ-convexity were introduced by Mulder in [18]. In [17], the Δ-convex sets, where they were called Δ-closed sets played an essential role in the characterization of quasi-median graphs. In [1], the authors studied the complexity of Δ-hull number of a graph and given an upper bound for the hull number. In this article, we study the Δ-convexity number and Δ-number of graphs and graph products. Denote $C(G)$ as the set of all cut vertices of G. A block graph is a graph where every 2-connected component is a complete graph.

For the three graph products (Cartesian, strong, lexicographic) of G and H, the vertex set is $V(G) \times V(H)$. Their edge sets are defined as follows. In the *Cartesian product* $G \,\square\, H$ two vertices are adjacent if they are adjacent in one coordinate and equal in the other. Two distinct vertices (g_1, h_1) and (g_2, h_2) are adjacent with respect to the *strong product* if (a) $g_1 = g_2$ and $h_1 h_2 \in E(H)$, or (b) $h_1 = h_2$ and $g_1 g_2 \in E(G)$ or (c) $g_1 g_2 \in E(G)$ and $h_1 h_2 \in E(H)$. Finally, two vertices (g, h) and (g', h') are adjacent in the *lexicographic product* $G \circ H$ (also $G[H]$) either if $gg' \in E(G)$ or if $g = g'$ and $hh' \in E(H)$. For $* \in \{\Box, \boxtimes, \circ\}$ we call the product $G * H$ *nontrivial* if both G and H have at least two vertices. For $h \in V(H)$, $g \in V(G)$, and $* \in \{\Box, \boxtimes, \circ\}$, call $G^h = \{(g, h) \in G * H : g \in V(G)\}$ a G *layer* in $G * H$, and call ${}^g H = \{(g, h) \in G * H : h \in V(H)\}$ an H *layer* in $G * H$. Note that the subgraph of $G * H$ induced on G^h is isomorphic to G and the subgraph of $G * H$ induced on ${}^g H$ is isomorphic to H for $* \in \{\Box, \boxtimes, \circ\}$. The map $p_G : V(G * H) \to V(G)$ defined with $p_G((g, h)) = g$ is called a *projection map on to G* for $* \in \{\Box, \boxtimes, \circ\}$. Similarly we can define the *projection map on to H*. We will also use for a graph G the standard notations $N^G(g)$ for the open neighbourhood $\{g' : gg' \in E(G)\}$, $N^G[g]$ for the closed neighbourhood $N^G(g) \cup \{g\}$.

2 Convexity Number of Block Graphs

If a graph G has a vertex x which is not in any K_3 of G, then $G \setminus \{x\}$ will be a convex set in G and in that case $C_\Delta(G) = |V(G)| - 1$. So we are interested in graphs with all the vertices lie on some triangles. Such graphs are called a Δ-*graph*. If a block graph is a Δ-graph, then we say that it is a block Δ-graph.

Theorem 1. *Let G be a block Δ-graph with $diam(G) \leq 3$. Then the convexity number $C_\Delta(G) = | V(G) \setminus V(B_p) | + 1$, where B_p is the smallest pendant block in G.*

Proof. By the definition of block graph, for each pendant block B of G, the convex hull of $(V(G) \setminus V(B)) \cup C(B)$ is a proper convex set in G. For any intermediate block B' of G, $(V(G) \setminus V(B')) \cup C(B')$ contains at least two elements of $C(B')$ and its convex hull contains the entire block B'. Hence for any intermediate block B' of G, $(V(G) \setminus V(B')) \cup C(B')$ will not be a proper convex set in G. Since $diam(G) \leq 3$, then G has only one intermediate block and from the above argument its removal (not removing its cut vertices) does not form a proper convex set in G. Intermediate block itself is a proper convex set, but it has at least two neighbouring blocks, so if we add one more block to it then it will also be a proper convex set. Hence the intermediate block itself will not be a maximum proper convex set in G. If there are only two blocks, then the largest block will be the maximum one. Assume G has at least three blocks. For getting a maximum proper convex set we have to add pendant blocks to the intermediate block and we have to remove at least one pendant block (except the cut vertex) to get a proper maximum convex set of G. That will be maximum only when the removing pendant block (except the cut vertex) of G is a minimum one. So for getting a maximum proper convex set we have to remove the smallest pendant block B_p of G and add the cut vertex of B_p. i.e., $(V(G) \setminus V(B_p)) \cup C(B_p)$ is the maximum proper convex set of G. Hence $C_\Delta(G) = | V(G) \setminus V(B_p) | + 1$. $\Box$

A cut vertex x is said to be an intermediate cut of a block graph G, if x is not in a pendant block. A minimum intermediate cut of a block graph G (each block of G contains at most two cut vertices) is an intermediate cut a of G with minimum $|N^G(a)|$.

Theorem 2. *Let G be a block Δ-graph with $diam(G) > 3$ and each block contains at most two cut vertices. Then the convexity number $C_\Delta(G) = max\{|V(G) \setminus V(B_p)| + 1, |(V(G) \setminus N^G[x]) \cup (N^G(x) \cap C(G))|\}$, where B_p is the smallest pendant block, x is the minimum intermediate cut of G and $C(G)$ is the set of all cut vertices of G.*

Proof. For any intermediate block B of G, $(V(G) \setminus V(B)) \cup C(B)$ contains at least two elements of $C(B)$ and its convex hull contains the entire block B. Hence for any intermediate block B of G, $(V(G) \setminus V(B)) \cup C(B)$ will not be a proper convex set in G. So if we remove a cut vertex x from an intermediate block, then we have to remove all the blocks which contains x, otherwise its first iteration produce $V(G)$. Now we have to find the minimum of $\{|N^G(x)| : x \in C(G), x$ is not a cut vertex of a pendant block$\}$. So assume x is a minimum intermediate cut, then the induced subgraph of $(V(G) \setminus N^G[x]) \cup (N^G(x) \cap C(G))$ will be a proper convex set in G, since no edge in the induced graph of $(V(G) \setminus N^G[x]) \cup (N^G(x) \cap C(G))$ form a triangle with a vertex in $N^G[x] \setminus C(G)$ and viceversa. If we remove a vertex x from an intermediate block of G other than a cut vertex, then the induced graph of $V(G) \setminus \{x\}$ will be a proper convex set only when we choose at most one vertex from the neighbouring blocks, otherwise the first iteration cover the entire blocks. So in this way we cannot get a maximum proper convex sets. From the proof of Theorem 1, for any pendant block B^* of G, the induced graph of $(V(G) \setminus V(B^*)) \cup C(B^*)$ is a proper convex set in G. These two convex sets have the chance to attain a maximum proper convex set. So the maximum proper convex set will be the maximum of $(V(G) \setminus N^G[x]) \cup (N^G(x) \cap C(G))$, where x is a minimum intermediate cut and $(V(G) \setminus V(B_p)) \cup C(B_p)$. Now we can conclude that $C_\Delta(G) = max\{|V(G) \setminus V(B_p)| + 1, |(V(G) \setminus N^G[x]) \cup (N^G(x) \cap C(G))|\}$. $\square$

If a graph G has hull number 2, then the proper convex sets in G are the set of pairwise nonadjacent vertices. i.e, the maximum independent sets in G will be the maximum proper convex sets. Therefore its convexity number is $\alpha(G)$, the independent number of G.

3   Convexity Number in Graph Products

If a graph G contains a vertex x which is not a part of any triangles in G, then the induced graph of $V(G) \setminus \{x\}$ will be a proper convex set in G. If such a vertex does not exist in G, then the convexity number will be less than $|V(G)| - 1$. i.e., $C_\Delta(G) = |V(G)| - 1$ if and only if there exists a vertex in G which is not a part of any K_3 in G. If G or H contains a vertex which does not lie on any K_3, then $C_\Delta(G \square H) = |V(G \square H)| - 1 = |V(G)||V(H)| - 1$. But in the case of strong and lexicographic product, all the vertices are on some triangles, hence $C_\Delta(G * H) < |V(G * H)| - 1$, $* \in \{\boxtimes, \circ\}$.

Theorem 3. *Let G and H be two connected Δ-graphs with orders m and n respectively, then $C_\Delta(G \,\square\, H) = max\{(n-1)m + C_\Delta(G), (m-1)n + C_\Delta(H)\}$.*

Proof. By the definition of Cartesian product of graphs for any two vertices $g_1, g_2 \in V(G)$, there does not exist a K_3 with one vertex in ^{g_1}H and the others in ^{g_2}H. This is true for all G-layers. So $I_\Delta(V(^gH)) = V(^gH)$ and $I_\Delta(V(G^h)) = V(G^h)$. Thus the two possibilities for getting a maximum proper convex set are the following:

1. Take $n-1$ complete G-layers and add in the remaining layer, the vertices of a maximum Δ-convex set $S \subseteq V(G)$, which brings additional $|S| = C_\Delta(G)$ vertices.
2. Take $m-1$ complete H-layers and add in the remaining layer, the vertices of a maximum Δ-convex set $S \subseteq V(H)$, which brings additional $|S| = C_\Delta(H)$ vertices.

Therefore $C_\Delta(G \,\square\, H) = max\{(n-1)m + C_\Delta(G), (m-1)n + C_\Delta(H)\}$. $\qquad\square$

Theorem 4. *Let G and H be two connected non-trivial graphs. Then the convex hull of any two vertices of an edge in $G * H$, $* \in \{\boxtimes, \circ\}$ is $V(G * H)$.*

Proof. First take an edge from an H layer, say ^gH. Let $(g, h_1)(g, h_2) \in E(G * H), * \in \{\boxtimes, \circ\}$. Then $I_\Delta((g, h_1)(g, h_2))$ contains $g' \times \{h_1, h_2\}$, where $g' \in N^G(g)$, since $gg' \in E(G)$ and $h_1h_2 \in V(H)$. The second iteration contains $N^G(g') \times \{h_1, h_2\}$. After a finite steps of iterations we get all the vertices of G_1^h and G_2^h. Now by continuing the iteration with the edges of G_1^h and G_2^h all the H- layers are produced and we get $V(G*H)$, $* \in \{\boxtimes, \circ\}$. The same thing will happen when we take an edge from any of the G-layers. Now let $(g_1, h_1)(g_2, h_2) \in E(G * H)$, $* \in \{\boxtimes, \circ\}$, where $g_1 \neq g_2$ and $h_1 \neq h_2$. Then $I_\Delta(g_1, h_1)(g_2, h_2)$ produce (g_1, h_2) and (g_2, h_1). Now we get two edges from the G-layer and from the first part, after a finite number of iterations we will get the whole vertices of $G * H$, $* \in \{\boxtimes, \circ\}$. $\qquad\square$

From Theorem 4 the possible proper convex sets of $G * H$, $* \in \{\boxtimes, \circ\}$ are the sets of pairwise non-adjacent vertices. Hence the proper convex sets in $G*H$ are the independent sets. We mention this as the following corollary.

Corollary 1. *Let G and H be two nontrivial graphs, then*

(a) $C_\Delta(G \boxtimes H) = \alpha(G \boxtimes H)$.
(b) $C_\Delta(G \circ H) = \alpha(G \circ H)$.

4 Realizing Δ-Number and Hull Number of Graphs

To understand the basic terminologies of Δ-number and hull number, in this section we illustrate some graphs. Also the following argument gives us a clear view of the inequality $h_n(G) \le \Delta(G)$ for an arbitrary graph G. A vertex u of a graph G is said to be a *simplicial vertex* if the subgraph induced by u and its neighbours induce a complete subgraph called an extreme subgraph denoted by $Ext(G)$.

Theorem 5. *For every pair (a, b) of integers $2 \leq a \leq b$, there exists a graph G such that $h_n(G) = a$ and $\Delta(G) = b$.*

Proof. **Case 1**: $a = b$.
The triangle free graphs on 'a' vertices has the desired property.
Case 2: $a < b$.
We construct a graph as follows. Let $m = b - a$ and $\ell = a - 2$. Let H be a disconnected graph containing two paths P_1 and P_2 which are on $m + 2$ vertices say $V(P_1) = \{u_1, u_2, \ldots, u_{m+2}\}$ and $m + 1$ vertices say $V(P_2) = \{v_2, v_3, \ldots, v_{m+2}\}$. Now make v_i adjacent to u_{i-1} and on u_i for $2 \leq i \leq m + 2$. Also add another $\ell + 1$ vertices say $\{v_1, w_1, w_2, \ldots, w_\ell\}$. Make v_1 adjacent to u_1 and u_2. Also add edges between w_i and v_1 for $1 \leq i \leq \ell$. The resulting graph G is given in the Fig. 1.

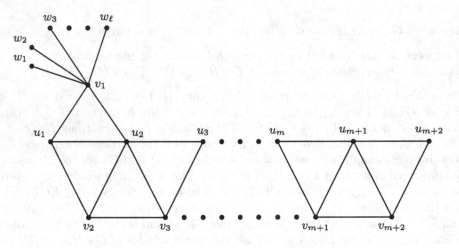

Fig. 1. Graph G with $h_n(G) = a$ and $\Delta(G) = b$.

Now $\langle \{u_1, v_1\} \rangle = V(G) \setminus \{w_1, w_2, \ldots w_\ell\}$. Hence we can see that the minimum hull set should contain $\{w_1, w_2, \ldots w_\ell\}$ and any two adjacent vertices from $V(P_1) \cup V(P_2) \cup \{v_1\}$. Hence $h_n(G) = \ell + 2 = a$. Now the first iteration of the set $S = \{u_1, u_2, \ldots, u_{m+2}\}$ is $\{u_1, u_2, \ldots, u_{m+2}, v_1, v_2, \ldots v_{m+2}\}$. Hence we can see that these S together with $\{w_1, w_2, \ldots, w_\ell\}$ is a minimum Δ-set of the graph. So $\Delta(G) = m + 2 + \ell = b - a + 2 + a - 2 = b$. $\qquad \square$

Theorem 6. *For every pair (a, b) of integers and $a, b \geq 2$ there exists a graph G such that $|Ext(G)| = a$ and $\Delta(G) = b = h_n(G)$.*

Proof. Take K_a and a path $P_{b-2} = u_1 u_2 \ldots u_{b-2}$, a path on $b - 2$ vertices. Let u be any vertex in K_a. Let G be a new graph obtained from K_a and P_{b-2} by joining an edge between u and u_1. We can see that G has exactly 'a' number of simplicial vertices namely $K_a \setminus \{u\}$ and u_{b-2}. Also $u_1 u_2 \ldots u_{b-2}$ together with u and any other vertex from K_a will constitute a Δ-set of G, so that $\Delta(G) = b - 2 + 2 = b$. Hence the theorem. $\qquad \square$

Theorem 7. *For integers* (n, d, Δ) *such that* $n \geq 4$, $2 \leq d \leq n - 2$, $2 \leq \Delta \leq n$ *and* $d \leq \Delta$, *there exists a graph* G *of order* n, *diameter* d *and* Δ-*number* Δ *with* $\Delta = h_n(G)$.

Proof. We first construct the graph as follows. First choose K_2, the complete graph on 2 vertices and label the vertices as a, b. Add $n - d$ vertices say, $v_1, v_2, \ldots v_{n-d}$. Now make the vertices $v_1, v_2, \ldots, v_{n-\Delta}$ adjacent to both a and b, and $v_{n-\Delta+1}, v_{n-\Delta+2}, \ldots v_{n-d}$ adjacent to either a or b. Now consider a path on $d - 2$ vertices, say, $u_2, u_3, \ldots u_{d-1}$ and make u_2 adjacent to $v_{n-\Delta}$. The resulting graph is given in the Fig. 2. Now we can see that it has diameter d, since the path $v_1 a v_{n-\Delta} u_2 \ldots u_{d-1}$ is of length $d - 2 + 2 = d$. The set $S = \{a, b, u_2, u_3, \ldots u_{d-1}, v_{n-\Delta+1}, v_{n-\Delta+2}, \ldots, v_{n-d}\}$ is a hull set and $I_\Delta(S) = V(G)$. Now $|S| = n - d - (n - \Delta) + d - 2 + 2 = \Delta$, which completes the proof. $\square$

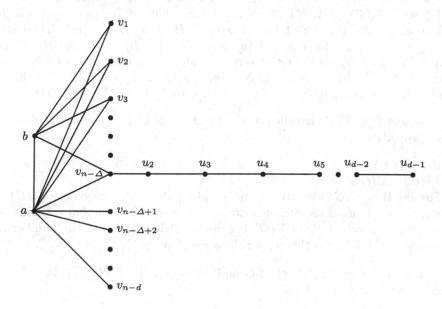

Fig. 2. Graph G with order n, diameter d and Δ-number $\Delta = h_n$.

Theorem 8. *Let* G *be a block graph. Then* $\Delta(G) \leq n_c + n_p$, *where* n_c *is the number of cut vertices and* n_p *is the number of pendant blocks of* G.

Proof. Since G is a block graph, every intermediate block has at least two cut vertices and every pendant block has exactly one cut vertex. Then $I_\Delta(C(G))$ will cover all the vertices of the intermediate blocks. Since each pendant block shares a cut vertex to an intermediate block, we need only one more vertex from each pendant blocks (except the cut vertices) to generate all the vertices of the pendant blocks. If we take such vertices with $C(G)$, then its first iteration will cover all the vertices of G. Hence $\Delta(G) \leq |C(G)| +$ number of pendant blocks, i.e, $\Delta(G) \leq n_c + n_p$.   □

5   Delta Number in Graph Products

Theorem 9. *Let G and H be two nontrivial graphs with orders m and n respectively. If m^* and n^* are the number of vertices which is not a part of any K_3 in G and H respectively, Then $\Delta(G \,\square\, H) \leq min\{n(\Delta(G) - m^*) + m^*\Delta(H), m(\Delta(H) - n^*) + n^*\Delta(G)\}$.*

Proof. Assume $S \subseteq V(G)$ and $I_\Delta(S) = V(G)$ with $|S| = \Delta(G)$. Let $g_1, g_2, \ldots, g_{m^*}$ be the vertices of G which are not in any K_3 in G. We get $\langle S \setminus \{g_1, g_2, \ldots, g_{m^*}\}\rangle = V(G) \setminus \{g_1, g_2, \ldots, g_{m^*}\}$. Then the convex hull of $(S \setminus \{g_1, g_2, \ldots, g_{m^*}\}) \times V(H)$ will cover all the vertices of $V(G \,\square\, H)$ except the H-layers ^{g_i}H, $i = 1, 2, \ldots, m^*$. Now let $R \subseteq V(H)$ and $|R| = \Delta(H)$ with $I_\Delta(R) = V(H)$. Then the convex hull of $\{g_1, g_2, \ldots, g_{m^*}\} \times R$ will cover all the vertices of the H-layers ^{g_i}H, $i = 1, 2, \ldots, m^*$. Therefore $\langle (S \setminus \{g_1, g_2, \ldots, g_{m^*}\}) \times V(H) \cup \{g_1, g_2, \ldots, g_{m^*}\} \times R\rangle = V(G \,\square\, H)$. Similarly if $h_1, h_2, \ldots, h_{n^*}$ is the set of vertices which are not in any K_3 of H, then $\langle V(G) \times (S \setminus \{h_1, h_2, \ldots, h_{m^*}\}) \cup S \times \{v_1, v_2, \ldots, v_{m^*}\}\rangle = V(G \,\square\, H)$. Hence $\Delta(G \,\square\, H) \leq min\{n(\Delta(G) - m^*) + m^*\Delta(H), m(\Delta(H) - n^*) + n^*\Delta(G)\}$.

A subset $U \subseteq V(G)$ is called a Steiner-Δ set of G if U have the following three properties:

1. $|U| = \Delta(G)$.
2. $I_\Delta(U) = V(G)$.
3. For any $W \subseteq V(G)$ with $|W| = \Delta(G)$ and $I_\Delta(W) = V(G)$, we have $d_S(U) \leq d_S(W)$, where d_S is the Steiner distance. (The smallest tree that contains all vertices of a subset W of $V(G)$ is called a Steiner tree and the number of edges of such a tree is the Steiner distance of W).

Let U be any Steiner-Δ set of G and $M = \{x \in V(G) | N^G[x] \cap U = \{x\}\}$, then we define $n_S(G) = |M|$.

Theorem 10. *Let G be a nontrivial graph and H a complete graph, then $\Delta(G \circ H) = \Delta(G \boxtimes H) = \Delta(G) + n_S(G)$.*

Proof. Let $U \subseteq V(G \boxtimes H)$ with $|U| = \Delta(G \boxtimes H)$. Then $I_\Delta(p_G(U)) = V(G)$. If not, $I_\Delta(U)$ will not cover $V(G \boxtimes H)$ (Since U is the minimum set which has the said properties).
Case 1: $M = \phi$ in G.

$M = \phi$ in G means there exists some $S \subseteq V(G)$ with $|S| = \Delta(G)$, $I_\Delta(S) = V(G)$ and $d_S(S) = |S| - 1$. Here $n_s(G) = 0$. Take S as $\{g_1, g_2, \ldots, g_r\}$. Then for any $h \in V(H)$, $I_\Delta(S \times \{h\}) = V(G \boxtimes H)$, because if $g \in I_\Delta(g_i, g_j)$ for $g_i, g_j \in S$, then $V(^g H) \subseteq I_\Delta((g_i, h), (g_j, h))$. If $g_k g_l \in E(G)$ and $g_k, g_l \in S$, then all the vertices of ^{g_k}H and ^{g_l}H form triangles with $(g_k, h)(g_l, h)$ in $G \boxtimes H$, i.e.,

$^{g_k}H \cup {}^{g_l}H \subseteq I_\Delta((g_k, h), (g_l, h)) \subseteq I_\Delta(S \times \{h\})$. Hence $I_\Delta(S \times h) = V(G \boxtimes H)$. Therefore in this case $\Delta(G \circ H) = \Delta(G \boxtimes H) = \Delta(G)$.

Case 2: $M \neq \phi$ in G.

If two vertices g_1, g_2 of S are adjacent, then from the proof of above case we can say that $^{g_1}H \cup {}^{g_2}H \subseteq I_\Delta(S \times \{h\})$ also for any $g \in I_\Delta(g_i, g_j)$ and $g_i, g_j \in S$, we have $V(^gH) \subseteq I_\Delta((g_i, h), (g_j, h)) \subseteq I_\Delta(S \times \{h\})$. Here $I_\Delta(S \times \{h\})$ cover $I_\Delta(S) \times V(H)$. So it remains to add vertices of the H-layers ^gH with $N_G(g) \cap S = \{g\}$ and the number of such vertices will be minimum when S is a Steiner-Δ set. We have denoted such vertex set as M and its cardinality is $n_s(G)$. Now let S be a Steiner-Δ set in G. If $g \in M \cap S$, then the layer ^gH will cover $I_\Delta(S)$ only when S contains a vertex of ^gH other than (g, h), since H is complete. So we have to add such vertices to S, hence $\Delta(G \circ H) = \Delta(G \boxtimes H) = \Delta(G) + n_s(G)$.
□

Theorem 11. *Let G and H be two non-complete graphs. Then $\Delta(G \boxtimes H) \leq min\{\alpha(G)\Delta(H), \Delta(G)\alpha(H)\}$.*

Proof. Let $S = \{g_1, g_2, \ldots, g_r\} \subseteq V(G)$ with $I_\Delta(S) = V(G)$ and $r = \Delta(G)$. If $g \in I_\Delta(g_i, g_j)$ for some $i, j = 1, 2, \ldots, r$. Then for any $h' \in N_H(h')$, $(g, h') \in I_\Delta((g_i, h), (g_j, h))$, since $g_i g, g_j g \in E(G)$ and $hh' \in E(H)$. Therefore $I_\Delta((g_i, h), (g_j, h))$ contains $I_\Delta(g_i, g_j) \times \{h\}$ and $\{g\} \times N(h)$. This is true for any $g_i, g_j \in S$. Then $I_\Delta((g_1, h), (g_2, h), \ldots, (g_r, h)) = I_\Delta(g_1, g_2, \ldots, g_r) \times \{h\} \cup V(G) \times N(h)$. Hence for any $h \in V(H)$, $I_\Delta(S \times \{h\})$ contains all the neighbouring H-layers. So if $h_1, h_2, \ldots, h_s$ be any independent set of H, then $I_\Delta(S \times \{h_1, h_2, \ldots, h_s\})$ cover all the vertices of $G \boxtimes H$. i.e., $\Delta(G \boxtimes H) \leq \{\Delta(G)\alpha(H)\}$. Similarly we can prove that $\Delta(G \boxtimes H) \leq \{\alpha(G)\Delta(H)\}$. Hence $\Delta(G \boxtimes H) \leq min\{\alpha(G)\Delta(H), \Delta(G)\alpha(H)\}$.
□

Theorem 12. *If G is a Δ-graph and H a non-complete graphs, then $\Delta(G \circ H) \leq 2\Delta(G)$.*

Proof. Let $S = \{g_1, g_2, \ldots, g_r\} \subseteq V(G)$ such that $I_\Delta(S) = V(G)$ and $r = \Delta(G)$. Let $h_1 h_2 \in E(H)$. Consider the set $S' = \{(g_1, h_1), (g_2, h_1), \ldots, (g_r, h_1), (g_1, h_2), (g_2, h_2), \ldots, (g_r, h_2)\}$. Our aim is to prove $I_\Delta(S') = V(G \circ H)$. If $g_l \in I_\Delta(g_i, g_j)$, for some $g_i, g_j \in S$ in G, then $V(^{g_l}H) \subseteq I_\Delta((g_i, h_1), (g_j, h_1)) \subseteq I_\Delta(S')$. Hence for any $g \in V(G) \setminus S$, $V(^gH) \subseteq I_\Delta(S')$. So it remains to prove $V(^{g_i}H) \subseteq I_\Delta(S')$ for all $g_i \in S$. Since G is a Δ-graph, for any $g_i \in S$, there exists some $g_j \in S$ such that $g_i g_j \in E(G)$. Since $h_1 h_2 \in E(H)$, $(g_i, h_1)(g_i, h_2), (g_j, h_1)(g_j, h_2) \in E(G \circ H)$. By the definition of lexicographic product both (g_i, h_1) and (g_i, h_2) are adjacent to $\{g_j\} \times V(H) = V(^{g_j}H)$, since $g_i g_j \in E(G)$. Then $I_\Delta((g_i, h_1), (g_i, h_2))$ will contain $\{g_j\} \times V(H) = V(^{g_j}H)$, because $(g_i, h_1)(g_i, h_2) \in E(G \circ H)$. Therefore $I_\Delta(S') = V(G \cup H)$ and $|S'| = 2r = 2\Delta(G)$.
□

Acknowledgements. We are grateful to the anonymous referees for their numerous valuable comments.

References

1. Anand, B.S., Anil, A., Changat, M., Dourado, M.C., Sain R.,S.: Computing the hull number in Δ-convexity (2018, submitted)
2. Anand, B.S., Changat, M., Klavžar, S., Peterin, I.: Convex sets in lexicographic products of graphs. Graphs Comb. **28**(1), 77–84 (2012)
3. Anand, B.S., Changat, M., Narasimha-Shenoi, P.G.: Helly and exchange numbers of geodesic and Steiner convexities in lexicographic product of graphs. Discrete Math. Algorithms Appl. **7**(04), 1550049 (2015)
4. Anand, B.S., Changat, M., Peterin, I., Narasimha-Shenoi, P.G.: Some Steiner concepts on lexicographic products of graphs. Discrete Math. Algorithms Appl. **6**(04), 1450060 (2014)
5. Changat, M., Mathew, J.: On triangle path convexity in graphs. Discrete Math. **206**(1–3), 91–95 (1999)
6. Changat, M., Mathew, J.: Induced path transit function, monotone and peano axioms. Discrete Math. **286**(3), 185–194 (2004)
7. Changat, M., Mulder, H.M., Sierksma, G.: Convexities related to path properties on graphs. Discrete Math. **290**(2–3), 117–131 (2005)
8. van De Vel, M.L.: Theory of Convex Structures, vol. 50. Elsevier, Burlington (1993)
9. Dourado, M.C., Penso, L.D., Rautenbach, D.: On the geodetic hull number of P_k-free graphs. Theoret. Comput. Sci. **640**, 52–60 (2016)
10. Dourado, M.C., Sampaio, R.M.: Complexity aspects of the triangle path convexity. Discrete Appl. Math. **206**, 39–47 (2016)
11. Duchet, P.: Convexity in combinatorial structures. In: Proceedings of the 14th Winter School on Abstract Analysis, pp. 261–293. Circolo Matematico di Palermo (1987)
12. Duchet, P.: Convex sets in graphs, II. Minimal path convexity. J. Comb. Theor. Ser. B **44**(3), 307–316 (1988)
13. Farber, M.: Bridged graphs and geodesic convexity. Discrete Math. **66**(3), 249–257 (1987)
14. Jiang, T., Pelayo, I., Pritikin, D.: Geodesic convexity and Cartesian products in graphs. Graphs Comb. (2004, submitted)
15. Kannan, B., Changat, M.: Hull numbers of path convexities on graphs. In: Proceedings of the International Instructional Workshop on Convexity in Discrete Structures, vol. 5, pp. 11–23 (2008)
16. Morgana, M.A., Mulder, H.M.: The induced path convexity, betweenness, and svelte graphs. Discrete Math. **254**(1–3), 349–370 (2002)
17. Mulder, H.M.: The Interval Function of a Graph. MC Tracts. Mathematisch Centrum, Amsterdam (1980)
18. Mulder, M.H.: Transit functions on graphs (and posets). Technical report (2007)
19. Pelayo, I.M.: Geodesic Convexity in Graphs. Springer, New York (2013). https://doi.org/10.1007/978-1-4614-8699-2
20. Polat, N., Sabidussi, G.: On the geodesic pre-hull number of a graph. Eur. J. Comb. **30**(5), 1205–1220 (2009)
21. Sierksma, G.: Convexity on unions of sets. Compositio Mathematica **42**(3), 391–400 (1980)

On Cartesian Products of Signed Graphs

Dimitri Lajou$^{(\boxtimes)}$

Univ. Bordeaux, Bordeaux INP, CNRS, LaBRI, UMR5800, 33400 Talence, France
dimitri.lajou@gmail.com

Abstract. In this paper, we study the Cartesian product of signed graphs as defined by Germina, Hameed and Zaslavsky (2011). Here we focus on its algebraic properties and look at the chromatic number of some products. One of our main result is the unicity of the prime factor decomposition of signed graphs. This leads us to present an algorithm to compute this decomposition in quasi-linear time. Both these results use their counterparts for ordinary graphs as building blocks. We also study the signed chromatic number of graphs with underlying graph of the form $P_n \square P_m$, of products of signed paths, of products of signed complete graphs and of products of signed cycles, that is the minimum order of a signed graph to which they admit a homomorphism.

1 Introduction

Signed graphs were introduced by Harary in [8]. Later, the notion of homomorphism of signed graphs was introduced by Guenin [7], and latter studied by Naserasr, Rollová and Sopena [14] and gave rise to a notion of signed chromatic number $\chi_s(G, \sigma)$ of a signed graph (G, σ) defined as the smallest order of a signed graph (H, π) to which (G, σ) admits a homomorphism.

In this paper, we are interested in the study of the Cartesian product of signed graphs, defined by Germina, Hameed and Zaslavsky in [6]. They mainly study the spectral properties of the Cartesian product.

The Cartesian product of two ordinary graphs G and H, noted $G \square H$, has been extensively studied. In 1957, Sabidussi [15] showed that $\chi(G \square H) = \max(\chi(G), \chi(H))$ where $\chi(G)$ is the chromatic number of the graph G. Another notable article on the subject by Sabidussi [16] shows that every connected graph G admits a unique prime decomposition, *i.e.*, there is a unique way to write a graph G as a product of some graphs up to isomorphism of the factors. This result was also independently discovered by Vizing in [17]. Another algebraic property, the cancellation property, which states that if $A \square B = A \square C$, then $B = C$, was proved by Imrich and Klavžar [10] using a technique of Fernández, Leighton and López-Presa [5]. On the complexity side, the main question associated with the Cartesian product is to be able to decompose a graph with the best possible complexity. The complexity of this problem has been improved successively in [1,3,4,18] to finally reach an optimal complexity of $O(m)$ in [11] where m is the number of edges of the graph.

© Springer Nature Switzerland AG 2020
M. Changat and S. Das (Eds.): CALDAM 2020, LNCS 12016, pp. 219–234, 2020.
https://doi.org/10.1007/978-3-030-39219-2_19

Our study of the Cartesian product of signed graphs is divided in several sections. First in Sect. 2, we present general definitions of graph theory and set our notation. In Sect. 3, we present some useful results on signed graphs and on the Cartesian product of graphs. In Sect. 4, we present the definition of the Cartesian product of signed graphs and give some first properties and easy consequences of the definition. We also prove the prime decomposition theorem for signed graphs and give an algorithm to decompose a signed product into its factors. We study the signed chromatic number of products of complete graphs in Sect. 5 and products of cycles in Sect. 6. Finally we present some concluding remarks in Sect. 7.

2   Definitions and Notation

All graphs we consider are undirected, simple and loopless. For classical graph definitions, we refer the reader to [2].

A *homomorphism* from G to H is a function φ from $V(G)$ to $V(H)$ such that for all $x, y \in V(G)$, $xy \in E(G)$ implies $\varphi(x)\varphi(y) \in E(H)$. In this case, we note $G \to H$. Note that $\chi(G)$, the *chromatic number of G* (see [2]), can also be defined as the smallest order of a graph H such that $G \to H$. An *isomorphism* from G to H is a bijection φ from $V(G)$ to $V(H)$ such that for all $x, y \in V(G)$, $xy \in E(G)$ if and only if $\varphi(x)\varphi(y) \in E(H)$. In this case, we note $G = H$.

A *walk* in a graph G is a sequence $s_0, \ldots, s_n$ of vertices of G such that $s_i s_{i+1} \in E(G)$. Its *starting vertex* is s_0 and its *end vertex* is s_n. A *closed walk* is a walk where its starting vertex s_0 and its end vertex s_n are identified (*i.e.* $s_0 = s_n$). If all elements of a walk are pairwise distinct, then the walk is a *path*. A closed walk where all elements are pairwise distinct (except s_0 and s_n) is a *cycle*. Suppose W is a walk $s_0, \ldots, s_n$, we define the *length* of W by its number of edges n (taken with multiplicity) and the *order* of W by its number of vertices (again taken with multiplicity). Note that a walk has order $n + 1$ while a closed walk has order n since we consider that s_0 and s_n to count for only one vertex.

A graph is *connected* if for all pairs of vertices $u, v \in V(G)$, there is a path between u and v. If $X \subseteq V(G)$, then the graph $G[X]$ is the subgraph of G induced by X. We say that $G[X]$ is an *induced subgraph* of G. The *complete graph* K_p is the graph of order p such that for all pair of distinct vertices of G, u and v, uv is an edge of K_p.

A *signed graph* (G, σ) is a graph G along with a function $\sigma : E(G) \to \{+1, -1\}$ called its *signature* and $\sigma(e)$ is the *sign* of the edge $e \in E(G)$. The edges in $\sigma^{-1}(+1)$ are the *positive* edges and the edges in $\sigma^{-1}(-1)$ are the *negative* edges of (G, σ). We often write a signed graph (G, σ) as (G, Σ) in place of (G, σ) where Σ is the set of negative edges $\sigma^{-1}(-1)$. These two ways to represent a signed graph are equivalent and will be used interchangeably. We note K_p^+ (resp. K_p^-) the complete graph $(K_p, \varnothing)$ (resp. $(K_p, E(K_p))$) of order p with only positive (resp. negative) edges.

Suppose that (G, σ) is a signed graph and W is a walk $s_0, \ldots, s_n$ in G. We say that W is a *balanced walk* if $\sigma(W) = \sigma(s_0 s_1)\sigma(s_1 s_2) \ldots \sigma(s_i s_{i+1}) \ldots \sigma(s_{n-1} s_n) = 1$ and an *unbalanced walk* otherwise. Similarly, this notion can be extended to closed walks, paths and cycles. We note an unbalanced path (resp. balanced path) of order k by UP_k (resp. BP_k) and an unbalanced cycle (resp. balanced cycle) of order k by UC_k (resp. BC_k). Similarly, we can define an unbalanced (resp. balanced) path UP_k (resp. BP_k) of order k. A signed graph where all closed walks are balanced (resp. unbalanced) is said to be *balanced* (resp. *antibalanced*). Generally, for the same ordinary graph G, there are several signatures σ such that (G, σ) is balanced. In particular it is the case for all signatures of a forest. These notions of balanced and antibalanced where introduced by Harary in [8].

This led Zaslavsky in [19] to define the notion of equivalent signed graphs. Two signed graphs (G, σ_1) and (G, σ_2) on the same underlying graph are *equivalent* if they have the same set of balanced closed walks. Note that this is equivalent to having the same set of balanced cycles. In this case we note $(G, \sigma_1) \equiv (G, \sigma_2)$. We also say that the two signatures σ_1 and σ_2 (resp. Σ_1 and Σ_2) are equivalent and we note $\sigma_1 \equiv \sigma_2$ (resp. $\Sigma_1 \equiv \Sigma_2$).

Let (G, σ) be a signed graph and v be a vertex of G. *Switching* (G, σ) at v creates the signed graph (G, σ') where $\sigma'(e) = -\sigma(e)$ if e incident to v and $\sigma'(e) = \sigma(e)$ otherwise. One can check that the switch operation does not modify the set of balanced closed walks as switching at a vertex of a closed walk does not change the sign of this walk. Moreover, two signed graphs are equivalent if and only if one can be obtained from the other by a sequence of switches [19]. This means that we can work with the balance of closed walks or with the switches depending on which notion is the easiest to use.

A *homomorphism* of a signed graph (G, σ) to a signed graph (H, π) is a homomorphism φ from G to H which maps balanced (resp. unbalanced) closed walks of (G, σ) to balanced (resp. unbalanced) closed walks of (H, π). Alternatively, a homomorphism from (G, σ) to (H, π) is a homomorphism from G to H such that there exists a signature σ' of G with $\sigma' \equiv \sigma$, such that if e is an edge of G, then $\pi(\varphi(e)) = \sigma'(e)$. When there is a homomorphism from (G, σ) to (H, π), we note $(G, \sigma) \longrightarrow_s (H, \pi)$ and say that (G, σ) *maps to* (H, π). Here (H, π) is the *target graph* of the homomorphism. When constructing a homomorphism, we can always fix a given signature of the target graph as proven in [14].

The *signed chromatic number* $\chi_s(G, \sigma)$ of a signed graph (G, σ) is the smallest k for which (G, σ) admits a homomorphism to a signed graph (H, π) of order k. Alternatively, a signed graph (G, σ) admits a k-*(vertex)-colouring* if there exists $\sigma' \equiv \sigma$ such that (G, σ') admits a proper vertex colouring $\theta : V(G) \rightarrow [\![k]\!]$ verifying that for every $i, j \in [\![k]\!]$, all edges uv with $\theta(u) = i$ and $\theta(v) = j$ have the same sign in (G, σ'). Then $\chi_s(G, \sigma)$ is the smallest k such that (G, σ) admits a k-vertex-colouring. The two definitions are equivalent, as with any colouring of a signed graph, we can associate a signed homomorphism which identifies the vertices with the same colour. The homomorphism is well defined as long as the target graph is simple, which is the case here by definition of a k-vertex-colouring.

Suppose that G and H are two ordinary graphs. The *Cartesian product* of G and H is the graph $G \square H$ whose vertex set is $V(G) \times V(H)$ and where (x, y) and (x', y') are adjacent if and only if $x = x'$ and y is adjacent to y' in H or $y = y'$ and x is adjacent to x' in G.

A graph G is *prime* if there are no graphs A and B on at least two vertices for which $G = A \square B$. A decomposition D of a graph G is a multi-set $\{G_1, \ldots, G_k\}$ such that the G_i's are graphs containing at least one edge and $G = G_1 \square \cdots \square G_k$. A decomposition is *prime* if all the G_i's are prime. The G_i's are called *factors* of G. A decomposition D is *finer* than another decomposition D' if $D' = \{G_1, \ldots, G_k\}$ and for each i there exists a decomposition D_i of G_i such that $D = \bigcup_i D_i$. Note that by definition, every decomposition is finer than itself.

Suppose that G is a graph and $D = \{G_1, \ldots, G_k\}$ is a decomposition of G such that $G = G_1 \square \ldots \square G_k$. A *coordinate system* for G under decomposition D is a bijection $\theta : V(G) \to \prod_{i=1}^{k} V(G_i)$ verifying that for each vertex v of G, the set of vertices which differ from v by the ith coordinate induces a graph noted G_i^v called a G_i-*layer* which is isomorphic to G_i by the projection on the ith coordinate. We say that a subgraph X of G is a copy of G_i if X is a G_i-layer of G. An edge uv of G is a copy of an edge ab of G_i if $\theta(u)$ and $\theta(v)$ differ only in their ith coordinate with $u_i = a$ and $v_i = b$.

3   Preliminary Results

The goal of this section is to present useful results on signed graphs and on the Cartesian product.

In [19], Zaslavsky gave a way to determine if two signed graphs are equivalent in linear time. In particular, all signed forests with the same underlying graph are equivalent. This theorem comes from the following observation.

Lemma 1 (Zaslavsky [19]). *If C is a cycle of a graph G, then the parity of the number of negative edges of C in (G, σ) is the same in all (G, σ') with $\sigma' \equiv \sigma$.*

This implies that we can separate the set of all cycles into four families BC_{even}, BC_{odd}, UC_{even} and UC_{odd}, depending on the parity of the number of negative edges (even for BC_{even} and BC_{odd} and odd for UC_{even} and UC_{odd}) and the parity of the length of the cycle (even for BC_{even} and UC_{even} and odd for BC_{odd} and UC_{odd}).

Theorem 1. *If (C, σ) is a signed cycle with $(C, \sigma) \in BC_{even}$, then $\chi_s(C, \sigma) = 2$. If $(C, \sigma) \in BC_{odd} \cup UC_{odd}$, then $\chi_s(C, \sigma) = 3$. Finally, if $(C, \sigma) \in UC_{even}$, then $\chi_s(C, \sigma) = 4$.*

Proof. By [13], we already have the upper bounds. A homomorphism of signed graphs is also a homomorphism of graphs thus $\chi(C) \leq \chi_s(C, \sigma)$. If $(C, \sigma) \equiv UC_{2q}$ suppose its signed chromatic number is less than 3. Then $(C, \sigma) \longrightarrow_s (H, \pi)$ where (H, π) is a triangle or a path. In each case, (H, π) can be switched to be all positive or all negative. This means that (C, σ) can be switched to be all positive or all negative, which is not the case, a contradiction. We get the desired lower bounds in each case. □

One of the first results on Cartesian products is a result from Sabidussi on the chromatic number of the product of two graphs.

Theorem 2 (Sabidussi [15]). *If G and H are two graphs, then $\chi(G \square H) = \max(\chi(G), \chi(H))$.*

Following this paper, Sabidussi proved one of the most important results on the Cartesian product: the unicity of the prime decomposition of connected graphs. This result was also proved by Vizing.

Theorem 3 (Sabidussi [16] and Vizing [17]). *Every connected graph G admits a unique prime decomposition up to the order and isomorphisms of the factors.*

Using some arguments of [5] and the previous theorem, Imrich and Klavžar proved the following theorem.

Theorem 4 (Imrich and Klavžar in [9] and [10]). *If A, B and C are three ordinary graphs such that $A \square B = A \square C$, then $B = C$.*

The unicity of the prime decomposition raises the question of the complexity of finding such a decomposition. The complexity of decomposing algorithms has been extensively studied. The first algorithm, by Feigenbaum *et al.* [4], had a complexity of $O(n^{4.5})$ where n is the order of the graph (its size is denoted by m). In [18], Winkler proposed a different algorithm improving the complexity to $O(n^4)$. Then Feder [3] gave an algorithm in $O(mn)$ time and $O(m)$ space. The same year, Aurenhammer *et al.* [1] gave an algorithm in $O(m \log n)$ time and $O(m)$ space. The latest result is this time optimal algorithm.

Theorem 5 (Imrich and Peterin [11]). *The prime factorization of connected graphs can be found in $O(m)$ time and space. Additionally a coordinate system can be computed in $O(m)$ time and space.*

4   Cartesian Products of Signed Graphs

We recall the definition of the signed Cartesian product.

Definition 1 ([6]). *Let (G, σ) and (H, π) two signed graphs. We define the signed Cartesian product of (G, σ) and (H, π), and note $(G, \sigma) \square (H, \pi)$, the signed graph with vertex set $V(G) \times V(H)$. It has for positive (resp. negative) edges the pairs $\{(u_1, v_1), (u_2, v_2)\}$ such that $u_1 = u_2$ and $v_1 v_2$ is a positive (resp. negative) edge of (H, π) or such that $v_1 = v_2$ and $u_1 u_2$ is a positive (resp. negative) edge of (G, σ). The underlying graph of $(G, \sigma) \square (H, \pi)$ is the ordinary graph $G \square H$.*

Using this definition, we can derive that this product is associative and commutative.

The following result shows that it is compatible with the homomorphism of signed graphs and in particular with the switching operation.

Theorem 6. If (G, σ), (G', σ'), (H, π), (H', π') are four signed graphs such that $(G, \sigma) \longrightarrow_s (G', \sigma')$ and $(H, \pi) \longrightarrow_s (H', \pi')$, then:

$$(G, \sigma) \,\square\, (H, \pi) \longrightarrow_s (G', \sigma') \,\square\, (H', \pi').$$

Proof. By commutativity of the Cartesian product and composition of signed homomorphisms, it suffices to show that $(G, \sigma) \,\square\, (H, \pi) \longrightarrow_s (G', \sigma') \,\square\, (H, \pi)$. Since $(G, \sigma) \longrightarrow_s (G', \sigma')$, there exists a set S of switched vertices and a homomorphism φ from G to G' such that if (G, σ_S) is the signed graph obtained from (G, σ) by switching at the vertices of S, then $\sigma'(\varphi(e)) = \sigma_S(e)$ for every edge e of G. We note $P = (G, \sigma) \,\square\, (H, \pi)$ and $X = \{(g, h) \in V(G \,\square\, H) \mid g \in S\}$. Let P' be the signed graph obtained from P by switching at the vertices in X.

If $(g, h)(g, h')$ is an edge of P, then in P' this edge was either switched twice if $g \in S$ or not switched if $g \notin S$. In both cases its sign did not change. If $(g, h)(g', h)$ is an edge of P, then in P' this edge was switched twice if $g, g' \in S$, switched once if $g \in S, g' \notin S$ or $g \notin S, g' \in S$, and not switched if $g, g' \notin S$. In each case its new sign is $\sigma_S(gg')$. Thus $P' = (G, \sigma_S) \,\square\, (H, \pi)$. Now define $\varphi_P(g, h) = (\varphi(g), h)$. It is a homomorphism from $G \,\square\, H$ to $G' \,\square\, H$ by definition. By construction, the target graph of φ_P is $(G', \sigma') \,\square\, (H, \pi)$ as the edges of H do not change and the target graph of φ is (G', σ'). $\qquad\square$

As mentioned, we can derive the following corollary from Theorem 6.

Corollary 1. If (G, σ), (G, σ'), (H, π), (H, π') are four signed graph such that $\sigma \equiv \sigma'$ and $\pi \equiv \pi'$, then:

$$(G, \sigma) \,\square\, (H, \pi) \equiv (G, \sigma') \,\square\, (H, \pi').$$

One first observation is that we can apply Theorem 6 to the case of forests.

Corollary 2. If (G, σ) is a signed graph and (F, π) is a signed forest with at least one edge, then:

$$\chi_s((G, \sigma) \,\square\, (F, \pi)) = \chi_s((G, \sigma) \,\square\, K_2).$$

In particular, for $n, m \geq 2$, $\chi_s((P_n, \sigma_1) \,\square\, (P_m, \sigma_2)) = 2$.

Note that there is a difference between considering the chromatic number of a signed product and the chromatic number of a signed graph with the product graph as underlying graph. For example, $BC_4 = K_2 \,\square\, K_2$ but $\chi_s(UC_4) \neq \chi_s(BC_4)$. Another example: $\chi_s(P_n \,\square\, P_m) = 2$, for any $n, m \in \mathbb{N}$ but the following theorem shows that not all signed grids have chromatic number 2.

Theorem 7. *If n, m are two integers and (G, σ) is a signed grid with $G = P_n \square P_m$, then $\chi_s(G) \leq 6$. If n or m is less than 4, then $\chi_s(G) \leq 5$. Moreover there exist grids with signed chromatic number 5.*

Due to page constraints, we do not include the proof of this result. Nonetheless, the proof is available in the full version of this paper which you can find on the author's web page. Note that we are not aware of a signed grid with signed chromatic number six.

Question 1. What is the maximal value of $\chi_s(G, \sigma)$ when (G, σ) is a signed grid? Is it 5 or 6?

Our goal is now to prove that each connected signed graph has a unique prime s-decomposition. Let us start with some definitions.

Definition 2. *A signed graph (G, σ) is said to be s-prime if and only if there do not exist two signed graphs (A, π_A) and (B, π_B) such that $(G, \sigma) \equiv (A, \pi_A) \square (B, \pi_B)$. An s-decomposition D of a signed connected graph (G, σ) is a multi-set $\{(G_1, \pi_1), \ldots, (G_k, \pi_k)\}$ such that:*

1. *the (G_i, π_i)'s are signed graphs containing at least one edge and*
2. *$(G, \pi) \equiv (G_1, \pi_1) \square \cdots \square (G_k, \pi_k)$.*

An s-decomposition D is prime *if all the (G_i, π_i)'s are s-prime. The (G_i, π_i)'s are called* factors *of D. An s-decomposition D is* finer *than another s-decomposition D' if $D' = \{(G_1, \pi_1), \ldots, (G_k, \pi_k)\}$ and for each $i \in [1, k]$ there exists an s-decomposition D_i of (G_i, π_i) such that $D = \bigcup_i D_i$. Recall that, if D is finer than D', then we may have $D = D'$.*

Suppose $D = \{G_1, \ldots, G_k\}$ is a decomposition of a graph G. We say that two copies X_1 and X_2 of G_i are adjacent by G_j *if and only if there exists an edge ab of G_j such that for all $u \in V(G_i)$ and u_1, u_2 its corresponding vertices in X_1 and X_2, $u_1 u_2$ is a copy of ab. In other words, the subgraph induced by the vertices of X_1 and X_2 is isomorphic to $G_i \square K_2$ where the edge K_2 corresponds to a copy of an edge of G_j.*

Note that if $G = A \square B$, then it is not always true that (G, σ) is the product of two signed graphs. For example, for $(G, \sigma) \equiv UC_4$, it is s-prime but C_4 is not a prime graph as $C_4 = K_2 \square K_2$.

We now present a useful tool to show that a signed graph is a product.

Lemma 2. *If (G, σ), (A, π_A) and (B, π_B) are three connected signed graphs with $G = A \square B$, then we have $(G, \sigma) \equiv (A, \pi_A) \square (B, \pi_B)$ if and only if:*

1. *all copies of A are equivalent to (A, π_A),*
2. *one copy of B is equivalent to (B, π_B) and*
3. *for each edge e of A, for each pair of distinct copies e_1, e_2 of e, if these two edges belong to the same square, then it is a BC_4.*

Proof. ($\Rightarrow$) This follows from the definition of the Cartesian product.

($\Leftarrow$) We will do the following switches: switch all copies of A to have the same signature π_A.

Now we claim that all copies of B have the same signature π'_B equivalent to π_B. Indeed take one edge xy of B and two copies of this edge x_1y_1 and x_2y_2 in G. Take a shortest path from x_1 to x_2 in the first copy of A. Now if u_1, u_2 are two consecutive vertices along the path and v_1 and v_2 are their copy in the second copy of A, then $u_1u_2v_2v_1$ is a BC_4 by the third hypothesis.

As u_1u_2 and v_1v_2 have the same sign by the previous switches, it must be that u_1v_1 and u_2v_2 have the same sign. Thus all copies of an edge of B have the same sign.

Thus $(G, \sigma) \equiv (A, \pi_A) \,\square\, (B, \pi'_B) \equiv (A, \pi_A) \,\square\, (B, \pi_B)$ by Theorem 6. $\square$

One of our main results is the prime decompositions theorem.

Theorem 8. *If (G, σ) is a connected signed graph and D is the prime decomposition of G, then (G, σ) admits a unique (up to isomorphism of the factors) prime s-decomposition D_s. Moreover if we see D_s as a decomposition, then D is finer than D_s.*

For its proof, we need the following lemma.

Lemma 3. *If (G, σ) is a connected signed graph that admits two prime s-decomposition D_1 and D_2, then there is a signed graph (X, π_X) such that $(G, \sigma) \equiv (X, \pi_X) \,\square\, (Y, \pi_Y)$ with $D_1 = \{(X, \pi_X)\} \cup D'_1$ and $D_2 = \{(X, \pi_X)\} \cup D'_2$ where D'_1 and D'_2 are two decompositions of (Y, π_Y).*

Proof. Suppose there exists a signed graph (G, σ) that admits two s-decompositions D_1 and D_2. Fix an edge e, e belongs to some factor Z of the prime decomposition of G. The edge e belongs to some copy of some signed graph (A, π_A) in D_1 and to some copy of (B, π_B) in D_2. The graph Z is a factor of A and B by unicity of the prime factor decomposition of G. Let X be the greatest common divisor of A and B, we have $e \in E(X)$ as $e \in E(Z)$. Now $G = X \,\square\, Y$ for some graph Y. Let us show that $(G, \sigma) \equiv (X, \pi_X) \,\square\, (Y, \pi_Y)$ for some signature π_X and π_Y of X and Y. We can suppose that $Y \neq K_1$, as otherwise the result is trivial.

First we want to show that all copies of X have equivalent signatures. Take two adjacent copies of X, if they are in different copies of A, then they are equivalent since they represent the same part of (A, π_A). If they are in the same copy of A, then they are in different copies of B since X is the greatest common divisor of A and B. The same argument works in this case. Thus two adjacent copies of X are two equivalent copies of some signed graph (X, π_X), and since there is only one connected component in Y, all copies of X have equivalent signatures.

Let π_Y be the signature of one copy of Y. Fix e an edge of X and X_1 and X_2 two copies of X. Now consider the square containing the two copies of this edge (if exists), if X_1 and X_2 are in different copies of A, then this is a BC_4 by Lemma 2 in G, otherwise it is a BC_4 as they are in different copies of B.

By Lemma 2, we can conclude that $(G, \sigma) \equiv (X, \pi_X) \,\square\, (Y, \pi_Y)$.

Now suppose that $A = X \,\square\, W$, then we can use Lemma 2 to show $(A, \pi_X) \equiv (X, \pi_X) \,\square\, (W, \pi_W)$ as all copies of X have equivalent signatures by $(G, \sigma) \equiv (X, \pi_X) \,\square\, (Y, \pi_Y)$ and all C_4 between two copies of an edge are BC_4 by the same argument. As (A, π_A) is s-prime this is absurd, so $(X, \pi_X) \equiv (A, \pi_A)$. Thus $(X, \pi_X) \equiv (A, \pi_A) \equiv (B, \pi_B)$ and this proves the lemma. $\square$

Proof of Theorem 8. Any signed graph has a prime s-decomposition by taking an s-decomposition of (G, σ) that cannot be refined. If we take a prime s-decomposition D_s of (G, σ), then it is a decomposition of G thus the prime decomposition of G is finer than D_s.

It is left to show that the prime s-decomposition of (G, σ) is unique. Suppose, to the contrary, that (G, σ) is a minimal counterexample to the unicity. Thus (G, σ) has two prime s-decompositions D_1 and D_2 and by Lemma 3, $(G, \sigma) \equiv (X, \pi_X) \,\square\, (Y, \pi_Y)$ with $D_1 = \{(X, \pi_X)\} \cup D_1'$ and $D_2 = \{(X, \pi_X)\} \cup D_2'$ where D_1' and D_2' are two decompositions of (Y, π_Y). By minimality of (G, σ), $D_1' = D_2'$ in Y. Thus $D_1 = D_2$, a contradiction. $\square$

Note that Theorem 8 implies the following result.

Theorem 9. *If $(A, \pi_A), (B, \pi_B)$ and (C, π_C) are three signed graphs with $(A, \pi_A) \,\square\, (B, \pi_B) \equiv (A, \pi_A) \,\square\, (C, \pi_C)$, then $(B, \pi_B) \equiv (C, \pi_C)$.*

The proof of this result is exactly the same as the proof for ordinary graphs presented in [10]. Indeed, we have all the necessary tools for the proof. The first one is Theorem 8, the other one is the semi-ring structure of signed graphs (quotiented by the equivalence relation) with the disjoint union and the Cartesian product which follows from the definition. See [10] for more details.

In the last part of this section, we propose an algorithm to decompose connected signed graphs. Decomposing a graph can be interpreted in multiple ways: finding a decomposition, identifying which edge of G belongs to which factor, or even better getting a coordinate system that is compatible with the decomposition. In [11], Imrich and Peterin gave an $O(m)$ time and space (m is the number of edges of G) algorithm for these three questions for ordinary graphs.

Our goal is to give an algorithm that does as much for signed graphs based on their algorithm. Due to size constraints, we do not give all the details of the algorithm but only the key ideas. Nonetheless, the proof is available in the full version of this paper which you can find on the author's web page.

A coordinate system for an s-decomposition D of (G, σ) is a coordinate system for D seen as an ordinary decomposition such that there exists a fixed equivalent signed graph (G, σ') of (G, σ) in which all layers have the same signature. This definition is very similar to the definition of a coordinate system for ordinary graphs but differ by the fact that we can switch our graph in order to have a fixed signature for which all isomorphisms are tested.

The idea of the algorithm is to start by partitioning (G, σ) with the prime decomposition of G and to merge some factors in order to get the prime

s-decomposition of (G, σ). If at one point of the algorithm we get an s-decomposition, then we can stop as all factors must be prime by the merging rules. As we remarked before, UC_4's are s-prime graphs thus in an s-decomposition of a signed graph, the UC_4 are included in the prime factors, this means that finding a UC_4 in (G, σ) between two distinct factors of our current decomposition implies that these two factors (of the ordinary graph) belong to the same s-prime factor of the signed graph.

Another reason to merge is when one factor X has two copies that do not represent the same signed graph. Then we need to find two adjacent copies X_1 and X_2 of X adjacent by factor Y, and we must merge the factor X with the factor Y. This can be tested efficiently in $O(m)$. Note that the proof of this fact amounts to finding a UC_4 between the two factors but it is faster to keep this remark as a separate rule. Note that it might be possible to use technics of [12] to improve the time complexity to an optimal $O(m)$ by avoiding to use the second merging rule.

With these two merging rules we can prove that once no merging is needed to be done, we get the prime s-decomposition of (G, σ).

Theorem 10. *Let (G, σ) be a connected signed graph of order n and size m. If k is the number of factors in the prime decomposition of G, then we can find in time $O(mk) = O(m \log n)$ and space $O(m)$ the prime s-decomposition of (G, σ) and a coordinate system for this decomposition.*

5   Signed Chromatic Number of Cartesian Products of Complete Graphs

In this section, we show a simple upper bound on the signed chromatic number of a product and compute the signed chromatic number of some special complete graphs. We start by defining a useful tool on signed graphs.

In what follows we define the notion of an s-redundant set in a signed graph. Informally, it is a set of vertices such that removing them does not remove any distance two constraint between the pairs of vertices left in the graph.

Definition 3. *Let (G, σ) be a signed graph and $S \subseteq V(G)$. We say that the set S is s-redundant if and only if, for every $x, y \in V(G) - S$ such that $xy \notin E(G)$, for every $z \in S$ and for all signatures σ' with $\sigma' \equiv \sigma$, in the signed graph (G, σ'), $xzy = UP_3$ implies that there exists $w \in V(G) - S$ such that $xwy = UP_3$.*

The following proposition is an alternative formulation of the definition which is useful in order to prove that a set is an s-redundant set.

Proposition 1. *If (G, σ) is a signed graph and $S \subseteq V(G)$, then S is s-redundant if and only if for every $z \in S$, and every $x, y \in N(z) \setminus S$ with $xy \notin E(G)$, there exists $w \in V(G) \setminus S$ such that $xwyz$ is a BC_4.*

Proof. Take $x, y \in V(G) - S$ such that $xy \notin E(G)$ and $z \in S$. If $xzy = UP_3$ in a signature $\sigma' \equiv \sigma$, then $x, y \in N(z)$. Now if S is a redundant set, then with the notation of the definition $xzyw$ is a BC_4 in (G, σ') and thus in (G, σ). If $xzyw$ is a BC_4 and xzy is a UP_3 in a given signature σ', then xwy is also a UP_3 as $xzyw$ is balanced. This proves the equivalence between the two statements. $\square$

The next theorem is the reason why we defined this notion. It allows us to compute an upper bound of the chromatic number of a graph given the chromatic number of one of its subgraphs. One example of utilisation of this notion is given by the proof of Theorem 12.

Theorem 11. *If (G, σ) is a signed graph and S is an s-redundant set of (G, σ), then:*

$$\chi_s(G, \sigma) \leq |S| + \chi_s((G, \sigma) - S).$$

Proof. Let φ be a homomorphism from $(G, \sigma) - S$ to a signed graph (H, π) of order $\chi_s((G, \sigma) - S)$. We construct a homomorphism φ' of (G, σ) as follows:

$$\varphi'(v) = \begin{cases} \varphi(v) & \text{if } v \notin S, \\ v & \text{otherwise.} \end{cases}$$

This homomorphism is well defined. To show this we need to prove that the image of φ' has no loops and no UC_2 (*i.e.* is simple). It does not create loops as φ does not. As φ is well defined, if there was a UC_2, it would come from identification of two vertices x and y in $(G, \sigma) - S$ and $z \in S$ such that xzy is a UP_3. By definition of S, if so, there would be a UC_2 in $(G, \sigma) - S$, which cannot be.

Thus φ is a well defined homomorphism. Thus $\chi_s(G, \sigma) \leq |S| + \chi_s((G, \sigma) - S)$.

$\square$

This result does not hold for any set S. For example, if $(G, \sigma) = UC_4$ and $S = \{v\}$ is a single vertex of G, then $\chi_s(G, \sigma) = 4$ but $\chi_s((G, \sigma) - v) \leq 3$.

As a direct corollary of Theorem 6, we get the following upper bound on the chromatic number of a product of signed graphs.

Corollary 3. *If $(G_1, \sigma_1), \ldots, (G_k, \sigma_k)$ are k signed graphs, then:*

$$\chi_s((G_1, \sigma_1) \square \cdots \square (G_k, \sigma_k)) \leq \prod_{1 \leq i \leq k} \chi_s(G_i, \sigma_i).$$

We study the product of balanced and antibalanced complete graphs in our next result. Recall that K_p^+ (resp. K_q^-) is the complete graph with only positive edges (resp. negative edges).

Theorem 12. *For every two integers p, q with $p, q \geq 2$, we have*

$$\chi_s(K_p^+ \square K_q^-) = \left\lceil \frac{pq}{2} \right\rceil.$$

Proof. Let us note $(P, \pi) = K_p^+ \,\square\, K_q^-$. First let us show that $\chi_s(P, \pi) \geq \lceil \frac{pq}{2} \rceil$.

Suppose it is not the case. By the pigeon hole principle, there exist x, y and z three vertices of the product that were identified. These vertices cannot be in the same lines or columns as these induce cliques. Consider the subgraph (H, σ) of the product composed of vertices which are in the same line as one of x, y, z and in the same column as one of x, y and z. We have $(H, \sigma) = K_3^+ \,\square\, K_3^-$ (see Fig. 1).

By assumption x, y and z of (H, σ) are identified. By the pigeon hole principle, two of x, y and z have been switched the same way. Without loss of generality suppose they are x and y. Then if a is one of their common neighbours in H, the edges xa and ya are of different signs, thus x and y cannot be identified. This is a contradiction.

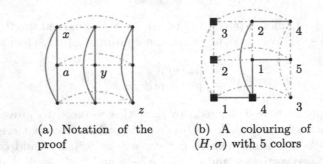

(a) Notation of the proof

(b) A colouring of (H, σ) with 5 colors

Fig. 1. The graph (H, σ) of the proof of the Theorem 12. The big squared vertices have been switched.

Now let us show that $\chi_s(P, \pi) \leq \lceil \frac{pq}{2} \rceil$. By symmetry suppose that $p \geq q$. We will prove this statement by induction. If $p = 2$, then $(P, \pi) \equiv BC_4$ and $\chi(P, \pi) = 2 \leq 2$. If $p = 3$ and $q = 2$, then $(P, \pi) \equiv BC_3 \,\square\, K_2$ whose chromatic number is $3 \leq 3$. If $p = 3$ and $q = 3$, then $(P, \pi) \equiv (H, \sigma)$ from the previous part of the proof. We have $\chi_s(P, \pi) = 5$, indeed Fig. 1 gives a 5-colouring of (P, π).

Now we can assume that $p \geq 4$. Suppose the vertices of P are labelled $v_{(i,j)}$ for $1 \leq i \leq p$ and $1 \leq j \leq q$ such that the labelling corresponds to the product. Now switch all vertices in $\{v_{(i,j)} \mid i = 1\}$. Now identify $v_{(1,j)}$ with $v_{(2,j+1)}$ (which are non adjacent) where the indices are taken modulo q to get the graph (P', π'). Let S be the set of identified vertices in (P', π'). We want to show that S is s-redundant to use the induction hypothesis. Take $z \in S$ and $x, y \in N(z) \setminus S$ such that $xy \notin E(P)$. If xzy is an unbalanced path of length 2, then x is some $v_{(i,j)}$ and y is some $v_{(k,j+1)}$ with $i, k \geq 3$. For $a = v_{(i,j+1)}$, $xayz$ is a BC_4.

By Proposition 1, S is s-redundant thus $\chi_s(P, \pi) \leq \chi_s(P', \pi') \leq |S| + \chi_s((P', \pi') - S)$ by Theorem 11. By induction hypothesis, as $(P', \pi') - S = K_{p-2}^+ \,\square\, K_q^-$, we get $\chi_s((P', \pi') - S) \leq \lceil \frac{(p-2)q}{2} \rceil$. Thus $\chi_s(P, \pi) \leq q + \lceil \frac{pq}{2} \rceil - q \leq \lceil \frac{pq}{2} \rceil$. $\qquad\square$

For this product the upper bound of Corollary 3 is pq. We thus have an example where the chromatic number is greater than half the simple upper bound.

Question 2. What is the supremum of the set of $\lambda \in [\frac{1}{2}, 1]$ such that there exist arbitrarily big signed graphs (G_1, σ_1) and (G_2, σ_2) such that $\chi_s((G_1, \sigma_1) \,\Box\, (G_2, \sigma_2)) \geq \lambda \cdot \chi_s(G_1, \sigma_1) \cdot \chi_s(G_2, \sigma_2)$?

Table 1. The signed chromatic number for each type of products of two cycles.

$(G, \sigma) \,\Box\, (H, \pi)$	BC_{even}	BC_{odd}	UC_{even}	UC_{odd}
BC_{even}	2	3	4	3
BC_{odd}	3	3	5	5
UC_{even}	4	5	4	5
UC_{odd}	3	5	5	3

6 Signed Chromatic Number of Cartesian Products of Cycles

The goal of this section is to determine the chromatic number of the product of two cycles. As there are four kind of cycles (balanced/unbalanced and of even/odd length), we have a number of cases to analyse. In most cases some simple observations are sufficient to conclude. For the other cases, we need the following lemma.

Lemma 4. *For every two integers* $p, q \in \mathbb{N}$:

$$\chi_s(UC_q \,\Box\, BC_{2p+1}) > 4.$$

Due to size constraints we do not include the proof of Lemma 4. Nonetheless, the proof is available in the full version of this paper which you can find on the author's web page. With this lemma, we can state this section's main result.

Theorem 13. *If* (G, σ) *and* (H, π) *are two signed cycles, then the signed chromatic number of* $(P, \rho) = (G, \sigma) \,\Box\, (H, \pi)$ *is given by Table 1.*

Proof. If G is a cycle of type BC_{even} (resp. BC_{odd}, UC_{even}, UC_{odd}), then $G \longrightarrow_s BC_2 = K_2$ (resp. BC_3, UC_4, UC_3). By computing the signed chromatic numbers of the products of (G, σ) and (H, π) when they belong to $\{K_2, BC_3, UC_4, UC_3\}$, we get an upper bound for each of the product type equal to the corresponding value in the table. These cases are represented in Fig. 2. Note that to color some graphs, we switched at some vertices.

For the lower bound, we can see that $\chi_s((G, \sigma) \,\Box\, (H, \pi)) \geq \max(\chi_s(G, \sigma), \chi_s(H, \pi))$. Note that $\chi_s(BC_{even}) = 2$, $\chi_s(BC_{odd}) = 3$, $\chi_s(UC_{even}) = 4$ and $\chi_s(UC_{odd}) = 3$ by Theorem 1. Lemma 4 allows us to conclude for the remaining cases as $\chi_s(UC_q \,\Box\, BC_{2p+1}) = \chi_s(UC_q \,\Box\, UC_{2p+1})$ by symmetry between the two edge types. $\quad\Box$

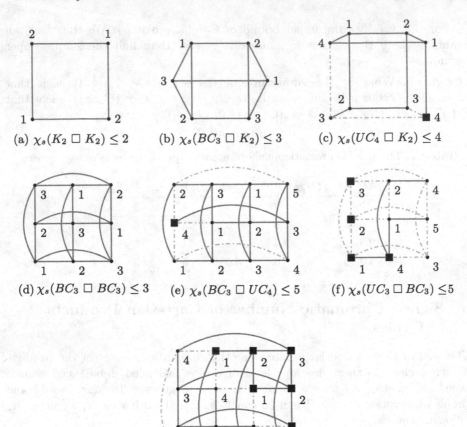

(a) $\chi_s(K_2 \,\square\, K_2) \le 2$ (b) $\chi_s(BC_3 \,\square\, K_2) \le 3$ (c) $\chi_s(UC_4 \,\square\, K_2) \le 4$

(d) $\chi_s(BC_3 \,\square\, BC_3) \le 3$ (e) $\chi_s(BC_3 \,\square\, UC_4) \le 5$ (f) $\chi_s(UC_3 \,\square\, BC_3) \le 5$

(g) $\chi_s(UC_4 \,\square\, UC_4) \le 4$

Fig. 2. All the cases for the product of cycles up to symmetry positive/negative. The large squared vertices have been switched from the product signature.

One further question would be to compute the signed chromatic number of an arbitrary number of signed cycles. An interesting remark is that $BC_3 \,\square\, BC_3 \longrightarrow_s BC_3$. This is also true for K_2, UC_3 and UC_4. Thus if we suppose that the length of the cycles does not impact the result outside of their parity, the only interesting case for upper bounds would be to compute $\chi_s(BC_3 \,\square\, UC_3 \,\square\, UC_4)$.

We now give the idea of the proof of Lemma 4. The proof is by contradiction. If $\chi_s(UC_q \,\square\, BC_{2p+1}) > 4$, then $(P, \rho) = UC_q \,\square\, BC_{2p+1} \longrightarrow_s (T, \theta)$ where (T, θ) is a complete graph of order 4. By looking at which (T, θ) are suitable targets, we get that (T, θ) must be the graph K_4 where only one edge is negative. (P, ρ) must be switched into another signed graph (P, ρ') which maps to (T, θ)

without switches. By counting the negative edges in (P, ρ') in different ways we get our contradiction.

In particular, we use topological arguments on (P, ρ') seen as a toroidal grid. We look at connected components of the negative edges and create a set of closed walks "surrounding" them. These closed walks are bipartite, this fact gives us constraints on how they can wrap around the torus. These constraints allows us to count the negative edges one way. The other way to count is directly given by the definition of the product.

7 Conclusion

To conclude, in this paper, we showed a number of results on Cartesian products of signed graphs. We proved a number of algebraic properties: Theorems 6, 8 and 9. We also presented an algorithm to decompose a signed graph into its factors in time $O(m \log n)$. This complexity is theoretically not optimal, thus we can ask the question: can we decompose a signed graph in linear time?

Finally, we computed the chromatic number of products: products of any graphs by a signed forest, products of signed paths, signed graphs with underlying graph $P_n \, \square \, P_m$, products of some signed complete graphs and products of signed cycles. We also presented a tool called a s-redundant set that helped us compute chromatic number of signed graphs. It would be interesting to know how close to the simple upper bound can be the chromatic number of the product of two signed graphs. It is not clear if there is a sequence of signed graphs approaching this bound or if there is a constant $\lambda < 1$ such that for all signed graphs (G, Σ) and (H, Π), $\chi_s((G, \Sigma) \, \square \, (H, \Pi)) \leq \lambda \max(\chi_s(G, \Sigma), \chi_s(H, \Pi))$. It would also be interesting to compute the chromatic number of more products.

Acknowledgements. We would like to thank Hervé Hocquard and Éric Sopena for their helpful comments through the making of this paper. We would also like to thank the reviewers for their comments and especially Reviewer 2 for pointing us to the techniques of [12] which could improve our algorithm.

References

1. Aurenhammer, F., Hagauer, J., Imrich, W.: Cartesian graph factorization at logarithmic cost per edge. Comput. Complex. **2**(4), 331–349 (1992)
2. Bondy, J.A., Murty, U.S.R.: Graph Theory, 1st edn. Springer, Dordrecht (2008). https://doi.org/10.1007/978-1-4020-6754-9
3. Feder, T.: Product graph representations. J. Graph Theory **16**(5), 407–488 (1992)
4. Feigenbaum, J., Hershberger, J., Schäffor, A.A.: A polynomial time algorithm for finding the prime factors of cartesian-product graphs. Discrete Appl. Math. **12**(2), 123–138 (1985)
5. Fernández, A., Leighton, T., López-Presa, J.L.: Containment properties of product and power graphs. Discrete Appl. Math. **155**(3), 300–311 (2007)
6. Germina, K.A., Hameed K, S., Zaslavsky, T.: On products and line graphs of signed graphs, their eigenvalues and energy. Linear Algebra Appl. **435**(10), 2432–2450 (2011). Special Issue in Honor of Dragos Cvetkovic

7. Guenin, B.: Packing odd circuit covers: a conjecture. Manuscript (2005)
8. Harary, F.: On the notion of balance of a signed graph. Michigan Math. J. **2**(2), 143–146 (1953)
9. Imrich, W., Klavžar, S.: Product Graphs, Structure and Recognition, January 2000
10. Imrich, W., Klavžar, S., Rall, D.F.: Cancellation properties of products of graphs. Discrete Appl. Math. **155**(17), 2362–2364 (2007)
11. Imrich, W., Peterin, I.: Recognizing cartesian products in linear time. Discrete Math. **307**(3), 472–483 (2007). Algebraic and Topological Methods in Graph Theory
12. Imrich, W., Peterin, I.: Cartesian products of directed graphs with loops. Discrete Math. **341**(5), 1336–1343 (2018)
13. Jacques, F., Montassier, M., Pinlou, A.: The chromatic number and switching chromatic number of 2-edge-colored graphs of bounded degree. Private communication
14. Naserasr, R., Rollová, E., Sopena, É.: Homomorphisms of signed graphs. J. Graph Theory **79**(3), 178–212 (2015)
15. Sabidussi, G.: Graphs with given group and given graph theoretical properties. Canad. J. Math **9**, 515–525 (1957)
16. Sabidussi, G.: Graph multiplication. Mathematische Zeitschrift **72**, 446–457 (1959/60)
17. Vizing, V.G.: The cartesian product of graphs (russian). Vycisl. Sistemy **9**, 30–43 (1963)
18. Winkler, P.: Factoring a graph in polynomial time. Eur. J. Comb. **8**(2), 209–212 (1987)
19. Zaslavsky, T.: Signed graphs. Discrete Appl. Math. **4**(1), 47–74 (1982)

List Distinguishing Number of pth Power of Hypercube and Cartesian Powers of a Graph

L. Sunil Chandran[1], Sajith Padinhatteeri[2]([✉]) [iD], and Karthik Ravi Shankar[3]

[1] Department of CSA, Indian Institute of Science, Bangalore, India
sunil@iisc.ac.in
[2] Department of ECE, Indian Institute of Science, Bangalore, India
sajithp@iisc.ac.in
[3] Department of ES, Indian Institute of Technology Hyderabad, Hyderabad, India
es16btech11014@iith.ac.in

Abstract. A graph G is said to be k-distinguishable if every vertex of the graph can be colored from a set of k colors such that no non-trivial automorphism fixes every color class. The distinguishing number $D(G)$ is the least integer k for which G is k-distinguishable. If for each $v \in V(G)$ we have a list $L(v)$ of colors, and we stipulate that the color assigned to vertex v comes from its list $L(v)$ then G is said to be $\mathcal{L}$-distinguishable where $\mathcal{L} = \{L(v)\}_{v \in V(G)}$. The list distinguishing number of a graph, denoted $D_l(G)$, is the minimum integer k such that every collection of lists $\mathcal{L}$ with $|L(v)| = k$ admits an $\mathcal{L}$-distinguishing coloring. In this paper, we prove that
- when a connected graph G is prime with respect to the Cartesian product then $D_l(G^r) = D(G^r)$ for $r \geq 3$ where G^r is the Cartesian product of the graph G taken r times.
- The pth power of a graph (Some authors use G^p to denote the pth power of G, to avoid confusion with the notation of Cartesian power of graph G we use $G^{[p]}$ for the pth power of G.) G is the graph $G^{[p]}$, whose vertex set is $V(G)$ and in which two vertices are adjacent when they have distance less than or equal to p. We determine $D_l(Q_n^{[p]})$ for all $n \geq 7, p \geq 1$, where $Q_n = K_2^n$ is the hypercube of dimension n.

Keywords: List distinguishing number · Cartesian power · Hypercube · Hypercube power

AMS Subject Classification (2010): 05C15 · 05C25 · 05C76 · 05C80

1 Introduction

Let G be a graph and $\text{Aut}(G)$ be the automorphism group of G. We denote the set $\{1, 2, \ldots, n\}$ by $[n]$ and the symmetric group defined over $[n]$ by S_n. A

Second author is supported by grant PDF/2017/002518, Science and Engineering Research Board, India.

r-vertex coloring of G is a map $f : V(G) \to \{1, 2, \ldots, r\}$, and the sets $f^{-1}(i)$ for $i \in \{1, 2 \ldots, r\}$ shall be referred to as the color classes of f. An automorphism $\sigma \in \mathrm{Aut}(G)$ is said to fix a color class C of f if $\sigma(C) = C$, where $\sigma(C) = \{\sigma(v) : v \in C\}$. A vertex coloring [1] of the graph G with the property that no non-trivial[2] automorphism of G fixes all the color classes is called a distinguishing coloring of the graph G.

Albertson and Collins [2] defined the distinguishing number of graph G, denoted $D(G)$, as the minimum r such that G admits a distinguishing r-vertex coloring. An interesting variant of the distinguishing number of a graph, due to Ferrara, Flesch, and Gethner [8] goes as follows. Given an assignment $\mathcal{L} = \{L(v)\}_{v \in V(G)}$ of lists of available colors to vertices of G, we say that G is $\mathcal{L}$-distinguishable if there is a distinguishing coloring f of G such that $f(v) \in L(v)$ for all v. The list distinguishing number of G, denoted $D_l(G)$, is the minimum positive integer k such that G is $\mathcal{L}$-distinguishable for any list assignment $\mathcal{L}$ with $|L(v)| = k$ for all v.

For example consider the graph C_4. One can do a proper coloring of this graph using two colors but any of the two colorings of C_4 is not a distinguishing coloring. This is because there exists a nontrivial automorphism of the graph which swaps the vertices in the same color class and hence preserve the color classes. It is shown in [2] that $D(C_4) = 3$. Usually, distinguishing coloring is done by keeping the vertices of a suitable subgraph in a single color class. But in list coloring this possibility is rare. Therefore in practice, finding a list distinguishing coloring of a graph is difficult than having a distinguishing coloring.

The list distinguishing number of cycles of size at least 6, Cartesian products of cycles, and for graphs whose automorphism group is a dihedral group D_{2n} is determined in [8]. The list distinguishing number of Trees [9], Interval graphs [11] and Kneser graphs [5] is also known. In all these cases the list distinguishing number of the graph is equal to the distinguishing number. In [8] authors raised the following question 'Is $D_l(G) = D(G)$ for all graphs G?' This question is still open and our results shows that $D_l(G) = D(G)$ holds for certain families of graphs. A necessary and sufficient condition for a graph G such that $D_l(G) = D(G)$ is given in [3].

Note that $D_l(G) \geq D(G)$, since in the special case where the assigned lists of size $D(G)$ for all the vertices are the same then the list distinguishing coloring turns out to be a distinguishing coloring of the graph.

In this article we deal with the list distinguishing number of Cartesian powers of a graph G and that of p th power of hypercube.

First, we recall some definitions and results which are necessary to understand the ensuing discussion. We follow [10] for the concepts related to the Cartesian product of graphs.

---

[1] Here the vertex coloring may not be a proper coloring always.

[2] The identity automorphism of a graph G is called as the trivial automorphism of G.

Definition 1. *The Cartesian product of graphs G and H is a graph $G \square H$ with vertex set $V(G) \times V(H)$. Two vertices (u, v) and (u', v') are adjacent in $G \times H$ if and only if $u = u'$ and $vv' \in E(H)$ or $uu' \in E(G)$ and $v = v'$. The r th Cartesian power of a graph G, denoted by G^r, is the Cartesian product of G with itself taken r times. That is $G^r = G \square G \square \ldots \square G$, r times.*

The graphs G and H are called factors of the product $G \square H$. A graph G is prime with respect to the Cartesian product if it is nontrivial and cannot be represented as the product of two nontrivial graphs[3]. The automorphism group of a connected graph is generated by automorphism of the factor graphs and the transpositions of the isomorphic prime factors.

Theorem 2. *Let ϕ be a automorphism of a connected graph G with prime factor decomposition $G = G_1 \square G_2 \square \ldots \square G_k$. Then there is a permutation π of $\{1, 2, \ldots, k\}$ and isomorphism $\phi_i : G_{\pi(i)} \to G_i$ for which*

$$\phi(x_1, x_2, \ldots, x_k) = (\phi_1(x_{\pi(1)}), \phi_2(x_{\pi(2)}), \ldots \phi_k(x_{\pi(k)}))$$

The following remark is a special cases of this theorem.

Remark 11. *Let G be a connected prime graph. Then G^k, being a connected graph, has a unique representation as a products of prime graphs. By our assumption the graph G is a prime graph and therefore the unique representation of G^k is $G \square G \square \ldots \square G$, k times, itself. This implies that for each $i \in [k]$, ϕ_i in Theorem 2 is an element of $Aut(G)$. Therefore π in Theorem 2 has $k!$ choices and each ϕ_i has $|Aut(G)|$ choices to form an automorphism of G^k. That is, we have $|Aut(G^k)| = k!|Aut(G)|^k$.*

A hyper cube Q_d of dimension d is the graph K_2^d, the dth Cartesian power of the complete graph K_2. It is shown in [6] that the distinguishing number of the hypercube Q_d is two when $d \geq 4$ and $D(Q_d) = 3$ if $d \in \{2, 3\}$. Later, in [1], Albertson has shown that when G is a connected prime graph then $D(G^r) = 2$ when $r \geq 4$. In the same article it is conjectured that the prime case is not necessary and for sufficiently large $r = R(G)$ the distinguishing number of G^r when G is a connected graph is two. This conjecture has been proved in [13] and a complete description of $D(G^r)$ for any graph G is given in [12].

The pth power of a graph G is the graph $G^{[p]}$, whose vertex set is $V(G)$ and in which two vertices are adjacent if the distance between them in the graph G is at most p. In [6] it is shown that $D(Q_n^{[2]}) = 2$ for $n \geq 4$, $D(Q_d^{[2]}) = 4$ when $d \in \{2, 3\}$, $D(Q_n^{[n]}) = 2^n$ and $D(Q_n^{[n-1]})$ is the minimum integer x such that $\binom{x}{2} \geq 2^{d-1}$. Later, M. Chan has determined the remaining cases in [7]. Chan has shown that $D(Q_n^{[p]}) = 2$ for all $2 < p < n - 1$.

---

[3] A trivial graph is the single vertex complete graph K_1.

In this article we discuss the list version of the distinguishing coloring of the Cartesian powers of connected prime graphs and that of the pth power of a hypercube. We show that in this case the list distinguishing number is equal to the distinguishing number. This also strengthens the belief that $D_l(G) = D(G)$ for all graphs. Moreover every connected graph has a unique representation as a product of prime graphs, up to isomorphism and the order of factors. Therefore to check the validity of the question, Is $D_l(G) = D(G)$, Cartesian powers of graphs is a suitable starting point.

The main results in this article are as follows:

Theorem 3. *Let G be a connected prime graph then*

- *If $|G| \neq 2$, then $D_l(G^r) = 2$ for $r \geq 3$.*
- *If $|G| = 2$ then $D_l(G^r) = 2$ for $r \geq 4$ and $D_l(G^r) = 3$ when $r \in \{2,3\}$.*

and

Theorem 4. *Let Q_n be the hypercube of dimension n then $D_l(Q_n^{[p]}) = D(Q_n^{[p]})$ for all $p \geq 1$, $n \geq 7$.*

In Sect. 2 we compute the list distinguishing number of hyper cubes. Section 3 discusses the list distinguishing number of Cartesian powers of an arbitrary connected prime graph. We determine the list distinguishing number of pth power of hypercube in Sect. 4.

2   List Distinguishing Number of the Hypercube

We shall show that $D_l(Q_d) = D(Q_d)$ for all $d \geq 2$. It is shown in [6] that the distinguishing number of the hypercube $D(Q_d) = 2$ for $d \geq 4$ and is 3 for $d \in \{2,3\}$. We shall show that the same is true for the list distinguishing number $D_l(Q_d)$ as well. We use a probabilistic argument to prove the case of $d \geq 6$, a more careful probabilistic argument for $d = 5$ and a combinatorial argument for $d \leq 4$.

Lemma 5. $D_l(Q_2) = D_l(Q_3) = 3$.

Proof. We consider the following cases:
Case $d = 2$: Recall that $D(Q_2) = 3$. Therefore $D_l(Q_2) \geq D(Q_2) = 3$. For each $v \in V(Q_2)$ let $l(v)$ be a given color list of size 3. If $| \cup_{v \in V(Q_2)} l_v| = 3$ then all the given lists are the same and in this case any valid 3 distinguishing coloring of the graph is a 3 list distinguishing coloring also. Therefore we assume $| \cup_{v \in V(Q_2)} l_v| \geq 4$ and color the vertices in $V(Q_2)$ as follows: assign different colors say a, b, c to vertices say v_1, v_2 and v_3 respectively. If the given list of the fourth vertex v_4 of the graph is $\{a, b, c\}$ then there is a color $d \notin \{a, b, c\}$ in the list of some vertex because of $| \cup_{v \in V(Q_2)} l_v| \geq 4$. Without loss of generality let $d \in l(v_2)$. Then recolor v_2 with color d and assign the color b to v_4. Since $|V(Q_2)| = 4$, Observe that such a coloring is always distinguishing.

Case $d = 3$: By following the similar arguments of the above case we can assume that $|\cup_{v \in V(Q_3)} l_v| \geq 4$. That is, since all the lists are not the same, there are two adjacent vertices, say u and v, which have different lists. Then we can ensure that the four vertices in $u \cup N(u)$ get different colors. This is done as follows: let $a \in l(u)$ be a color that not present in $l(v)$ and $b \in l(v)$ be the color not present in $l(u)$. Then assign a to u and b to v. Avoid the colors a and b from the third neighbor of u and give a color c. If the list of the fourth neighbor of u is exactly $\{a, b, c\}$, then assign a to this vertex and change the color of u to $d \neq a, b, c$. We color the remaining vertices arbitrarily by avoiding the color given to u. We show this to be a distinguishing coloring. Let ϕ be a color preserving nontrivial automorphism. Then ϕ must fix u, and cannot permute the neighbors of u as they have different colors. Without loss of generality let $u = (0, 0, 0)$ and $N(u) = \{(1, 0, 0), (0, 1, 0), (0, 0, 1)\}$. Since $\phi(0, 0, 0) = (0, 0, 0)$ by Theorem 2 each $\phi_i = I$ for $i \in [3]^4$. Now observe that since $\phi(1, 0, 0) = (1, 0, 0)$ the permutation π fixes 1. By similar argument $\pi(2) = 2$ and $\pi(3) = 3$. Therefore ϕ is the identity automorphism.

Lemma 6. $D_l(Q_4) = 2$.

Proof. Observe that Q_4 is the Cartesian product $C_4 \square C_4$, and it is shown in theorem 4 of [8] that $D_l(C_n \square C_m) = D(C_n \square C_m) \quad \forall n, m \geq 3$. Therefore, since $D(Q_4) = 2$, we have $D_l(Q_4) = 2$.

Before the next lemma we introduce some notations and explain the probabilistic techniques used in [5] to calculate the list distinguishing number of a graph. This is a reformulation of the motion lemma (See [1,13,15]) suitable for calculating the list distinguishing number of graphs discussed in this article.

Let G be a graph of order n with $\mathcal{L} = \cup_{i \in [n]} L(v_i)$, where $L(v_i)$ is the list of two different colors assigned for the vertex v_i. We choose randomly (uniformly) and independently for each vertex v, a color from its list $L(v)$. We call this coloring as random coloring and denote it by μ. Observe that the probability $P(\mu$ is not distinguishing$) < 1$ implies its complement $P(\mu$ is distinguishing$) > 0$, that is, there exists a distinguishing coloring of the graph G. Let $\mathcal{G} := \text{Aut}(G) \setminus I$, where I is the identity automorphism of the graph. Given an automorphism ϕ of G, for a vertex $v \in G$ let $O_v^\phi := \{v, \phi(v), \ldots \phi^t(v)\}$ with $\phi^{t+1}(v) = v$ denote the orbit of ϕ containing v and O^ϕ denote the class of all distinct orbits. Observe that O^ϕ is a partition of $V(G)$ and let the elements of O^ϕ be denoted by O_i^ϕ. The number of distinct orbits for a given ϕ is denoted by θ_ϕ and $\theta := \max_{\phi \in \mathcal{G}} \theta_\phi$.

Observe that μ is not distinguishing if and only if there exists a $\phi \in \mathcal{G}$ such that $\forall v \in V(G)$ we have $\mu(\phi(v)) = \mu(v)$. This implies there exists $\phi \in \mathcal{G}$ such that $|\mu(O_i^\phi)| = 1 \forall i \in [\theta_\phi]$. That is all the orbits of the automorphism, are

---

[4] This is because each ϕ_i is an automorphism of K_2 (since $Q_d = K_2^d$) and $\phi_i(0) = 0$ implies $\phi_i(1) = 1$.

monochromatic. We now bound the probability of occurrence of monochromatic orbits:

$$P(\mu \text{ is not distinguishing}) = P(\exists \phi \in \mathcal{G} \text{ such that } \phi \text{ is color preserving wrt } \mu)$$
$$\leq \sum_{\phi \in \mathcal{G}} P(\phi \text{ is color preserving wrt } \mu)$$
$$= \sum_{\phi \in \mathcal{G}} \prod_{i=1}^{\theta_\phi} P(O_i^\phi \text{ is monochromatic})$$

Since each of the lists in $\mathcal{L}$ has two colors, for a fixed automorphism $\phi \in \mathcal{G}$, we have,

$$P(O_i^\phi \text{ is monochromatic}\}) \leq 2^{1-|O_i^\phi|}.$$

Observe that this probability is equal to zero if there is no common color in the given lists of vertices in O_i^ϕ.

Therefore

$$P(\mu \text{ is not distinguishing}) \leq \sum_{\phi \in \mathcal{G}} \prod_{i=1}^{\theta_\phi} 2^{1-|O_i|}$$
$$= \sum_{\phi \in \mathcal{G}} 2^{\theta_\phi - n}$$
$$\leq \frac{|\text{Aut}(G)|}{2^{n-\theta}} \qquad (1)$$

where $n = |G|$ and $\theta = \max_{\phi \in \mathcal{G}} \theta_\phi$ as defined above. That is, if we know the value (or an upper bound) of θ then we can apply the above technique to check for the existence of a two list distinguishing coloring of the graph. A usual idea to bound θ is to determine the maximum number of fixed points that any non trivial automorphism can have.

Let $F_\phi := \{v \in V(G) \text{ such that } \phi(v) = v\}$ where F_ϕ is the set of all fixed point of ϕ. Let $f_\phi := |F_\phi|$ and $f := \max_{\phi \in \mathcal{G}} f_\phi$. Now observe that the fixed points of ϕ form orbits of ϕ of size one, and all other orbits of ϕ have size at least two. Therefore $\theta \leq \frac{n-f}{2} + f = \frac{n+f}{2}$. Again observe that $\theta - n \leq \frac{n+f}{2} - n = \frac{f-n}{2}$. Therefore by (1) we have

$$P(\mu \text{ is not distinguishing}) \leq \frac{|\text{Aut}(G)|}{2^{\frac{n-f}{2}}} \qquad (2)$$

However this may not always give a good upper bound on the probability. If we have better information about the automorphisms and their orbits the bound in (1) could be improved. That is if we know the maximum number of orbits of size $k \geq 1$ and the number of automorphisms that make orbits of size k, then we could fine-tune (2) and get better bound for the probability (See Lemma 9).

Using these techniques we determine the list distinguishing number of Q_d for $d \geq 5$. Before the proof, we discuss the following lemma which gives an upper bound of the maximum number of fixed points f in the graph Q_d.

Lemma 7. *The maximum number of fixed points, f in the graph Q_d is at most* 2^{d-1}.

Proof. Let S be the set of fixed points under a given nontrivial automorphism ϕ. Since ϕ is not the identity, there must be a u such that $\phi(u) = v$, $v \neq u$. Now for each $x \in S$ we see that the distance $d(x, u) = d(\phi(x), \phi(u)) = d(x, v)$, because distances are preserved under automorphisms. Since $u \neq v$ they must differ in some coordinate, let them differ on the ith coordinate. Then we claim that the vertex $\bar{x}$ obtained by flipping the ith coordinate of x is not in S. This follows because $d(\bar{x}, u) \neq d(\bar{x}, v)$ since one side must be $d(x, u) + 1$ while the other must be $d(x, u) - 1$. As for each $x \in S$ there is a unique $\bar{x} \notin S$, S can have at most 2^{d-1} elements.

Lemma 8. *For every $d \geq 6$, $D_l(Q_d) = 2$.*

Proof. We use the probabilistic argument discussed above to show that given any assignment of lists of size 2 to the vertices of Q_d, we can always find a distinguishing coloring.

Since Q_d is isomorphic to K_2^d, from Remark 11 we have $|\text{Aut}(Q_d)| = 2^d d!$. By Lemma 7 we have $f \leq 2^{d-1}$. Then by substituting in (2) and noting that $d! < 2^{(2^{(d-2)} - d)}$ holds for all $d \geq 6$ we have $P(\mu$ is not distinguishing$) < 1$ for $d \geq 6$. This implies $P(\mu$ is distinguishing$) > 0$ and there exists a list distinguishing coloring for the graph.

Observe that when $d = 5$ the bound given in (2) is bigger than one and hence the probabilistic argument fails to provide a list distinguishing coloring of Q_5. The following lemma uses a more careful bound on θ and this allows the same technique to be used for $d = 5$ as well.

Lemma 9. $D_l(Q_5) = 2$.

Proof. From Theorem 2, we observe that an easy way to characterize the automorphisms of Q_d is by saying that for $x = (x_1, x_2, \ldots, x_d) \in \{0, 1\}^d$ and $\sigma \in S_d$ we get a unique $\phi \in \text{Aut}(Q_d)$ which takes

$$v = (v_1, v_2 \ldots, v_d) \in V(Q_d) \quad \text{to} \quad \phi(v) = x \oplus (v_{\sigma(1)}, v_{\sigma(2)} \ldots, v_{\sigma(d)})$$

where $\oplus$ stands for the XOR of the two d length vectors, and all the $\phi \in \text{Aut}(Q_d)$ are characterized precisely in this manner. Now we calculate the number of automorphisms with $f = 2^{d-1}$. To see this, given a $\phi \in \text{Aut}(Q_d)$ represent it by (x, σ) for some $x \in \{0, 1\}^d$ and $\sigma \in S_d$. Let $C_1 C_2 \ldots C_t$ be the disjoint cycle decomposition of σ. Since x and the cycles of σ are fixed, if $v = (v_1, v_2 \ldots, v_d)$ is a fixed point of ϕ then we have

$$\phi(v_1, v_2 \ldots, v_d) = (v_1, v_2 \ldots, v_d) \iff x \oplus (v_{\sigma(1)}, v_{\sigma(2)} \ldots, v_{\sigma(d)}) = (v_1, v_2 \ldots, v_d)$$

Without loss of generality, let $C_1 = (i_1, i_2 \ldots, i_k)$ where $k \geq 1$. That is $\sigma(i_1) = i_2, \sigma(i_2) = i_3, \ldots, \sigma(i_k) = i_1$. For a fixed point $v = (v_1, v_2, \ldots, v_k)$ we need that $v_k = [\phi(v)]_{i_k} = x_k + v_{i_{k-1}}$ for all k. From this it is clear that if we assign 0 or 1 to v_k, then the value of $v_{k-1}, v_{k-2}, \ldots, v_1$ are fixed accordingly. Then $(v_{i_1}, v_{i_2}, \ldots, v_{i_k})$ have only two possible values in any fixed vertex v. The same is true in the cases of other cycles $C_2, C_3, \ldots C_t$ also. It follows that the number of fixed vertices with respect to ϕ is at most $2^{\#cycles(\phi)}$, where $\#cycles(\phi)$ is the number of cycles in the disjoint cycle decomposition of σ in the representation of ϕ. Then if an automorphism ϕ has 2^{d-1} fixed points means $2^{\#cycles(\phi)} = 2^{d-1}$. That is $\#cycles(\phi) = d - 1$. This occurs only if σ is a transposition. Let σ be (ij), then x will be forced to take $x_k = 0 \forall k \notin \{i, j\}$ and $x_i = x_j$ since for v to be a fixed point under π, $v_i = v_j \oplus x_i$ and $v_j = v_i \oplus x_j$, and this implies $x_i \oplus x_j = 0$. That is (x_i, x_j) should be either $(1, 1)$ or $(0, 0)$.

Therefore it follows that there are $\binom{d}{2} * 2$ automorphisms which fix 2^{d-1} points and the rest must fix at most 2^{d-2} points. We look next, at those automorphisms which do not fix any vertices. In fact if S is a subset of $[d]$ such that $\sigma|_S$ is the identity, then it must be the case that $\forall i \in S$ $\;\; x_i = 0$, if there are fixed points. Therefore at least $!(d - |S|) * (2^{|S|} - 1) * 2^{d - |S|}$ automorphisms will not have any fixed points, where the first term comes from taking derangements of remaining $d - |S|$ coordinates, the second term from preventing x_i from being 0 for $i \in S$, and the third term, from freely choosing the remaining values of x.

For each S let A_S be the automorphisms given above, which do not fix any points. Then for $S \neq S'$, $A_S \cap A_{S'} = \phi$. Hence summing over all subsets of $[d]$ we shall get a lower bound on the number of automorphisms which have no fixed points as follows:

$$\sum_{S \subset [d]} |A_S| = \sum_{m=1}^{d} \binom{d}{m} !(d - m) * (2^m - 1) * 2^{d-m}$$

For the case of $d = 5$, solving the above sum gives us a value of 1511. Therefore at least 1511 out $3840(= |Aut(Q_5)|)$ automorphisms have no fixed points. Now we try to bound (1) better by splitting it up.

$$\sum_{\pi \in Aut(Q_5)} 2^{-(2^5 - \theta_\pi)} \leq \frac{1511}{2^{2^4 - 0}} + \frac{20}{2^{2^4 - \frac{2^4}{2}}} + \frac{3840 - 1511 - 20}{2^{2^4 - \frac{2^3}{2}}} = 0.665 < 1$$

This proves $P(\mu$ is not distinguishing$) < 1$ in this case as well.

By Lemmas 8, 5, 6, 9 we have proved the main result of the section, stated below.

Theorem 10. $D_l(Q_d) = D(Q_d)$ $\forall d \geq 2$.

3   List Distinguishing Number of Products of Arbitrary Graphs

Let G be a connected prime graph with $|G| = n$. Then we show that $D_l(G^r) = D(G^r)$ for all $r \geq 3$. The proof is based on the methods used in [1] and [13].

Albertson [1] uses the motion lemma [15] to prove the existence of a distinguishing two coloring for G^r when $r \geq 4$. He also shows that when $|G| \geq 5$ we have $D(G^r) = 2$ for $r \geq 3$. The missing cases other than $r = 2$ have been solved in [13] again by using a refinement of the motion lemma provided $|G| \geq 3$. In both the cases they assign a color, uniformly and independently for each vertex and show that the probability that a nontrivial automorphism fixing all the color classes is strictly less than one.

In our case, the only difference is that we are provided with a list of colors for each vertex. For each vertex v, if we choose uniformly and independently a color from the corresponding lists, the probability bound discussed in (1) is same as that in [1] and [13]. Therefore one can use the probabilistic technique used in [1] and [13] in the list distinguishing coloring also. Therefore we have

Theorem 11. *Let G be a connected prime graph then $D_l(G^r) = D(G^r)$ for all $r \geq 3$.*

Proof. It is shown that(Corollary 1.2 in [1]) when G is a connected prime graph of order n then $D(G^r) \leq D(K_n^r)$. This is because every automorphism of G is an automorphism of K_n and by Theorem 2 it follows that $\mathrm{Aut}(G^r) \subseteq \mathrm{Aut}(K_n^r)$. Therefore any coloring that destroys every automorphism of K_n^r must also destroys every automorphism of G^r. Observe that this is true in the list case also. Hence we have $D_l(G^r) \leq D_l(K_n^r)$.

Since Q_d is isomorphic to K_2^d, by Theorem 10 we have $D_l(G^r) = D(G^r)$ for a connected prime graphs with $n = 2$. When $r \geq 4$ or $n \geq 5$ and $r = 3$, calculations similar to [1] (Theorem 4, Theorem 5 and proof of Theorem 2 in [1]) proves $D_l(G^r) = D(G^r)$. For the remaining cases similar arguments to those in Corollary 3.2 of [13] proves $D_l(G^r) = D(G^r)$ when $r, n \geq 3$. (See Appendix for details). ∎

4   List Distinguishing Number for p^{th} Power of the Hypercube

The pth power of a graph G is the graph $G^{[p]}$, whose vertex set is $V(G)$ and in which two vertices are adjacent if the distance between them in the graph G is at most p. In [14], the automorphism group of $Q_d^{[p]}$ is shown to be the same as $\mathrm{Aut}(Q_d^{[p \bmod 2]})$, when $2 \leq p < d - 1$ with $Q_d^{[0]}$ taken to be $Q_d^{[2]}$. Note that for graphs G, H with $V(G) = V(H)$ and $\mathrm{Aut}(G) \subseteq \mathrm{Aut}(H)$, we have $D_l(G) \leq D_l(H)$. This is because a coloring which destroys all the automorphism of H also destroys the automorphisms of G. Therefore $D_l(Q_d^{[t]}) = D_l(Q_d)$ for any odd number t and we have already determined $D_l(Q_d)$ in Sect. 2. Therefore it remains only to calculate $D_l(Q_d^{[2]})$. Moreover we have $D_l(Q_d^{[t]}) = D_l(Q_d^{[2]})$ for all even number t.

To understand the structure of the automorphisms of $Q_d^{[p]}$, we restate the result from [14] which characterize the automorphism group of hyper cubes.

Theorem 12. *1. For $n \geq 2$, the automorphism group $Aut(Q_n^{[n-1]})$ is isomorphic to a semi-direct product of $Z_2^{2^{n-1}}$ with $S_{2^{n-1}}$, of order $2^{2^{n-1}}(2^{n-1})!$.*

2. For $n \geq 4$ the automorphism group $Aut(Q_n^{[2]})$ is isomorphic to a semi-direct product of Z_2^n with S_{n+1} of order $2^n(n+1)!$.

3. For $2 \leq k < n-1$, we have

$$Aut(Q_n^{[k]}) = \begin{cases} Aut(Q_n^{[2]}), & \text{if } k \text{ is even} \\ Aut(Q_n), & \text{if } k \text{ is odd} \end{cases}$$

Moreover the same article describes the action of $\pi \in Aut(Q_n^{[2]})$ in the following way: For $x \in Q_n^{[2]}$ let $\phi_k(x) = x + e_k$ with $e_k \in Q_n^{[2]}$ consisting entirely of zeros except for a single one in the k-th coordinate position. Let N be the group generated by $\{\phi_k : 1 \leq k \leq n\}$. Let $G_0^{[2]}$ be the group generated by the permutation group S_n and τ_1 where

$$\tau_1(x) = \begin{cases} x + e_1, & \text{if } d(0,x) \text{ is even and one in the first coordinate position of } x \\ x + e_1, & \text{if } d(0,x) \text{ is odd and zero in the first coordinate position of } x \\ x, & \text{otherwise} \end{cases}$$

with $d(0,x)$ as the hamming weight of the vertex x. In short for $\pi \in Aut(Q_n^{[2]})$ there exists a unique $\phi \in N$ and $\alpha \in G_0^2$ such that $\pi(x) = \phi\alpha(x)$.

The following lemma shows that any nontrivial automorphism of the graph $Q_d^{[2]}$ can have at most 2^{d-1} fixed points.

Lemma 13. *The maximum number of fixed points, f, that any nontrivial automorphism can have in the graph $Q_d^{[2]}$ is at most 2^{d-1}.*

Proof. We observed that an automorphism π of $Q_d^{[2]}$ could be represented as $\phi\alpha$ where $\phi \in N$ and $\alpha \in G_0^{[2]}$. Moreover, α can be represented as σ or $\tau_1\sigma$ or $\sigma\tau_1$ for some $\sigma \in S_d$

Case 1: τ_1 is identity, that is $\alpha \in S_d$:

In this case the automorphism π is $\phi\sigma$ and that has the same structure as an automorphism of the hypercube Q_n. By similar arguments as in the proof of Lemma 9, we have the number of fixed vertices with respect to π to be at most 2^{d-1}.

Case 2: α has the form $\sigma\tau_1$:

Denote the union of vertices of the form $(1, a_2, \ldots, a_n)$ with even parity and vertices of the form $(0, a_2, \ldots, a_n)$ with odd parity as Class I and the remaining vertices as Class II.

Note that τ_1 acts as identity for Class II vertices and toggles the first position for vertices in Class I. Let S be the set of fixed vectors under the action of $\pi = \phi\sigma\tau_1$. Let $S_1 = S \cap$ Class I and $S_2 = S \cap$ Class II. We consider the following sub cases:

Sub case 2(1): σ fixes the first position, that is $\sigma(1) = 1$.

Suppose $[\phi]_1 = 1$. Then ϕ toggles the first bit of $v = (v_1, v_2, \ldots, v_d)$. When v is fixed point then we have $[\phi\sigma\tau_1(v)]_1 = v_1$. Since $\sigma(1) = 1$, we have $v \in S$ implies τ_1 is changing the first bit. Since τ_1 act as identity for the vertices in Class II, in this case a fixed point should be in Class I. That is we have $S \subseteq$ Class I and $|S| \leq |$Class I$| = 2^{d-1}$. Now suppose $[\phi]_1 = 0$. Then τ_1 should not toggle the first bit of any $v \in S$. Therefore in this case $S \subseteq$ Class II and $|S| \leq |$Class II$| = 2^{d-1}$ Sub case 2(2): $\sigma(1) \neq 1$:

Without loss o generality, let C_1 be a cycle containing 1 in the cycle decomposition of the permutation σ. Let $C_1 = (1, i_2, i_3, \ldots i_k)$ with $k \geq 2$ and $i_r \in [d]$. We claim that $|S \cap$ Class I$| \leq 2^{d-2}$ and $|S \cap$ Class II$| \leq 2^{d-2}$.

To see this let $v \in S \cap$ Class I. Then the choice of v_1 fixes the values of $v_{i_2}, v_{i_3}, \ldots, v_{i_k}$. That is $(v_1, v_{i_2}, \ldots, v_{i_k})$ has two choices depending on the choice of v_1. The choices of remaining bits v_t, $t \in [d] \setminus \{1, i_2, \ldots, i_k\}$ depends on the parity of $(v_1, v_{i_2}, \ldots, v_{i_k})$ and the choice of v_1. That is if $v_1 = 0$ and $(0, v_{i_2}, \ldots, v_{i_k})$ has odd parity then to make $v \in$ Class I the parity of the remaining entries v_t, $t \in [d] \setminus \{1, i_2, \ldots, i_k\}$ in v should be of even parity. Similarly if $(1, v_{i_2}, \ldots, v_{i_k})$ has odd parity the remaining entries in v should have odd parity to make v an element of Class I. We have 2^{d-k} possible patterns of the remaining entries of which 2^{d-k-1} are of odd parity and 2^{d-k-1} are of even parity. This implies that the total number of choices that v can make is at most $2 * 2^{d-k-1} = 2^{d-k}$. Since $k \geq 2$ we have $|S \cap$ Class I$| \leq 2^{d-2}$. Similarly we have $|S \cap$ Class II$| \leq 2^{d-2}$. Therefore $|S| \leq 2 * 2^{d-2} = 2^{d-1}$.

Case 3: α has the form $\sigma\tau_1$:

By the similar arguments as in Case 2 an automorphism π of this form can fix at most 2^{d-1} vertices.

Theorem 14. $D_l(Q_d^{[p]}) = D(Q_d^{[p]})$ for all $p \geq 1$ and $n \geq 7$.

Proof. By Theorem 10 in Sect. 2 we have $D_l(Q_d) = D(Q_d)$. By Theorem 12 and the discussions in the beginning of this section, we have $D_l(Q_d^{[t]}) = D_l(Q_d)$ and $D(Q_d^{[t]}) = D(Q_d)$ for $2 \leq t < d-1$. Therefore it follows that $D_l(Q_d^{[t]}) = D_l(Q_d) = D(Q_d) = D(Q_d^t)$ for all odd t. For the case when t is even, we prove $D_l(Q_n^{[2]}) = D(Q_n^{[2]})$ using the probabilistic arguments discussed in (2). It is shown that $|\text{Aut}(Q_d^{[2]})| = (d+1)! 2^d$ in [14]. By the above lemma (Lemma 13) we have $f \leq 2^{d-1}$. Substituting these values in (2), we have

$$P(\mu \text{ is not distinguishing}) \leq \frac{(d+1)! 2^d}{2^{2d-2}}. \text{ To get } P(\mu \text{ is not distinguishing}) < 1,$$

we need to show that $\frac{(d+1)! 2^d}{2^{2d-2}} < 1$. That is $(d+1)! < 2^{2^{d-2}-d}$. This holds for $d \geq 7$ and hence there exists a list distinguishing coloring of the graph when $n \geq 7$.

For $Q_n^{[n]}$, it being a K_{2^n}, we get $D_l(Q_n^{[n]}) = D(Q_n^{[n]}) = 2^n$. For $Q_n^{[n-1]}$, its complement being a perfect matching, we observe that each edge in the matching can either get swapped (That is for $uv \in E(G)$ u maps to v and v maps to u) or get assigned to another edge or any combination of the two. Therefore whenever $u, v \in V(G)$ are adjacent, they should get different colors, and for every two edges uv and xy, the set of colors assigned to uv should differ from that assigned to xy. Therefore when there are m $(= 2^{n-1})$ edges, l colors are necessary, where l is the smallest integer satisfying $\binom{l}{2} \geq m$. Clearly if we have lists of size l, a coloring satisfying the previous criteria can be chosen from it.

Appendix

Details of the proof of Theorem 11

There is a slight difference between the motion lemma discussed in this article (in [5] also) and that uses in [1] and [13]. Here we bound the number of orbits by bounding maximum number of fixed points that any non trivial automorphism could have. But in [1] and [13] they use the minimum number of vertices that any non trivial automorphism could move to bound the maximum number of orbits θ (See Sect. 2). Observe that the minimum number of moving points, denoted by m or $m(G)$, is equal to $|G|-$ the maximum number of fixed points. Therefore $2^{\frac{n-f}{2}}$ in (2) becomes $2^{\frac{m}{2}}$ in [1] and [13]. That is, the bounds for m in their discussions are bounds for $n - f$ in our discussions.

Keeping this in mind we discuss the details of the proof of Theorem 11. In [1], Theorem 4 calculates $|\mathrm{Aut}(K_n^r)|$ and Theorem 5 in the same paper determines $m(K_n^r)$, the minimum number of vertices that any non trivial automorphism must move. By substituting these values in (2), we have

$$P(\mu \text{ is not distinguishing}) \leq \frac{r!(n!)^r}{2^{n^{r-1}}}.$$

Again, in the proof of Theorem 2 of [1], it is shown that $r!(n!)^r < 2^{n^{r-1}}$ when $r \geq 4$ or $n \geq 5, r = 3$. Similarly the bounds and the refinement of motion lemma given in [13] could be used to prove the list version of the distinguishing coloring also.

Lemma 3.1 in [13] is a refinement of the motion lemma which states that.

Lemma 15 ([13]). *Let G be a graph with $|G| = n$. Suppose $Aut(G)$ acting on $V(G)$ has k orbits and $d \geq 2$ is an integer. If $n - m(G) \geq 3$ and*

$$\left(|Aut(G)| - \frac{kAut(G) - n}{n - m(G)} - 1\right)d^{-n/2} + \frac{kAut(G) - n}{n - m(G)}d^{-m(G)/2} < 1$$

then $D(G) \leq d$

The proof of this lemma is based on the calculation of the probability that a non trivial automorphism fixes all the color classes in a random d coloring of the vertices. We could use the same method in list coloring also. That is this lemma is valid in our case as well. The next result, Corollary 3.2 in [13] states that

Corollary 16 ([13]). *Suppose G is a vertex transitive graph with n vertices and with $n - m(G) \geq 3$. If $d \geq 2$ is an integer and*

$$|Aut(G)| \leq \frac{(n - m(G))d^{m(G)/2}}{(n - m(G))d^{(m(G)-n)/2} + 1}$$

then $D(G) \leq d$.

The proof of this Corollary depends on the previous lemma and vertex transitive property of the graph. Both are independent of the list and non list version of the vertex coloring. Now, the values of $m(K_n^r)$ and $|Aut(K_n^r)|$ given in [1] and the above corollary gives the desired result.

References

1. Albertson, M.O.: Distinguishing Cartesian powers of graphs. Electron. J. Combin. **12**, #N17 (2005)
2. Albertson, M.O., Collins, K.L.: Symmetry breaking in graphs. Electron. J. Comb. **3**, #R18 (1996)
3. Alikhani, S., Soltani, S.: Characterization of graphs with distinguishing number equal list distinguishing number. https://arxiv.org/abs/1711.08887
4. Balachandran, N., Padinhatteeri, S.: $\chi_D(G), |Aut(G)|$ and a variant of the Motion Lemma. Ars Math. Contemp. **12**(1), 89–109 (2016)
5. Balachandran, N., Padinhatteeri, S.: The list distinguishing number of Kneser graphs. Discrete Appl. Math. **236**, 30–41 (2018)
6. Bogstad, B., Cowen, L.: The distinguishing number of the hypercube. Discrete Math. **283**, 29–35 (2004)
7. Chan, M.: The distinguishing number of the augmented cube and hypercube powers. Discrete Math. **308**(11), 2330–2336 (2008)
8. Ferrara, M., Flesch, B., Gethner, E.: List-distinguishing coloring of graphs. Electron. J. Comb. **18**, #P161 (2011)
9. Ferrara, M., Gethner, E., Hartke, S., Stolee, D., Wenger, P.: List distinguishing parameters of trees. Discrete Appl. Math. **161**, 864–869 (2013)
10. Hammack, R., Imrich, W.: Sandi Klavžar: Handbook of Product Graphs. Discrete Mathematics and its Applications, 2nd edn. Taylor & Francis Group, LLC, Boca Raton (2011)
11. Immel, P., Wenger, P.S.: The list distinguishing number equals the distinguishing number for interval graphs. Discussiones Mathematicae Graph Theory **37**(1), 165–174 (2017). http://arxiv.org/abs/1509.04327v1
12. Imrich, W., Klavžar, S.: Distinguishing Cartesian powers of graphs. J. Graph Theory **53**(3), 250–260 (2006)
13. Klavžar, S., Zhu, X.: Cartesian powers of graphs can be distinguished by two labels. Eur. J. Comb. **28**, 303–310 (2007)
14. Miller, Z., Perkel, M.: A stability theorem for the automorphism groups of powers of the n-cube. Australas. J. Comb. **10**, 17–28 (1994)
15. Russell, A., Sundaram, R.: A Note on the asymptotics and computational complexity of graph distinguishability. Electron. J. Comb. **5**, #R23 (1998)

On Algebraic Expressions
of Two-Terminal Directed Acyclic Graphs

Mark Korenblit[1]($\boxtimes$) ⓘ and Vadim E. Levit[2] ⓘ

[1] Holon Institute of Technology, Holon, Israel
korenblit@hit.ac.il
[2] Ariel University, Ariel, Israel
levitv@ariel.ac.il

Abstract. The paper investigates relationship between algebraic expressions and graphs. Our intent is to simplify graph expressions and eventually find their shortest representations. We describe the decomposition method for generating expressions of complete st-dags (two-terminal directed acyclic graphs) and estimate the corresponding expression complexities. Using these findings, we present an $2^{O\left(\log^2 n\right)}$ upper bound of a length of the shortest expression for every st-dag of order n.

1 Introduction

A *graph* G consists of a *vertex set* $V(G)$ and an *edge set* $E(G)$, where each edge corresponds to a pair (v, w) of vertices. If the edges are ordered pairs of vertices (i.e., the pair (v, w) is different from the pair (w, v)), then we call the graph *directed* or *digraph*; otherwise, we call it *undirected*. If (v, w) is an edge in a digraph, we say that (v, w) *leaves* vertex v and *enters* vertex w. A vertex in a digraph is a *source* if no edges enter it, and a *target* if no edges leave it. A *path* from vertex v_0 to vertex v_k in a graph G is a sequence of its vertices $[v_0, v_1, v_2, \ldots, v_{k-1}, v_k]$ such that $(v_{i-1}, v_i) \in E(G)$ for $1 \leq i \leq k$. G is an *acyclic graph* if there is no closed path $[v_0, v_1, v_2, \ldots, v_k, v_0]$ in G. A two-terminal directed acyclic graph (*st-dag*) has only one source s and only one target t. In an st-dag, every vertex lies on some path from s to t.

A graph G' is a *subgraph* of G if $V(G') \subseteq V(G)$ and $E(G') \subseteq E(G)$. A graph G is *homeomorphic* to a graph G' if G can be obtained by subdividing edges of G' with new vertices.

Given a graph G, an *edge labeling* is a function $E(G) \longrightarrow R$, where R is a ring equipped with two binary operations + (addition or disjoint union) and · (multiplication or concatenation, also denoted by juxtaposition when no ambiguity arises). In what follows, elements of R are called *labels*, and a *labeled graph* refers to an edge-labeled graph with all labels distinct.

A path between the source and the target of an st-dag is called *spanning*. We define the sum of edge label products corresponding to all possible spanning paths of an st-dag G as the *canonical expression* of G. The label order in every product (from the left to the right) is identical to the order of corresponding edges

M. Changat and S. Das (Eds.): CALDAM 2020, LNCS 12016, pp. 248–259, 2020.
https://doi.org/10.1007/978-3-030-39219-2_21

in the path (from the source to the target). An algebraic expression is called an *st-dag expression* (a *factoring of an st-dag* in [2]) if it is algebraically equivalent to the canonical expression of an st-dag. An st-dag expression consists of labels, the two ring operators $+$ and $\cdot$, and parentheses. For example, clearly, the algebraic expression $ab + bc$ is not an st-dag expression. We denote an expression of an st-dag G by $Ex(G)$.

We define the total number of labels in an algebraic expression as its *complexity*. An *optimal representation of the algebraic expression F* is an expression of minimum complexity algebraically equivalent to F. Our intention is to simplify an st-dag expression to its optimal representation or, at least, to the expression with polynomial complexity in relation to the st-dag's order.

A *series-parallel graph* is defined recursively so that a single edge is a series-parallel graph and a graph obtained by a parallel or a series composition of series-parallel graphs is series-parallel. A series-parallel graph expression has a representation in which each label appears only once [2,13] (a *read-once formula* [7] in which Boolean operations are replaced by their arithmetic counterparts). This representation is an optimal representation of the series-parallel graph expression. For example, the canonical expression of the series-parallel graph presented in Fig. 1 is $abd + abe + acd + ace + fe + fd$ and it can be reduced to $(a(b + c) + f)(d + e)$.

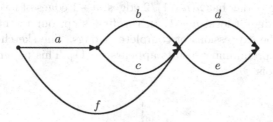

Fig. 1. A series-parallel graph.

As shown in [3], an st-dag is series-parallel if and only if it does not contain a subgraph homeomorphic to the *forbidden graph* illustrated in Fig. 2. Possible optimal representations of its expression are $a_1 (a_2 a_3 + b_2) + b_1 a_3$ or $(a_1 a_2 + b_1) a_3 + a_1 b_2$. For this reason, an expression of a non-series-parallel st-dag cannot be represented as a read-once formula. However, generating the optimum factored form for expressions which cannot be reduced to read-once formulae is **NP**-complete [23].

Problems related to computations on graphs whose edges are associated with additional data have applications in various areas. Specifically, some flow [22], scheduling [6], reliability [20], economical [19] problems are either intractable or have complicated solutions, in general, while they have efficient solutions for series-parallel graphs. Optimization problems on vertex-labeled graphs are considered in [5].

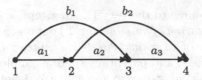

Fig. 2. The forbidden graph.

Interrelations between graphs and expressions are discussed in [1,2,4,8,10, 12,13,15–19,21], and other works. In particular, some algorithms developed in order to obtain good factored forms are described in [8,15]. In [13] we presented an algorithm, which generates the expression of $O\left(n^2\right)$ complexity for an n-vertex *Fibonacci graph* [9] that gives a generic example of non-series-parallel graphs. More complicated, *rhomboidal* graphs are considered in [12]. The total numbers of labels in expressions derived for these n-vertex graphs are $O\left(n^{\log_2 6}\right)$.

This paper is devoted to a more general problem. We estimate an upper bound of the optimal representation's complexity of the expression for each n-vertex st-dag.

To this end, we investigate a *complete st-dag* that has vertices $\{1,2,3,\ldots,n\}$ and edges $\{(v,w) \mid v = 1,2,\ldots,n-1 \mid w > v\}$ (see the example in Fig. 3). Edge $(v,v+l)$ in the graph is labeled by $e_v^{(l)}$ and is called an *edge of level l*. An n-vertex complete st-dag has $n(n-1)/2$ edges: $n-1$ edges of level 1, $n-2$ edges of level 2, ..., 1 edge of level $n-1$. In the first step, our intent is to generate and to simplify the expressions of complete st-dags. The sketch of the relevant study along with preliminary results appears in [11]. This paper contains exact formulae and proofs.

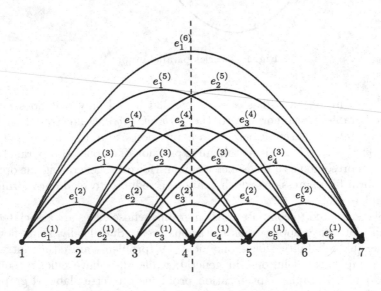

Fig. 3. A 7-vertex complete st-dag.

2 Generating Expressions for Complete St-Dags

For constructing expressions of complete st-dags we use a decomposition method. This method is based on recursive revealing subgraphs in the initial graph. The resulting expression is produced by a special composition of subexpressions describing these subgraphs.

Consider an n-vertex complete st-dag. Denote by $F(p,q)$ a subexpression related to its subgraph (which is a complete st-dag as well) having a source p ($1 \leq p \leq n$) and a target q ($1 \leq q \leq n$, $q \geq p$). If $q - p \geq 2$, then we choose a *decomposition vertex* i in the middle of a subgraph (i is $\underline{m}$ or $\overline{m}$, where $\underline{m} = \lfloor \frac{q+p}{2} \rfloor$, $\overline{m} = \lceil \frac{q+p}{2} \rceil$), and, in effect, split it at this vertex (vertex 4 in Fig. 3). Otherwise, we assign final values to $F(p,q)$. As follows from the structure of a complete st-dag, any path from vertex p to vertex q passes through vertex i or avoids it via an edge of level l ($l = 2, 3, ..., n - 1$) that connects revealed subgraphs (*connecting edge*). Therefore, $F(p,q)$ is generated by the following recursive procedure (*decomposition procedure*):

1. $F(p,p) \leftarrow 1$
2. $F(p,p+1) \leftarrow e_p^{(1)}$
3. $F(p,q) \leftarrow F(p,i)F(i,q) + F(p,i-1)e_{i-1}^{(2)}F(i+1,q)$

$$+ F(p,i-2)e_{i-2}^{(3)}F(i+1,q) + F(p,i-1)e_{i-1}^{(3)}F(i+2,q)$$
$$+ \ldots \ldots \ldots \ldots \ldots$$
$$+ F(p,i-\underline{m}+p)\,e_{i-\underline{m}+p}^{(\underline{m}-p+1)}F(i+1,q)$$
$$F(p,i-\underline{m}+p+1)\,e_{i-\underline{m}+p+1}^{(\underline{m}-p+1)}F(i+2,q)$$
$$+ \ldots + F(p,i-1)e_{i-1}^{(\underline{m}-p+1)}F(i+\underline{m}-p,q)$$
$$+ F(p,p)\,e_p^{(\underline{m}-p+2)}F(\underline{m}+2,q)$$
$$+ F(p,p+1)\,e_{p+1}^{(\underline{m}-p+2)}F(\underline{m}+3,q)$$
$$+ \ldots + F(p,\overline{m}-2)\,e_{\overline{m}-2}^{(\underline{m}-p+2)}F(q,q)$$
$$+ \ldots \ldots \ldots \ldots \ldots$$
$$+ F(p,p)e_p^{(q-p-1)}F(q-1,q) + F(p,p+1)e_{p+1}^{(q-p-1)}F(q,q)$$
$$+ F(p,p)e_p^{(q-p)}F(q,q)$$

Lines 1 and 2 contain conditions of exit from the recursion. The special case when a subgraph consists of a single vertex is considered in line 1. It is clear that such a subgraph can be connected to other subgraphs only serially. For this reason, it is accepted that its subexpression is 1, so that when it is multiplied by another subexpression, the final result is not influenced. Line 2 describes a subgraph consisting of a single edge. The corresponding subexpression consists of a single label. The general case is processed in line 3. Subgraphs described by subexpressions $F(p,i)$ and $F(i,q)$ include all paths from vertex p to vertex q passing through vertex i. Subgraphs described by other subexpressions include all paths from vertex p to vertex q passing through corresponding connecting edges.

The decomposition procedure is initially invoked by substituting parameters 1 and n instead of p and q, respectively.

Denote the number of connecting edges of level l by E_l. One can see that $E_2 = 1$. In an n-vertex complete st-dag, for odd n, $E_l = E_{l-1} + 1$ $\left(l = 3, 4, ..., \left\lceil \frac{n}{2} \right\rceil \right)$ and $E_l = E_{l-1} - 1$ $\left(l = \left\lceil \frac{n}{2} \right\rceil + 1, \left\lceil \frac{n}{2} \right\rceil + 2, ..., n - 1 \right)$. For even n, $E_l = E_{l-1} + 1$ $\left(l = 3, 4, ..., \frac{n}{2} \right)$, $E_{\frac{n}{2}+1} = E_{\frac{n}{2}}$, and $E_l = E_{l-1} - 1$ $\left(l = \frac{n}{2} + 2, \frac{n}{2} + 3, ..., n - 1 \right)$. Therefore, for odd and even n, all edges of level l $\left(l = \left\lceil \frac{n+1}{2} \right\rceil, \left\lceil \frac{n+1}{2} \right\rceil + 1, ..., n - 1 \right)$ are connecting edges.

For example, the expression of the 7-vertex complete st-dag (Fig. 3) derived in accordance with the decomposition procedure is

$$
\begin{aligned}
&\left(e_1^{(1)} \left(e_2^{(1)} e_3^{(1)} + e_2^{(2)} \right) + e_1^{(2)} e_3^{(1)} + e_1^{(3)} \right) \left(e_4^{(1)} \left(e_5^{(1)} e_6^{(1)} + e_5^{(2)} \right) + e_4^{(2)} e_6^{(1)} + e_4^{(3)} \right) \\
&+ \left(e_1^{(1)} e_2^{(1)} + e_1^{(2)} \right) e_3^{(2)} \left(e_5^{(1)} e_6^{(1)} + e_5^{(2)} \right) \\
&+ e_1^{(1)} e_2^{(3)} \left(e_5^{(1)} e_6^{(1)} + e_5^{(2)} \right) + \left(e_1^{(1)} e_2^{(1)} + e_1^{(2)} \right) e_3^{(3)} e_6^{(1)} \qquad\qquad (1) \\
&+ e_1^{(4)} \left(e_5^{(1)} e_6^{(1)} + e_5^{(2)} \right) + e_1^{(1)} e_2^{(4)} e_6^{(1)} + \left(e_1^{(1)} e_2^{(1)} + e_1^{(2)} \right) e_3^{(4)} \\
&+ e_1^{(5)} e_6^{(1)} + e_1^{(1)} e_2^{(5)} + e_1^{(6)}.
\end{aligned}
$$

It contains 47 labels.

The expression in step 3 of the decomposition procedure can be simplified by putting repeated subexpressions outside the brackets. Specifically, putting outside the brackets subexpressions $F(p, i - 1)$, $F(p, i - 2)$, ..., $F(p, i - \underline{m} + p + 1)$ and $F(p, p)$, $F(p, p + 1)$, ..., $F(p, \overline{m} - 3)$ gives the following statement:

$$
\begin{aligned}
F(p, q) \leftarrow\ & F(p, i) F(i, q) \qquad\qquad\qquad\qquad\qquad\qquad (2) \\
&+ F(p, i - 1) \left(e_{i-1}^{(2)} F(i + 1, q) + e_{i-1}^{(3)} F(i + 2, q) \right. \\
&\qquad \left. + ... + e_{i-1}^{(\underline{m}-p+1)} F(i + \underline{m} - p, q) \right) \\
&+ F(p, i - 2) \left(e_{i-2}^{(3)} F(i + 1, q) + e_{i-2}^{(4)} F(i + 2, q) \right. \\
&\qquad \left. + ... + e_{i-2}^{(\underline{m}-p+1)} F(i + \underline{m} - p - 1, q) \right) \\
&+ .... \\
&+ F(p, i - \underline{m} + p + 1) \left(e_{i-\underline{m}+p+1}^{(\underline{m}-p)} F(i + 1, q) \right. \\
&\qquad \left. + e_{i-\underline{m}+p+1}^{(\underline{m}-p+1)} F(i + 2, q) \right) \\
&+ F(p, i - \underline{m} + p) e_{i-\underline{m}+p}^{(\underline{m}-p+1)} F(i + 1, q) \\
&+ F(p, p) \left(e_p^{(\underline{m}-p+2)} F(\underline{m} + 2, q) + e_p^{(\underline{m}-p+3)} F(\underline{m} + 3, q) \right)
\end{aligned}
$$

$$+ \ldots + e_p^{(q-p-1)} F(q-1,q) + e_p^{(q-p)} F(q,q) \Big)$$

$$+ F(p,p+1) \Big(e_{p+1}^{(m-p+2)} F(\underline{m}+3,q) + e_{p+1}^{(m-p+3)} F(\underline{m}+4,q) $$

$$+ \ldots + e_{p+1}^{(q-p-2)} F(q-1,q) + e_{p+1}^{(q-p-1)} F(q,q) \Big)$$

$$+ \ldots \ldots \ldots \ldots$$

$$+ F(p,\overline{m}-3) \Big(e_{\overline{m}-3}^{(m-p+2)} F(q-1,q) + e_{\overline{m}-3}^{(m-p+3)} F(q,q) \Big)$$

$$+ F(p,\overline{m}-2) e_{\overline{m}-2}^{(m-p+2)} F(q,q).$$

Remark 1. In a similar manner, right repeated subexpressions instead of left ones can be put outside the brackets.

Actually, there are identical subexpressions in (2). For example, $F(p, i - \underline{m} + p)$ is equal to $F(p,p)$ for $i = \underline{m}$. Given this fact, (2) may be additionally simplified as follows:

$$F(p,q) \leftarrow F(p,p) \Big(e_p^{(i-p+1)} F(i+1,q) + e_p^{(i-p+2)} F(i+2,q) \tag{3}$$

$$+ \ldots + e_p^{(q-p-1)} F(q-1,q) + e_p^{(q-p)} F(q,q) \Big)$$

$$+ F(p,p+1) \Big(e_{p+1}^{(i-p)} F(i+1,q) + e_{p+1}^{(i-p+1)} F(i+2,q) $$

$$+ \ldots + e_{p+1}^{(q-p-2)} F(q-1,q) + e_{p+1}^{(q-p-1)} F(q,q) \Big)$$

$$+ \ldots \ldots \ldots \ldots$$

$$+ F(p,i-2) \Big(e_{i-2}^{(3)} F(i+1,q) + e_{i-2}^{(4)} F(i+2,q) $$

$$+ \ldots + e_{i-2}^{(i-p+1+\Delta)} F(q-1,q) + e_{i-2}^{(i-p+2+\Delta)} F(q,q) \Big)$$

$$+ F(p,i-1) \Big(e_{i-1}^{(2)} F(i+1,q) + e_{i-1}^{(3)} F(i+2,q) $$

$$+ \ldots + e_{i-1}^{(i-p+1+\Delta)} F(q,q) \Big)$$

$$+ F(p,i) F(i,q),$$

where $\Delta = 0$ for odd $p - q + 1$ (number of vertices in a subgraph), $\Delta = 1$ for even $p - q + 1$ and $i = \underline{m}$, $\Delta = -1$ for even $p - q + 1$ and $i = \overline{m}$.

Specifically, expression (1) after this simplification looks as follows:

$$e_1^{(4)} \Big(e_5^{(1)} e_6^{(1)} + e_5^{(2)} \Big) + e_1^{(5)} e_6^{(1)} + e_1^{(6)}$$

$$+ e_1^{(1)} \Big(e_2^{(3)} \Big(e_5^{(1)} e_6^{(1)} + e_5^{(2)} \Big) + e_2^{(4)} e_6^{(1)} + e_2^{(5)} \Big)$$

$$+ \Big(e_1^{(1)} e_2^{(1)} + e_1^{(2)} \Big) \Big(e_3^{(2)} \Big(e_5^{(1)} e_6^{(1)} + e_5^{(2)} \Big) + e_3^{(3)} e_6^{(1)} + e_3^{(4)} \Big)$$

$$+ \Big(e_1^{(1)} \Big(e_2^{(1)} e_3^{(1)} + e_2^{(2)} \Big) + e_1^{(2)} e_3^{(1)} + e_1^{(3)} \Big) \Big(e_4^{(1)} \Big(e_5^{(1)} e_6^{(1)} + e_5^{(2)} \Big) + e_4^{(2)} e_6^{(1)} + e_4^{(3)} \Big).$$

The new representation contains 39 labels.

Proposition 1. *For an n-vertex complete st-dag G_c, the total number of labels $L(n)$ in the expression $Ex(G_c)$ derived in accordance with (3) may be evaluated as follows:*

$$L(1) = 0, \quad L(2) = 1$$

$$L(n) < L\left(\left\lceil \frac{n}{2} \right\rceil + 1\right) + L\left(\left\lceil \frac{n}{2} \right\rceil\right) + \left(\left\lfloor \frac{n}{2} \right\rfloor + 1\right) \sum_{j=2}^{\left\lfloor \frac{n}{2} \right\rfloor} L(j) + \left\lfloor \frac{n}{2} \right\rfloor^2$$

$$(n > 2).$$

Proof. Initial statements for $L(1)$, $L(2)$ follow directly from steps 1, 2 of the decomposition procedure. Left subexpressions $F(p, p)$, $F(p, p+1)$, ..., $F(p, i-1)$ have $L(1)$, $L(2)$, ..., $L\left(\left\lfloor \frac{n}{2} \right\rfloor\right)$ labels, respectively, for odd n. For even n, these subexpressions contain $L(1)$, $L(2)$, ..., $L\left(\frac{n}{2} - 1\right)$ labels if $i = \underline{m}$ or $L(1)$, $L(2)$, ..., $L\left(\frac{n}{2}\right)$ labels if $i = \overline{m}$. Each of sums multiplied by one of the left subexpressions has the same number ($\left\lfloor \frac{n}{2} \right\rfloor$ or $\left\lfloor \frac{n}{2} \right\rfloor - 1$) of addends each of which includes a label of a connecting edge and a right subexpression ($F(i+1, q), F(i+2, q)$, ..., $F(q, q)$). These right subexpressions have $L\left(\left\lfloor \frac{n}{2} \right\rfloor\right)$, $L\left(\left\lfloor \frac{n}{2} \right\rfloor - 1\right)$, ..., $L(1)$ labels, respectively, for odd n. For even n, they contain $L\left(\frac{n}{2}\right)$, $L\left(\frac{n}{2} - 1\right)$, ..., $L(1)$ labels if $i = \underline{m}$ or $L\left(\frac{n}{2} - 1\right)$, $L\left(\frac{n}{2} - 2\right)$, ..., $L(1)$ if $i = \overline{m}$. Subexpressions $F(p, i)$ and $F(i, q)$ of the last product consist both of $\left\lceil \frac{n}{2} \right\rceil$ labels for odd n. For even n, the number of labels is $\frac{n}{2}$ in one of the subexpressions and $\frac{n}{2} + 1$ in another one. Taking into consideration that $L(1) = 0$, we obtain

$$L(n) < L(1) + \sum_{j=1}^{\left\lfloor \frac{n}{2} \right\rfloor} L(j) + \left\lfloor \frac{n}{2} \right\rfloor + L(2) + \sum_{j=1}^{\left\lfloor \frac{n}{2} \right\rfloor} L(j) + \left\lfloor \frac{n}{2} \right\rfloor$$

$$+ \ldots + L\left(\left\lfloor \frac{n}{2} \right\rfloor\right) + \sum_{j=1}^{\left\lfloor \frac{n}{2} \right\rfloor} L(j) + \left\lfloor \frac{n}{2} \right\rfloor$$

$$+ L\left(\left\lceil \frac{n}{2} \right\rceil\right) + L\left(\left\lceil \frac{n}{2} \right\rceil + 1\right)$$

$$= \left\lfloor \frac{n}{2} \right\rfloor \sum_{j=2}^{\left\lfloor \frac{n}{2} \right\rfloor} L(j) + \sum_{j=2}^{\left\lfloor \frac{n}{2} \right\rfloor} L(j) + L\left(\left\lceil \frac{n}{2} \right\rceil\right) + L\left(\left\lceil \frac{n}{2} \right\rceil + 1\right) + \left\lfloor \frac{n}{2} \right\rfloor^2$$

$$= L\left(\left\lceil \frac{n}{2} \right\rceil + 1\right) + L\left(\left\lceil \frac{n}{2} \right\rceil\right) + \left(\left\lfloor \frac{n}{2} \right\rfloor + 1\right) \sum_{j=2}^{\left\lfloor \frac{n}{2} \right\rfloor} L(j) + \left\lfloor \frac{n}{2} \right\rfloor^2.$$

$\square$

It may be easily shown by induction on n that asymptotically $L\left(\left\lceil\frac{n}{2}\right\rceil+1\right) < 2L\left(\left\lceil\frac{n}{2}\right\rceil\right)$. Thus

$$L\left(\left\lceil\frac{n}{2}\right\rceil+1\right) + L\left(\left\lceil\frac{n}{2}\right\rceil\right) + \left(\left\lfloor\frac{n}{2}\right\rfloor+1\right)\sum_{j=2}^{\left\lfloor\frac{n}{2}\right\rfloor}L(j) + \left\lfloor\frac{n}{2}\right\rfloor^2$$

$$< 2L\left(\left\lceil\frac{n}{2}\right\rceil\right) + L\left(\left\lceil\frac{n}{2}\right\rceil\right) + \left(\left\lfloor\frac{n}{2}\right\rfloor+1\right)\sum_{j=2}^{\left\lfloor\frac{n}{2}\right\rfloor}L(j) + \left\lfloor\frac{n}{2}\right\rfloor^2$$

$$< 3L\left(\left\lceil\frac{n}{2}\right\rceil\right) + \left(\left\lfloor\frac{n}{2}\right\rfloor+1\right)\left(\left\lfloor\frac{n}{2}\right\rfloor-1\right)L\left(\left\lceil\frac{n}{2}\right\rceil\right) + \left\lfloor\frac{n}{2}\right\rfloor^2$$

$$= \left(\left\lfloor\frac{n}{2}\right\rfloor^2+2\right)L\left(\left\lceil\frac{n}{2}\right\rceil\right) + \left\lfloor\frac{n}{2}\right\rfloor^2 \le \left(\left\lfloor\frac{n^2}{4}\right\rfloor+2\right)L\left(\left\lceil\frac{n}{2}\right\rceil\right) + \left\lfloor\frac{n}{2}\right\rfloor^2.$$

We interpret $\frac{n}{2}$ as either $\left\lfloor\frac{n}{2}\right\rfloor$ or $\left\lceil\frac{n}{2}\right\rceil$, and thus the upper bound of $L(n)$ can be estimated asymptotically as follows:

$$\overline{L}(n) = \frac{n^2}{4}\overline{L}\left(\frac{n}{2}\right) + \Theta\left(n^2\right). \tag{4}$$

Proposition 2. *The explicit solution of recurrence (4) is*

$$\overline{L}(n) = O\left(2^{\lceil\log_2 n\rceil^2 - \lceil\log_2 n\rceil}\right).$$

Proof. Denote $k = \lceil\log_2 n\rceil$. Iterating (4) yields

$$\overline{L}(n) = \frac{n^2}{4}\overline{L}\left(\frac{n}{2}\right) + n^2$$

$$= \frac{n^2}{4}\left(\frac{(n/2)^2}{4}\overline{L}\left(\frac{n}{4}\right) + \left(\frac{n}{2}\right)^2\right) + n^2$$

$$= \frac{\left(n^2\right)^2}{4^2\cdot 2^2}\overline{L}\left(\frac{n}{4}\right) + \frac{\left(n^2\right)^2}{4\cdot 2^2} + n^2$$

$$= \frac{\left(n^2\right)^2}{4^2\cdot 2^2}\left(\frac{(n/4)^2}{4}\overline{L}\left(\frac{n}{8}\right) + \left(\frac{n}{4}\right)^2\right) + \frac{\left(n^2\right)^2}{4\cdot 2^2} + n^2$$

$$= \frac{\left(n^2\right)^3}{4^3\cdot 2^{2+4}}\overline{L}\left(\frac{n}{8}\right) + \frac{\left(n^2\right)^3}{4^2\cdot 2^{2+4}} + \frac{\left(n^2\right)^2}{4\cdot 2^2} + n^2$$

$$= \frac{\left(n^2\right)^3}{4^3\cdot 2^{2+4}}\left(\frac{(n/8)^2}{4}\overline{L}\left(\frac{n}{16}\right) + \left(\frac{n}{8}\right)^2\right) + \frac{\left(n^2\right)^3}{4^2\cdot 2^{2+4}} + \frac{\left(n^2\right)^2}{4\cdot 2^2} + n^2$$

$$= \frac{\left(n^2\right)^4}{4^4\cdot 2^{2+4+6}}\overline{L}\left(\frac{n}{16}\right) + \frac{\left(n^2\right)^4}{4^3\cdot 2^{2+4+6}} + \frac{\left(n^2\right)^3}{4^2\cdot 2^{2+4}} + \frac{\left(n^2\right)^2}{4\cdot 2^2} + n^2$$

$$\cdots\cdots\cdots\cdots\cdots\cdots$$

$$= \frac{\left(n^2\right)^{k-1}}{4^{k-1} \cdot 2^{2(1+2+3+\ldots+k-2)}} \overline{L}\left(\frac{n}{2^{k-1}}\right) + \frac{\left(n^2\right)^{k-1}}{4^{k-2} \cdot 2^{2(1+2+3+\ldots+k-2)}}$$

$$+ \frac{\left(n^2\right)^{k-2}}{4^{k-3} \cdot 2^{2(1+2+3+\ldots+k-3)}} + \ldots + \frac{\left(n^2\right)^3}{4^2 \cdot 2^{2(1+2)}} + \frac{\left(n^2\right)^2}{4 \cdot 2^2} + n^2$$

$$= \frac{\left(n^2\right)^{k-1}}{4^{k-1} \cdot 2^{(k-2)(k-1)}} \overline{L}\left(2\right) + \frac{\left(n^2\right)^{k-1}}{4^{k-2} \cdot 2^{(k-2)(k-1)}} + \frac{\left(n^2\right)^{k-2}}{4^{k-3} \cdot 2^{(k-3)(k-2)}}$$

$$+ \ldots + \frac{\left(n^2\right)^3}{4^2 \cdot 2^{2\cdot 3}} + \frac{\left(n^2\right)^2}{4 \cdot 2^2} + n^2$$

$$= \frac{n^{2(k-1)}}{2^{k(k-1)}} + \sum_{j=0}^{k-2} \frac{n^{2(k-j-1)}}{2^{(k-j-2)(k-j+1)}}.$$

We need to estimate the asymptotic bound of the derived function.

$$\frac{n^{2(k-1)}}{2^{k(k-1)}} = \frac{n^{2(\lceil \log_2 n \rceil - 1)}}{2^{\lceil \log_2 n \rceil (\lceil \log_2 n \rceil - 1)}} \leq \frac{\left(2^{\lceil \log_2 n \rceil}\right)^{2(\lceil \log_2 n \rceil - 1)}}{2^{\lceil \log_2 n \rceil (\lceil \log_2 n \rceil - 1)}} = \left(2^{\lceil \log_2 n \rceil}\right)^{\lceil \log_2 n \rceil - 1}.$$

$$\frac{n^{2(k-j-1)}}{2^{(k-j-2)(k-j+1)}} = \frac{n^{2(\lceil \log_2 n \rceil - j - 1)}}{2^{(\lceil \log_2 n \rceil - j - 2)(\lceil \log_2 n \rceil - j + 1)}} \leq \frac{\left(2^{\lceil \log_2 n \rceil}\right)^{2(\lceil \log_2 n \rceil - j - 1)}}{2^{(\lceil \log_2 n \rceil - j - 2)(\lceil \log_2 n \rceil - j + 1)}}$$

$$= \frac{\left(2^{\lceil \log_2 n \rceil}\right)^{2(\lceil \log_2 n \rceil - j - 1)}}{2^{\lceil \log_2 n \rceil^2 - j\lceil \log_2 n \rceil - 2\lceil \log_2 n \rceil - j\lceil \log_2 n \rceil + j^2 + 2j + \lceil \log_2 n \rceil - j - 2}}$$

$$= \frac{\left(2^{\lceil \log_2 n \rceil}\right)^{2(\lceil \log_2 n \rceil - j - 1)}}{\left(2^{\lceil \log_2 n \rceil}\right)^{\lceil \log_2 n \rceil} \cdot 2^{-(2j+1)\lceil \log_2 n \rceil} \cdot 2^{j^2 + j - 2}}$$

$$= \frac{\left(2^{\lceil \log_2 n \rceil}\right)^{2(\lceil \log_2 n \rceil - j - 1) - \lceil \log_2 n \rceil + 2j + 1}}{2^{j^2 + j - 2}} = \frac{\left(2^{\lceil \log_2 n \rceil}\right)^{\lceil \log_2 n \rceil - 1}}{2^{j^2 + j - 2}}.$$

Therefore,

$$\overline{L}\left(n\right) \leq \left(2^{\lceil \log_2 n \rceil}\right)^{\lceil \log_2 n \rceil - 1} + \sum_{j=0}^{k-2} \frac{\left(2^{\lceil \log_2 n \rceil}\right)^{\lceil \log_2 n \rceil - 1}}{2^{j^2 + j - 2}}$$

$$= \left(1 + \sum_{j=0}^{k-2} \frac{1}{2^{j^2 + j - 2}}\right) \left(2^{\lceil \log_2 n \rceil}\right)^{\lceil \log_2 n \rceil - 1}$$

$$= \left(1 + 4 + 1 + \sum_{j=2}^{k-2} \frac{1}{2^{j^2 + j - 2}}\right) 2^{\lceil \log_2 n \rceil^2 - \lceil \log_2 n \rceil}$$

$$< 7 \cdot 2^{\lceil \log_2 n \rceil^2 - \lceil \log_2 n \rceil} = O\left(2^{\lceil \log_2 n \rceil^2 - \lceil \log_2 n \rceil}\right).$$

$\square$

Corollary 1. *When n is power of two ($n = 2^k$ for some positive integer $k \geq 1$), the explicit solution of recurrence (4) is*

$$\overline{L}(n) = O\left(n^{\log_2 n - 1}\right).$$

Proof. By Proposition 2, when $n = 2^k$

$$\overline{L}(n) = O\left(\left(2^{\lceil \log_2 n \rceil}\right)^{\lceil \log_2 n \rceil - 1}\right) = O\left(\left(2^{\log_2 n}\right)^{\log_2 n - 1}\right) = O\left(n^{\log_2 n - 1}\right).$$

$\square$

Thus the expression of an n-vertex complete st-dag has a representation with complexity that may be estimated as $O\left(2^{\lceil \log_2 n \rceil^2 - \lceil \log_2 n \rceil}\right)$ (specifically, $O\left(n^{\log_2 n - 1}\right)$ for n that is power of two).

2.1 Complexities of St-Dag Expressions

Recall that $Rp_{opt}(G)$ denotes the optimal representation of the algebraic expression describing a graph G and denote the complexity of the algebraic expression F by $L(F)$.

The following statement (so called *Monotonicity Lemma*) is proved in [14].

Lemma 1. *If an st-dag G_2 is a subgraph of an st-dag G_1, then $L\left(Rp_{opt}(G_2)\right) \leq L\left(Rp_{opt}(G_1)\right)$.*

It is clear that each n-vertex st-dag is a subgraph of an n-vertex complete st-dag. Hence, by Lemma 1, for every n-vertex st-dag G and an n-vertex complete st-dag G_c, $L\left(Rp_{opt}(G)\right) \leq L\left(Rp_{opt}(G_c)\right)$. Together with Proposition 2, it concludes with the following.

Theorem 1. *For each n-vertex st-dag G the upper bound of $L\left(Rp_{opt}(G)\right)$ is $O\left(2^{\lceil \log_2 n \rceil^2 - \lceil \log_2 n \rceil}\right)$.*

3 Conclusions

We have described the decomposition method for generating expressions of complete st-dags and have estimated the corresponding expression complexities in a quasi-polynomial manner. Based on the above, it has been shown that the expression $Ex(G)$ of an n-vertex st-dag G has a representation with complexity bounded by $O\left(2^{\lceil \log_2 n \rceil^2 - \lceil \log_2 n \rceil}\right)$. The question left is whether the decomposition method provides the optimal representation of $Ex(G_c)$ for an n-vertex complete st-dag G_c and, generally, what is a tight bound for the optimal representation of $Ex(G_c)$. The answer to this question will help to understand whether it is possible to improve the obtained $2^{O(\log^2 n)}$ upper bound for n-vertex st-dags.

References

1. Anderson, D.F., Levy, R., Shapiro, J.: Zero-divisor graphs, von Neumann regular rings, and Boolean algebras. J. Pure Appl. Algebra **180**, 221–241 (2003)
2. Bein, W.W., Kamburowski, J., Stallmann, M.F.M.: Optimal reduction of two-terminal directed acyclic graphs. SIAM J. Comput. **21**(6), 1112–1129 (1992)
3. Duffin, R.J.: Topology of series-parallel networks. J. Math. Anal. Appl. **10**, 303–318 (1965)
4. Fernau, H.: Algorithms for learning regular expressions from positive data. Inf. Comput. **207**, 521–541 (2009)
5. Fernau, H., Ryan, J.F., Sugeng, K.A.: A sum labelling for the generalised friendship graph. Discrete Math. **308**, 734–740 (2008)
6. Finta, L., Liu, Z., Milis, I., Bampis, E.: Scheduling UET-UCT series-parallel graphs on two processors. Theoret. Comput. Sci. **162**(2), 323–340 (1996)
7. Golumbic, M.C., Gurvich, V.: Read-once functions. In: Crama, Y., Hammer, P.L. (eds.) Boolean Functions: Theory, Algorithms and Applications, pp. 519–560. Cambridge University Press, New York (2011)
8. Golumbic, M.C., Mintz, A., Rotics, U.: Factoring and recognition of read-once functions using cographs and normality and the readability of functions associated with partial k-trees. Discrete Appl. Math. **154**(10), 1465–1477 (2006)
9. Golumbic, M.C., Perl, Y.: Generalized Fibonacci maximum path graphs. Discrete Math. **28**, 237–245 (1979)
10. Gulan, S.: Series parallel digraphs with loops. Graphs encoded by regular expression. Theory Comput. Syst. **53**, 126–158 (2013)
11. Korenblit, M.: Efficient computations on networks. Ph.D. Thesis, Bar-Ilan University, Israel (2004)
12. Korenblit, M.: Decomposition methods for generating algebraic expressions of full square rhomboids and other graphs. Discrete Appl. Math. **228**, 60–72 (2017)
13. Korenblit, M., Levit, V.E.: On algebraic expressions of series-parallel and Fibonacci graphs. In: Calude, C.S., Dinneen, M.J., Vajnovszki, V. (eds.) DMTCS 2003. LNCS, vol. 2731, pp. 215–224. Springer, Heidelberg (2003). https://doi.org/10.1007/3-540-45066-1_17
14. Korenblit, M., Levit, V.E.: Estimation of expressions' complexities for two-terminal directed acyclic graphs. Electron. Not. Discrete Math. **63**, 109–116 (2017)
15. Mintz, A., Golumbic, M.C.: Factoring Boolean functions using graph partitioning. Discrete Appl. Math. **149**(1–3), 131–153 (2005)
16. Mundici, D.: Functions computed by monotone boolean formulas with no repeated variables. Theoret. Comput. Sci. **66**, 113–114 (1989)
17. Mundici, D.: Solution of Rota's problem on the order of series-parallel networks. Adv. Appl. Math. **12**, 455–463 (1991)
18. Naumann, V.: Measuring the distance to series-parallellity by path expressions. In: Mayr, E.W., Schmidt, G., Tinhofer, G. (eds.) WG 1994. LNCS, vol. 903, pp. 269–281. Springer, Heidelberg (1995). https://doi.org/10.1007/3-540-59071-4_54
19. Oikawa, M.K., Ferreira, J.E., Malkowski, S., Pu, C.: Towards algorithmic generation of business processes: from business step dependencies to process algebra expressions. In: Dayal, U., Eder, J., Koehler, J., Reijers, H.A. (eds.) BPM 2009. LNCS, vol. 5701, pp. 80–96. Springer, Heidelberg (2009). https://doi.org/10.1007/978-3-642-03848-8_7
20. Satyanarayana, A., Wood, R.K.: A linear time algorithm for computing k-terminal reliability in series-parallel networks. SIAM J. Comput. **14**(4), 818–832 (1985)

21. Savicky, P., Woods, A.R.: The number of boolean functions computed by formulas of a given size. Random Struct. Algorithms **13**, 349–382 (1998)
22. Tamir, A.: A strongly polynomial algorithm for minimum convex separable quadratic cost flow problems on two-terminal series-parallel networks. Math. Program. **59**, 117–132 (1993)
23. Wang, A.R.R.: Algorithms for multilevel logic optimization. Ph.D. Thesis, University of California, Berkeley (1989)

The Relative Oriented Clique Number of Triangle-Free Planar Graphs Is 10

Soura Sena Das[1], Soumen Nandi[2(✉)], and Sagnik Sen[3]

[1] Indian Statistical Institute, Kolkata, India
[2] Institue of Engineering and Management, Kolkata, India
soumen2004@gmail.com
[3] Indian Institute of Technology, Dharwad, India

Abstract. A vertex subset R of an oriented graph $\overrightarrow{G}$ is a relative oriented clique if each pair of non-adjacent vertices of R is connected by a directed 2-path. The relative oriented clique number $\omega_{ro}(\overrightarrow{G})$ of $\overrightarrow{G}$ is the maximum value of $|R|$ where R is a relative oriented clique of $\overrightarrow{G}$. Given a family $\mathcal{F}$ of oriented graphs, the relative oriented clique number is $\omega_{ro}(\mathcal{F}) = \max\{\omega_{ro}(\overrightarrow{G})|\overrightarrow{G} \in \mathcal{F}\}$. For the family $\mathcal{P}_4$ of oriented triangle-free planar graphs, it was conjectured that $\omega_{ro}(\mathcal{P}_4) = 10$. In this article, we prove the conjecture.

Keywords: Oriented graph · Relative clique number · Triangle-free planar graph

1 Introduction and Main Result

Oriented colroing and chromatic number was introduced by Courcelle [1] in 1994 and then, following the works of Raspaud and Sopena [7], the decipline gained popularity [4–6,9–11].

Recently two parameters, namely, relative oriented clique number and absolute oriented clique number, associated to the notions of oriented coloring is being studied. Our focus is the former which was introduced by Nandy, Sopena and Sen [6].

A vertex subset R of an oriented graph $\overrightarrow{G}$ is a *relative oriented clique* if each pair of non-adjacent vertices of R is connected by a *directed 2-path* (an oriented 2-path uvw having arcs uv and vw) in $\overrightarrow{G}$. The *relative oriented clique number* $\omega_{ro}(\overrightarrow{G})$ of $\overrightarrow{G}$ is the maximum value of $|R|$ where R is a relative oriented clique of $\overrightarrow{G}$. Given a family $\mathcal{F}$ of oriented graphs, the relative oriented clique number is given by $\omega_{ro}(\mathcal{F}) = \max\{\omega_{ro}(\overrightarrow{G})|\overrightarrow{G} \in \mathcal{F}\}$.

Finding the value of $\omega_{ro}(\mathcal{P}_4)$, where $\mathcal{P}_4$ is the family of triangle-free oriented planar graphs, was mentioned as an open problem in the recent survey on

This work is partially supported by the IFCAM project "Applications of graph homomorphisms" (MA/IFCAM/18/39) and ANR project HOSIGRA (ANR-17-CE40-0022).

oriented coloring by Sopena [9]. Furthermore, the value $\omega_{ro}(\mathcal{P}_4) = 10$ was conjectured by Sen [8]. The best known result on this topic is $10 \leq \omega_{ro}(\mathcal{P}_4) \leq 15$ [2]. In this article, we close the open problem by proving the conjecture.

Theorem 1. *For the family $\mathcal{P}_4$ of triangle-free planar graphs, $\omega_{ro}(\mathcal{P}_4) = 10$.*

The proof of the above result is presented in the following section.

2   Proof of Theorem 1

The proof implies from the observations and lemmata proved in the following.

Let $\overrightarrow{H}$ be a minimal counter example of Theorem 1, with respect to lexicographic ordering of $(|V(H)|, |E(H)|)$. Let R be a maximum relative oriented clique of $\overrightarrow{H}$. Thus $|R| \geq 11$. Also let $S = V(H) \setminus R$. The vertices of R are *good* vertices while that of S are *helpers*. Moreover, assume a particular planar embedding of $\overrightarrow{H}$ for the rest of this section unless otherwise stated.

The set S of vertices is an independent set, as otherwise we can remove an edge with both end points in S in order to obtain a smaller counter example contradicting the minimality of $\overrightarrow{H}$. We need some terminologies and notations before going further.

The set of vertices and arcs of an oriented graph $\overrightarrow{H}$ is denoted by $V(\overrightarrow{H})$ and $A(\overrightarrow{H})$, respectively. The underlying simple graph of $\overrightarrow{H}$ is H. The set of neighbors $N(v)$ of a vertex v is the set of all adjacent vertices of v. Given an arc $uv \in A(\overrightarrow{G})$, u is an *in-neighbor* of v and v is an *out-neighbor* of u. The set of all in-neighbors and out-neighbors of v is denoted by $N^-(v)$ and $N^+(v)$, respectively. Moreover the *degree*, *in-degree* and *out-degree* of a vertex v is given by $d(v) = |N(v)|$, $d^-(v) = |N^-(v)|$ and $d^+(v) = |N^+(v)|$, respectively.

Two vertices u, v *agree* on a third vertex w if $w \in N^\alpha(u) \cap N^\alpha(v)$ for some $\alpha \in \{+, -\}$. Also u, v *disagree* on w if $w \in N^\alpha(u) \cap N^\beta(v)$ for some $\{\alpha, \beta\} = \{+, -\}^1$.

We want to show that any helper h in $\overrightarrow{H}$ has $d(h) = 2$. To begin with, note that $\overrightarrow{H}$ is connected due to its minimality. Moreover if $d(h) = 1$, then even after deleting the vertex h the set R remains a relative oriented clique. Thus $d(h) \geq 2$.

Furthermore, given a planar graph and its embedding we will denote the region of the plane corresponding to a face f of the graph by R_f. The notation is unambiguous always but for some exceptions. In those exceptional cases, we will use a different way to describe a face/region. Thus there is no scope of ambiguity.

To begin with, we will improve the above result. We will need some nomenclatures for the proof. A vertex v *sees* a vertex u if they are adjacent or they are connected by a directed 2-path. If u, v are connected by a directed 2-path with the thrid vertex of the directed 2-path being w, we say that u sees v *via* w.

---

[1] We use this notation frequently to denote $\alpha, \bar{\alpha} \in \{+, -\}$ and $\alpha \neq \bar{\alpha}$. Our notation is a set theoretic equation whose solutions arc the values that $\alpha, \bar{\alpha}$ may take.

The proof of the following result was implicit inside the proof of Theorem 11 in the paper by Klostermeyer and MacGillivray [3]. However, we reprove it for the sake of completeness.

Lemma 1. *Five good vertices $v_1, v_2, \cdots, v_5$ cannot agree on a vertex v.*

Proof. As H is triangle-free, $v_1, v_2, \cdots, v_5$ are independent vertices. Assume that $v_1, v_2, \cdots, v_5$ are arranged in a clockwise order around v in the planar embedding of H and that the five vertices agree with each other on v.

Note that v_2 must see v_4 via some vertex h_1. Therefore, v_3 is forced to see v_1 and v_5 via h_1. This implies that either v_1 does not see v_4 or v_2 does not see v_5. □

Using triangle-freeness of the planar graph, we can do a bit better than above.

Lemma 2. *Four good vertices v_1, v_2, v_3, v_4 cannot agree with each other on a vertex v.*

Proof. Assume that v_1, v_2, v_3, v_4 are arranged in a clockwise order around v in the planar embedding of H and that they are out-neighbors of v.

Note that v_2 must see v_4 via some h_1 as H is triangle-free. Therefore, v_1 is forced to see v_3 via h_1. Without loss of generality we may assume that $v_1, v_2 \in N^-(h_1)$ and that $v_3, v_4 \in N^+(h_1)$.

Observe that v_1 must see v_2 via some h_2 and v_3 must see v_4 via some h_3. As h_2 cannot see v_4, h_2 is a helper. Similarly, as h_3 cannot see v_2, h_2 is a helper.

Any good vertex containing in $R_{v_1 h_2 v_2 h_1 v_1}$ or in $R_{v_3 h_3 v_4 h_1 v_3}$ cannot see v. On the other hand, for similar reasons, any good vertex contained in $R_{vv_1 h_2 v_2 v}$ or in $R_{vv_3 h_3 v_4 v}$ is an in-neighbor of v in order for seeing v_3 or v_2 via v. However, if each of $R_{vv_1 h_2 v_2 v}$ and $R_{vv_3 h_3 v_4 v}$ contains at least one good vertex, they cannot see each other. Therefore, without loss of generality $R_{vv_3 h_3 v_4 v}$ does not contain any good vertex.

Any good vertex contained in $R_{vv_2 h_1 v_3 v}$ or in $R_{vv_1 h_1 v_4 v}$ cannot, respectively, see v_1, v_4 or v_2, v_3 via h_1 alone. Thus they must see some of them via v. Hence any good vertex contained in $R_{vv_2 h_1 v_3 v}$ or in $R_{vv_1 h_1 v_4 v}$ is also an in-neighbor of v. As $|R \setminus (N(v) \cup \{v, h_1\})| \geq 5$, we have at least five good vertices agreeing on v. This contradicts Lemma 1. □

Using the above result, we prove the following.

Lemma 3. *It is not possible to have three good vertices v_1, v_2, v_3 disagree with a fourth good vertex v_4 on a vertex v.*

Proof. Assume that v_1, v_2, v_3, v_4 are arranged in a clockwise order around v in the planar embedding of H and that $v_1, v_2, v_3 \in N^-(v)$ while $v_4 \in N^+(v)$.

Note that v_1 must see v_3 via some h_1. Without loss of generality assume that $v_1 \in N^-(h_1)$ and $v_3 \in N^+(h_1)$.

Let the face $vv_1 h_1 v_3 v$ divides the plane into two connected regions: A is the one containing v_2 and B is the one containing v_4. Assume that each of A and B

contains a good vertex. Suppose a good vertex w_1 is contained in A and another good vertex w_2 is contained in B.

Observe that w_1 cannot see w_2 via v, as otherwise four vertices will agree on v contradicting Lemma 2. Therefore, w_1 must see w_2 via h_1. Thus each vertex of $R \setminus (N(v) \cup \{v, h_1\})$ must be adjacent to h_1. However as $|R \setminus (N(v) \cup \{v, h_1\})| \geq 5$, the vertex h_1 has at least seven good neighbors. Hence by pigeonhole principle at least four good vertices agree on h_1, a contradiction to Lemma 2.

Therefore, all the vertices of $R \setminus (N(v) \cup \{v, h_1\})$ must be contained either in A or in B.

If they are all inside A, then they have to see v_4. The only options are via v or h_1. If one of them sees v_4 via v, then v agrees on at least four good vertices contradicting Lemma 2. Thus all the vertices of $R \setminus (N(v) \cup \{v, h_1\})$ sees v_4 via h_1, again contradicting Lemma 2 as $|R \setminus (N(v) \cup \{v, h_1\})| \geq 5$.

On the other hand, if they are all inside B, then they have to see v_2. The only options are via v or h_1. If three of them sees v_2 via v, then v agrees on at least four good vertices contradicting Lemma 2. Thus at most two of the vertices can see v_2 via v. As $|R \setminus (N(v) \cup \{v, h_1\})| \geq 5$, at least three vertices of $R \setminus (N(v) \cup \{v, h_1\})$ sees v_2 via h_1, yet again contradicting Lemma 2. □

Now we focus on proving that a vertex v cannot have two good in-neighbors and two good out-neighbors. The proof is divided into two cases. The first case follows.

Lemma 4. *Let v_1, v_2, v_3, v_4 be good neighbors of a vertex v arranged in a clockwise order around v. It is not possible to have $v_1, v_3 \in N^+(v)$ and $v_2, v_4 \in N^-(v)$.*

Proof. Assume that v_1 sees v_3 via some h_1. Then v_2 is forced to see v_4 via h_1 as well. Let w_1 be a good vertex contained inside $R_{vv_i h_1 v_{i+1} v}$, where $i \in \{1, 2, 3, 4\}$ and $+$ operation is taken modulo 4. Note that w_1 have to see v_{i+2} ($+$ is taken modulo 4) via v or h_1. In either case, v (or h_1) becomes adjacent to four good vertices among which three disagree with the fourth one on v or (h_1). This is a contradiction to Lemma 3. □

Now we present the second case.

Lemma 5. *Let v_1, v_2, v_3, v_4 be good neighbors of a vertex v arranged in a clockwise order around v. It is not possible to have $v_1, v_2 \in N^+(v)$ and $v_3, v_4 \in N^-(v)$.*

Proof. Assume that v_1 sees v_2 via some h_1. Also suppose that v_3 sees v_4 via the same h_1. Let w_1 be a good vertex contained inside $R_{vv_i h_1 v_{i+1} v}$, where $i \subset \{1, 2, 3, 4\}$ and $+$ operation is taken modulo 4. Note that w_1 have to see v_{i+2} ($+$ is taken modulo 4) via v or h_1. In either case, v (or h_1) becomes adjacent to four good vertices among which three disagree with the fourth one on v or (h_1). This is a contradiction to Lemma 3.

Thus v_3 must see v_4 via a different vertex h_2. If the region $R_{vv_1 h_1 v_2 v}$ contain a vertex w, then w must see v_3, v_4 via h_1 contradicting Lemma 3. Thus $R_{vv_1 h_1 v_2 v}$ does not contain a good vertex. Similarly, $R_{vv_3 h_1 v_4 v}$ also does not contain a good vertex.

Let $W = R \setminus \{v_1, v_2, v_3, v_4, v, h_1, h_2\} = \{w_1, w_2, \cdots, w_t\}$. Thus $|W| \geq 4$. Note that a w_i must be contained in $R_{vv_1h_1v_2vv_3h_2v_4v}$ and can see at most two of the four good vertices v_1, v_2, v_3, v_4 via a particular vertex due to Lemma 3. Now let us consider some cases.

- If w_1 sees v_1, v_2 via h_3 and v_3, v_4 via h_4, then both w_2 and w_3 (irrespective of their placement) is forced to see v_1 or v_2 via h_3 and v_3 or v_4 via h_4 creating a vertex with three good neighbor disagreeing on it with a fourth one. This is a contradiction to Lemma 3.

- If w_1 sees v_1, v_3 via h_3 and v_2, v_4 via h_4, then both w_2 and w_3 (irrespective of their placement) is forced to see v_1 or v_3 via h_3 and v_2 or v_4 via h_4 creating a vertex with three good neighbor disagreeing on it with a fourth one. This is a contradiction to Lemma 3.

- If w_1 sees v_1, v_2 via h_3 and v_3 via h_4 (or by being adjacent to it) and v_4 via h_5 (or by being adjacent to it), then w_2 is not able to see at least one of v_1, v_2, v_3, v_4 irrespective of its placement.

- If w_1 sees v_1 via h_3 (or by being adjacent to it) and v_2 via h_4 (or by being adjacent to it) and v_3 via h_5 (or by being adjacent to it) and v_4 via h_6 (or by being adjacent to it), then w_2 is not able to see at least two of v_1, v_2, v_3, v_4 irrespective of its placement.

Thus we are done.   □

Therefore, we can conclude that the graph $\vec{H}$ does not have any vertex v with at least four good neighbors. However, we want to improve this and show that $|N(v) \cap R| \leq 2$ for any $v \in V(\vec{H})$. Indeed, the previous results will be used to prove so.

Lemma 6. *Three good vertices v_1, v_2, v_3 cannot agree with each other on a vertex v.*

Proof. Let v_1 see v_2 via some h_1. Let v_3 also see v_2 via h_1.

Let w_1 be a good vertex contained inside $R_{vv_ih_1v_{i+1}v}$, where $i \in \{1, 2, 3\}$ and $+$ operation is taken modulo 3. Note that w_1 have to see v_{i+2} ($+$ is taken modulo 3) via v or h_1. In either case, v (or h_1) becomes adjacent to four good vertices among which three disagree with the fourth one on v or (h_1). This is a contradiction to Lemma 3.

Hence v_2 must see v_3 via h_2 and v_3 must see v_1 via h_3. Let w_1 be a good vertex contained inside $R_{vv_ih_iv_{i+1}v}$, where $i \in \{1, 2, 3\}$ and $+$ operation is taken modulo 3. Note that w_1 have to see v_{i+2} ($+$ is taken modulo 3) via v or h_i. In either case, v (or h_i) becomes adjacent to four good vertices among which three disagree with the fourth one on v or (h_i). This is a contradiction to Lemma 3.

Therefore, every vertex of $W = R \setminus \{v, v_1, v_2, v_3, h_1, h_2, h_3\}$ must be contained in $R_{v_1h_1v_2h_2v_3h_3v_1}$ and $|W| \geq 4$.

Note that a w_i must be contained in $R_{v_1h_1v_2h_2v_3h_3v_1}$ and can see at most two of the three good vertices v_1, v_2, v_3 via a particular vertex due to Lemma 3. Now let us consider some cases.

- If w_1 sees v_1, v_2 via h_4 and v_3 via h_5, then w_2 (irrespective of their placement) is forced to see v_1 or v_2 via h_4 creating a vertex with four good neighbors, a contradiction.
- If w_1 sees v_1 via h_4 (or by being adjacent to it) and v_2 via h_5 (or by being adjacent to it) and v_3 via h_6 (or by being adjacent to it), then w_2 is not able to see at least one of v_1, v_2, v_3 irrespective of its placement.

Thus we are done.   □

The final lemma in the similar direction follows.

Lemma 7. *It is not possible to have two good vertices v_1, v_2 disagree with a third good vertex v_3 on a vertex v.*

Proof. Let v_1 see v_2 via some h_1. The cycle $vv_1h_1v_2v$ divides the plane into two connected regions: A containing v_3 and B not containing v_3.

Let w_1 be a good vertex contained inside B. Note that w_1 have to see v_3 via v or h_1. In either case, v (or h_1) becomes adjacent to four good vertices among which three disagree with the fourth one on v or (h_1). This is a contradiction to Lemma 3.

Therefore, every vertex of $W = R \setminus \{v, v_1, v_2, v_3, h_1\}$ must be contained in to A. Thus $|W| \geq 6$.

Note that a w_i must be contained in A and can see at most two of the three good vertices v_1, v_2, v_3 via a particular vertex. Now let us consider some cases.

- If w_1 sees v_1, v_2 via h_2 and v_3 via h_3, then w_2 (irrespective of their placement) is forced to see v_1 or v_2 via h_2 creating a vertex with four good neighbors, a contradiction.
- If w_1 sees v_1 via h_4 (or by being adjacent to it) and v_2 via h_5 (or by being adjacent to it) and v_3 via h_6 (or by being adjacent to it), then w_2 is not able to see at least one of v_1, v_2, v_3 irrespective of its placement.

Thus we are done.   □

This implies that the graph $\overrightarrow{H}$ does not have any vertex v with at least three good neighbors.

Lemma 8. *It is not possible for a vertex to have at least three good neighbors.*

Proof. Follows directly from Lemma 6 and 7.   □

In particular, for any helper h we have $d(h) \leq 2$. Thus, using our earlier observation that the degree of h is at least 2, we can conclude that $d(h) = 2$.

Observe that two helpers h_1 and h_2 cannot have $N(h_1) = N(h_2) = \{u, v\}$. The reason is, both h_1 and h_2 contributes in u seeing v. Therefore, even if we delete h_2, the set R still remains a relative oriented clique contradicting the minimality of $\overrightarrow{H}$.

Now construct a graph H^* from H as follows: delete each helper and add an edge between its neighbors. Observe that H^* is planar, not neccesarily triangle-free, with $V(H^*)$ being the set of good neighbors of $\overrightarrow{H}$. Also the degree of a vertex v in H^* is greater than equal to the degree of v in H. As H^* is a planar graph, it must have a vertex x with degree at most five. Therefore, we can say that there exists a good vertex x in $\overrightarrow{H}$ having degree at most five. We fix the name of this vertex x for the rest of this section.

Proof of Theorem 1. Let x be a good vertex of $\overrightarrow{H}$ having degree at most five whose existance follows from the above paragraph. Let $X = R \setminus (N(x) \cup \{x\})$. We know due to Lemma 8 that $|R \cap N(x)| \leq 2$.

As $|R| \geq 11$, we have $|X| \geq 8$. Note that each vertex of X must see x via one of its neighbors. Therefore, by pigeonhole principle at least one of the neighbors x_1 (say) of x will have two good neighbors from X. Thus x_1 has three good neighbors, contradicting Lemma 8. $\square$

References

1. Courcelle, B.: The monadic second order logic of graphs VI: on several representations of graphs by relational structures. Discrete Appl. Math. **54**(2), 117–149 (1994)
2. Das, S., Prabhu, S., Sen, S.: A study on oriented relative clique number. Discrete Math. **341**(7), 2049–2057 (2018)
3. Klostermeyer, W.F., MacGillivray, G.: Analogues of cliques for oriented coloring. Discussiones Mathematicae Graph Theory **24**(3), 373–388 (2004)
4. Kostochka, A.V., Sopena, É., Zhu, X.: Acyclic and oriented chromatic numbers of graphs. J. Graph Theory **24**(4), 331–340 (1997)
5. Marshall, T.H.: Homomorphism bounds for oriented planar graphs. J. Graph Theory **55**, 175–190 (2007)
6. Nandy, A., Sen, S., Sopena, É.: Outerplanar and planar oriented cliques. J. Graph Theory **82**(2), 165–193 (2016)
7. Raspaud, A., Sopena, É.: Good and semi-strong colorings of oriented planar graphs. Inf. Process. Lett. **51**(4), 171–174 (1994)
8. Sen, S.: A contribution to the theory of graph homomorphisms and colorings. Ph.D. thesis, Bordeaux University, France (2014)
9. Sopena, É.: Homomorphisms and colourings of oriented graphs: an updated survey. Discrete Math. **339**(7), 1993–2005 (2016)
10. Sopena, É.: The chromatic number of oriented graphs. J. Graph Theory **25**, 191–205 (1997)
11. Sopena, É.: Oriented graph coloring. Discrete Math. **229**(1–3), 359–369 (2001)

Combinatorial Optimization

On the Minimum Satisfiability Problem

Umair Arif[2], Robert Benkoczi[2], Daya Ram Gaur[2(✉)],
and Ramesh Krishnamurti[1]

[1] School of Computing Science, Simon Fraser University, Burnaby, BC, Canada
ramesh@sfu.ca
[2] Department of Math and Computer Science, University of Lethbridge,
Lethbridge, AB, Canada
umair.arif@alumni.uleth.ca, {benkoczi,gaur}@cs.uleth.ca

Abstract. We characterize the optimal solution to the LP relaxation of the standard formulation for the minimum satisfiability problem. Based on the characterization, we give a $O(nm^2)$ combinatorial algorithm to solve the fractional version of the minimum satisfiability problem optimally where $n(m)$ is the number of variables (clauses). As a by-product, we obtain a $2(1 - 1/2^k)$ approximation algorithm for the minimum satisfiability problem where k is the maximum number of literals in any clause. We also give a simple linear time 2 approximation algorithm.

Keywords: Primal-dual algorithm · Minimum satisfiability ·
Approximation algorithm

1 Introduction

Since the decision version of the satisfiability problem was shown to be the first NP-complete problem (Cook 1971), the satisfiability problem has been studied extensively. Here we consider weighted MINSAT, a satisfiability problem in which each clause has a non-negative weight, and the goal is to obtain a truth assignment that minimizes the total weight of the satisfied clauses. MINSAT was first studied by Kohli et al. (1994), who showed that the problem is NP-complete even when every clause has at most two literals. They also showed that a greedy algorithm, similar to one given by Johnson (1974) for MAXSAT, has a performance ratio of k, where k is the maximum number of literals in a clause. Kohli et al. (1994) also provide a probabilistic greedy algorithm with a performance ratio of 2.

Bertsimas et al. (1999) provide a randomized approximation algorithm with a performance ratio of $2 - 1/2^{k-1}$, where k is the number of literals in any clause. The algorithm is based on rounding the LP solution using a technique called dependent rounding. Marathe and Ravi (1996) provide an approximation preserving reduction and show that approximating vertex cover with maximum degree Δ is equivalent to approximating MINSAT with at most Δ literals in each clause. Berman and Fujito (1999) and Halperin (2002) studied the vertex cover problem with a maximum degree of Δ.

© Springer Nature Switzerland AG 2020
M. Changat and S. Das (Eds.): CALDAM 2020, LNCS 12016, pp. 269–281, 2020.
https://doi.org/10.1007/978-3-030-39219-2_23

Bar-Yehuda and Even (1981, 1985), Hochbaum (1982, 1983), and Monien and Speckenmeyer (1985) provide 2 factor approximation algorithms for the vertex cover problem, the best known so far. Dinur and Safra (2005) show that it is NP-hard to approximate vertex cover within a factor of 1.3606. On the flip side, several algorithms with approximation ratios better than 1/2 are known for MAXSAT. Approximation algorithms with performance ratio 3/4 for MAXSAT are due to Yannakakis (1992) and Goemans and Williamson (1994). Currently, the best known, the 0.878 factor approximation for MAXSAT due to Goemans and Williamson (1994) is based on rounding solutions to a semidefinite program.

MINSAT, therefore, seems to be harder than MAXSAT in the sense of approximations, given the reduction in (Marathe and Ravi 1996). Given an instance of MINSAT, all the clauses are falsified by some assignment if and only if all the variables occur either as positive or only as negative literals (Marathe and Ravi 1996). Marathe and Ravi (1996) gave the following reduction from MINSAT to vertex cover. Given an instance of MINSAT, they construct a graph called an auxiliary graph. In the auxiliary graph, there is a node for each clause and an edge between nodes v_i and v_j (corresponding to clauses c_i and c_j) if and only if there is a variable which occurs as a positive (negative) literal in c_i and as a negative (positive) literal in c_j. The edge in the auxiliary graph implies that no assignment can simultaneously falsify both the clauses. This inturn implies that given an assignment that satisfies k clauses, we can find a vertex cover of size k in the auxiliary graph and vice-versa. One can now compute a 2-approximate vertex cover using the well known primal-dual algorithm for vertex cover (Bar-Yehuda and Even 1981). In the unweighted case, the dual solution corresponds to a maximal matching in the auxiliary graph, and the primal solution corresponds to selecting all the matched nodes (both the endpoints of each edge in the matching). It is generally believed that vertex cover is hard to approximation with a constant better than 2.

There is a conditional result which states that it might not be possible to approximate vertex cover within a factor of $2 - \epsilon$. The LP formulation that Bertsimas et al. (1999) use for their dependent rounding has at most 2 variables in each constraint, and can be solved using Tardos's algorithm (Tardos 1986) in $O(n^5)$ time. The dependent rounding step is randomized, and a de-randomization is not given in their paper.

Hochbaum (1998) show that the optimal solution to the LP relaxation is half-integral, can be computed in time $O(m^2 n^2 \log (m + n))$ and can be rounded deterministically to obtain a 2 approximation for MINSAT. This is achieved by rewriting the constraints and solving the resulting LP problem using flow or cut algorithms.

To obtain a faster, deterministic and combinatorial approximation algorithm for MINSAT, we characterize the optimal solution to the LP relaxation. Our proof is nonconstructive. Such characterizations are known for the maximum matching problem (Edmonds 1965), the vertex cover problem (Bourjolly and Pulleyblank 1989), the maximum charge problem (Krishnamurti et al. 2006), and the linear ordering problem (Iranmanesh 2016) to name a few. We show that the absence of appropriately defined structures (called augmenting paths and augmenting lassos) in the bipartite graph associated with a dual solution implies that the current dual solution is optimal.

Using our characterization, we give a combinatorial algorithm with running time $O(nm^2)$ to solve the LP relaxation of the minimum satisfiability problem where n is the number of variables, and m is the number of clauses. The running time for K-MINSAT is $O(km^2)$. We also show that there exists an optimal solution to the LP relaxation that is half-integral. Using the fact that we have a half-integral solution, we give a simple randomized rounding scheme with performance ratio $2(1 - 1/2^k)$ for K-MINSAT. The scheme can be de-randomized using the method of conditional expectation. We also give a simple $O(nm)$ primal-dual approximation algorithm with a performance ratio of 2.

1.1   Our Contributions

In Theorem 1 we give a simple, primal-dual approximation algorithm for the MINSAT problem with a performance ratio of 2. Our primal-dual scheme for MINSAT is also an alternative proof of the result that the size of an optimal vertex cover is at most twice the size of a maximal matching (Bar-Yehuda and Even 1981).

In Theorem 4, we characterize the optimal solution to the LP relaxation of the minimum satisfiability problem and show that it is half-integral. This characterization gives us a $O(nm^2)$ combinatorial algorithm for solving the LP (see Theorem 2). We then give a simple randomized rounding scheme with performance ratio $2(1 - 1/2^k)$ for MINSAT with at most k literals in each clause (see Theorem 3). Our rounding scheme can be de-randomized easily. This gives us a faster $O(km^2)$ deterministic approximation algorithm compared to the one due to Hochbaum (1998) with the same performance ratio for approximating K-MINSAT. Our approximation algorithm for MINSAT is faster in the following sense (i) our algorithm does not require doubling of the number of variables and the constraints as in (Hochbaum 1998) (ii) the faster $O(m^2n + mn^2)$ algorithm needed in (Hochbaum 1998) to solve the max flow problem (due to (Orlin 2013)) has an extra time $O(mn^2)$ term. We give a $O(m^2n)$ algorithm which should be faster in practice.

Given the conditional results of Khot and Regev (2008) and the relationship of MINSAT with vertex cover (Marathe and Ravi 1996), any improvement over a bound of $2(1-1/2^k)$ looks improbable. Therefore the best that one can hope for is a faster deterministic combinatorial algorithm, which is the main contribution in this paper. Our algorithm has a running time of $O(nm^2)$ compared to previous ones $O(m^2n^2 \log{(m + n)})$ in (Hochbaum 1998) and $O(m^2n + n^2m)$ if we use the recent max flow algorithm (Orlin 2013). We also give a simpler and much faster $O(mn)$ ($O(mk)$ for K-MINSAT) 2-factor approximation algorithm using the primal-dual schema. The nonconstructive proof technique used in Theorem 4. to characterize the structure of the optimal solution to the LP relaxation might be of independent interest.

2 A Primal-Dual Approximation

We begin by describing a simple, primal-dual approximation algorithm. We use the integer linear program (ILP) formulation due to Bertsimas et al. (1999) for the MINSAT problem as the starting point. We assume that no clauses contains x and the negation of x for some variable x. Associated with each clause, c_i is an indicator variable z_i. The indicator variable is 1 if the clause is satisfied and 0 otherwise. The clauses are indexed by i, and the variables are indexed by j. V is the set of variables and C is the set of clauses. P_i is the set of positive literals in clause i. Similarly, N_i is the set of negative literals in clause i.

The integer linear program (ILP) for MINSAT and the LP dual are given below.

$$ILP : \min \sum_{i=1}^{n} w_i z_i$$
$$z_i \geq x_j, \forall j \in P_i$$
$$z_i \geq 1 - x_j, \forall j \in N_i$$
$$z_i, x_j \in \{0, 1\}, \ i \in C, j \in V$$

$$Dual : \max \sum_{i \in C, j \in N_i} y_{ij}$$
$$\sum_{j \in P_i \cup N_i} y_{ij} \leq w_i, \ i \in C$$
$$- \sum_{i \mid j \in P_i} y_{ij} + \sum_{i \mid j \in N_i} y_{ij} \leq 0, \ j \in V$$
$$y_{ij} \geq 0, \ i \in C, j \in V$$

We will describe a procedure DUAL that will compute a feasible solution to the dual. Furthermore, if the dual feasible solution has value k, then we construct a primal solution to the ILP, which is feasible and has value at most $2k$.

Theorem 1. *There exists a $O(nm)$ primal dual algorithm for the minimum satisfiability problem with performance ratio 2.*

Proof. A bipartite graph can be associated with the dual in the following way: the variables and the clauses are the vertices. There is an edge (v, c) if variable v occurs in clause c. We further classify this edge as a positive edge if v occurs as a positive literal in c. If v occurs as a negative literal in c, then the edge is classified as a negative edge.

Given a dual solution y, a clause i is saturated if $\sum_{i \in P_i \cup N_i} y_{ij} = w_i$. For each clause i, define slack $s_i = w_i - \sum_{i \in P_i \cup N_i} y_{ij}$. Slack $s_i = 0$ for saturated clauses and $s_i > 0$ for unsaturated clauses given any dual feasible solution. Define $P(v) = \sum_{i \mid v \in P_i} s_i$ ($N(v) = \sum_{i \mid v \in N_i} s_i$) as the sum of the slack on the clauses that contain v as a positive (negative) literal. assume without loss of generality that $P(v) \geq N(v)$. If $P(v) < N(v)$ then we can rename v with $\bar{u}$. If ϕ is the original formula and ϕ_u is the formula in which v is replaced with $\bar{u}$ then any clause that is satisfied in ϕ with v set to TRUE, is satisfied in ϕ_u with u set to FALSE. Similarly, any clause that is falsified in ϕ with v set to FALSE, is still falsified in ϕ_u with u set to TRUE. This relabeling may be needed in every iteration.

DUAL:

1. Let the set U of unsaturated clauses be C. Set $V = \phi$.
2. If there exists a variable $v \notin V$ such that $P(v) \geq N(v) = S > 0$
   (a) Pick all the k negative edges in $N(v)$ and assign each edge incident on clause i a value of s_i where i is the unsaturated clause that contains v as a negative literal ($y_{vi} = y_{vi} + s_i$). Note that we always saturate each of these clauses in this case.
   (b) Next, pick a subset of positive edges such that the total slack on the positive edges is more than S. Assign an additional total value of S on the positive edges, without violating the capacity constraints on the edges. This can be done in a greedy manner.
3. Update the slack on each clause and remove the edges that are incident on any saturated clause.
4. If for all variables $v \notin V$, $N(v) \leq 0$ then stop, else GOTO Step 2.

Upon termination, (i) there are only positive edges incident on clauses that are not saturated, (ii) the total weight of the saturated clauses is at most twice the weight of the dual solution (the negative edges picked). The unsaturated clauses can all be falsified by setting the variables to FALSE.

Note that the values on the edges are a feasible solution to the dual. We use the following procedure to construct an assignment for the primal problem. Variables that are picked in Step 2 are assigned a value of FALSE. The clauses that contain $\bar{v}$ where $v \in V$ are satisfied under this assignment. Clauses that contain $v \in V$ have at least one literal set to FALSE under this assignment. We show that the set of remaining clauses in U can be falsified. Take any clause $c \in U$ upon termination of the procedure DUAL. This clause has only positive edges incident on it from variables not in V (the set of variables not picked in Step 2). If there is a negative edge from some variable not in V, then the procedure would not terminate. If there was a negative edge incident from one of the variables in V, then the clause would be saturated and removed from U. All the positive edges set some literal to FALSE. Therefore c has a subset of literals set to FALSE, and the rest of the literals are unassigned. In fact, for each clause in U, the set of unassigned literals are all positive. We can falsify every clause in U by setting all the variables not in V to FALSE. The weight of satisfied clauses is at most twice the weight of negative edges picked in Step 2 of the DUAL procedure. The weight of negative edges picked is the value of the objective function in the dual. Therefore procedure DUAL is a 2-approximation algorithm. The running time of the procedure DUAL is proportional to the number of edges in the bipartite graph, which is $O(nm)$ ($O(mk)$ for K-MINSAT) where n is the number of variables, and m is the number of clauses in the minimum satisfiability instance.

Next, we give an example that shows that the analysis is tight. Consider the three clauses, $c_1 = (\bar{a}, b), c_2 = (\bar{b}, c), c_3 = (\bar{c}, a)$. The vertices in the bipartite graph are a, b, c, c_1, c_2, c_3 and the edges are $(a, c_1), (a, c_3), (b, c_2), (b, c_1), (c, c_3), (c, c_2)$. The negative edges are $(a, c_1), (b, c_2), (c, c_3)$ and the rest are positive edges. The DUAL procedure first considers variable c and picks edges $(c, c_3), (c, c_2)$. This saturates the two clauses c_2, c_3. The next variable considered is a, and since $P(a) < N(a)$, a is replaced with $\bar{a}$. Now the unsaturated clause c_1 has only positive edges incident on it. The clause c_1 can be falsified by setting $\bar{a}$ to FALSE, thereby setting a to TRUE, and b, c to FALSE. There are two satisfied clauses and the value of the feasible dual solution is 1. Therefore the analysis of an upper bound of 2 on the approximation ratio is tight.

3   A Combinatorial Algorithm for Solving the Dual

In this section, we give a primal-dual algorithm to compute an optimal solution to the LP relaxation of MINSAT. For background on the primal-dual method, please refer to the book by Papadimitriou and Steiglitz (1982) or (Arif 2017) (pages 19–24). In fact, we will solve the dual optimally and recover the primal optimal solution using complementary slackness conditions. We begin our presentation by defining the structures that are used to augment the current dual solution. We then give an efficient algorithm to find such structures. We also prove that if no augmenting structures exist, then the current dual solution is optimal. Our characterization also implies that there is an optimal solution to the LP relaxation that is half-integral.

We assume that (i) each variable occurs as a positive and as a negated literal in some clause, (ii) each clause contains at least two literals. We first describe the dual of the restricted primal (DRP). We then give a method based on breadth-first search (BFS) that computes a solution with a non-zero value to the DRP (if one exists). The method increases the value of the dual solution by at least $1/2$ in each iteration for the unweighted case. The total number of calls to the method to solve the DRP is linear in the number of clauses, and each invocation of the method is linear in the number of edges in the bipartite graph. Therefore, the method takes $O(nm^2)$ time, where n is the number of variables and m, is the number of clauses ($O(km^2)$ for K-MINSAT). We also comment on why the running time is $O(nm^2)$ for the weighted MINSAT as well.

Given y', a feasible solution to the dual, we say an edge (i, j) is saturated if $y'_{ij} = 0$, a clause i is saturated if $\sum_{j \in N_i, P_i} y'_{ij} = w_i$, and a variable j is saturated if $\sum_{i:i \in N_i} y'_{ij} = \sum_{i:i \in P_i} y'_{ij}$. Let E^*, C^*, V^* be sets of saturated edges, clauses, vertices respectively. Given y', we now state the dual of the restricted primal,

$$DRP : \max \sum_{i \in C, j \in N_i} y_{ij}$$

$$st \sum_{j \in N_i, P_i} y_{ij} \leq 0, \forall i \in C^*$$

$$\sum_{i:j \in N_i} y_{ij} - \sum_{i:j \in P_i} y_{ij} \leq 0, \forall j \in V^*$$

$$y_{ij} \geq 0, \forall (i,j) \in E^*$$

$$y_{ij} \leq 1, \forall (i,j) \in E$$

For a detailed derivation of the DRP, please see (Arif 2017) (pages 31–35). We start with a feasible dual solution, $y = 0$. At the start, all the variables and the edges are saturated, and none of the clauses is saturated. In each iteration, we find a feasible solution to the DRP using the procedure below, and augment the current dual solution. The update is such that the variable vertices remain saturated. The structure that we find using modified BFS is either a path that starts and ends at two different unsaturated clauses and the number of negative edges is odd, or an even cycle with a path hanging off the cycle. All the intermediate vertices are saturated. The start of the path is an unsaturated clause vertex. In both cases, the number of negative edges (counting the multiplicities) is odd. The structure of the first type is called an augmenting path, and the structure of the second type is called an augmenting lasso. The first vertex happens to be the last vertex in an augmenting lasso.

Examples of the two types of structures are shown in Fig. 1. The clauses are labelled $1, 2, 3, 4$ and the variables are $5, 6, 7$. The negative edges are dashed; the positive edges are solid. The unsaturated clause vertices are shown in solid black. The augmenting path is $1, 5, 2, 6, 3, 7, 4$, and the augmenting lasso is $1, 5, 3, 6, 4, 7, 2, 5, 1$. The dual solution is shown in brackets. The values on the edges correspond to a feasible solution to the DRP with a non-zero value. In the lasso, the label $+1$ occurs twice on edge $1,5$ as it is traversed twice. A feasible solution to the DRP for the lasso can be obtained by halving all the values. In both the examples, the objective function value is $1/2$.

Now that we have sufficient intuition let us define the augmenting structures. A walk W is a sequence of vertices $c_1, v_1, c_2, v_2, \ldots, v_k, c_k$ where v_i is variable and c_j is a clause. A walk in which no vertex is repeated is called a path. Two edges are said to be of the same type if both are positive edges or both are negative edges; otherwise, they are of a different type. We label the edges in the increasing order of distance from the source as follows: the first edge is labelled $+1$, at every variable vertex if the incoming and the outgoing edges are of different types then they have the same label, else the label on the outgoing edge is negative of the label on the incoming edge. At every clause vertex, if the incoming edge has label a then the outgoing edge has label $-a$.

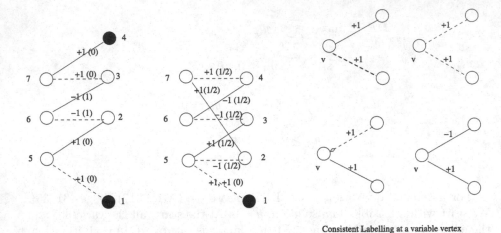

Consistent Labelling at a variable vertex

Fig. 1. Augmenting path and lasso

A labelling is called consistent wrt to a dual solution y if every edge that gets a label of -1 is not saturated in the dual solution y, i. e. $y_e > 0$. The labelling l is a feasible solution to the DRP, and $y' + \theta l$ for $\theta \in \{1/2, 1\}$ is a feasible solution to the dual with a larger value.

- A walk is augmenting, if the following conditions hold: (i) the first and the last vertices (clauses) are unsaturated and every intermediate vertex is saturated, (ii) the number of negative edges is odd (iii) and the labelling l (as defined above) is consistent.
- An augmenting walk in which there is no repeated vertex is called an augmenting path.
- An augmenting walk, which can be decomposed into a path incident on a cycle called an augmenting lasso. In a lasso, the edges on the path are visited twice, and the edges on the cycles are visited once. The first and the last vertex is the same clause vertex unsaturated by negative edges.

Figure 1 illustrates the two types of augmenting walks. The example above does not illustrate the case when the edges incident on some variable vertex are of the same type.

Theorem 2. *An augmenting path or an augmenting lasso can be discovered in $O(nm)$ time if one exists, where n is the number of variables and m is the number of clauses. The running time of the algorithm for K-MINSAT is $O(km)$.*

Proof. We show how to find an augmenting path or an augmenting lasso in the bipartite graph, if one exists using a variant of BFS, given a dual solution y. Let us first describe the case of augmenting paths. We start at an unsaturated clause vertex, and use the following strategy (dependent on the node) to select the next outgoing edge;

1. at a variable vertex: If the incoming negative edge has a label $a \in \{+1, -1\}$ then all the positive edges with label a, and all the negative edges with label $-a$ are the candidate edges. An edge with label -1 cannot be saturated. Therefore if the outgoing edge is to have a label -1 then the candidate edge has to be unsaturated. Similarly, if the incoming edge is a positive edge with label a then all the negative edges with label a, and the positive edges with label $-a$ are candidates. Only the candidates that lead to consistent labelling are to be considered, and only the un-visited edges are possible candidates.
2. at a clause vertex: If the incoming edge has label $+1$, then the possible candidates are un-visited unsaturated edges with label -1. If the incoming edge has label -1 then the possible candidates are un-visited edges with label $+1$. Note that if the clause vertex is the first vertex then there is an outgoing edge with label $+1$ and we use the same.

BFS terminates when an unsaturated clause vertex is encountered, and the number of negative edges is odd. To discover an augmenting lasso, we modify the termination criteria and the candidate criteria slightly. An un-visited edge (incident on a variable or a clause) is a not a candidate if it leads to a visited vertex and the total number of negative edges is even (counting the multiplicities). We terminate the search when a vertex is revisited, and the number of negative edges in the walk, counting the multiplicities is odd. The total time to discover an augmenting path/lasso is linear in the number of edges.

Given the definition of augmenting walk, we know that the number of negative edges is odd, and the first negative edge has a label of $+1$. The labels on the negative edges alternate in sign. Therefore, the sum of labels on the negative edges >0. We also note that the dual solution is half-integral, the solution to the DRP is in $\{0, +1, -1\}$, and $\theta \in \{1/2, 1\}$, so after the update, the new dual solution is still half-integral. The dual objective has increased by at least $1/2$ in value in the unweighted case.

We use the algorithm above to repeatedly increase the objective function value of the dual until no improvement is possible and recover the primal solution using complementary slackness conditions. The following theorem gives a simple randomized rounding scheme to compute an approximate solution.

Theorem 3. *If z^*, x^* is a half integral optimal solution to the primal then there exists a deterministic rounding scheme with performance ratio $2(1 - 1/2^k)$.*

Proof. Given a dual optimal solution y^*, we compute z_i using the complementary slackness conditions as follows: if the ith constraint has a slack then $z_i = 0$, if $y_{ij} > 0$ then $z_i = x_j$ if $j \in P_i$ else $z_i = 1 - x_j$ if $j \in N_i$. This gives us $n + m$ equation in $n + m$ unknowns. For all i such that $z_i = 0$ we set x_j as 0 or 1 depending on the equation above. Finally, we are left with equations of the type $z_i = x_j$ or $z_i = 1 - x_j$, which are satisfied by setting all the remaining z_i and x_j to $1/2$. Complementary slackness guarantees that such a feasible solution to the primal exists and it is half-integral. This procedure is linear in the number of edges.

We use the values of the literals in the primal solution for rounding. We set literal x_j to 1 with probability x_j^* where x_j^* is the optimal primal solution. The probability that a clause z_i with k literals is unsatisfied is $1/2^k$. So each clause i with $z_i^* = 1/2$ is satisfied with probability at most $(1 - 1/2^k)$. Let z_1 be the number of clauses satisfied ($z_i = 1$) in the optimal solution to the LP, and $z_{1/2}$ be the number of clauses with $z_i = 1/2$ in the optimal solution. The expected number of satisfied clauses is $z_1 + z_{1/2}(1 - 1/2^k)$ and the optimal solution has value $z_1 + z_{1/2}$. Therefore, the performance ratio is

$$\frac{z_1 + z_{1/2}(1 - 1/2^k)}{z_1 + (1/2)z_{1/2}} \leq 2(1 - 1/2^k).$$

The rounding scheme can be de-randomized using the method of conditional expectation.

We need to show that if there are no augmenting paths or lassos, then the dual solution is optimal. We will show the contrapositive that if the dual solution is not optimal, then there exists an augmenting path or a lasso. The following argument is based on the idea in (Krishnamurti et al. 2006).

Theorem 4. *Let y be a dual solution such that all the variable vertices are saturated. Let y^* be an optimal solution to the dual such that $\sum_{e \in E} |y_e^* - y_e|$ is the minimum possible, and all the variable vertices are saturated. If y is not an optimal solution to the dual, then there exists an augmenting path or an augmenting lasso.*

Proof. Let $P(v)(N(v))$ be the set of positive (negative) edges incident on a node v. Let us assign the label to the edge (i, j) in the graph as $sgn(y_{ij}^* - y_{ij}) \times 1$ where sgn returns the sign of the argument. Edges for which $y_{ij} = y_{ij}^*$ are assigned a label of 0, edges for which $y_{ij} < y_{ij}^*$ are assigned a label of $+1$, the rest of the edges are assigned a label of -1.

Since y is not an optimal solution to the dual, $\sum_{c \in C} \sum_{e \in N(c)} y_e < \sum_{c \in C} \sum_{e \in N(c)} y_e^*$. Furthermore, because every variable node is saturated, $\sum_{c \in C} \sum_{e \in N(c)} y_e = \sum_{c \in C} \sum_{e \in P(c)} y_e$. From the inequality $\sum_{c \in C} \sum_{e \in N(c)} y_e < \sum_{c \in C} \sum_{e \in N(c)} y_e^* <= \sum_{c \in C} \sum_{e \in P(c)} y_e^*$, it follows that $\sum_{c \in C} \sum_{e \in P(c)} y_e < \sum_{c \in C} \sum_{e \in P(c)} y_e^*$. Therefore, it follows that $\sum_{c \in C} \sum_{e \in N(c),P(c)} y_e < \sum_{c \in C} \sum_{e \in N(c),P(c)} y_e^*$. Thus, there is a clause c such that $\sum_{e \in N(c),P(c)} y_e < \sum_{e \in N(c),P(c)} y_e^*$. This implies there is either a positive or negative edge $e = (v, c)$ incident on c such that $y_e^* - y_e > 0$ with label $+1$. We start a walk at c, pick the edge (c, v) with label $+1$. The next outgoing edge (among the edges not used so far) is chosen according to the following rules:

(i) at a variable node: if the incoming edge is a negative (positive) edge with label $a \in \{+1, -1\}$ then the outgoing edge is a positive (negative) edge with label a. If no positive (negative) edge with label a exists then select a negative (positive) edge with label $-a$. We will show later that such a positive (negative) edge always exists. In other words, if the outgoing edge and the incoming edge are of the same type, then the label is different. If the outgoing and the incoming edges are of different types, then the labels are the same. We need to show that an appropriate outgoing edge always exists. Suppose we are at a variable node, and the incoming edge is negative with label $+1$, and every positive edge in $P(v)$ has a label -1 or 0 then

$$\sum_{e \in P(v)} (y_e^* - y_e) \leq 0.$$

Furthermore, if every negative edge in $N(v)$ has a label of $+1$ or 0 then,

$$\sum_{e \in N(v)} (y_e^* - y_e) > 0.$$

This implies that

$$\sum_{e \in N(v)} y_e < \sum_{e \in N(v)} y_e^* \leq \sum_{c \in P(v)} y_e^* \leq \sum_{c \in P(v)} y_e.$$

But we know that v is saturated in y, implying $\sum_{e \in N(v)} y_e = \sum_{e \in P(v)} y_e$, leading to a contradiction. Therefore, either there is a positive edge with label $+1$ or there is a negative edge with label -1. Similarly, if the incoming edge is negative with label -1, either there is a positive edge with label -1 or there is a negative edge with label $+1$. The case when the incoming edge is positive is similar.

(ii) at a clause node: if the incoming edge has a label a then select any un-visited edge with label $-a$.

We stop when either we revisit a vertex, or visit a clause vertex c for which $\sum_{e \in N(c), P(c)} y_e < \sum_{e \in N(c), P(c)} y_e^*$.

Let us now consider the case when the incoming edge has label $+1$, and we are at a clause c. As the label is $+1$, $y_e^* > y_e$ on this edge. If $y_e^* \geq y_e$ on the rest of the edges (except the incoming edge) then $\sum_{e \in N(c), P(c)} y_e^* > \sum_{e \in N(c), P(c)} y_e$, and we stop. Otherwise, for some edge $y_e^* < y_e$, and we continue.

Suppose, the label on the incoming edge is -1, then $y_e^* < y_e$ for the incoming edge incident on clause c. If for all other edges e incident on c, $y_e^* \leq y_e$, we stop the search. We need to show that this case cannot happen. We rely on perturbing the optimal solution to obtain a contradiction, as shown in this paragraph. We note that in any walk, the labels on the negative edges alternate in sign. Suppose the incoming edge is negative then clearly the number of negative edges is even. If the incoming edge is positive with label -1 then the last negative edge also has a label -1. Once again, the number of negative edges is even. This means that the number of negative edges in the walk is even. For every edge e with label -1 we add ϵ to y_e^*, and for every edge with label $+1$ we subtract ϵ from the edge, where $\epsilon = \min_{e \in E}\{|(y_e^* - y_e)|\}$ over all the edges e in the walk. It can be verified that this perturbation does not violate the constraints in the dual. This modification to y^* gives another optimal solution y' (with the same value) such that $\sum_{e \in E} |y_e' - y_e|$ is smaller. Therefore the search terminates only at a clause c for which $\sum_{e \in N(c), P(c)} y_e < \sum_{e \in N(c), P(c)} y_e^*$.

We need to show that there are an odd number of negative edges in the walk. Suppose there are an even number of negative edges in the walk, then perturbing the optimal solution y^* as above leads to another optimal solution with a smaller value of $\sum_{e \in E} |y_e'' - y_e|$, therefore the number of negative edges have to be odd. It can be verified that no two negative edges that are consecutive in the subsequence both have a label of -1. Therefore, the value of $y + \theta w$ is strictly larger than the value of y. This is a nonconstructive proof to show the existence of an augmenting path or lasso. We use modified BFS to discover augmenting structures if one exists.

For an instance of MINSAT in which all the clauses have uniform weight w, the progress of the primal-dual method can be measured by an increase in the value of any single variable. We can bound the number of iterations needed as follows: the optimal is at most mw and in each iteration the value of the dual increases by $w/2$, so the total number of iterations is m. However, if the weights are non-uniform, then we need a different measure of progress. It is sufficient to note that when we augment the dual solution along a path or a lasso, the clauses that are saturated remain saturated after augmentation. Additionally, at least one additional clause is saturated. Upon termination, at every clause $\sum_{e \in N(c), P(c)} y_e \geq \sum_{e \in N(c), P(c)} y_e^*$. Therefore the total number of iterations is bounded by m, and each iteration of modified breadth-first search takes $O(nm)$ time. This gives us a running time of $O(nm^2)$ for weighted MINSAT as well. For weighted , K-MINSAT, the running time of the algorithm is $O(km^2)$.

4   Conclusions

We give a combinatorial algorithm for determining a half-integral solution to the LP relaxation of the minimum satisfiability problem. This along with a simple rounding scheme implies a deterministic approximation algorithm with performance ratio $2(1 - 1/2^k)$ for MINSAT with at most k literals in each clause. Our results improve the running time of the previously best-known approximation algorithm. We also give a simple linear time approximation algorithm with a performance ratio of 2. The proof technique used to characterize the structure of the optimal solution to the LP relaxation might be of independent interest.

Acknowledgments. Robert Benkoczi, Daya Gaur, Ramesh Krishnamurti would like to acknowledge the support from NSERC in the form of individual discovery grants. The authors would like thank the reviewers for the comments.

References

Arif, U.M.: On primal-dual schema for the minimum satisfiability problem. Master's thesis, University of Lethbridge, Canada (2017)

Bar-Yehuda, R., Even, S.: A linear-time approximation algorithm for the weighted vertex cover problem. J. Algorithms **2**(2), 198–203 (1981). ISSN 0196–6774

Bar-Yehuda, R., Even, S.: A local-ratio theorem for approximating the weighted vertex cover problem. North-Holland Math. Stud. **109**, 27–45 (1985)

Berman, P., Fujito, T.: On approximation properties of the independent set problem for low degree graphs. Theory Comput. Syst. **32**(2), 115–132 (1999)

Bertsimas, D., Teo, C., Vohra, R.: On dependent randomized rounding algorithms. Oper. Res. Lett. **24**(3), 105–114 (1999)

Bourjolly, J.-M., Pulleyblank, W.R.: König-Evervári graphs, 2-bicritical graphs and fractional matchings. Discrete Appl. Math. **24**(1–3), 63–82 (1989)

Cook, S.A.: The complexity of theorem-proving procedures. In: Proceedings of the Third Annual ACM Symposium on Theory of Computing, STOC 1971, pp. 151–158, New York, NY, USA. ACM (1971)

Dinur, I., Safra, S.: On the hardness of approximating minimum vertex cover. Ann. Math. **162**, 439–485 (2005)

Edmonds, J.: Paths, trees, and flowers. Can. J. Math. **17**(3), 449–467 (1965)

Goemans, M.X., Williamson, D.P.: New $\frac{3}{4}$-approximation algorithms for the maximum satisfiability problem. SIAM J. Discrete Math. **7**(4), 656–666 (1994)

Halperin, E.: Improved approximation algorithms for the vertex cover problem in graphs and hypergraphs. SIAM J. Comput. **31**(5), 1608–1623 (2002)

Hochbaum, D.S.: Instant recognition of half integrality and 2-approximations. In: Jansen, K., Rolim, J. (eds.) APPROX 1998. LNCS, vol. 1444, pp. 99–110. Springer, Heidelberg (1998). https://doi.org/10.1007/BFb0053967

Hochbaum, D.S.: Approximation algorithms for the set covering and vertex cover problems. SIAM J. Comput. **11**(3), 555–556 (1982)

Hochbaum, D.S.: Efficient bounds for the stable set, vertex cover and set packing problems. Discrete Appl. Math. **6**(3), 243–254 (1983)

Iranmanesh, E.: Algorithms for Problems in Voting and Scheduling. Ph.D. thesis, Simon Fraser University (2016)

Johnson, D.S.: Approximation algorithms for combinatorial problems. J. Comput. Syst. Sci. **9**(3), 256–278 (1974)

Khot, S., Regev, O.: Vertex cover might be hard to approximate to within $2 - \varepsilon$. J. Comput. Syst. Sci. **74**(3), 335–349 (2008)

Kohli, R., Krishnamurti, R., Mirchandani, P.: The minimum satisfiability problem. SIAM J. Discret. Math. **7**(2), 275–283 (1994)

Krishnamurti, R., Gaur, D.R., Ghosh, S.K., Sachs, H.: Berge's theorem for the maximum charge problem. Discrete Optim. **3**(2), 174–178 (2006)

Marathe, M., Ravi, S.: On approximation algorithms for the minimum satisfiability problem. Inf. Process. Lett. **58**(1), 23–29 (1996)

Monien, B., Speckenmeyer, E.: Ramsey numbers and an approximation algorithm for the vertex cover problem. Acta Informatica **22**(1), 115–123 (1985)

Orlin, J.B.: Max flows in O(nm) time, or better. In: Proceedings of the Forty-Fifth Annual ACM Symposium on Theory of Computing, pp. 765–774. ACM (2013)

Papadimitriou, C.H., Steiglitz, K.: Combinatorial Optimization: Algorithms and Complexity. Courier Corporation, North Chelmsford (1982)

Tardos, E.: A strongly polynomial algorithm to solve combinatorial linear programs. Oper. Res. **34**(2), 250–256 (1986)

Yannakakis, M.: On the approximation of maximum satisfiability. In: Proceedings of the Third Annual ACM-SIAM Symposium on Discrete Algorithms, SODA 1992, pp. 1–9, Philadelphia, PA, USA (1992). ISBN 0-89791-466-X

Waiting for Trains: Complexity Results

Bjoern Tauer[1,2]([envelope]) [iD], Dennis Fischer[1]([envelope]) [iD], Janosch Fuchs[1]([envelope]) [iD],
Laura Vargas Koch[2]([envelope]) [iD], and Stephan Zieger[3]([envelope]) [iD]

[1] Department of Computer Science, RWTH Aachen University, Aachen, Germany
{tauer,fischer,fuchs}@algo.rwth-aachen.de
[2] Department of Business and Economics, RWTH Aachen University,
Aachen, Germany
{tauer,laura.vargas}@oms.rwth-aachen.de
[3] Institute of Transport Science, RWTH Aachen University, Aachen, Germany
zieger@via.rwth-aachen.de

Abstract. We introduce a model for train routing on railway systems. Trains route through a network over time from a start to an end depot. They occupy consecutive nodes and edges corresponding to their length and block each other. We study the case where the depots are part of the network (internal) and the case where the depots are not part of the network (external).

The problem is a generalization of packet routing without buffers. We consider two different kinds of optimization problems. In the first, trains are only allowed to wait on predefined paths and in the second, trains are additionally allowed to shunt, i.e., change direction. In both cases, we are interested in minimizing the overall makespan.

For waiting instances, we find NP-hardness results even on unidirectional paths. We also show $W[1]$-hardness and lower bounds on the running time using the Exponential Time Hypothesis. For shunting instances, we show PSPACE-completeness results on honeycomb graphs and transfer the previously shown NP-hardness results. We present a polynomial time algorithm for a special subclass of unidirectional paths.

Keywords: Train scheduling · Wormhole routing · Packet routing · Parameterized complexity

1 Introduction

Transportation issues are one of the crucial challenges of our time. Consequently, to perform efficient routing of goods, persons or data messages various models are studied intensively. The model approach has to fit the specific needs of the transportation system depending on the considered asset.

We want to thank the Deutsche Forschungsgemeinschaft (DFG, German Research Foundation) – UnRAVeL-Research Training Group 2236/1 funding this research and connecting the involved authors from the Department of Computer Science, the School of Business and Economics as well as the Faculty of Civil Engineering.

© Springer Nature Switzerland AG 2020
M. Changat and S. Das (Eds.): CALDAM 2020, LNCS 12016, pp. 282–303, 2020.
https://doi.org/10.1007/978-3-030-39219-2_24

Store and Forward Packet Routing. In case of a single item, e.g., an item that can be represented by a packet, one extensively studied approach is the store and forward packet routing (see e.g., Leightons survey [13]). Store and forward packet routing can be solved efficiently if all packets share the same initial and target position (single commodity instance) by calculation of a maximum flow over time [14]. If the locations differ, the problem becomes NP-hard [5]. In a game theoretic variant the hardness landscape behaves accordingly. Here, trains represented as packets route selfishly through the network. If capacity conflicts between two packets occur various mechanism schemes are analyzed [6,9,10].

Message Routing. In message routing each message consists of a certain amount of packets of unit size. A message is sent packet per packet along an edge. One packet is allowed to traverse an edge per time step until the whole message reached the end of the edge. After all packets are collected the next edge can be entered. In the special case where every message contains only one packet, this problem reduces to store and forward packet routing mentioned above, and thus the hardness results can be transferred. However, for arbitrary message sizes, the problem is NP-hard even on a directed path [14].

Wormhole Routing. A non-preemptive version of message routing is wormhole routing where data is sent as worms through the network. The body of the worm consists of a sequence of fixed sized units called flits, trailed after its head unit. During traversing the network, a contiguous sequence of edges along a certain path is blocked where every flit occupies one edge. One fundamental aspect of wormhole routing is the detection of deadlocks [4].

Train Routing. We modify the wormhole routing approach such that it is applicable to train networks. Doing so requires some subtle model modifications that have a huge impact on the behavior of the model while some known results can be adapted easily. Similar to the worms, trains are not allowed to be split. Furthermore, no two trains can traverse the same node at the same time, i.e., not only edge capacities are considered; additionally, node capacities have to be taken into account (see Fig. 8 in Appendix A of the full version for an example).

Moreover, trains are permanently present in the network, so they are not just injected when needed and deleted as soon as they arrived at their target location. Thus, the total number of wagons in the network is constant and occupies a certain amount of nodes and edges of the graph which could cause additional difficulties.

Our Contributions
We study hardness of train routing on different variations of the problem as well as on selected graph classes. In a *waiting* instance, the paths of the trains are already completely defined. The optimization problem is to determine waiting times that minimize the arrival time of the last train. Whereas, in a *shunting* instance subpaths are given, but trains can be routed along other edges, too, to let other trains pass and improve the total latest arrival time. Moreover, we distinguish the case where trains are equipped with external individual depots in contrast to internal start and end links which can be used by all other trains.

We show that the waiting and shunting problem with external depots is already NP-complete on unidirectional paths. Moreover, if all trains have length one, the waiting problem is also NP-hard on degree-three-bounded planar graphs. For general graphs the waiting problem becomes $W[1]$-hard parameterized in the number of trains. Shunting problems are already PSPACE-complete on honeycomb graphs.

On the positive side, we give an FPT-algorithm on trees with external depots for which the number of trains is the parameter. Although trees are a simple graph class, for train scheduling this instance is highly relevant. At least from a local point of view with fixed track orientation, railway networks can be assumed to possess a tree-structure. Due to the hardness results, a global optimal solution for the train problem is unlikely to be found and thus it is relevant to search for solutions that are locally optimal.

Furthermore, we give a polynomial time algorithm minimizing the makespan on a unidirectional path with non-nested depots, where at least one node is traversed by every train. Our complexity results are condensed in Table 1.

Table 1. Overview of complexity results for train routing on various graph classes. Extern/Intern refers to external and internal depots.

Instance	Directed path		Tree	Max. degree ≤ 3		General
	Non-nested shared node	General		Honey-comb	Planar, length 1	Equal train length
Waiting (extern)	P Theorem 6	NP Theorem 1 FPT Theorem 3	NP Theorem 1 FPT Theorem 3	NP Theorem 1	NP Theorem 2	$W[1]$ Theorem 4
Shunting (intern)	P Proposition 2	P Proposition 2	NP Corollary 1	PSPACE Theorem 5		PSPACE Theorem 5
Shunting (extern)		NP Corollary 2	NP Corollary 2	NP Corollary 2	NP Corollary 3	$W[1]$ Corollary 4

2 Preliminaries

As a first relevant simplification of a real train setting, undirected graphs are considered. In Figs. 1, 2 and 3 the abstraction from a real track layout towards a graph model is presented. A track layout has certain restrictions on the movement of the trains, e.g., at a real crossing a train is unable to switch from a horizontal direction to a vertical one and vice versa. Furthermore, at a switch not all combinations of incoming and outgoing edges are possible. We overcome this gap between the graph model and the track layout by introducing black-lists to forbid illegal movements. To increase readability we only draw angles for switches, e.g., for node v_1 in Fig. 3.

In Sect. 2.1, we formalize the input and the problem, with respect to the above mentioned restrictions. Moreover, we distinguish four different variations

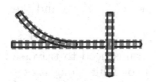 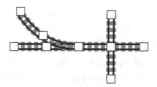 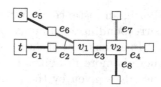

Fig. 1. Switch and cross-ing that do not allow arbi-trary directions.

Fig. 2. Slicing the track into a graph with unit edge length.

Fig. 3. A train may tra-verse from e_2 to e_6 or e_4 to e_7 without a blacklist.

of our model: Waiting/Shunting instances with internal/external depots. As in reality, in our model deadlocks impose a challenging task. Hence, we discuss how our model can cope with this.

2.1 Model

In a train routing problem, we are given an undirected graph G which consists of a finite set of nodes V and a finite set of edges $E \subseteq V \times V$. We assume that every edge has unit length and unit capacity and nodes have unit capacity. In our model, time is discretized and each train is allowed to move up to one node and edge per time step.

Due to railway system properties, every node is equipped with a list of infeasible edge combinations a train cannot traverse consecutively, i.e., node v has the so called *blacklist* $b_v = \{\{e_i, e_j\}, \{\cdot, \cdot\}, \ldots\}$ for $e_i \cap e_j \cap v \neq \emptyset$. If $\{e_i, e_j\} \in b_v$ it means that trains cannot traverse v by coming from e_i and going to e_j and vice versa. Further, there is a finite set of trains $N = \{1, \ldots, n\} = [n]$ where each train $i \in N$ is associated with an initial position $E_{s_i} \subset E$ and target position $E_{t_i} \subset E$.

We denote $E_i := E_{s_i} \cup E_{t_i}$ and the union over all initial and target positions with $\mathcal{E} = \bigcup_{i \in N} E_i$. For a feasible instance it holds that $E_{s_i} \cap E_{s_j} = \emptyset$ and $E_{t_i} \cap E_{t_j} = \emptyset$ for all $i, j \in N$, $i \neq j$. If other trains are allowed to enter edges in E_i for all $i \in N$ we denote it as an instance with internal depots. If E_i is reserved only for train i we call it an instance with external depots. For external depots, the graph we consider as input is formally $G \setminus \mathcal{E}$. If it is oblivious, we denote the initial or target position just by the node where the depot is adjacent to instead of the edge set itself. Additionally, each train has a certain weight $w_i \in \mathbb{N}_{\geq 1}$ which corresponds to its length (and thus $w_i = |E_{s_i}| = |E_{t_i}|$). Since trains do not vanish, at each point in time a train i blocks w_i edges and $w_i - 1$ nodes.

Further, we distinguish instances with respect to the ground set of feasible solutions. In a waiting instance we are additionally given a path P_i for every train $i \in N$ and we optimize over possible waiting times θ_v at intermediate nodes $v \in P_i$ of the path, i.e., waiting times before entering an edge of the path.

Hence, an instance of waiting can be formalized as a tuple $(G, N, \mathcal{E}, P)$ and the corresponding solution as $(\theta_1, \ldots, \theta_{|P_i|})_{i \in N}$.

In an instance of shunting, a path P_i for every train $i \in N$ is given as well, but there is more flexibility in constructing a solution. The trains need to traverse the edges given by the paths, but now they are allowed to shunt along theses edges. Additionally, an edge set $S_i \subseteq E$ is provided, which also can be used for shunting. Further waiting time for every intermediate node of the chosen path can be defined, i.e., a solution to the instance is a path P_i' for every train $i \in N$, such that $P_i \subseteq P_i' \subseteq P_i \cup S_i$ and additionally a vector of waiting times $(\theta_i, \ldots, \theta_{|P_i'|})$ for every train $i \in N$.

2.2 Feasible Train Movement

Throughout the paper, we use the head of a train to describe the position of the complete train, since the distance of a specific wagon to the head is constant. The movement of a train is completely described by a path through the network from its start to its target depot. There are three different actions a train is able to perform in every time step. Either it keeps its current position, it is forwarded to the next edge of the path or there is a change in direction κ in the path, which means that instead of the currently first wagon, the currently last wagon is the new head. To describe this formally, we extend the classical path definition such that a path can contain the action κ additionally to traversing edges. Furthermore, we introduce a function W_i for every train i such that $W_i(j, \theta)$ returns the position of the jth wagon of train i (counted from the head onwards) at time θ.

In a *feasible path*, for two consecutive edges e_k, e_{k+1}, it classically needs to hold that $e_k \cap e_{k+1} \neq \emptyset$. Moreover, it is necessary that no edge combination which is forbidden by the blacklist is chosen. If two edges are separated by a κ, i.e., (e_k, κ, e_{k+2}) the situation is slightly more complicated. At time θ the current engine enters edge e_k and the last wagon is at position $W_i(w_i, \theta)$, since this wagon enters an adjacent edge in the next step $W_i(w_i, \theta) \cap e_{k+2} \neq \emptyset$. For example, in Fig. 3 the only valid path from s to t regarding the original train track includes a change in direction κ after traversing node v_1. We denote the corresponding path by $P = (e_5, e_6, e_3, e_4, \kappa, e_2, e_1)$. For two consecutive edges (e_k, e_{k+1}) of a path it needs to hold that $\{e_k, e_{k+1}\} \notin b_{e_k \cap e_{k+1}}$ and for two edges separated by a directional change (e_k, κ, e_{k+2}) it needs to hold, if e_k is entered at θ, that $\{W_i(w_i, \theta), e_{k+2}\} \notin b_{W_i(w_i, \theta) \cap e_{k+2}}$.

The possibilities we get from shunting enable us to find feasible solutions (see deadlock handling of Fig. 4 in Appendix B of the full version), or, if a waiting instance already had a feasible solution, to reach a smaller makespan for it. The example given in Appendix C of the full version proves the following proposition.

Proposition 1. *There are instances of the train routing problem in which the optimal shunting solution is lower than the optimal waiting solution.*

2.3   Deadlock

Given a feasible path for every train, we get a solution P. Such a solution is called feasible if every edge and every node is used only by one wagon of a train per time unit and if all chosen times can be satisfied. If this is not the case we call the situation a deadlock. There are instances in which there is no feasible solution at all and there are instances where only some solutions end up with a deadlock. In Fig. 4 we can see an example where there is no feasible solution for trains of length one, i.e. the trains always end up in a deadlock situation; see Example 1 in Appendix D of the full version. On contrary, if an instance contains a subgraph as depicted in Fig. 5, it might happen that a deadlock occurs if both trains start at the same time, but there are also deadlock free solutions, like in Example 2 in Appendix D of the full version.

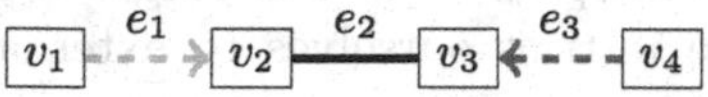

Fig. 4. Two unit length trains with $E_{s_1} = E_{t_2} = e_1$ and mirrored $E_{s_2} = E_{t_1} = e_3$. No train can circumvent the other train.

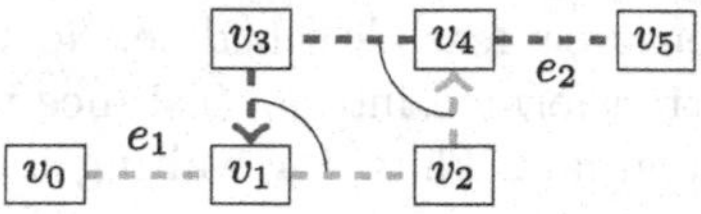

Fig. 5. The blue train wants to travel to v_0 and the red one to v_5. The red train can only leave edge e_1 if it enters e_2 simultaneously while the blue one has to enter e_1 to free e_2. (Color figure online)

To overcome this trivial blocking, we want to investigate a special variant of the problem, where each train has an individual start depot and target depot. The assumption relates to injection buffer and delivery buffers used for example in [2]. Each depot is a path of w_i consecutive edges, such that in the initial configuration the trains head is pointing to its original node. Example 3 in Appendix D of the full version showcases the transformation of the situation depicted in Fig. 4 into a solvable instance.

If the graph contains an individual depot tuple for every train, a sorting algorithm guarantees a deadlock-free feasible solution. Sort the trains according to their name and delay every train until all trains preceding in the ordering reached their destination depot completely. Thus, there is always only one train traveling through the network and every choice of a feasible path is conflict-free. This gives a polynomial worst-case solution and thus the corresponding optimization problem with external depots is contained in NP.

Obviously, this approach is not efficient since only a small fraction of edges is used at every point in time. For a well-defined problem it is necessary that we can check if a given solution is feasible. This is possible for a solution of our problem, since we can embed the trains along their paths according to their waiting times and check if an edge or a node is used by two or more trains at the same time. This is possible by embedding the trains time step per time step (and update only if some train moves at the respective time). By using this procedure we can also determine the arrival time of the last train, i.e., the makespan.

Before we start presenting complexity results, we first want to reconsider classic packet routing classifications. Often single commodity instances are analyzed. In those special cases easier solutions exist. Due to the node capacity, this classification is not helpful for the train routing problem as only one train at a time can traverse the shared first node and thus no further interaction takes place. In the remaining part of this paper, we consider multi commodity instances only.

3   Complexity Results

This section gives an overview of the complexity landscape of the waiting and the shunting problem. Table 1 summarizes all results. As stated in Sect. 2.3, external depots ensure a feasible schedule. Note, that every instance with external depots can be transformed into an instance with internal depots by considering $G \cup \mathcal{E}$ as input graph. Thus, for waiting instances, we restrict ourselves to external depots.

3.1   Waiting

Unexpectedly, the problem turned out to be very hard. Even the task to compute an optimal schedule with minimal makespan on a *unidirectional path* is NP-complete. In a unidirectional path, w.l.o.g. the target depots are always positioned to the right hand side of the starting depots. This means, that all trains travel from left to right.

Theorem 1. *In a waiting instance on a unidirectional path it is* NP-*complete to decide if a schedule with makespan of at most k exists.*

Proof. We will show hardness by a reduction to unary 3-Partition [7]. Consider an instance of 3-partition of $3z$ integers $A = (a_1, a_2, \ldots, a_{3z})$ encoded in unary such that $\sum_{k \in A} a_i = zB$ and $\frac{B}{4} < a_k < \frac{B}{2}$ for all $k \in [3z]$. We construct $z + 1$ trains of unit length b_k for $k \in 0, \ldots, z$ which represent the boundaries of each set. Moreover, we add one train for each integer a_k of weight $w_{a_k} = a_k$, so there are $n = 4z + 1$ trains in total. The graph G is a path of $2z(B+1) + 1$ consecutive edges and both, edges and nodes, are numbered ascending. We arrange the depots for the trains b_k as follows: The start depot of train b_k is located at $v_{k(B+1)}$ and its target depot is located at $v_{(z+k)(B+1)+1}$. All start depots of the integer trains are located at $v_{z(B+1)}$ and the target depots at $v_{z(B+1)+1}$, see also Fig. 6 for a visualization. Each set train b_k needs at least $z(B + 1) + 2$ time units to reach its destination. On the other hand, edge $e_{z(B+1)}$ is traversed by $z + 1$ set trains and $3z$ integer trains of total length zB. So, $z(b + 1) + 2$ is a lower bound on the makespan again. To reach this makespan all set trains have to start at time zero and traverse their complete path without being delayed at all. Due to the congestion on edge e_{zB+1} this edge has to be used for every time step, thus the

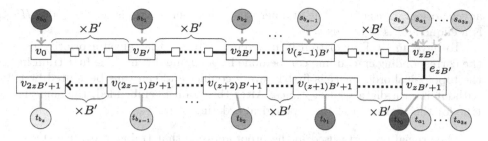

Fig. 6. Reduction to unidirectional path of total length $2zB' + 1$ where $B' := (B+1)$. B is the sum of the elements in each set. There are $z + 1$ unit length trains (red) to symbolize the set boundaries and $3z$ trains with identical depots and length according to the size of the corresponding integers (blue). (Color figure online)

gaps between the set trains have to be filled exactly by the integer trains. Thus, there exists a schedule with makespan $z(B+1) + 2$ if and only if there exist a 3-partition of A into z sets of size B.

After showing hardness for the simple case of unidirectional paths, we consider the case where the trains have all minimal length, i.e., one. To show hardness, the graph class becomes slightly more complicated.

Theorem 2. *In a waiting instance on planar graphs with degree bounded by three it is* NP-*complete to decide if a schedule with makespan of at most k exists – even for trains of size one.*

The proof is a reduction to Simple Planar 4-bounded exactly 3–SAT contained in Appendix E of the full version.

We showed that the problem is NP-hard on a unidirectional path as well as for trains of length 1. Thus, we consider fixed parameter tractability of the problem on trees. The idea is to identify a set of fixed parameters in which the algorithm is allowed to have superpolynomial running time. This means, for a fixed parameter n a problem is in FPT if there is an algorithm with running time $\mathcal{O}(f(n) \cdot poly(input))$ [1].

Theorem 3. *Computing an optimal solution for a waiting instance on trees, parameterized by the total number of trains n, is possible with running time $\mathcal{O}(n! \cdot n \cdot |E|)$. Thus, the problem is in* FPT.

Proof. First, we construct a conflict graph $H - (N, E(H))$ such that there exist one node for every train $i \subset N$. Consider the nodes of the original graph G and add an edge $(i, j), i, j \in N$ to $E(H)$ if and only if $P_i \cap P_j \neq \emptyset$. Since G does not contain any cycles a priority list of all trains is sufficient to handle all potentially occurring conflicts. Observe that an acyclic orientation of H corresponds to a priority list. Here, every train i is prioritized over j if there is a directed path from j to i. Thus, to identify the optimal priority list it is sufficient to check all acyclic orientations of H. The number of acyclic orientations is maximized if H is a clique and is $n!$ in this case.

To have an FPT algorithm in n it remains to see that given a priority list the corresponding embedding can be done in polynomial time. The first train in the topological order of H is first in every conflict and can thus be routed first without intermediate waiting times. After this train is embedded, the same is true for the next train in the topological ordering of H.

The crucial property used in the proof above is that trains interact at most once. This is the reason why an optimal priority order, automatically corresponds to an optimal schedule. As soon as the graph contains a cycle this is not true anymore as we can see in Example 4 in Appendix F of the full version. Indeed, it turns out that the makespan minimization problem is $W[1]$ hard on these graph classes which means it is most likely not in FPT parameterized in the number of trains.

Theorem 4. *It is $W[1]$-hard to compute the optimal solution of a waiting instance parameterized by the number of trains n. Thus, there is no algorithm that solves the problem in $f(n) |E|^{o(n/\log n)}$ time for some function f unless the ETH fails.*

Proof. We reduce our problem from the Multi Way Number Partition problem [3] like presented in Fig. 7.

For every bin $j \in [n]$ we introduce a train with external depots. We generate $n \times (k+1)$ nodes, where k is the number of objects. Additionally, we add a path with a_l nodes in front of v_1^l for all $l \in [k]$. We connect each node v_j^{l-1} with the first node of the path in front of v_1^l and v_j^l, for all $j \in [n]$. The starting depot of train j is adjacent to node v_j^0. Since the depot is defined as path, we add an additional node per final depot of a train and connect those with all v_j^k for $j \in [n]$. Each train has a length of $w_j = A + k$ where $A := \sum_{j \in [k]} a_j$ and thus $A \geq k$.

Then, the Multi Way Number Partition is solvable if a schedule exists such that the makespan is $C^* = A(1 + \frac{1}{n}) + 2k + 1$. Note that all trains have the same length and the shortest path of each train has length $k + 1$. The only way to realize the makespan C^* is that no train is be delayed. This implies by the construction of the graph and the train length that every train traverses exactly one node v_j^l for all $l \in [k]$. The train that traverses v_1^j also traverses the path of length a_j in front of it. Thus, to minimize the makespan an optimal schedule divides the additional path length equally between the trains. This results in an additional $\frac{A}{n}$ travel time for each train. Summarizing, an optimal schedule corresponds to a solution of the Multi Way Number Partition Problem and vice versa.

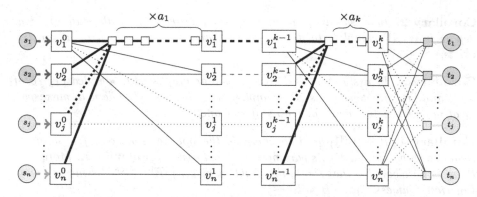

Fig. 7. The graph constructed from an instance of Multi Way Number Partition with k objects and n bins, corresponding to trains, each one with length $w_j = k + \sum a_k$. All thick paths between v_{i-1} and v_i consist out of $a_i + 1$ edges for all $i \in [k]$. The pathwidth is bounded by $pw(G) \le n+1$. Each node has two switches, one for all edges from the left and one for all edges on the right.

3.2   Shunting

For shunting, we distinguish between internal and external depots.

Internal Depots. For trees, we can use our results from Theorem 1. The basic idea is to consider the path and add the external depots into the graph. This, obviously, results in a tree. In this modified instance, the result from Theorem 1 still holds. Since, one edge is permanently used and there are trains that do not wait, shunting cannot reduce the makespan.

Corollary 1. *In a shunting instance on a tree with internal depots it is NP-complete to decide if a schedule with makespan of at most k exists.*

If the graph class becomes more complex, we can enhance the hardness result to PSPACE-completeness. The graph we consider is the honeycomb graph which is a subgraph of a grid. Note that the maximum degree of honeycomb graphs is three.

Theorem 5. *Finding a feasible schedule for a shunting instance with internal depots on a honeycomb graph is* PSPACE-*complete – even if all trains have equal length.*

We reduce our problem from Rush Hour Puzzle. The proof can be found in Appendix G of the full version.

External Depots. Note that, in all the reductions in Sect. 3.1 shunting never helps to improve the makespan. Thus, the optimal solution for waiting corresponds to the optimal solution for shunting in these instances and therefore the hardness results transfer to shunting problems.

Corollary 2. *In a shunting instance on a unidirectional path with external depots it is NP-complete to decide if a schedule with makespan of at most k exists.*

Corollary 3. *In a shunting instance on planar graphs with bounded degree of three and external depots it is NP-complete to decide if a schedule with makespan of at most k exists – even for trains of size one.*

Corollary 4. *It is $W[1]$-hard to compute the optimal solution of a shunting instance with external depots parameterized by the number of trains n. Moreover, there is no algorithm that solves the problem in $f(n)\,|E|^{o(n/\log n)}$ time for some function f unless the ETH fails.*

4 Polynomial Time Algorithms for Paths

On a path with internal depots, the trains cannot overtake each other since they are permanently present. Thus, it is necessary that the order in which they start is equal to the ordering of the target depots. Moreover, in this case no train needs to wait for another one. Thus, an algorithm checks the order of the starting depots and compares it to the order of the target depots. If they are equal, the makespan is the longest distance between start and target depot, otherwise there is no feasible schedule.

Proposition 2. *In a shunting or waiting instance with internal depots on a path there is a linear time algorithm that computes the optimal makespan.*

So, in the following we focus on external depots. The problem of optimizing a waiting problem turned out to be hard already on a unidirectional path. Therefore, we present a polynomial time algorithm for waiting instances on a special subclass of unidirectional paths.

We characterize an instance as *non-nested* if the start-destination intervals of trains are not included in each other. To illustrate the definition, assume that all trains move from left to right and are numbered according to their start nodes. Then, the destination nodes appear in the same ordering, i.e. for $s_i < s_j$ it holds that $t_i < t_j$.

Observe that for $i < j$ the only possibility of train i to traverse a shared edge before train j is that train j waits in its depot until train i traversed s_j. Furthermore, train i will be in front of j on every shared edge afterwards. In Appendix H of the full version, we present two examples. One in which it is necessary to delay a train such that a preceding train can pass and another in which the preceding train moves first. Both variants demonstrate that the delaying decision does not only depend on the position of the depots (see Example 5 in Appendix H of the full version).

A class for which we can optimize the waiting time is a non-nested instance on a unidirectional path where all trains share one node, i.e., $\bigcap_{i \in N} P(s_i, t_i) \neq \emptyset$. The idea of the algorithm is starting with the rightmost train. Then, we always start the next train to the left such that it can directly follow the preceding train. If the distance between two subsequent trains is large enough, we can safely start the train at the left start node at time 0 without interfering, too, and repeat the same procedure for all trains to the left of it. The pseudo code of the algorithm can be found in Appendix I of the full version, Algorithm 1.

Theorem 6. *Consider a waiting instance on a unidirectional path where all depots are non-nested and fulfill $\bigcap_{i \in N} P(s_i, t_i) \neq \emptyset$. Algorithm 1 (Rightmost-Train-First – RTF) computes a schedule minimizing the makespan.*

Algorithm 1 does not return the optimal solution as soon as the instance does not fulfill $\bigcap_{i \in N} P(s_i, t_i) = \emptyset$ – even if trains have the same length. Moreover, the Farthest-Destination-First Algorithm (FDF), which is optimal for a non-nested instance in the message routing problem [12], does not solve the train routing problem to optimally as shown in Appendix J of the full version.

Lemma 1. *If at least one node v on a non-nested unidirectional path exists that is contained in the path of every train, then there exists an optimal schedule where the first train traverses v last.*

Proof. Assume for contradiction that no such schedule exists. Consider an optimal schedule where train one traverses the shared node latest among all optimal schedules. Then, there exists a train j that directly follows train one, because otherwise train one can wait for one time unit. At some point in time t train one traverses s_j and from this point onward train j follows train one. Together they block $w_j + w_1$ consecutive edges. So at point t train j can leave the depot and train one follows train j. Since all trains use shared edges after s_j, this does not further delay later trains and the makespan does not get worse because the remaining distance for train one is shorter than the remaining distance for train j. This is a contradiction.

Proof. (Theorem 6) With use of Lemma 1, we can compute an optimal schedule recursively by finding an optimal schedule for trains two to n and then scheduling train one last, but as early as possible. This is exactly the way Algorithm 1 works.

5 Conclusion

In this work, we introduced a model for train routing problems on railway networks. Unfortunately, problems get hard very fast in this model – even for trains of length one and on paths. Thus, it is an interesting challenge finding good heuristics and approximation algorithms. In classical packet routing problems, randomization is a powerful tool, but in this case due to the possibility of deadlocks it is not easy to use.

A   Difference Between Wormhole and Train Routing

In Fig. 8, one elemental difference between wormhole and train routing is depicted. In contrast to classical wormhole routing, in train routing node capacities and blacklists exist. Both constraints restrict the solution space.

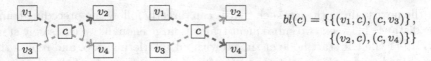

Fig. 8. Two track intersections for the train routing with two trains/worms of length 2 (train 1 red, train 2 blue) are shown. In train routing only one train is allowed to traverse c at a time, while in whormhole routing both worms could traverse c simultaneously. Moreover, trains are only allowed to enter and leave c along pahts that are not forbidden by the blacklist $bl(c)$. (Color figure online)

B   Difference Between Waiting and Shunting Instance

In Fig. 4 a deadlock instance is presented. Adding edge e_4 (see Fig. 9) and allowing at least on train to shunt along this edge resolves the deadlock situation. For example, the optimal solution for the instance with $N = \{red, blue\}$,

$$P_{red} = (e_2, e_3),\ P_{blue} = (e_2, e_1)\ \text{and}\ S_{blue} = \{e_4\}\ \text{is}$$

$$P'_{red} = (e_2, e_3)\ \text{with}\ \theta_{red} = (0, 0)\ \text{and}\ P'_{blue} = (e_4, e_2, e_1)\ \text{with}\ \theta_{blue} = (0, 1, 0).$$

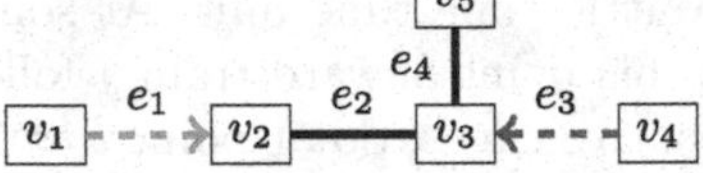

Fig. 9. The instance is only feasible if one allows at least on train to shunt.

C   Proof of Proposition 1

Proposition 1. *There are instances of the train routing problem in which the optimal shunting solution is lower than the optimal waiting solution.*

Proof. Consider the network depicted in Fig. 10 with trains of weight vector $w = (2, 1, 1)$. The number on the connection of two nodes denotes the number of edges in the path connecting the two nodes. In the shunting problem as well as in the waiting time problem the given paths for the trains are the unique shortest paths and no further edges are offered, thus $S_i = \emptyset$ for all $i \in [3]$.

The optimal makespan of this instance as a shunting problem is 43. The strategy for train 1 is to pass node v_2 such that train 2 can leave the depot. After train 2 traversed v_2 train 1 is shunted back to let train 3 enter into depot t_3. Only afterwards train 1 continues to t_1. If train 1 cannot shunt, there are three routing possibilities. If train 1 is allowed to traverse v_2 and v_3 first, train 3 arrives at time 58. If train 1 only traverse v_2 first and waits for train 3, train 2 arrives at time 45. On contrary, if train 2 is in front of train 1 it only arrives at time 52 and thus no waiting solution attains the optimal shunting makespan.

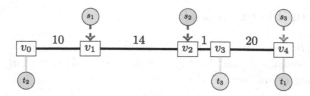

Fig. 10. The optimal shunting solution is lower than the optimal waiting solution.

D   Deadlock Examples

Example 1. Even for a simple path with only two trains of unit length, no feasible solution exists if the start edge of train 1 (red) is the target edge of train 2 (blue) and vice versa as depicted in Fig. 4. There is no track that allows one train to traverse the other, regardless of any possible intermediate waiting or multiple use of edges.

Example 2. In Fig. 5, there are two trains of length three. The blue one is starting at node v_5 and a red one starting at node v_0, both in external depots. The path of the blue train is $P_{blue} = \{\{v_5, v_4\}, \{v_4, v_3\}, \{v_3, v_1\}, \{v_1, v_0\}\}$ and the path of the red train is $P_{red} = \{\{v_0, v_1\}, \{v_1, v_2\}, \{v_2, v_4\}, \{v_4, v_5\}\}$. When both trains start at the same time and proceed without waiting they are on the first three edges of their path, blocking each other such that no train can continue, as depicted in Fig. 5. Thus, if they are not allowed to shunt this is a deadlock situation. If one of the trains waits for five time units, the other train reached its depot and the train can travel without waiting at intermediate nodes.

Example 3. The setup depicted in Fig. 4 is transformed to a solvable instance presented in Fig. 11. We introduce two depots for each train. Edges and nodes that are related to only one train are colored in the train specific color while the trains are initially orientated towards G.

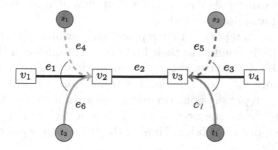

Fig. 11. A two-train instance with $s_1 = v_2$ and $t_1 = v_3$ and thus $E_{s_1}^{red} = \{e_4\}$, $E_{t_1}^{red} = \{e_7\}$ as well as $s_2 = v_3$ and $t_2 = v_2$ and thus $E_{s_2}^{blue} = \{e_5\}$, $E_{t_2}^{blue} = \{e_6\}$. While one train is parking in one of its depots the other train can travel from one of its depots to the other. (Color figure online)

E   Proof of Theorem 2

Theorem 2. *In a waiting instance on planar graphs with degree bounded by three it is NP-complete to decide if a schedule with makespan of at most k exists – even for trains of size one.*

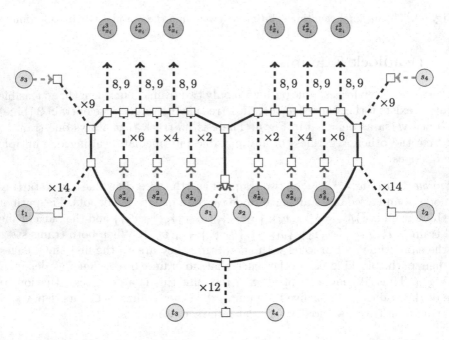

Fig. 12. A variable component.Either the trains from s_5, s_6, s_7 are delayed by s_1 or the trains from s_8, s_9, s_{10} are delayed by s_2.

Proof. The proof uses a reduction to Simple Planar 4-bounded exactly 3–SAT [11] with 22 trains per variable and one train per clause. Note that in Simple Planar 4-bounded exactly 3–SAT, a variable occurs at most four times and at least once negated and non-negated.

Figure 12 shows a component that replaces one variable x_i. The trains, starting at s_1 and s_2, are in conflict at their first edge. The choice of the first moving train represents the decision to set x_i to true or false. If we priories s_1 it will delay the trains from s_{x_i}. Additionally, the trains from $s_{\bar{x}_i}$ will not be delayed. Each train from s_{x_i} represents one occurrence of x_i in a clause. Vice versa, each train from $s_{\bar{x}_i}$ represents one occurrence of $\bar{x}_i$ in a clause. Delaying the trains from s_{x_i} means setting x_i to false. Thus, if the train from s_2 is prioritized x_i is set to true.

The distance from s_1 and s_2 to their depot is 23. At least one of these trains is delayed and a second delay increases the makespan to 25 which is not optimal if the 3-SAT formula is satisfiable. A delayed train is in conflict with the train from s_3 or s_4. Because these trains have also a distance of 23 to their depot, they are allowed to be delayed at most once, too. Moreover, they are in conflict with each other. Thus, at most one of them is allowed to be delayed. Therefore, only one of the trains from s_1 and s_2 is allowed to be delayed once.

The dashed edges from Fig. 12 labeled with $8, 9$ are continued in Fig. 13. The label represents the possible arrival times for the trains from Fig. 12. The presented construction ensures that delayed trains are delayed two more times. The trains s_{11} and s_{12} have a distance of 24 to their depots. Thus, if they are delayed once and the 3-SAT formula is satisfiable it increases the makespan.

Figure 14 shows a component that replaces one clause c. The clause train from s_c to t_c is delayed three times if all of the three crossing trains were delayed. In this case, it arrives at t_c in time step 25. So, if one of the crossing trains is early, the clause train arrives earlier and the clause train arrives in a time step smaller than 25. If in each clause construction at least one variable train is early (true), the makespan is less than 25. The crossing variable trains have a remaining distance of 9 to their depots and cannot be delayed if they were already delayed.

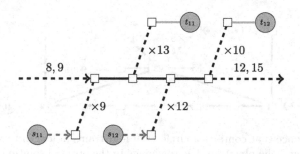

Fig. 13. A delay component that delays a train that was delayed once in the variable component.

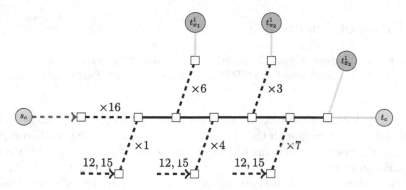

Fig. 14. A component that represents a clause $(x_1 \lor x_2 \lor \bar{x}_3)$ where each variable appears for the first time.

F   Counter-Example: FPT Approach Cannot Be Generalized

In the following example, we show that the approach presented in Theorem 3 can not be generalized to a broader class of graphs. As soon as there are cycles present a global priority order is not enough to find the optimal makespan.

Example 4. Consider Fig. 15. To achieve the optimal makespan of $C^* = 8$ train 1 has to traverse v_1 in front of train 2. Due to the congestion at node v_3 it is optimal to route train 1 via the longer horizontal path. During this time train 2 can traverse v_1, use the idle time between train 4 and train 3 to traverse v_3 and finally traverse v_2 in front of train 1. This is the only schedule where none of the trains is directly delayed. Since the shortest path of train 2, 3 and 4 are 6 their paths are fixed. Train 1 can not deviate without delaying at least one other train thus no other schedule is optimal. Conclusively, the priority between train 1 and 2 has to be changed to achieve the optimal solution.

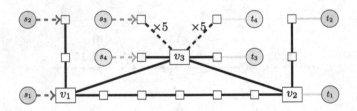

Fig. 15. An instance that contains a circle with four trains of length two. The priority between the red and the blue train has to change in the optimal solution. (Color figure online)

G   Proof of Theorem 5

Theorem 5. *Finding a feasible schedule for a shunting instance with internal depots on a honeycomb graph is PSPACE-complete – even if all trains have equal length.*

Proof. We reduce the problem from Rush-Hour puzzle, by the construction described below. Even though the Rush-Hour puzzle itself is PSPACE-complete [8], we use the restricted version where all blocks are of size 1×2, which is still PSPACE-complete.

Given an instance of Rush-Hour puzzle (see Fig. 16a), we transform the puzzle into a honeycomb graph. Therefore, we introduce an edge for every field of the puzzle. We connect each two horizontal field neighbors as well as each vertical field neighbors with an edge such that honeycombs are created as depicted in Fig. 16c. Moreover, we introduce one additional node for every horizontal block

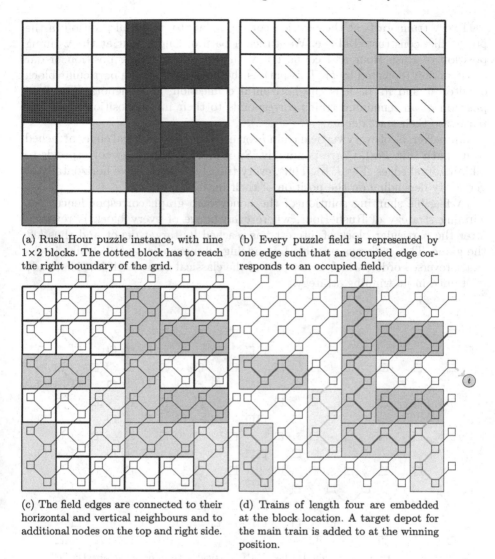

(a) Rush Hour puzzle instance, with nine 1×2 blocks. The dotted block has to reach the right boundary of the grid.

(b) Every puzzle field is represented by one edge such that an occupied edge corresponds to an occupied field.

(c) The field edges are connected to their horizontal and vertical neighbours and to additional nodes on the top and right side.

(d) Trains of length four are embedded at the block location. A target depot for the main train is added to at the winning position.

Fig. 16. Transformation of a Rush Hour puzzle into a grid track instance.

as well as one node for every vertical block and connect them to the neighboring block via an edge.

We introduce a train of length four for every block of the puzzle. Therefore, every train is located on the two edges that represent the fields of the puzzle. Additionally, one unit is placed on the edge that connects both of them. The fourth edge which is occupied by the train is the one attached at the right hand side or on top, depending if it is a horizontal or vertical one (see Fig. 16d).

Every train starts at the position corresponding to its initial position in the puzzle like constructed above. We attach a path of length four at the winning position of Rush Hour and connect it with the corresponding position of our construction (depicted by t). This path is the final depot of the particular block of interest and its path is the horizontal connection of its depots. The final position of all remaining trains corresponds to their initial position and their paths are their start depots.

Moreover, S_i of every vertical train i corresponds to all vertical edges attached to its path (Fig. 17a). Correspondingly, S_j of every horizontal j corresponds to all horizontal edges (Fig. 17b). Thus, every train can either move horizontally or vertically depending on the position of their initial depot.

A feasible shunting solution of the honeycomb graph correspondence to a winning strategy of Rush Hour, where every move of every block is reversed after the particular block of interest has reached its target depot and thus left the game. This is always possible by undoing all movements of all trains in the exact reversed order, which is not part of the classical game itself but is already contained in its winning strategy.

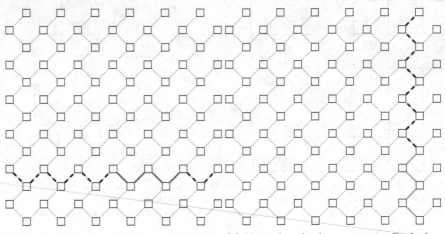

(a) Horizontal paths of pink train. Dashed edges correspond to S_{pink} and magenta edges are $E_{s_{pink}} = E_{t_{pink}}$.

(b) Vertical path of orange train. Dashed edges correspond to S_{orange} and orange edges are $E_{s_{orange}} = E_{t_{orange}}$.

Fig. 17. Subpaths along trains can be routed. Each zig-zag path corresponds to the horizontal or vertical movement of the blocks in the puzzle.

H   Delaying on a Path

Example 5. It might be necessary to delay a train until a train that starts on the left hand side has passed. Figure 18a shows an instance in which it is optimal to delay train 2 in its depot until the first train has passed. Observe that the location of the depots alone is not sufficient to decide which train goes first. This can be seen in Fig. 18b, which extends the previous instance by several copies of the second train. Now the optimal order of the first two trains is reversed.

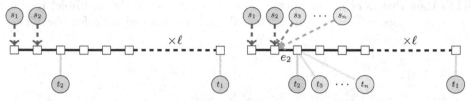

(a) Handle conflicts only if train heads interact is not optimal. Delay train 2 without immediate conflict is necessary to achieve the optimal makespan of $C^* = 6 + l$.

(b) Due to the increased congestion on edge e_2 it is optimal to delay the first train one time unit such that train 2 traverses e_2 immediately and avoids idle time on e_2.

Fig. 18. Delaying does not only depend on the position of the depots.

I Algorithm to Compute Waiting Times on an Uni-directional Path on Which All Trains Share One Node

---

Algorithm 1. Rightmost Train First - (RTF)

Input: Path P and a set N of n trains with weight w_i and external non-nested depots, adjacent to s_i (start depot) and t_i (target depot) such that $\bigcap_{i \in N} P(s_i, t_i) \neq \emptyset$.

Output: Arrival time of every train in a solution minimizing the makespan.

1   $\tau \leftarrow 0;\ k \leftarrow n;$
2   **for** $i = n$ **to** 1 **do**
3      $dist(i) \leftarrow |P(s_i, s_k)|;$
4   **if** $\tau - dist(i) \geq 0$ **then**
5         $start(i) \leftarrow \tau - dist(i);$
6         $\tau \leftarrow \tau + w_i;$
7   **else**
8         $k \leftarrow i;\ start(i) \leftarrow 0;\ \tau \leftarrow w_i;$
9   **end**
10     $C_i \leftarrow start(i) + |P(s_i, t_i)| + w_i;$
11 **end**
12 **return** $(C_i)_{i \in N}$

J   Counter-Example: RTF-Algorithm Cannot be Generalized if Trains do not Share a Node

Example 6. In Fig. 19 we present an example where trains travel on a unidirectional path and are non-nested but they do not share one node. Since the RTF-Algorithm does not compute an optimal solution in this example, it is a necessary property of an instance that all trains share one node for the RTF-Algorithm to work.

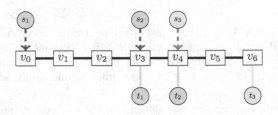

Fig. 19. Non-nested instance with three trains of length five. Its optimal to start the second train first to achieve a minimal makespan of 13 even though the last train has the longer remaining path $P(v_4, v_7)$ in contrast to train 2.

References

1. Cygan, M., et al.: Parameterized Algorithms, 1st edn. Springer, Heidelberg (2015). https://doi.org/10.1007/978-3-319-21275-3
2. Cypher, R., Meyer auf der Heide, F., Scheideler, C., Vöcking, B.: Universal algorithms for store-and-forward and wormhole routing. In: Miller, G.L. (ed.) Proceedings of the Twenty-Eighth Annual ACM Symposium on the Theory of Computing, Philadelphia, Pennsylvania, USA, 22–24 May 1996, pp. 356–365. ACM (1996)
3. Dreier, J., et al.: The complexity of packing edge-disjoint paths. CoRR abs/1910.00440 (2019). http://arxiv.org/abs/1910.00440
4. Duato, J.: A necessary and sufficient condition for deadlock-free adaptive routing in wormhole networks. IEEE Trans. Parallel Distrib. Syst. **6**(10), 1055–1067 (1995). https://doi.org/10.1109/71.473515
5. Hall, A., Hippler, S., Skutella, M.: Multicommodity flows over time: efficient algorithms and complexity. Theor. Comput. Sci. **379**(3), 387–404 (2007)
6. Harks, T., Peis, B., Schmand, D., Tauer, B., Koch, L.V.: Competitive packet routing with priority lists. ACM Trans. Econ. Comput. **6**(1), 4:1–4:26 (2018)
7. Hartmanis, J.: Computers and intractability: a guide to the theory of NP-completeness (michael r. garey and david s. johnson). SIAM Rev. **24**(1), 90 (1982)
8. Hearn, R.A., Demaine, E.D.: Pspace-completeness of sliding-block puzzles and other problems through the nondeterministic constraint logic model of computation. Theor. Comput. Sci. **343**(1–2), 72–96 (2005)
9. Hoefer, M., Mirrokni, V.S., Röglin, H., Teng, S.: Competitive routing over time. Theor. Comput. Sci. **412**(39), 5420–5432 (2011)

10. Ismaili, A.: Routing games over time with FIFO policy. In: Devanur, N.R., Lu, P. (eds.) WINE 2017. LNCS, vol. 10660, pp. 266–280. Springer, Cham (2017). https://doi.org/10.1007/978-3-319-71924-5_19
11. Jansen, K., Müller, H.: The minimum broadcast time problem for several processor networks. Theor. Comput. Sci. **147**(1–2), 69–85 (1995)
12. Koch, R., Peis, B., Skutella, M., Wiese, A.: Real-time message routing and scheduling. In: Dinur, I., Jansen, K., Naor, J., Rolim, J. (eds.) APPROX/RANDOM -2009. LNCS, vol. 5687, pp. 217–230. Springer, Heidelberg (2009). https://doi.org/10.1007/978-3-642-03685-9_17
13. Leighton, F.T.: Methods for message routing in parallel machines. In: Kosaraju, S.R., Fellows, M., Wigderson, A., Ellis, J.A. (eds.) Proceedings of the 24th Annual ACM Symposium on Theory of Computing, 4–6 May 1992, Victoria, British Columbia, Canada, pp. 77–96. ACM (1992)
14. Leung, J.Y., Tam, T.W., Wong, C.S., Young, G.H.: Routing messages with release time and deadline constraint. J. Parallel Distrib. Comput. **31**(1), 65–76 (1995)

Distributed Algorithms

Oriented Diameter of Star Graphs

K. S. Ajish Kumar[1]([📧])[iD], Deepak Rajendraprasad[2][iD], and K. S. Sudeep[3][iD]

[1] Department of Electronics and Communication Engineering, National Institute of Technology Calicut, Kozhikode, India
ajishsreerangathu@gmail.com
[2] Department of Computer Science and Engineering, Indian Institute of Technology Palakkad, Palakkad, India
[3] Department of Computer Science and Engineering, National Institute of Technology Calicut, Kozhikode, India

Abstract. An *orientation* of an undirected graph G is an assignment of exactly one direction to each edge of G. Converting two-way traffic networks to one-way traffic networks and bidirectional communication networks to unidirectional communication networks are practical instances of graph orientations. In these contexts minimising the diameter of the resulting oriented graph is of prime interest.

The n-star network topology was proposed as an alternative to the hypercube network topology for multiprocessor systems by Akers and Krishnamurthy [IEEE Trans. on Computers (1989)]. The *n-star graph* S_n consists of $n!$ vertices, each labelled with a distinct permutation of $[n]$. Two vertices are adjacent if their labels differ exactly in the first and one other position. S_n is an $(n-1)$-regular, vertex-transitive graph with diameter $\lfloor 3(n-1)/2 \rfloor$. Orientations of S_n, called *unidirectional star graphs* and distributed routing protocols over them were studied by Day and Tripathi [Information Processing Letters (1993)] and Fujita [The First International Symposium on Computing and Networking (CANDAR 2013)]. Fujita showed that the (directed) diameter of this unidirectional star graph $\overrightarrow{S_n}$ is at most $\lceil 5n/2 \rceil + 2$.

In this paper, we propose a new distributed routing algorithm for the same $\overrightarrow{S_n}$ analysed by Fujita, which routes a packet from any node s to any node t at an undirected distance d from s using at most $\min\{4d+4, 2n+4\}$ hops. This shows that the (directed) diameter of $\overrightarrow{S_n}$ is at most $2n + 4$. We also show that the diameter of $\overrightarrow{S_n}$ is at least $2n$ when $n \geq 7$, thereby showing that our upper bound is tight up to an additive factor.

Keywords: Strong orientation · Oriented diameter · Star graphs

1 Introduction

Let $G = (V, E)$ be an undirected graph with vertex set V and edge set E. An orientation $\overrightarrow{G}$ of G is a directed graph obtained by assigning exactly

© Springer Nature Switzerland AG 2020
M. Changat and S. Das (Eds.): CALDAM 2020, LNCS 12016, pp. 307–317, 2020.
https://doi.org/10.1007/978-3-030-39219-2_25

one direction to each edge of G. An orientation is called a *strong orienta-tion* if the resulting directed graph is strongly connected. A directed graph $\overrightarrow{G}$ is said to be strongly connected, if there exists at least one directed path from every vertex of $\overrightarrow{G}$ to every other vertex. There can be many strong orientations for G. The smallest diameter among all possible strong orienta-tions of G is called the *oriented diameter* of G, denoted by $\overrightarrow{diam}(G)$. That is, $\overrightarrow{diam}(G) = \min\{\text{diam}(\overrightarrow{G}) | \ \overrightarrow{G} \text{ is a strong orientation of G}\}$.

The research on strong orientations dates back to 1939 with Robbins [2], solving the *One Way Street* problem. Given the road network of city, the One Way Street problem poses the following question: Is it possible to implement one way traffic in every street without compromising the accessibility of any of the junctions of the network? Robbins proved that the necessary and sufficient condition for the existence of a strong orientation of a graph G is the 2-edge connectivity of G. A 2-edge connected graph is one that cannot be disconnected by removal of a single edge. The research on orientations that minimise the resulting distances was initiated by Chvátal and Thomassen in 1978 [3]. They proved that, for every undirected graph G there exists an orientation $\overrightarrow{G}$ such that for every edge (u, v) which belongs to a cycle of length k, either (u, v) or (v, u) belong to a cycle of length $h(k)$ in $\overrightarrow{G}$, where $h(k) = (k-2)2^{\lfloor \frac{(k-1)}{2} \rfloor} + 2$. They also showed that every 2-edge connected undirected graph of diameter d will possess an orientation with diameter at most $2d^2 + 2d$. Further, they proved that it is NP-hard to decide whether an undirected graph possesses an orientation with diameter at most 2.

Fomin et al. [9] continues the algorithmic study on oriented diameter on chordal graphs. They show that every chordal graph G has an oriented diameter at most $2 \, \text{diam}(G) + 1$. This result proves that the oriented diameter problem is $(2, 1)$-approximable for chordal graphs. A polynomial time algorithm for finding the oriented diameter of planar graphs was given by Eggemann [5]. Fomin *et al.* [10] have proved that the oriented diameter of every AT-free bridgeless connected graph G is at most $2 \, \text{diam}(G) + 11$ and for every interval graph G, it is at most $\frac{5}{4} \, \text{diam}(G) + \frac{29}{2}$. Dankelmann et al., [8] proved that every n-vertex bridgeless graph with maximum degree Δ has oriented diameter at most $n - \Delta + 3$. For balanced bipartite graphs (a bipartite graph with equal number of vertices on both halves of the bipartition), they prove a better bound of $n - 2\Delta + 7$. The problems of finding strong orientations that minimize the parameters such as diameter, distance between pairs of vertices *etc.*, have been investigated for other restricted subclasses of graphs like n-dimensional hypercube [11], torus [6], star graph [4,13], and (n,k)-star graph [7].

Oriented diameter problem finds a significant application in parallel comput-ing. In interconnection networks of parallel processing systems, the processing elements are connected together using fibre optic links that support high band-width, high speed and long distance data communication. However, the optical transmission medium suffers from the drawback that the links are inherently unidirectional [11]. In the case of optical links, a naive strategy to achieve bidi-rectional communication is to use two separate optical links between every pair of communication entities. But, such a naive approach increases the hardware complexity and cost of the network. On the other hand, unidirectional communication links are simple and cost effective but require more number of intermediate communication hops to establish bidirectional communication. Thus, the average interprocessor communication delay is generally more in the case of unidirectional interconnection networks. However, unidirectional interconnection networks might be the best choice if we can trade off communication delay with cost and hardware complexity of the network.

1.1 The n-star Graph (S_n)

In [1], Akers and Murthy presented a group theoretic model called *Cayley Graph Model* for designing symmetric interconnection networks. In parallel computing the interconnection networks provide an efficient communication mechanism among the processors and the associated memory. For a finite group Γ and a set S of generators of Γ, the Cayley Graph $D = D(\Gamma, S)$ is the directed graph defined as follows. The vertex-set of D is Γ. There is an arc from a vertex u to a vertex v in D, if and only if there exist a generator g in S such that $ug = v$. Further, if the inverse of every element in S is also in S, the two directed edges between u and v are replaced by a single undirected edge, resulting in an undirected graph. In [12], Akers and Murthy proposed a new symmetric graph, called *Star Graph, S_n*. Let G be a group with elements being all permutations of the set $\{1, 2, \ldots, n\}$ and group operation being composition. The star graph S_n is a Cayley graph on G with generator set $S = \{g_2, g_3, \ldots, g_n\}$, where g_i is the permutation obtained by swapping the first and i^{th} value of the identity permutation. It is easy to see that, S_n has degree $n - 1$, and it has been shown that the diameter of S_n is $\lfloor 3(n-1)/2 \rfloor$ [12]. The star graph has many desirable properties of a good interconnection network such as symmetry (vertex transitivity), small diameter, small degree and large connectivity. A symmetric interconnection network allows the use of same routing algorithm for every node, while a small degree reduces the cost of the network. Further, a small diameter reduces overall communication delay and large connectivity offers good fault tolerance.

Two different strong orientation schemes have been proposed for S_n. The first one was by Day and Tripathi [4]. They showed that the diameter of their orientation is at most $5(n - 2) + 1$. The second orientation scheme was proposed by Fujita [13]. The diameter of this orientation scheme was shown to be at most $\lceil 5n/2 \rceil + 2$. We observe that these two orientation schemes are essentially the same. Both the schemes partition the set of generators into nearly equal halves. The edges due to first set of generators are oriented from the odd permutation to the even permutation and those due to the second set of generators are oriented in the opposite direction. The difference between the two orientation schemes lies in the way by which the two schemes partition the set of generators. The Day-Tripathi scheme splits the set of generators based on the parity of i of a generator g_i, i.e., generators with odd parity for i belong to the first set and even parity for i belong to the second set. In the case of Fujita's orientation, the first partition consists of generators from g_2 to g_k, $k = \lfloor (n-1)/2 \rfloor + 1$, whereas, the second partition consists of generators from g_{k+1} to g_n. The details of the two orientation schemes described above are depicted in Fig. 1, for two nodes with labels 12345 and 21345, and their neighbours in S_5.

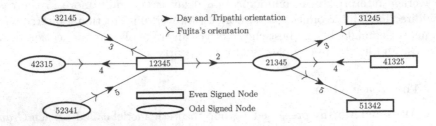

Fig. 1. Day-Tripathi and Fujita orientation schemes for S_5

In this paper, we propose a new distributed routing algorithm for the same $\overrightarrow{S_n}$ analysed by Fujita. We show that the proposed algorithm routes a packet in $\overrightarrow{S_n}$ from any node s to any other node t using at most $\min\{4d + 4, 2n + 4\}$ hops, where d is the distance between s and t in S_n. In particular, this shows that the (directed) diameter of $\overrightarrow{S_n}$ is at most $2n + 4$, which is an improvement over Fujita's upper bound. We also show that the diameter of $\overrightarrow{S_n}$ is at least $2n$ when $n \geq 7$, thereby showing that our upper bound is tight up to an additive factor. We do not believe that either of the above orientations of S_n are optimal in terms of achieving the minimum (directed) diameter. In fact, we believe that the oriented diameter of S_n is $3n/2 + O(1)$.

2   Preliminaries

2.1   Graph Terminology

Some of the basic definitions in graph theory which are required to understand the details of this work are explained in this section. Let $G = (V, E)$ be any undirected graph with vertex-set V and edge-set E. Two vertices of G are called *neighbours* when they are connected by an edge. The *degree* of a vertex u is the number of neighbours of u. If all the vertices of G have the same degree, G is called *regular*. The *distance* between two nodes u and v, denoted by $d(u, v)$, is the number of edges along a shortest path between u and v. The *diameter* of G, denoted by $\operatorname{diam}(G)$, is the maximum of $d(u, v)$ among all $u, v \in V$. An *automorphism* of G is a permutation π of V such that for every pair of vertices $u, v \in V$, $\{u, v\}$ is an edge in E, if and only if $\{\pi(u), \pi(v)\}$ is an edge in E. Two vertices u and v of G are said to be *similar* if there is an automorphism π of G with $\pi(u) = v$. G is *vertex-transitive* when every pair of vertices in G are similar.

Let $D = (V, E)$ be any directed graph with vertex-set V and edge-set E. If (u, v) is an edge (arc) in E, then u is called an *in-neighbour* of v and v is called an *out-neighbour* of u. The *in-degree* and *out-degree* of a vertex u are, respectively, the number of in-neighbours and out-neighbours of u. The *distance* from a node u to a node v, denoted by $\overrightarrow{d}(u,v)$, is the number of edges along a shortest directed path from u to v. The *diameter* of D (diam(D)), is the maximum of $\overrightarrow{d}(u,v)$ among all $u, v \in V$. Automorphism and Vertex-transitivity among directed graphs are defined similar to that of undirected graphs.

2.2 Cycle Structure of Permutations

Let π be a permutation of $\{1, \ldots, n\}$. The *sign* of π, denoted by Sign(π), is defined as the parity of the number of inversions in π, that is $x, y \in \{1, \ldots, n\}$, such that $x < y$ and $\pi(x) > \pi(y)$. A *cycle* $(a_0, \ldots, a_{k-1})$ is a permutation π of $\{a_0, \ldots, a_{k-1}\}$ such that $\pi(a_i) = a_{i+1}$ where addition is modulo k. Two cycles are disjoint if they do not have common elements. Every permutation of $[n]$ has a unique decomposition into a product of disjoint cycles. The sign of a permutation turns out to be the parity of the number of even-length cycles in that permutation.

One hop in an n-star graph corresponds to moving from a permutation σ to another permutation π, by exchanging the value $\sigma(1)$ with a value $\sigma(k)$, $k \in \{2, \ldots n\}$. We would like to make some observations about the cycle structure of π and σ. In the case when 1 and k belong to the same cycle of σ, this cycle gets broken into two disjoint cycles in π (Fig. 2). Notice that, if $\sigma(1) = k$, then one of the resulting cycles is a singleton. In the case when 1 and k belong to different cycles of σ, these two cycles merge and form a single cycle in π (Fig. 3).

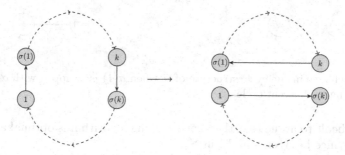

Fig. 2. The change in the cycle structure of σ, when $\sigma(1)$ is swapped with $\sigma(k)$, when 1 and k belong to the same cycle.

Given two permutations π and t, we call the cycles of $\pi \circ t^{-1}$ as the cycles of π *relative* to t. The above observations about the cycle structure of two permutations π and σ which differ by a single swap between 1 and k will apply in this case to the relative cycle structure of π and σ with respect to t.

2.3   Routing in Undirected Star Graph

In this section, we describe the routing algorithm for the undirected star graph S_n presented in [1]. Assume that a node labelled c forwards a packet P from a source s to a destination t. The destination label t is available in the packet. Upon receiving the packet, c accepts P if c is same as t. Otherwise, when $c(1) \neq t(1)$, the node c forwards P through the link labelled i, where i is the position of $c(1)$ in t. We call such a move a *settling move*. A value is called *settled* if it is in the same location in c and t, and *unsettled* otherwise. When $c(1) = t(1)$ (but $c \neq t$), the node c forwards P through a link i, where i is the position of an unsettled value. We call such move a *seeding move*. Notice that, during the course of routing P from s to t, the number of seeding moves is same as the number of non-singleton cycles in s relative to t. Also, no move disturbs an already settled value. Therefore, we can observe that the total number of steps required to settle all unsettled values in s, denoted by d, is at most $m(s,t) + c(s,t)$, where $m(s,t)$ is the number of mismatched values (i.e., values that are not in their correct position with respect to t) and $c(s,t)$ is the number of non-singleton cycles in s relative to t. More closer analysis yield [1] the following result.

$$ d = \begin{cases} m(s,t) + c(s,t), & \text{if } s(1) = t(1) \\ m(s,t) + c(s,t) - 2, & \text{otherwise.} \end{cases} \tag{1} $$

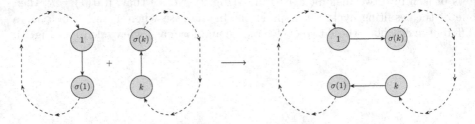

Fig. 3. The change in the cycle structure of σ when $\sigma(1)$ is swapped with $\sigma(k)$, when 1 and k belong to different cycle.

It is not difficult to argue that the above routing algorithm is optimal and hence d is the distance between s and t in S_n.

3   The Proposed Routing Algorithm

There are two different ways in which one can describe and analyse a routing algorithm on a star graph. In the first view, which we call the "network view", we consider each vertex of S_n as a communication node whose *address* is the permutation labelling that vertex. Depending on the sign of the address of a node, we classify it as an *even node* or an *odd node*. We consider each arc of $\overrightarrow{S_n}$ as a unidirectional communication link and label it by the unique position in $\{2, \ldots, n\}$, where the addresses of the endpoints of the arc differ. Hence every node has $n - 1$ links attached to it with unique labels from $\{2, \ldots, n\}$. For an even node, the links labelled 2 to $\lceil (n - 1)/2 \rceil + 1$ are outgoing links and the remaining are incoming. The situation is reversed for odd nodes. Every packet that is to be routed along the network will have the destination address in its header. We describe the algorithm by which a node, on receiving a packet not destined for itself, selects the outgoing link along which to relay that packet. This selection is based on the address of the current node and the destination address.

In the second view, which we call the "sorting view", we consider each vertex of S_n as a permutation of $[n]$. Thus a routing is viewed as a step-by-step procedure to sort the permutation labelling the source to the permutation labelling the destination. Each step in this sorting is restricted to be a transposition $(1, i)$, where $i \in \{2, \ldots, \lceil (n-1)/2 \rceil + 1\}$, if the current permutation is even, and $i \in \{\lceil (n-1)/2 \rceil + 2, \ldots, n\}$, if the current permutation is odd. Hence a directed path in $\overrightarrow{S_n}$ will correspond to an alternating sequence of right half and left half transpositions. This is the view with which we will analyse our routing algorithm in Sect. 4.

In a given permutation, let us call the positions 2 to $\lceil (n-1)/2 \rceil + 1$ as the *left half*, and the positions $\lceil (n-1)/2 \rceil + 2$ to n as the *right half*. First, we analyse the case of sorting a permutation π in which all the left values are in a derangement in the left half itself, and all the right values are in a derangement in the right half itself. For every $n \geq 5$, an example for π is the permutation obtained by cyclically shifting the left-half and right-half by one position each. That is, the cycle decomposition of π is $(1)(2, \ldots, k)(k+1, \ldots, n)$, where $k = \lceil (n-1)/2 \rceil + 1$. This analysis serves two purposes. Firstly, it establishes a lower bound on the diameter of $\overrightarrow{S_n}$. Secondly, it illustrates a typical run of our proposed algorithm to be described later. Let $\pi = \pi^0, \ldots, \pi^l = id$ be the nodes of a shortest directed path from π to id in $\overrightarrow{S_n}$. Notice that, in π, every value except 1 is not in its "correct" position (with respect to the identity permutation) and hence needs to be moved. This requires a transposition $a = (1, \pi^{-1}(i))$ to remove i from its present position and a transposition $b = (1, i)$ to place i in its final position. Let α and β be, respectively, the permutations in $\{\pi^1, \ldots, \pi^l\}$ which appear immediately after the transposition a and immediately before the transposition b. Notice that $\alpha(1) = \beta(1) = i$. The key observation is that α and β cannot be the same permutation. This is because, for every $i \in \{2, \ldots, n\}$, both $\pi^{-1}(i)$ and i are in the same half and the directions in $\overrightarrow{S_n}$ constraints one to alternate between left half and right half transpositions. Hence for every $i \in \{2, \ldots, n\}$, there exists at least two distinct permutations in $\{\pi^0, \ldots, \pi^l\}$ which has i in the first position. Moreover $\pi^0(1) = \pi^l(1) = 1$ (i.e., the value 1 appears in the first position for at least two permutations). Thus $l + 1 \geq 2n$ and hence the length of the path is at least $2n - 1$. If π was an even permutation, we could have improved the lower bound by 1, since the distance between two even permutations has to be even. This is indeed the case when n is odd. When n is even and $n \geq 8$ (and thereby $k \geq 5$, we can choose π to be $(1)(2, 3)(4, \ldots, k)(k+1, \ldots, n)$. This improvement does not work for $n = 6$, and it is indeed established by computer simulation that diam $\overrightarrow{S_n} = 2n - 1$ when $n = 6$ [4]. Hence we conclude

Theorem 1. *For every $n \geq 5$ the diameter of $\overrightarrow{S_n}$ is at least $2n - 1$. Further if $n \neq 6$, the diameter of $\overrightarrow{S_n}$ is at least $2n$.*

Now let us see a way to sort the permutation $\pi = (1)(2, \ldots, k)(k+1, \ldots, n)$ for an odd $n \geq 5$ and $k = \lceil (n-1)/2 \rceil + 1$. We do not attempt to rigorously justify the claims made in the following discussion as they are proved in more generality in Sect. 4. We do the sorting in two phases. In the first phase (the *crossing phase*), we obtain a permutation γ in which all the values in $\{2, \ldots, k\}$ (the *small values*) are in the right half and all the values in $\{k+1, \ldots, n\}$ (the *big values*) are in the left half. This can be done in $n + 1$ steps; the first step places 1 in the left half (seeding move) and all the subsequent steps either places a small value in the right half or a large value in the left half (crossing moves). Only thing one has to be careful about is to remove 1 from the left half only in the last transposition. For example, one can attack the positions $2, k+1, 3, k+2, \ldots, k-1, n, 2$ in that order to arrive at $\gamma = (1)(2, k+1)(3, k+2) \cdots (n-1, n)$. In the second phase (the *settling phase*), when a value i, $i \neq 1$ appears in the first position for the first time, in the very next step we will settle it, i.e., place it in position i. This will be possible since, $\gamma^{-1}(i)$ and i are in different halves for all $i \in 2, \ldots .n$. This phase could have been completed in n steps provided the elements $\{2, \ldots, n\}$ formed a singe cycle in γ. Otherwise, after placing all the elements in a cycle of γ to their correct positions, 1 will return to the first position. This results in one extra move (a *seeding move*) per non-singleton cycle of γ. An extreme example of this can be seen by analysing the case when γ is as above, wherein one requires $\lceil n/2 \rceil - 1$ seeding moves. Hence the number of moves in the settling phase is $n - 1 + c(\gamma)$ where $c(\gamma)$ is the number of non-singleton cycles in γ. Since the number of non-singleton cycles in any permutation of $[n]$ is at most $\lfloor n/2 \rfloor$, one quickly sees that π can be sorted in a total of $\lfloor 5n/2 \rfloor$ steps. One can then easily extend this analysis to an arbitrary permutation in place of π and show that the diameter of $\overrightarrow{S_n}$ is at most $5n/2 + O(1)$, reproving the bound of Fujita [13]. But we show that we have enough freedom while building γ to ensure that γ consists of at most two non-singleton cycles. This is done by showing that during all but the final two transpositions of the crossing phase, we can select the swaps so as not to complete a new cycle among the crossed values. This is what helps us in achieving the bound of $2n + O(1)$ on the diameter of $\overrightarrow{S_n}$.

One drawback of the above method is that, even if the source permutation π is very close to the identity permutation in terms of distance in S_n like $\pi = (1)(2, 3)(4)(5) \cdots (n)$, this method may take $2n$ steps. Hence, we modify the above method by making sure that, if π has m small values and m large values which are already in their correct positions, then those $2m$ values are not disturbed during the sorting. We then analyse this strategy to show that any permutation π can be sorted in at most $4d + 4$ steps, where d is the distance between π and id in S_n.

These attempts to reduce the number of cycles in γ and to disturb as few settled values in π as possible is what makes the crossing phase of the routing algorithm slightly complex. Moreover, when $\pi(1) \neq 1$, we have two possibilities. If π is even and $\pi(1)$ is a large value, we continue as if we are in the crossing phase. We do the same when π is odd and $\pi(1)$ is a small value. In the other two cases (π even, $\pi(1)$ small and π odd, $\pi(1)$ large), we start by settling $\pi(1)$ and continue in the settling phase till either 1 appears in the first position or one of the two cases mentioned above occurs. Then we go into the crossing phase, complete it, and enter the settling phase for a second time. Hence one cycle of the settling phase can happen before the crossing phase. With this high-level idea, we formally state our proposed routing algorithm.

Definition 2. For a permutation s of $[n]$, we call $L(s) = \{s(i) : 2 \leq i \leq \lceil (n-1)/2 \rceil + 1\}$ and $R(s) = \{s(i) : \lceil (n-1)/2 \rceil + 2 \leq i \leq n\}$ as the sets of *left values* and *right values* of s, respectively.

Given two permutations s and t of $[n]$ for some n, we define $S(s,t) = \{s(i) : s(i) = t(i), 1 \leq i \leq n\}$, and $U(s,t) = [n] \setminus S(s,t)$ respectively, as the sets of *settled* and *unsettled* values between s and t. We partition $U(s,t) \setminus \{s(1), t(1)\}$ into four sets

$$ULL(s,t) = U(s,t) \cap L(s) \cap L(t),$$
$$URR(s,t) = U(s,t) \cap R(s) \cap R(t),$$
$$ULR(s,t) = U(s,t) \cap L(s) \cap R(t),$$
$$URL(s,t) = U(s,t) \cap R(s) \cap L(t).$$

We also partition $S(s,t)$ into two sets

$$SL(s,t) = S(s,t) \cap L(t), \quad SR(s,t) = S(s,t) \cap R(t).$$

Let us call $X(s,t) = ULR(s,t) \cup URL(s,t)$ as the set of *crossed values* between s and t. A cycle of s relative to t is called *alternating* if it has size at least two, and the successive elements of the cycle alternate between $L(s)$ and $R(s)$. Finally, $\chi(s,t)$ will denote the number of alternating cycles of s with respect to t.

The processing done by an even node is given in Algorithm 1. The processing done by an odd node is similar (the roles of "left" and "right" are reversed) and hence omitted. In every move, $c(1)$ is exchanged with $c(i)$ for some $i \in \{2, \ldots, n\}$. We classify these moves into three types. If $c(1) = t(1)$, the move is called a *seeding move*. If $c(1) \neq t(1)$ and $c(1) = t(i)$, i.e., $c(1)$ moves to its correct location in t, it is called a *settling move*. If $c(1) \in L(t)$ and it moves to the right half or if $c(1) \in R(t)$ and it moves to the left half, the move is called a *crossing move*.

4   Analysis of the Proposed Routing Algorithm

In this section, we are going to prove the following upper bound on the number of hops that Algorithm 1 uses to reach from a node s to a node t based on the

Algorithm 1. Processing done by an even node labelled c upon receiving a packet P destined for a node labelled t.

1: **procedure** ROUTEEVEN(PACKET P)
2:     Receive packet P, extract the destination address t.
3:     If the address of the current node c is the same as t, accept P and return.
4:     Let $L_t = L(t)$, $R_t = R(t)$, $ULL = ULL(c,t)$, $URR = URR(c,t)$, $SL = SL(c,t)$, $URL = URL(c,t)$.
5:     *Case 1 (Settling Move):* $c(1) \in L_t$.
6:         Let i be the position of $c(1)$ in the permutation t.
7:     *Case 2 (Crossing/Seeding Move):* $c(1) \notin L_t$ and $|ULL| + |URR| > 0$.
             (*Crossing* when $c(1) \in R_t$, *Seeding* when $c(1) = t(1)$)
8:         The forwarding link i is selected based on the cycle structure of c with respect to t.
9:         *Case 2.1:* When ULL contains a value that is not part of the cycle containing $c(1)$.
10:            Pick i as the c-index of that value.
11:        *Case 2.2:* When all values in ULL are part of the cycle containing $c(1)$.
12:            Pick i as the c-index of the value in ULL that comes first on traversing this cycle backward from $c(1)$.
13:        *Case 2.3:* When ULL is empty
14:            Pick i as the c-index of any value from SL.
15:    *Case 3(Crossing Move):* $c(1) \in R_t$ and $|ULL| + |URR| = 0$.
16:    *Case 3.1(Final Crossing Move):* When $|URL| > 0$
17:            Pick i as the c-index of a value from URL. If possible, select i from an alternating cycle in c.
18:    *Case 3.2(Final/Pre-Final Crossing Move):* When $|URL| = 0$
             (*Final Crossing Move* when $t(1)$ is picked, *Pre-Final Crossing Move* when a settled value is picked)
19:            Pick i as the c-index of $t(1)$ if possible, otherwise pick a settled value.
20:    *Case 4 (Seeding Move):* $c(1) = t(1)$ and $|ULL| + |URR| = 0$.
21:            Pick i as the c-index of a value from URL.
22:    Send P along the edge labelled i and terminate.
23: **end procedure**

relative structure of the permutations that label s and t. The proofs are omitted due to constraints on space and are available in the full version of the paper [14].

Theorem 3. *Let s and t be the permutations labelling any two nodes of the oriented star graph $\overrightarrow{S_n}$. Then, Algorithm 1 will send a packet from s to t in at most*

$$|X(s,t)| + \max\{6, y\}$$

steps, where

$$y = 4\max\{|ULL(s,t)|, |URR(s,t)|\} + \chi(s,t) + 4.$$

Corollary 1. *Let d be the distance between any two nodes s and t in the unoriented star graph S_n. Then $\overrightarrow{d}$, the distance between s and t in the oriented star* *graph $\overrightarrow{S_n}$ oriented using scheme in [13] is upper bounded as,*

$$\overrightarrow{d} \leq 4d + 4 \tag{2}$$

Corollary 2. *The diameter of the oriented star graph $\overrightarrow{S_n}$,*

$$\text{diam}(\overrightarrow{S_n}) \leq \begin{cases} 2n + 2, & \text{when } n \text{ is odd,} \\ 2n + 4, & \text{otherwise.} \end{cases} \tag{3}$$

Day and Tripathi have numerically computed the diameter of $\overrightarrow{S_n}$ for n in the range 3 to 9 [4]. Our bounds on the diameter of $\overrightarrow{S_n}$ agrees with their computation when $n \leq 8$. For $n = 9$ our upper bound is 20 while their computation reports 24.

References

1. Akers, S.B., Krishnamurthy, B.: A group-theoretic model for symmetric interconnection networks. IEEE Trans. Comput. **38**(4), 555–566 (1989)
2. Robbins, H.E.: A theorem on graphs, with an application to a problem of traffic control. Am. Math. Monthly **46**(5), 281–283 (1939)
3. Chvátal, V., Thomassen, C.: Distances in orientations of graphs. J. Comb. Theory Ser. B **24**(1), 61–75 (1978)
4. Day, K., Tripathi, A.: Unidirectional star graphs. Inf. Process. Lett. **45**(3), 123–129 (1993)
5. Eggemann, N., Noble, S.D.: Minimizing the oriented diameter of a planar graph. Electron. Notes Discrete Math. **34**, 267–271 (2009)
6. Konig, J.-C., Krumme, D.W., Lazard, E.: Diameter-preserving orientations of the torus. Networks **32**(1), 1–11 (1998)
7. Cheng, E., Lipman, M.J.: Unidirectional (n, k)-star graphs. J. Interconnect. Netw. **3**(01n02), 19–34 (2002)
8. Dankelmann, P., Guo, Y., Surmacs, M.: Oriented diameter of graphs with given maximum degree. J. Graph Theory **88**(1), 5–17 (2018)
9. Fomin, F.V., Matamala, M., Rapaport, I.: Complexity of approximating the oriented diameter of chordal graphs. J. Graph Theory **45**(4), 255–269 (2004)
10. Fomin, F.V., Matamala, M., Prisner, E., Rapaport, I.: AT-free graphs: linear bounds for the oriented diameter. Discrete Appl. Math. **141**(1–3), 135–148 (2004)
11. Chou, C.-H., Du, D.H.C.: Uni-directional hypercubes. In: Proceedings of Supercomputing 1990, pp. 254–263. IEEE (1990). https://doi.org/10.1109/SUPERC.1990.130028
12. Akers, S.B.: The star graph: an attractive alternative to the n-cube. In: Proceedings of International Conference on Parallel Processing (1987)
13. Fujita, S.: On oriented diameter of star graphs. In: 2013 First International Symposium on Computing And Networking (CANDAR), pp. 48–56. IEEE (2013). https://doi.org/10.1109/CANDAR.2013.16
14. Ajish Kumar, K.S., Rajendraprasad, D., Sudeep, K.S.: Oriented diameter of star graphs, arXiv preprint arXiv:1911.10340

Gathering over Meeting Nodes
in Infinite Grid

Subhash Bhagat, Abhinav Chakraborty, Bibhuti Das$^{(\boxtimes)}$,
and Krishnendu Mukhopadhyaya

ACM Unit, Indian Statistical Institute, Kolkata, India
subhash.bhagat.math@gmail.com, abhinav.chakraborty06@gmail.com,
dasbibhuti905@gmail.com, krishnendu.mukhopadhyaya@gmail.com

Abstract. The *gathering on meeting points* problem requires the robots to gather at one of the pre-defined meeting points. This paper investigates a discrete version of the problem where the robots and meeting nodes are deployed on the nodes of an anonymous infinite square grid. The robots are identical, autonomous, anonymous, and oblivious. They operate under an asynchronous scheduler. Robots do not have any agreement on a global coordinate system. Initial configurations, for which the problem is unsolvable, have been characterized. A deterministic distributed algorithm has been proposed to solve the problem, for the rest of the configurations.

Keywords: Swarm robotics · Asynchronous · Oblivious · Gathering · Infinite grid · Meeting nodes

1 Introduction

Swarm robotics in the discrete domain is an emerging field of research. A swarm of robots is a multi-robot system, consisting of small and inexpensive robots working together in a cooperative environment to achieve some goal. Robots are autonomous, anonymous, homogeneous, and oblivious. Robots are deployed on the nodes of an anonymous graph in the discrete domain. Robots do not have any explicit means of communication. They do not have any agreement on a global coordinate system.

Gathering problem by a swarm of robots is one of the essential tasks in the field of distributed computing. The *gathering problem* requires a set of n robots, initially placed at different locations to gather at some location, not known a priori, within finite time. When a robot becomes active, it operates according to Look-Compute-Move (LCM) cycle. In the *look* phase, a robot takes a snapshot of the current configuration. In the *compute* phase, it computes a destination point, which may be its current node also. It moves towards its destination point in the *move* phase. The movement of a robot is instantaneous, i.e., during the *look* phase robots are always detected on the nodes of the input graph. Cycles are performed asynchronously for each robot. We have considered infinite grid graph

© Springer Nature Switzerland AG 2020
M. Changat and S. Das (Eds.): CALDAM 2020, LNCS 12016, pp. 318–330, 2020.
https://doi.org/10.1007/978-3-030-39219-2_26

with robots placed at distinct nodes initially. In addition, we have also considered some pre-defined nodes referred to as *meeting nodes* [1]. These meeting nodes are visible to every robot. We assume that robots have *global weak* multiplicity detection capability, by which a robot can detect whether a node is occupied by more than one robots.

Gathering over meeting nodes problem in an infinite grid requires the robots to gather at one of the *meeting nodes*. This is a variant of the *gathering on meeting points* problem [1], where the robots are deployed on the Euclidean plane.

2 Earlier Works

Gathering is one of the most active research topics in the domain of the multi-robot systems. *Gathering* has been extensively studied in the continuous domain [2]. In the discrete domain, the robots are deployed on the nodes of the input graph. The problem has been largely studied on ring topologies by Klasing et al. in [3,4] and Angelo et al. in [5,6]. In [3], Klasing et al. proved that *gathering* in the ring is impossible without any multiplicity detection capability. With the weak multiplicity detection capability, they solved the problem for all configurations with an odd number of robots and all the asymmetric configurations with an even number of robots. Klasing et al. studied symmetric configurations with an even number of robots, and the problem was solved for more than eighteen robots [4]. Kamei et al. studied the problem using local multiplicity detection capability [7,8].

In [9], Angelo et al. studied *gathering* on trees and finite grids. It was shown that even with global-strong multiplicity detection capability a configuration remains ungatherable, if it is periodic or symmetric with the line of symmetry passing through the edges of the grid. The problem was solved for all other remaining configurations without assuming any multiplicity detection capability. In [10], Stefano et al. studied the basic results for the optimal *gathering* of robots in graphs. They have considered *gathering* on finite graphs with respect to both its feasibility and the minimum number of asynchronous moves performed by robots. In [11], Stefano et al. studied the *gathering* problem on infinite grids. This problem has been studied by introducing optimal algorithms in terms of the total number of moves. In [12], Bose et al. studied the *gathering* problem in hypercubes. Their proposed algorithm is optimal with respect to the total number of moves executed by the robots.

In [1], Cicerone et al. studied a variant of the *gathering* problem, where robots require to gather at one of the pre-determined points, referred to as meeting nodes. They solved the *gathering* on meeting points problem with respect to two objective functions, by minimizing the total distance travelled by all robots and by minimizing the maximum distance travelled by a single robot.

3   Our Contributions

This paper considers the *gathering on meeting points* problem on an infinite grid by asynchronous oblivious mobile robots. The infinite grid is a natural discretization of the plane. It has been shown that even with the global weak-multiplicity detection, some configurations remain ungatherable. If the configuration is symmetric w.r.t. a single line of symmetry and the line of symmetry does not contain any robot or *meeting node*, then the problem is unsolvable. The configuration is ungatherable even if the configuration admits a rotational symmetry without having any robot or *meeting nodes* on the center of rotation. For the rest of the configurations, a deterministic distributed algorithm has been proposed for the *gathering* of robots on *meeting nodes*.

4   Model and Definitions

Robots are assumed to be dimensionless, autonomous, anonymous, homogeneous, and oblivious. They do not have any explicit means of communication. They have unlimited visibility range, i.e., they can sense the entire grid. The robots do not have any agreement on a global coordinate system and chirality. Robots are either in an active state or in an inactive state. They execute Look-Compute-Move (LCM) cycle under asynchronous scheduler when they become active. Initially, robots are assumed to be on distinct nodes of the input grid, and no robots are deployed on the *meeting nodes*. Robots are equipped with global-weak multiplicity detection capability. A robot can move to one of its adjacent nodes along the grid lines. We assume that the movement is instantaneous, i.e., the robot can be seen only on nodes. We use some of the notions as defined in [11].

- **System Configuration:** Let $P = (\mathbb{Z}, E')$ denote the infinite path graph where $E' = \{i, i + 1 : i \in \mathbb{Z}\}$. Consider the cartesian product $P \times P$ as an input grid graph. Let $d(u, v)$ denote the Manhattan distance between nodes u and v. $R = \{r_1, r_2, \ldots, r_n\}$ is the set of robots deployed on the nodes of the grid. Suppose $r_i(t)$ denotes the position of the robot r_i at time t. $R(t) = \{r_1(t), r_2(t), \ldots, r_n(t)\}$ is the set of robot positions at time t. Let $M = \{m_1, m_2, \ldots, m_r\}$ be the set of *meeting nodes*. Let $C(t) = (R(t), M)$ denote the configuration at time t. In an *initial* configuration, all the robots occupy distinct nodes of the grid, and no robot is deployed on any *meeting node*. We call a configuration *final* if all the robots are on a single meeting node.
- **Symmetry:** An automorphism of a graph $G = (V, E)$ is a bijection $\phi: V \to V$ such that u and v are adjacent iff $\phi(u)$ and $\phi(v)$ are adjacent. Automorphism of graphs can be extended similarly to define automorphism of configurations.

Let $l : V \rightarrow \{0, 1, 2, 3, 4, 5\}$ be a function, where

$$
l(v) = \begin{cases}
0 & v \text{ is an empty node} \\
1 & v \text{ is a } meeting\ node \\
2 & v \text{ is a robot position on a } meeting\ node \\
3 & v \text{ is a robot position} \\
4 & v \text{ is a robot multiplicity} \\
5 & v \text{ is a robot multiplicity on a } meeting\ node
\end{cases}
$$

An automorphism of a configuration $(C(t), l)$ is an automorphism ϕ such that $l(v) = l(\phi(v))$ for all $v \in V$. The set of all automorphisms for a configuration forms a group which is denoted by $Aut(C(t), l)$. Next, we define symmetry in terms of an automorphism of a configuration. If $|Aut(C(t), l)| = 1$, then the configuration is asymmetric. Otherwise, the configuration is symmetric. We assume that the grid is embedded in the Cartesian plane. The grid can have only three types of symmetry, namely, translation, reflection, and rotation. Since the number of robots and *meeting node* is only finite in number, translational symmetry is not possible. In case the configuration has reflectional symmetry, only horizontal, vertical, or diagonal is considered. The axis of symmetry can pass through nodes or edges. In the case of rotational symmetry, the angle of rotation can be 90° or 180°. The center of rotation can be a node or center of an edge or center of a unit square [11].

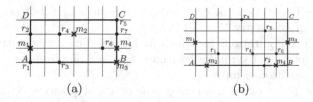

(a)   (b)

Fig. 1. In this figure, cross represents *meeting node* and black circles represent robot positions. The lexicographic smallest string in this figure is $s_{DA} = 03130000030301000000000303311$. (a) $\mathcal{I}_1$-configuration. (b) $\mathcal{I}_2$-configuration.

- **Configuration view:** Let $\mathcal{MER} = ABCD$ denote the minimum enclosing rectangle of $R \cup M$. Let the dimension of $\mathcal{MER}$ be $n \times m$ where $|AB| = n$ and $|AD| = m$. We first assume that $\mathcal{MER}$ is a rectangle with $n > m$. We associate a senary string of length mn to each corner of $\mathcal{MER}$. The two strings associated with corner A will be denoted by s_{AD} and s_{AB}. Since $\mathcal{MER}$ is a non-square rectangle, we associate the string s_{AD} or s_{AB} for corner A, in the direction of the smallest side. We have defined the string s_{AD} as follows: Starting from the corner A, we proceed in the direction parallel to AD and scan the grid from A to D and sequentially consider all grid lines parallel to AD. The direction parallel to AD is considered as the *string direction* from corner A. We associate 0 for empty nodes, 1 for *meeting nodes*, 2 for robot positions on *meeting nodes*, 3 for robot positions, 4 for a robot multiplicity node and 5 for a robot multiplicity on a *meeting node*. Similarly, we associate strings for each other corners. In case of a square grid, between the two strings associated with a corner, we select the string which is lexicographically minimum. Note that both the strings associated with a corner are equal if $\mathcal{MER}$ is symmetric with respect to the diagonal passing through that corner. In this case, we select any one of the strings. If the configuration is asymmetric, then we would always get a unique smallest lexicographic string (Fig. 1). Without loss of generality, let s_{AD} be the smallest lexicographic string. Then we refer A as the *key corner*. The configuration view of a node is defined as the tuple (d', x) where d' denotes the distance of a node from the *key corner* in its *string direction* and $x \in \{f, m, a, r, s, w\}$ whether the node is free, a *meeting node*, a robot position on a *meeting node*, a robot position, a robot multiplicity node or a robot multiplicity on a *meeting node*, respectively. We have used the idea proposed in [11].

- **Partitioning of the initial configurations:** All the configurations can be partitioned into the following disjoint classes:

  1. $\mathcal{I}_1-$ All configurations for which M is asymmetric (Fig. 1(a)).
  2. $\mathcal{I}_2-$ All configurations for which M is symmetric and $R \cup M$ is asymmetric (Fig. 1(b)).
  3. $\mathcal{I}_3-$ All configurations for which M is symmetric with a single line of symmetry, and $R \cup M$ is symmetric w.r.t. the line of symmetry for M. This can be further partitioned into:

     $\mathcal{I}_3^a-$ There exists at least one *meeting node* or robot position on the line of symmetry (Fig. 2(a) and (b)).

     $\mathcal{I}_3^b-$ There does not exist any *meeting node* or robot position on the line of symmetry (Fig. 2(c)).
  4. $\mathcal{I}_4-$: All configurations for which M has rotational symmetry. Also, $R \cup M$ has rotational symmetry without any line of symmetry. This can be further partitioned into:

     $\mathcal{I}_4^a-$ There exists a *meeting node* or robot position on the center of rotational symmetry (Fig. 3(a) and (b)).

     $\mathcal{I}_4^b-$ There does not exist any *meeting node* or robot position on the center of rotational symmetry (Fig. 3(c)).
  5. $\mathcal{I}_5-$: All configurations for which M has rotational symmetry. Also, $R \cup M$ has rotational symmetry with multiple lines of symmetry.

     $\mathcal{I}_5^a-$ There exists a *meeting node* or robot position on the center of rotational symmetry (Fig. 4(a) and (b)).

     $\mathcal{I}_5^b-$ There does not exist any *meeting nodes* or robot positions on the center of rotational symmetry (Fig. 4(c)).

- In case M admits a single line of symmetry, let L be the line of symmetry. Note that L can be a horizontal, vertical, or diagonal line of symmetry. Let c be the center of rotation, in case M admits rotational symmetry.
- **Problem Definition**: Given a configuration $C(t) = (R(t), M)$, the *gathering over meeting nodes problem* requires the robots to gather on one of the *meeting nodes* within finite time.

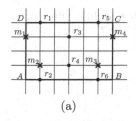 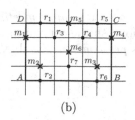 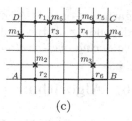

(a)	(b)	(c)

Fig. 2. (a) $\mathcal{I}_3^a$-configuration without *meeting nodes* on L. (b) $\mathcal{I}_3^a$-configuration with *meeting nodes* on L. (c) $\mathcal{I}_3^b$-configuration with no robots and *meeting nodes* on L.

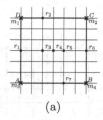 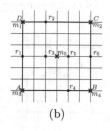 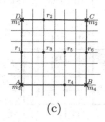

(a)	(b)	(c)

Fig. 3. (a) $\mathcal{I}_4^a$-configuration with a robot on c. (b) $\mathcal{I}_4^a$-configuration with a *meeting nodes* on c. (c) $\mathcal{I}_4^a$-configuration without robot or *meeting node* on c.

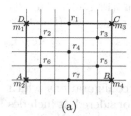 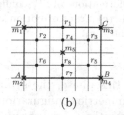 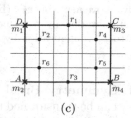

(a)	(b)	(c)

Fig. 4. (a) $\mathcal{I}_5^a$-configuration with a robot on c. (b) $\mathcal{I}_5^a$-configuration with a *meeting node* on c. (c) $\mathcal{I}_5^b$-configuration without robot or *meeting node* on c.

5   Impossibility Results

Observation 1. *If the initial configuration is either asymmetric or admits a single line of symmetry, then initially, all the robots can have an agreement on the same north-south direction.*

Theorem 1. *If $C(0) \in \mathcal{I}_3^b \cup \mathcal{I}_4^b \cup \mathcal{I}_5^b$, then the gathering over meeting nodes problem is unsolvable.*

Proof. If possible, let algorithm $\mathcal{A}$ solve the *gathering* problem. Let $\gamma(r)$ denote the orbit of r [11]. Consider the scheduler to be semi-synchronous. We assume that all the robots of $\gamma(r)$ are activated at the same time. We also assume that the distance travelled by each robot in $\gamma(r)$ is the same. Under this setup, the configuration symmetry is preserved. Since the problem requires the robots to gather at one of the *meeting nodes*, the *target meeting node* should lie on c or L, depending on the configuration. As the configuration $C(0) \in \mathcal{I}_3^b \cup \mathcal{I}_4^b \cup \mathcal{I}_5^b$, the *gathering over meeting nodes problem* is unsolvable. $\qquad \square$

6   Algorithm

In this section, we propose a deterministic distributed algorithm *Gathering()* for solving the *gathering over meeting nodes problem* in an infinite grid. Our proposed algorithm solves the problem for configurations having at least six robots. It mainly consists of the following phases:

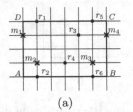 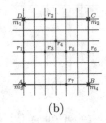 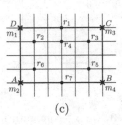

        (a)   (b)   (c)

Fig. 5. (a) Transformation of $\mathcal{I}_3^a$-configuration (Fig. 2(a)) into $\mathcal{I}_2$-configuration by movement of r_3. (b) Transformation of $\mathcal{I}_4^a$-configuration (Fig. 3(a)) into $\mathcal{I}_2$-configuration by movement of r_4. (c) Transformation of $\mathcal{I}_5^a$-configuration (Fig. 4(a)) into $\mathcal{I}_3^a$-configuration by movement of r_4.

1. **Symmetry Breaking:** In this phase, the symmetric configurations which can be transformed into asymmetric configurations are considered. It includes the configurations $C(t) \in \mathcal{I}_3^a$ without having any *meeting nodes* on the line of symmetry and $C(t) \in \mathcal{I}_4^a \cup \mathcal{I}_5^a$ without having any *meeting nodes* on the center of rotational symmetry. First, consider the case when $C(t) \in \mathcal{I}_3^a$ without having any *meeting nodes* on the line of symmetry. Let L be the line of symmetry for $R \cup M$ and r be the northernmost robot on L (Robots agree on north-south directions). Robot r is moved to one of the adjacent nodes away from L (Fig. 5(a)). Next, consider the case when the configuration $C(t) \in \mathcal{I}_4^a \cup \mathcal{I}_5^a$ without having any *meeting nodes* on the center of rotational symmetry. Let c be the center of symmetry for $R \cup M$ and r be the robot on c. Robot r is moved to one of the adjacent nodes away from c (Fig. 5(b) and (c)).

2. **Guard Selection and Placement:** Robots execute *GuardSelection()* procedure to select some robot positions as *guards*. First, we consider the case when $R \cup M$ is asymmetric (i.e., $C(t) \in \mathcal{I}_1 \cup \mathcal{I}_2$), and the *key corner* contains a robot position. Let r be the robot on the *key corner*. We move r to the adjacent node in the *string direction*. Robot r and the robots on the other three corners are selected as guards. Next, consider the case when $R \cup M$ is asymmetric, and the *key corner* does not contain a robot position (Figs. 6(a) and 7(a)). Let r be the first robot position along the *string direction* from the

key corner. Robot r is selected as one of the guards. For each of the other three corners, the first robot along the *string direction* (Note that the first robot for *key corner* is not considered) is moved to their corners, respectively. All three robots on other corners are also selected as guards. In case, a single robot is the first robot along the *string direction* for two corners, then the corner opposite to the *key corner* is given preference. If the configuration $C(t) \in \mathcal{I}_3^a$ having only *meeting nodes* on the line of symmetry (Fig. 8(a)) and $C(t) \in \mathcal{I}_4^a$ having *meeting node* on the center of rotational symmetry (Fig. 9(a)), then we will have two *key corners*. Without loss of generality, let A and B be the *key corners*. In this case, if the *key corners* contain robot positions, then the robots on the *key corners*, are moved to their adjacent nodes respectively (The adjacent node for movement is selected such that the minimum enclosing rectangle $\mathcal{MER}$ remains invariant). The first robot along the string direction from *key corner*, along with the robots on the *non-key corners*, are selected as guards. Next, consider the case when the *key corners* do not contain robot positions. Let r_A and r_B be the first robots along the *string directions* from *key corners* A and B, respectively. The first robot along the *string directions* from the *non-key corners* are moved to their respective corners. Robots r_A and r_B along with the two robots on the *non-key corners* are selected as guards. In case the configuration $C(t) \in \mathcal{I}_5^a$ having a *meeting node* on the center of rotational symmetry (Fig. 10(a)) without robot positions, then we will have four *key corners*. In this case, for each corner, the first robot along the *string directions*, is selected as a guard.

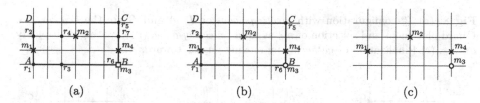

(a)   (b)   (c)

Fig. 6. (a) $\mathcal{I}_1$-configuration (Fig. 1(a)). D is the *key corner*. Guards placement and selection of m_3 as *target meeting node*. (b) *Gathering* of all robots other than guards on m_3. (c) Finalization of *gathering* by moving guards towards m_3. Square represents robot r_6 on m_3. The empty circle represents multiplicity.

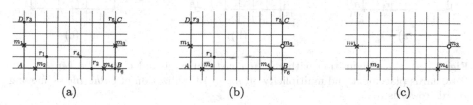

(a)   (b)   (c)

Fig. 7. (a) $\mathcal{I}_2$-configuration (Fig. 1(b)). A is the *key corner*. Guards placement and selection of m_3 as *target meeting node*. (b) Creating multiplicity on m_3. (b) Finalization of *gathering* by moving guards towards m_3.

3. **Creating Multiplicity on Target Meeting Node:** In this phase, robots move to create a unique multiplicity point at a meeting node by executing *MakeMultiplicity()*. First, consider the case when $R \cup M$ is asymmetric. Let m be the last *meeting node* while traversing $\mathcal{MER}$ from the *key corner* along the *string direction*. In case the configuration $C(t) \in \mathcal{I}_3^a$ and has *meeting nodes* on the line of symmetry, m is the northernmost *meeting node*. Let m be the *meeting node* on c, when $C(t) \in \mathcal{I}_4^a \cup \mathcal{I}_5^a$ and has a *meeting node* on the center of rotational symmetry. In each case, m is selected as the *target meeting node*. All robots other than guards move towards the *target meeting node* (Figs. 6(b), 7(b), 8(b), 9(b) and 10(b)). Since the guards do not move during this movement, the key corner remains invariant. As a result the *target meeting node* remains uniquely identifiable during this movement. Note that the robots may create multiplicity on a meeting node other than the *target meeting node* during this movement. Since the *target meeting node* is uniquely identifiable, eventually all robots other than guards create a multiplicity on it.

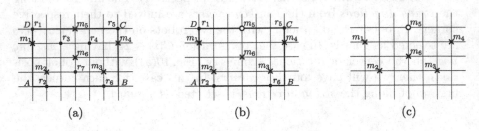

(a)   (b)   (c)

Fig. 8. (a) $\mathcal{I}_3^a$-configuration with *meeting nodes* on L. A and B are the *key corners*. Guard placement and selection of m_5 as *target meeting node*. (b) Creating multiplicity on m_5. (c) Finalization of *gathering* by moving guards towards m_5.

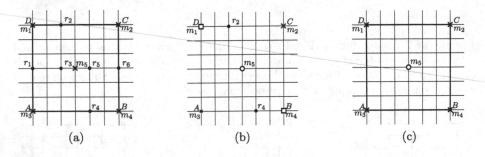

(a)   (b)   (c)

Fig. 9. (a) $\mathcal{I}_4^a$-configuration with *meeting node* on c. A and C are the *key corners*. (b) Guards placement and multiplicity on m_5. (c) Finalization of *gathering* by moving guards towards m_5.

4. **Finalization of Gathering:** In this phase, the robots on the guard positions move to the final gathering point m by executing *GuardMovement()*

procedure. The robots on the guard positions start moving only when the configuration contains at most five distinct robot positions (Figs. 6(c), 7(c), 8(c), 9(c) and 10(c)). Since the robots have multiplicity detection capability and the target gathering point m is the unique meeting point with multiple robots on it, the robots can easily indentify the final gathering point. Note that during this phase there can be at most six distinct robots which does not lie on the *target* meeting node m. Let S denote the set of robots which are not on m and not on a robot multiplicity point. If $r \in S$, it starts moving towards m. When S is empty, robots on a multiplicity point start moving towards m. During their movement, they do not create any multiplicity on a *meeting node* other than m. To ensure this, each robot selects a shortest path towards m, which does not contain *meeting node* other than m. By the choice of m there always exists one half-line from m in the grid which does not contain *meeting nodes*. Note that during this movement, robots identify m as the unique *meeting node* containing robot multiplicity. Eventually, they finalize the *gathering* on m.

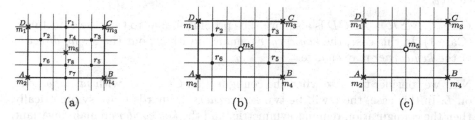

Fig. 10. (a) $\mathcal{I}_5^a$-configuration with *meeting nodes* on c. A, B, C, and D are the *key corners*. (b) Multiplicity on *meeting node* m_5. (c) Finalization of *gathering* by moving the guards towards m_5.

7   Correctness

Lemma 1. *If $C(t) \in \mathcal{I}_1 \cup \mathcal{I}_2$ or $C(t) \in \mathcal{I}_3^a$ with meeting nodes on L, then during the execution of GuardSelection(), the key corner remains invariant.*

Proof. First we consider the case when configuration $C(t) \in \mathcal{I}_1 \cup \mathcal{I}_2$. We have the following cases:

Case 1. $\mathcal{MER} = ABCD$ is not a square. Without loss of generality, let $s_{AD} = a_1 a_2 \ldots a_{mn}$ be the smallest lexicographic string. Consider the smallest string attached to the corner $\lambda \in \{B, C, D\}$ as $s_\lambda = \lambda_1 \lambda_2 \ldots \lambda_{mn}$. Since $s_{AD} < s_\lambda$, assume k to be the first position where $a_k < \lambda_k$. Let i be the first robot position in s_λ. We have the following subcases:

Subcase 1. Corner A contains a robot position i.e., $a_1 = 3$. We have $\lambda_1 = 3$ for all the other three corners. During the execution of the algorithm *GuardSelection*(), the robot on A is moved to the adjacent node along the *string direction*. Due to this movement, we have $a_1 < \lambda_1 = 3$. This implies that $s_{AD} < s_\lambda$, i.e., corner A remains the *key corner*. If the configuration is in $\mathcal{I}_3^a$ without any robot position on the line of symmetry, we have two key corners. The robots on the key corners are moved to adjacent nodes such that $\mathcal{MER}$ remains invariant. Note that in this case also we have $a_1 < \lambda_1 = 3$.

Subcase 2. Corner A does not contain robot position. Suppose the robot on ith position in s_λ is moved to jth position in s_λ during the execution of the algorithm *GuardSelection*(). Note that $j < i$, since robot moves towards the corner. Next, we consider different cases for this robot's movement. If $i < k$ then (i) if $a_j = \lambda_j = 0$ initially we have $\lambda_j = 3 > a_j$ or (ii) if $a_j = \lambda_j = 1$ initially we have $\lambda_j = 2 > a_j$. This implies that $s_{AD} < s_\lambda$. If $i > k$ and $j > k$ then A remains the *key corner* as $a_k < \lambda_k$. Otherwise, if $i > k$ and $j \leq k$ then (i) if $a_j = \lambda_j = 0$ initially, we have $\lambda_j = 3 > a_j$ or (ii) if $a_j = \lambda_j = 1$ initially, we have $\lambda_j = 2 > a_j$. This implies that $s_{AD} < s_\lambda$. Hence corner A remains the *key corner*.

Case 2. $\mathcal{MER} = ABCD$ is a square. The proof is similar to the case of rectangle (Case 1). In this case, the *key corner* remains invariant, but the *string direction* of the *key corner* may change.

Next we consider the case when the configuration $C(t) \in \mathcal{I}_3^a$ with *meeting nodes* on L. In this case, there will be two *key corners*. If guards move symmetrically, then the configuration remains symmetric, and the *key corners* remain invariant. The proof is similar to the $\mathcal{I}_1 \cup \mathcal{I}_2$ case (Case 1). Otherwise, if the configuration becomes asymmetric, then there will be a unique *key corner* similar to as in Case 1.   □

Observation 2. *If $C(t) \in \mathcal{I}_1 \cup \mathcal{I}_2$, then the target meeting node remains invariant during the execution of MakeMultiplicity*().

Observation 3. *If $C(t) \in \mathcal{I}_4^a \cup \mathcal{I}_5^a$ with a meeting node on c, then the target meeting node remains invariant during the execution of MakeMultiplicity*().

Lemma 2. *If $C(t) \in \mathcal{I}_3^a$ with meeting nodes on L, the target meeting node remains invariant during the execution of the algorithm MakeMultiplicity*().

Proof. Since the configuration is symmetric about L, there are two *key corners* which are symmetric w.r.t. L. Let m be the northernmost *meeting node* on L, which is selected as the *target meeting node*. During the execution of algorithm *MakeMultiplicity*(), even if $R \cup M$ becomes asymmetric, robots can identify L as the line of symmetry for M and all the *meeting nodes* on L. We need to show that robots can identify m as the northernmost *meeting node* among all the *meeting node* on L. Since, during the execution of *MakeMultiplicity*(), guards do not move, the *key corners* remain invariant. As the northernmost *meeting node* on

L is decided by the position of *key corners*, m remains the *target meeting node*. Hence, during the execution of the algorithm $MakeMultiplicity()$, the *target meeting node* remains invariant. □

Let m be the *target meeting node*, selected after guards placement at time t. Let $d(t) = \sum_{r_i \in R(t)} d(r_i(t), m)$.

Theorem 2. *Execution of Algorithm Gathering() eventually solves the gathering over meeting nodes problem.*

Proof. First, we consider the execution of algorithm $MakeMultiplicity()$. Let t' be an arbitrary point of time after guards selection and placement. We also assume that at t' time at least one robot which is not a guard has completed its LCM cycle. We have $d(t') = \sum_{r_i \in R(t')} d(r_i(t'), m)$. If there are at most six robots which are not on m at time t', then execution of $GuardMovement()$ is started. Otherwise, let r be the robot which has computed its LCM cycle at time t'. Since r has moved at least one node closer to m, we have $d(t') < d(t)$. This implies that eventually, all the non-guard robots will reach m and execution of $GuardMovement()$ will be started when there are at most six robots which are not on m.

Next, we consider the execution of algorithm $GuardMovement()$. In this phase each robot selects the shortest path, which does not contain any *meeting nodes* among all the shortest paths during their movement. Since in each LCM cycle, robots not on m move at least one node towards m along the selected path, within a finite amount of time, they will reach m. This implies that eventually, robots will finalize *gathering* on m.

Hence, execution of algorithm $Gathering()$ eventually solves the *gathering over meeting nodes problem* within finite time. □

8   Conclusion

In this work, we have studied the *gathering over meeting nodes problem* in infinite grids with global-weak multiplicity detection. We have proved some configurations to be ungatherable, and for the rest of the configurations, we have proposed a deterministic distributed algorithm for the *gathering* of robots over *meeting nodes*. The future interest would be to consider randomized algorithm for breaking the symmetry when there are no robots or *meeting nodes* on the line of symmetry and the center of rotation. Another direction of future interest would be to consider the optimal algorithms in terms of the number of moves for *gathering*.

References

1. Cicerone, S., Stefano, G.D., Navarra, A.: Gathering of robots on meeting-points: feasibility and optimal resolution algorithms. Distrib. Comput. **31**(1), 1–50 (2018)

2. Flocchini, P., Prencipe, G., Santoro, N. (eds.): Distributed Computing by Mobile Entities, Current Research in Moving and Computing. LNCS, vol. 11340. Springer, Cham (2019). https://doi.org/10.1007/978-3-030-11072-7
3. Klasing, R., Markou, E., Pelc, A.: Gathering asynchronous oblivious mobile robots in a ring. Theoret. Comput. Sci. **390**(1), 27–39 (2008)
4. Klasing, R., Kosowski, A., Navarra, A.: Taking advantage of symmetries: gathering of many asynchronous oblivious robots on a ring. Theoret. Comput. Sci. **411**(34–36), 3235–3246 (2010)
5. D'Angelo, G., Stefano, G.D., Navarra, A.: Gathering six oblivious robots on anonymous symmetric rings. J. Discrete Algorithms **26**, 16–27 (2014)
6. D'Angelo, G., Stefano, G.D., Navarra, A.: Gathering on rings under the look-compute-move model. Distrib. Comput. **27**(4), 255–285 (2014)
7. Izumi, T., Izumi, T., Kamei, S., Ooshita, F.: Mobile robots gathering algorithm with local weak multiplicity in rings. In: Patt-Shamir, B., Ekim, T. (eds.) SIROCCO 2010. LNCS, vol. 6058, pp. 101–113. Springer, Heidelberg (2010). https://doi.org/10.1007/978-3-642-13284-1_9
8. Kamei, S., Lamani, A., Ooshita, F., Tixeuil, S.: Gathering an even number of robots in an odd ring without global multiplicity detection. In: Rovan, B., Sassone, V., Widmayer, P. (eds.) MFCS 2012. LNCS, vol. 7464, pp. 542–553. Springer, Heidelberg (2012). https://doi.org/10.1007/978-3-642-32589-2_48
9. D'Angelo, G., Stefano, G.D., Klasing, R., Navarra, A.: Gathering of robots on anonymous grids and trees without multiplicity detection. Theoret. Comput. Sci. **610**, 158–168 (2016)
10. Stefano, G.D., Navarra, A.: Optimal gathering of oblivious robots in anonymous graphs and its application on trees and rings. Distrib. Comput. **30**(2), 75–86 (2017)
11. Stefano, G.D., Navarra, A.: Gathering of oblivious robots on infinite grids with minimum traveled distance. Inf. Comput. **254**, 377–391 (2017)
12. Bose, K., Kundu, M.K., Adhikary, R., Sau, B.: Optimal gathering by asynchronous oblivious robots in hypercubes. In: Gilbert, S., Hughes, D., Krishnamachari, B. (eds.) ALGOSENSORS 2018. LNCS, vol. 11410, pp. 102–117. Springer, Cham (2019). https://doi.org/10.1007/978-3-030-14094-6_7

0-1 Timed Matching in Bipartite Temporal Graphs

Subhrangsu Mandal$^{(\boxtimes)}$ and Arobinda Gupta

Indian Institute of Technology Kharagpur, Kharagpur 721302, West Bengal, India
{subhrangsum,agupta}@cse.iitkgp.ac.in

Abstract. Temporal graphs are introduced to model dynamic networks where the set of edges and/or nodes can change with time. In this paper, we define *0-1 timed matching* for temporal graphs, and address the problem of finding the *maximum 0-1 timed matching* for bipartite temporal graphs. We show that the problem is NP-Complete for bipartite temporal graphs even when each edge is associated with exactly one time interval. We also show that the problem is NP-Complete for rooted temporal trees even when each edge is associated with at most three time intervals. Finally, we propose an $O(n^3)$ time algorithm for the problem on a rooted temporal tree with n nodes when each edge is associated with exactly one time interval.

Keywords: 0-1 timed matching · Temporal matching · Temporal graph

1 Introduction

Temporal graphs [14] have been used to model dynamic network topologies where the edge set and/or node set vary with time. Some examples of such dynamic networks in practice include delay tolerant networks, vehicular ad-hoc networks, social networks etc. For many graph related problems, the usual definitions and algorithms for the problem on static graphs do not apply directly to dynamic graphs. Several works have addressed graph related problems such as finding paths and trees [7,13], computing dominating sets [15], traveling salesman problem [17] etc. on temporal graphs.

Finding maximum matching for a bipartite graph [12] is a well-studied problem in static graphs due to its wide application. In this paper, we investigate the problem of matching in bipartite temporal graphs. In particular, we define a type of matching called the *0-1 timed matching* for a temporal graph, and investigate the complexity and algorithms for finding *maximum 0-1 timed matching* for bipartite temporal graphs and trees. Note that an edge in a temporal graph may not exist for the lifetime of the graph, and thus can be represented by labelling it with discrete time intervals for which the edge exists. We first show that the problem of finding a maximum 0-1 timed matching in a bipartite temporal graph is NP-Complete even when each edge is labelled with exactly one interval. We

M. Changat and S. Das (Eds.): CALDAM 2020, LNCS 12016, pp. 331–346, 2020.
https://doi.org/10.1007/978-3-030-39219-2_27

next show that the problem is NP-Complete for rooted temporal trees, a special case of bipartite temporal graphs, even when each edge is labelled with at most three time intervals. Finally, we present a $O(n^3)$ time algorithm for finding a maximum 0-1 timed matching in a rooted temporal tree with n nodes when each edge is labelled with exactly one time interval.

The rest of this paper is organised as follows. Section 2 describes some related work in the area. Section 3 describes the system model. Section 4 formally defines the problem. Section 5 presents the NP-completeness results. Section 6 describes the algorithm for finding a maximum 0-1 timed matching for a rooted temporal tree. Finally Sect. 7 concludes the paper.

2   Related Work

The problem of finding a maximum matching for a static graph has been extensively studied for both general graphs and different special classes of graphs [8–10,12,16,18]. Many algorithms have also been proposed to maintain an already computed maximum matching for a graph under addition and/or deletion of edges and/or nodes [3,5,6,19]. However, none of these works address the problem of computing any type of matching for temporal graphs.

There have been a few recent works addressing the problem of matching in temporal graphs. Given a temporal graph, Michail et al. [17] consider the decision problem of finding if there exists a maximum matching M in the underlying graph (the static graph formed from the temporal graph by considering all edges that exist for at least one timestep) which can be made temporal by assigning a single distinct timestep to each edge of M such that the edge exists in the temporal graph at the timestep assigned to it. They show that the problem is NP-Hard. In [2], Baste et al. have defined another version of temporal matching called γ-matching. γ-edges are defined as edges which exist for at least γ consecutive timesteps. The maximum γ-matching is defined as a maximum cardinality subset of γ-edges such that no two γ-edges in the subset share any node at any timestep. They proved that the problem of finding the maximum γ-matching is NP-Hard when $\gamma > 1$, and propose a 2-approximation algorithm to address the problem.

In this paper we propose another version of matching on temporal graph called *0-1 timed matching*. To the best of our knowledge, there is no prior work which have addressed this problem for temporal graphs.

3   System Model

We represent a temporal graph by the *evolving graphs* [11] model. In this model, a temporal graph is represented as a finite sequence of static graphs, each static graph being an undirected graph representing the graph at a discrete timestep. The total number of timesteps is called the *lifetime* of the temporal graph. In this paper, we assume that the node set of the temporal graph remains unchanged throughout the lifetime of the temporal graph; only the edge set changes with time. All the changes in the edge set are known a priori. Also,

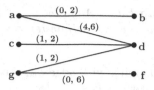

Fig. 1. A bipartite temporal graph

there are no self-loops and at most one edge exists between any two nodes at any timestep. Thus, a temporal graph is denoted by $\mathcal{G}(\mathcal{V}, \mathcal{E}) = \{G_0(\mathcal{V}, \mathcal{E}_0), G_1(\mathcal{V}, \mathcal{E}_1), \cdots, G_{\mathcal{T}-1}(\mathcal{V}, \mathcal{E}_{\mathcal{T}-1})\}$ where $\mathcal{V}$ is the node set, $\mathcal{E} = \bigcup_{i=0}^{\mathcal{T}-1} \mathcal{E}_i$ is the edge set and $\mathcal{T}$ is the lifetime of the temporal graph. Each G_i denotes the static graph at timestep i with node set $\mathcal{V}$ and set of edges $\mathcal{E}_i$ that exist at timestep i. As only the edge set changes with time, each edge in $\mathcal{E}$ of a temporal graph $\mathcal{G}$ can be equivalently represented by specifying the time intervals for which the edge exists. Thus an edge $e \in \mathcal{E}$ between nodes u and v can be represented as $e(u, v, (s_1, f_1), (s_2, f_2), \cdots, (s_k, f_k))$, where $u, v \in \mathcal{V}, u \neq v$, $f_k \leq \mathcal{T}$ and a pair (s_i, f_i) indicates that the edge exists for the time interval $[s_i, f_i)$, where $0 \leq s_i < f_i \leq \mathcal{T}$ (and hence the edge exists in all the static graphs $G_{s_i}, G_{s_i+1}, \cdots, G_{f_i-1}$). Also, if an edge e has two such pairs (s_i, f_i) and (s_j, f_j), $s_i \neq s_j$ and if $s_i < s_j$ then $f_i < s_j$. Thus the maximum number of time intervals for an edge can be $\lfloor \frac{\mathcal{T}}{2} \rfloor$. An edge at a single timestep is called an instance of that edge. For simplicity, we also denote the edge e between nodes $u, v \in \mathcal{V}$ by e_{uv} when the exact time intervals for which e exists are not important. The corresponding instance of e_{uv} at time t is represented as e_{uv}^t.

4   0-1 Timed Matching

In this section, we formally define the 0-1 timed matching problem on temporal graphs. For a given temporal graph $\mathcal{G}(\mathcal{V}, \mathcal{E})$ with lifetime $\mathcal{T}$, the underlying graph of $\mathcal{G}$ is defined as $\mathcal{G}_U(\mathcal{V}, \mathcal{E}_U)$, where $\mathcal{E}_U = \{(u, v) \mid \exists t \text{ such that } e_{uv}^t \text{ is an instance of } e_{uv} \in \mathcal{E}\}$. A temporal graph $\mathcal{G}(\mathcal{V}, \mathcal{E})$ is said to be a *bipartite temporal graph* if the underlying graph of $\mathcal{G}$ is a bipartite graph. Similarly, a temporal graph $\mathcal{G}(\mathcal{V}, \mathcal{E})$ is said to be a *temporal tree* if the underlying graph of $\mathcal{G}$ is a tree. A *rooted temporal tree* $\mathcal{G}(\mathcal{V}, \mathcal{E})$ rooted at node r is a temporal tree with one node $r \in \mathcal{V}$ chosen as root of the tree.

Note that the underlying graph of a rooted temporal tree is also a rooted tree. For any node v, let $P(v)$ denote the parent node of v and $child(v)$ denote the set of children nodes of v in the underlying graph of the rooted temporal tree. For the root node r, $P(r) = \phi$. *Depth* of a node v in a rooted temporal tree $\mathcal{G}$ rooted at r is the path length from r to v in $\mathcal{G}_U$. *Height* of a rooted temporal tree $\mathcal{G}$ is the maximum depth of any node in $\mathcal{G}$.

Definition 1. *Overlapping Edge: Any edge $e_{vw} \in \mathcal{E}$ is said to be overlapping with another edge e_{uv} if there exists a timestep t such that both e^t_{vw} and e^t_{uv} exist.*

Note that if e_{uv} is overlapping with edge e_{vw}, then e_{vw} is also overlapping with e_{uv}. We refer to such pair of edges as *edges overlapping with each other*. In Fig. 1, e_{gd} is an overlapping edge with e_{gf} because both edges are incident on g and both e^1_{gd} and e^1_{gf} exist. On the other hand, edges e_{ab} and e_{ad} are incident on the same node a, but there is no timestep t when both e^t_{ab} and e^t_{ad} exist. Thus e_{ab} and e_{ad} are *non-overlapping with each other*. For any two sets of edges E_1, E_2, if $E_1 \subseteq E_2$ and no two edges in E_1 are overlapping with each other, then E_1 is called a *non-overlapping subset* of E_2.

Definition 2. *0-1 Timed Matching: A 0-1 timed matching M for a given temporal graph $\mathcal{G}(\mathcal{V}, \mathcal{E})$ is a non-overlapping subset of $\mathcal{E}$.*

Definition 3. *Maximum 0-1 Timed Matching (MAX-0-1-TM): The maximum 0-1 timed matching for a given temporal graph is the 0-1 timed matching with the maximum cardinality.*

For the bipartite temporal graph shown in Fig. 1, the maximum 0-1 timed matching M is $\{e_{ab}, e_{ad}, e_{cd}, e_{fg}\}$. Note that the edges in a 0-1 timed matching for a given temporal graph may not be a matching for its underlying graph. In the next section, we investigate the hardness of the problem of finding the maximum 0-1 timed matching for bipartite temporal graphs.

5   Complexity of Finding Maximum 0-1 Timed Matching in Bipartite Temporal Graphs

In this section, we first show that the problem of finding a maximum 0-1 timed matching in a bipartite temporal graph even when each edge is associated with a single time interval (referred to as the MAX-0-1-TMB-1 problem) is NP-Complete. We next show that the problem is NP-Complete for rooted temporal trees even when there are at most three time intervals associated with any edge (referred to as the MAX-0-1-TMT-3 problem).

5.1   NP-Completeness of MAX-0-1-TMB-1

We define the decision version of the MAX-0-1-TMB-1 problem (referred to as D-MAX-0-1-TMB-1) as follows:

Definition 4. *D-MAX-0-1-TMB-1: Given a bipartite temporal graph $\mathcal{G}(\mathcal{V}, \mathcal{E})$ with lifetime $\mathcal{T}$ where each edge in $\mathcal{E}$ is associated with a single time interval, and a positive integer r, does there exist a 0-1 timed matching M for $\mathcal{G}$ such that $|M| = r$?*

We prove NP-Completeness of the D-MAX-0-1-TMB-1 problem by show-ing that there exists a polynomial time reduction from the decision version of the *scheduling jobs with fixed start and end times on non-identical machines* problem (referred to as the D-SJFSETNM problem) [1], which is known to be NP-Complete. The D-SJFSETNM problem is defined as follows.

Definition 5. *D-SJFSETNM: Let $J := \{J_1, J_2, \cdots, J_m\}$ be a set of m jobs such that each job $J_i \in J$ has start time s_i and end time t_i. Let $P := \{P_1, P_2, \cdots, P_k\}$ be a set of k non-identical machines, and S be a set of job-machine mappings between J and P, where each job $J_i \in J$ is mapped with one or more machines in P. In the mapping S, each machine $P_j \in P$ can be mapped with zero or more jobs in J. The D-SJFSETNM problem is to find whether there exists a subset $C \subseteq S, |C| = m$ such that all m jobs in J are executed once and no machine executes more than one job at the same time.*

Theorem 1. *The D-MAX-0-1-TMB-1 problem is NP-Complete.*

Proof. We first show that the D-MAX-0-1-TMB-1 problem is in NP. Consider a certificate $\langle \langle \mathcal{G}(\mathcal{V}, \mathcal{E}), r \rangle, M \rangle$, where $\mathcal{G}$ is a bipartite temporal graph with lifetime $\mathcal{T}$ and each edge is associated with a single time interval, r is a given integer and M is a given set of edges. We consider each edge $e_{uv} \in M$ at a time and compare associated time interval of e_{uv} with associated time intervals of all the other edges in M to find any edges overlapping with each other. This can be done in $O(|\mathcal{E}|^2)$ time. Checking $|M| = r$ and $M \subseteq \mathcal{E}$ can also be done in $O(|\mathcal{E}|)$ time. Hence the D-MAX-0-1-TMB-1 problem is in NP.

Next, we prove that there exists a polynomial time reduction from the D-SJFSETNM problem to the D-MAX-0-1-TMB-1 problem. Consider an instance $\langle J, P, S \rangle$ of the D-SJFSETNM problem where $J := \{J_1, J_2, \cdots, J_m\}$ is a set of m jobs such that each job $J_i \in J$ has start time s_i and end time t_i. $P := \{P_1, P_2, \cdots, P_k\}$, k is a set of non-identical machines. $S \subseteq J \times P$ (here $\times$ denotes the cartesian product) is a set of job-machine mappings between J and P, where each job $J_i \in J$ is mapped to one or more machines in P. Each machine $P_j \in P$ can be mapped to zero or more jobs in J. Given this instance of the D-SJFSETNM problem, we construct an instance of the D-MAX-0-1-TMB-1 problem as follows. We define the bipartite temporal graph $\mathcal{G}(\mathcal{V}, \mathcal{E})$ with the node set $\mathcal{V} := A \cup B$, where $A := \{J_i \mid J_i \in J\}$ and $B := \{P_i \mid P_i \in P\}$, and the edge set $\mathcal{E} := \{e(J_i, P_j, (s_i, t_i)) \mid (J_i, P_j) \in S\}$. The lifetime $\mathcal{T}$ is set to t_{max}, where $t_{max} := max\{t_i \mid 1 \le i \le m\}$. The positive integer r is set to m. Hence the bipartite temporal graph $\mathcal{G}(\mathcal{V}, \mathcal{E})$ has $m + k$ nodes. Figure 2 shows an example of the construction of the bipartite temporal graph.

We prove that if there is a solution for the instance of the D-MAX-0-1-TMB-1 problem for $\mathcal{G}$, there is a solution for the instance of the D-SJFSETNM problem. For a 0-1 timed matching $M, |M| = r$, for $\mathcal{G}$, we construct a job-machine mapping $C \subseteq S$ as follows. Consider the job-machine mapping $C = \{(J_i, P_j) \mid e_{J_i P_j} \in M\}$. As $|M| = r$ and $r = m$, $|C| = m$. We prove that C is a job-machine mapping such that each job in J gets executed once by following C and no machine in P executes two jobs at the same timestep. At first, we prove that following C, all

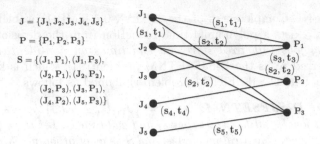

Fig. 2. Construction of a bipartite temporal graph from an instance of SJFETNM

the jobs in J gets executed once. Assume that, this is not true and there is one job $J_k \in J$ which is not executed. As $|C| = m$, then by pigeon hole principle, there is a job $J_i \in J$ which gets executed at least twice. This implies that there are two edges in M, incident on the same node J_i. According to the construction, each edge incident on J_i in $\mathcal{G}$ exists at the same time interval (s_i, t_i). Thus M is not a 0-1 timed matching for $\mathcal{G}$, which is a contradiction. Next, we show that no machine in P executes two different jobs at the same timestep. Let there be a timestep t_k and a machine $P_j \in P$ such that P_j executes two different jobs $J_h, J_i \in J$ at t_k. This implies that there exists two edges $e_{J_h P_j}, e_{J_i P_j}$ in M, which are overlapping with each other. This also contradicts the fact that M is a 0-1 timed matching for $\mathcal{G}$.

Next, we show that, if there is no solution for the instance of the D-MAX-0-1-TMB-1 problem, then there is no solution for the instance of the D-SJFSETNM problem. To show this, we prove that if there is a solution for the instance of the D-SJFSETNM problem, then there is a solution for the instance of the D-MAX-0-1-TMB-1 problem. For a solution C, $|C| = m$ for the instance of the D-SJFSETNM problem, we construct a 0-1 timed matching M for $\mathcal{G}$ as follows. Consider the set $M = \{e_{J_i P_j} \mid (J_i, P_j) \in C\}$. As $|C| = m$ and $m = r$, hence $|M| = r$. Next, we prove that M is a 0-1 timed matching for $\mathcal{G}$. Assume that, M is not a 0-1 timed matching. This implies that, there are at least two edges in M, which are overlapping with each other. Following C each job is executed only once. Thus, only one edge incident on each node in A is selected in M. Thus, those two overlapping edges $e_{J_h P_j}, e_{J_i P_j}$ are incident on the same node $P_j \in B$. This implies that there exists a machine P_j which executes two or more jobs at the same timestep, which is a contradiction.

Hence, the D-MAX-0-1-TMB-1 problem is NP-Complete.   □

5.2   NP-Completeness of MAX-0-1-TMT-3

We define the decision version of MAX-0-1-TMT-3 (referred to as the D-MAX-0-1-TMT-3 problem) as follows.

Definition 6. *D-MAX-0-1-TMT-3: Given a rooted temporal tree $\mathcal{G}(\mathcal{V}, \mathcal{E})$ with lifetime $\mathcal{T}$, where each edge in $\mathcal{E}$ is associated with at most 3 time intervals* *and a positive integer p, does there exist a 0-1 timed matching M for $\mathcal{G}$ such that $|M| \geq p$?*

We show that there is a polynomial time reduction from the decision version of the problem of finding the *maximum independent set when the degree of each node of the input graph is bounded by 3* (referred to as the D-MAX-IS-3 problem) [4] to the D-MAX-0-1-TMT-3 problem. The D-MAX-IS-3 problem is known to be NP-Complete [4]. In this subsection, we refer to a graph G such that the degree of each node of G is bounded by 3 as *BDG-3*. The D-MAX-IS-3 problem is defined as follows.

Definition 7. *D-MAX-IS-3:* *Given a BDG-3 $G(V, E)$ and a positive integer k, does there exist a set $I \subseteq V$ such that no two nodes in I are connected by an edge in E and $|I| \geq k$?*

Theorem 2. *The D-MAX-0-1-TMT-3 problem is NP-Complete.*

Proof. We first show that the problem is in NP. Consider a certificate $\langle\langle\mathcal{G}(\mathcal{V}, \mathcal{E}), p\rangle, M\rangle$, where $\mathcal{G}$ is a rooted temporal tree with lifetime $\mathcal{T}$ with each edge associated with at most 3 time intervals, p is a given integer and M is a given set of edges. We consider each edge $e_{uv} \in M$ at a time and compare associated time intervals of e_{uv} with associated time intervals of all the other edges in M to find any edges overlapping with each other. This can be done in $O(|\mathcal{E}|^2)$ time. Checking $|M| \geq p$ and $M \subseteq \mathcal{E}$ can be done in $O(|\mathcal{E}|)$ time. Hence the D-MAX-0-1-TMT-3 problem is in NP.

Next, we prove that there exists a polynomial time reduction from the D-MAX-IS-3 problem to the D-MAX-0-1-TMT-3 problem. Consider an instance $\langle G(V, E), k\rangle$ of the D-MAX-IS-3 problem where $G(V, E)$ is a BDG-3 with $|V| = n$, and $|E| = m$, and k is a positive integer. For our reduction, we assume that each edge in E is labelled with a distinct integer from 0 to $m-1$. We also assume that the set $V_0 \subseteq V$ of nodes with degree 0 is given, and $|V_0| = n_0$. If V_0 is not given, it can be easily computed in polynomial time.

From the given instance of the D-MAX-IS-3 problem, we construct an instance of the MAX-0-1-TMT-3 problem as follows. We construct the temporal graph $\mathcal{G}(\mathcal{V}, \mathcal{E})$ with node set $\mathcal{V} := (V \setminus V_0) \cup \{r\}$, where $r \notin V$ and $\mathcal{E} = \{e(v, r, (a_v^0, a_v^0 + 1), \cdots, (a_v^l, a_v^l + 1)) \,|\, v \in (V \setminus V_0),\, a_v^i$ is the label of the ith edge incident on v, $l \leq 2\}$. The lifetime $\mathcal{T}$ is set to m and p is set to $k - n_0$. Thus, the temporal graph $\mathcal{G}$ has $n - n_0 + 1$ nodes and $n - n_0$ edges. As the degree of each node in V is bounded by 3, the number of intervals associated with each $e_{vr} \in \mathcal{E}$ is at most 3. As the degree of each node in $V \setminus V_0$ is at least 1, there is an edge between each node in $\mathcal{V} \setminus \{r\}$ and r. Thus $\mathcal{G}$ is a temporal tree rooted at r. Figure 3 shows the construction of $\mathcal{G}$ from a given BDG-3.

We first show that, if there is a solution for the instance of the D-MAX-0-1-TMT-3 problem, then there is a solution for the instance of the D-MAX-IS-3 problem. For a solution M, $|M| \geq p$, for the D-MAX-0-1-TMT-3 problem on $\mathcal{G}$, we construct a solution I, $|I| \geq k$ for the D-MAX-IS-3 problem on G as follows. Consider the set of nodes $I = \{v \,|\, e_{rv} \in M\} \cup V_0$. We show that I is

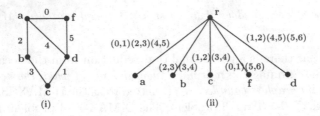

Fig. 3. (i) A BDG-3 with edges labelled by integers, (ii) Corresponding temporal tree

an independent set for G and $|I| \geq k$. As $|M| \geq p$ and $|V_0| = n_0$, then $|I| \geq k$. Assume that I is not an independent set for G. This implies that there are at least two nodes $u, v \in I$ such that $(u, v) \in E$. As $u, v \in I$ and degree of each node in V_0 is 0, then e_{ur} and e_{vr} both are in M. Each edge in E is labelled with an integer. Let a_v^i be the integer associated with the edge (u, v). Thus, the time interval $(a_v^i, a_v^i + 1)$ is associated with both e_{ur} and e_{vr}, and both are incident on r. This implies that M is not a 0-1 timed matching for G. This is a contradiction.

We next show that, if there is no solution for the instance of the D-MAX-0-1-TMT-3 problem, then there is no solution for the instance of the D-MAX-IS-3 problem. To show this, we prove that if we have a solution for the instance of the D-MAX-IS-3 problem, then we have a solution for the instance of the D-MAX-0-1-TMT-3 problem. For a solution I, $|I| \geq k$, for the instance of the D-MAX-IS-3 problem on G, we construct a solution M for the instance of the D-MAX-0-1-TMT-3 on G as follows. Consider the set $M = \{e_{vr} \mid v \in (I \setminus V_0)\}$. We prove that M is a 0-1 timed matching for G and $|M| \geq p$. As $|I| \geq k$, $|V_0| = n_0$ and for all nodes in $I \setminus V_0$ we add an edge to M, $|M| \geq p$. Assume that M is not a 0-1 timed matching for G. This implies that there are at least two edges e_{ur} and e_{vr} such that both are incident on r and there is a time interval $(a_v^i, a_v^i + 1)$ when both of them exist. This implies that there is an edge $(u, v) \in E$ which is labelled with a_v^i and $u, v \in I$. This implies that I is not an independent set for G. This is a contradiction.

Hence, the D-MAX-0-1-TMT-3 problem is NP-Complete.   □

6   Finding Maximum 0-1 Timed Matching for Rooted Temporal Tree with Single Time Interval Per Edge

In this section, we present a dynamic programming based algorithm for finding the maximum 0-1 timed matching for a rooted temporal tree $G(V, \mathcal{E})$ with root $r \in V$ where each edge of G is associated with a single time interval. In the rest of this paper, T_v denotes the temporal subtree rooted at a node $v \in V$ and M_v denotes the maximum 0-1 timed matching for T_v.

The algorithm orders the nodes in non-increasing order of depths, and then computes the maximum 0-1 timed matching for the subtrees rooted at each node in this order. For any leaf node u_i, T_{u_i} has no edges, and hence $M_{u_i} = \emptyset$. To compute M_v for T_v where v is a non-leaf node, two cases are possible: (i) no

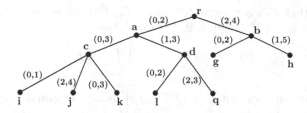

Fig. 4. A temporal tree rooted at r

edge incident on v in T_v is included in M_v, or (ii) one or more edges incident on v in T_v are included in M_v. Let $TM1[v]$ and $TM2[v]$ denote the maximum 0-1 timed matching for T_v which does not include any edge incident on v (Case (i)), and which includes at least one edge incident on v (Case (ii)) respectively. Note that for a leaf node, $TM1[v] = \emptyset$ and $TM2[v] = \emptyset$. Then, it is easy to see that

$$M_v := cardMax(TM1[v], TM2[v]) \tag{1}$$

where the function $cardMax(X, Y)$ returns the maximum cardinality set among two sets X and Y.

We first describe the method for computing $TM1[v]$ for T_v when for all $u_i \in child(v)$, M_{u_i} for T_{u_i} are already computed. As $TM1[v]$ does not include any edge e_{vu_i} for any $u_i \in child(v)$,

$$TM1[v] := \bigcup_{\forall u_i \in child(v)} M_{u_i} \tag{2}$$

For example, in Fig. 4, while computing M_a, $M_c = \{e_{ci}, e_{cj}\}$ and $M_d = \{e_{dl}, e_{dq}\}$ where c and d are two child nodes of a. Thus by Eq. 2, $TM1[a] = \{e_{ci}, e_{cj}, e_{dl}, e_{dq}\}$.

Next, we describe the method to compute $TM2[v]$ for T_v when for all $u_i \in child(v)$, M_{u_i} for T_{u_i} are already computed. We first define the following sets.

Definition 8. *Maximum Allowable Set for* T_v*: The maximum allowable set for* T_v*, denoted by* A_v*, is a maximum cardinality set of edges incident on* v*, such that* $A_v \cup (\bigcup_{\forall u_i \in child(v)} M_{u_i})$ *is a 0-1 timed matching for* T_v*.*

Note that there can be more than one possible maximum allowable sets for T_v. We give preference to a particular type of these sets, as defined below.

Definition 9. *Maximum Feasible Set for* T_v*: The maximum feasible set for* T_v*, denoted by* F_v*, is a maximum allowable set such that* $F_v \cup \{e_{vP(v)}\}$ *is a 0-1 timed matching for* $T_{P(v)}$*. If there is no such maximum allowable set (i.e, for any maximum allowable set* A_v*,* $A_v \cup \{e_{vP(v)}\}$ *is not a 0-1 timed matching for* $T_{P(v)}$*), then* F_v *is set to an arbitrary maximum allowable set.*

Assuming that F_v's are computed for all T_v, $TM2[v]$ is computed as follows. If $F_v = \emptyset$, then we set $TM2[v] := \emptyset$. If $F_v \neq \emptyset$, then

$$TM2[v] := F_v \cup (\bigcup_{\forall u_i \in child(v)} M_{u_i}) \tag{3}$$

We illustrate the computation of $TM2[v]$ using the graph shown in Fig. 4. While computing M_a, $M_c = \{e_{ci}, e_{cj}\}$ and $M_d = \{e_{dl}, e_{dq}\}$. Note that if edge e_{ac} is included in a 0-1 timed matching for T_a, we need to remove both e_{ci} and e_{cj} to maintain the properties of 0-1 timed matching. Similarly, e_{ad} cannot be included along with e_{dl} and e_{dq} in any 0-1 timed matching for T_a. Thus for T_a, both A_a and F_a are $\emptyset$. Again while computing M_b, $\{e_{bg}\}$ and $\{e_{bh}\}$ both are A_b. But only $\{e_{bg}\} \cup \{e_{rb}\}$ is a 0-1 timed matching for T_r. Thus $F_b = \{e_{bg}\}$. It can be observed that $M_a \cup \{e_{ra}\}$ and $M_b \cup \{e_{rb}\}$ are both 0-1 timed matching for T_r, and e_{ra}, e_{rb} are non-overlapping with each other and $P(r) = \phi$. Thus, $F_r = \{e_{ra}, e_{rb}\}$. As $F_a = \emptyset$ for T_a, hence $TM2[a] = \emptyset$. Again, as $F_r = \{e_{ra}, e_{rb}\}$ for T_r, then $TM2[r] = \{e_{ra}, e_{rb}, e_{ci}, e_{cj}, e_{dl}, e_{dq}, e_{bg}\}$.

In Fig. 4, as illustrated above, $TM1[a] = \{e_{ci}, e_{cj}, e_{dl}, e_{dq}\}$ and $TM2[a] = \emptyset$, hence $M_a = \{e_{ci}, e_{cj}, e_{dl}, e_{dq}\}$. Similarly, $TM1[r] = \{e_{ci}, e_{cj}, e_{dl}, e_{dq}, e_{bg}\}$ and $TM2[r] = \{e_{ra}, e_{rb}, e_{ci}, e_{cj}, e_{dl}, e_{dq}, e_{bg}\}$. Hence $M_r = \{e_{ra}, e_{rb}, e_{ci}, e_{cj}, e_{dl}, e_{dq}, e_{bg}\}$.

Algorithm 1 describes the pseudocode of the proposed algorithm. The algorithm calls a function $createLevelList$ to put all the nodes in $\mathcal{G}$ in different lists, with two nodes put in the same list if their depth in $\mathcal{G}$ are the same. After that, for all leaf nodes u_i, it assigns $TM1[u_i] := \emptyset$ and $TM2[u_i] := \emptyset$. Then it processes rooted temporal subtrees according to the non-increasing order of the depths of their root node in $\mathcal{G}$, starting from the temporal subtree rooted at the node with the maximum depth. For each non-leaf node v, the algorithm computes $TM1[v]$, $TM2[v]$ and M_v following Eqs. 2, 3 and 1 respectively. F_v is computed using the function $computeFeasibleSet$.

We next describe how to compute F_v for all T_v using the function $compute$-$FeasibleSet$ when, for all $u_i \in child(v)$, M_{u_i} is already computed. For computing F_v, we first define the following.

Definition 10. Minimum Removal Set for e_{vu_i} on M_{u_i}: *The minimum removal set for e_{vu_i} on M_{u_i}, denoted as $R_{M_{u_i}}(e_{vu_i})$, is a minimum cardinality set of edges that needs to be removed from M_{u_i} such that $\{e_{vu_i}\} \cup (M_{u_i} \setminus R_{M_{u_i}}(e_{vu_i}))$ is a 0-1 timed matching for T_v, where $u_i \in child(v)$.*

In Fig. 4, for e_{ac} on T_a, $R_{M_c}(e_{ac}) = \{e_{ci}, e_{cj}\}$. Here $M_c = \{e_{ci}, e_{cj}\}$ and we need to remove $\{e_{ci}, e_{cj}\}$ from M_c such that $\{e_{ac}\} \cup (M_c \setminus R_{M_c}(e_{ac}))$ is a 0-1 timed matching for T_a.

To compute F_v, we first compute $R_{M_{u_i}}(e_{vu_i})$ for each edge e_{vu_i} incident on v. Let $\tilde{\mathcal{R}}_0(v)$ be the set of edges incident on v in T_v such that for any edge $e_{vu_i} \in \tilde{\mathcal{R}}_0(v)$, $R_{M_{u_i}}(e_{vu_i}) = \emptyset$. Let $\tilde{\mathcal{R}}_0^P(v) = \tilde{\mathcal{R}}_0(v) \cup \{e_{vP(v)}\}$. Note that for the root r, $\tilde{\mathcal{R}}_0(r) = \tilde{\mathcal{R}}_0^P(r)$ as it has no parent. Then the algorithm computes $MaxNOS(\tilde{\mathcal{R}}_0(v))$ and $MaxNOS(\tilde{\mathcal{R}}_0^P(v))$, where $MaxNOS(S)$

Algorithm 1. dp0-1TimedMatching

Input: $\mathcal{G}(\mathcal{V}, \mathcal{E})$, root node r
Output: $M \subseteq \mathcal{E}$, the maximum 0-1 timed matching
1: **if** $r = NULL$ **or** $child(r) = \emptyset$ **then**
2:     $M := \emptyset$; $return(M)$
3: **for all** leaf nodes u_i **do**
4:     $TM1[u_i] := \emptyset$; $TM2[u_i] := \emptyset$
5: $nList :=$ createLevelList(r)
6: **for** $level = max\_depth \rightarrow 0$ **do** $\triangleright max\_depth =$ maximum depth of a node
7:     **while** $(v := nList[level].extractNode())! = \emptyset$ **do**
8:         $TM1[v] := \bigcup_{u_i \in child(v)} M_{u_i}$; $TM2[v] := \emptyset$
9:         $F_v :=$ computeFeasibleSet$(v, child(v), \forall u_i \in child(v) M_{u_i})$
10:         **if** $F_v \neq \emptyset$ **then**
11:             $TM2[v] := (\bigcup_{u_i \in child(v)} M_{u_i}) \cup F_v$
12:         $M_v = cardMax(TM1[v], TM2[v])$
13: $return(M_r)$

---

denotes the maximum cardinality non-overlapping subset of any set of temporal edges S. If $|MaxNOS(\tilde{\mathcal{R}}_0^P(v))| > |MaxNOS(\tilde{\mathcal{R}}_0(v))|$ then F_v is set to $MaxNOS(\tilde{\mathcal{R}}_0^P(v)) \setminus \{e_{vP(v)}\}$, else F_v is set to $MaxNOS(\tilde{\mathcal{R}}_0(v))$ as F_v.

As an illustration, in Fig. 4, while computing F_b, both M_g and M_h are $\emptyset$. As both $R_{M_g}(e_{bg})$ and $R_{M_h}(e_{bh})$ are $\emptyset$, $\tilde{\mathcal{R}}_0(b) = \{e_{bg}, e_{bh}\}$ and $\tilde{\mathcal{R}}_0^P(b) = \{e_{bg}, e_{bh}, e_{rb}\}$. Therefore $MaxNOS(\tilde{\mathcal{R}}_0(b))$ can be both $\{e_{bh}\}$ and $\{e_{bg}\}$, but $MaxNOS(\tilde{\mathcal{R}}_0^P(b)) = \{e_{bg}, e_{rb}\}$. Since $|MaxNOS(\tilde{\mathcal{R}}_0^P(b))| > |MaxNOS(\tilde{\mathcal{R}}_0(b))|$, therefore $F_b = \{e_{bg}\}$. Similarly, while computing F_c, $\tilde{\mathcal{R}}_0(c) = \{e_{ci}, e_{cj}, e_{ck}\}$ and $\tilde{\mathcal{R}}_0^P(c) = \{e_{ac}, e_{ci}, e_{cj}, e_{ck}\}$. In this case $MaxNOS(\tilde{\mathcal{R}}_0(c))$ and $MaxNOS(\tilde{\mathcal{R}}_0^P(c))$ are both equal to $\{e_{ci}, e_{cj}\}$. Thus, $F_c = \{e_{ci}, e_{cj}\}$.

Algorithm 2 describes the details of computing F_v. Algorithm 2 internally invokes the functions $maxNonOverlap(S)$ which returns the maximum cardinality non-overlapping subset of edges from a set of temporal edges S and $interSect(e_{vu_i}, S)$ which returns the number of edges overlapping with e_{vu_i} in S. Both of these functions are easy to implement in polynomial time.

6.1   Proof of Correctness

Lemma 1. *Suppose that for each $u_i \in child(v)$, M_{u_i} is already computed and there is no 0-1 timed matching M'_{u_i} for T_{u_i} such that $|M'_{u_i}| = |M_{u_i}|$ and $|R_{M'_{u_i}}(e_{vu_i})| = 0$ but $|R_{M_{u_i}}(e_{vu_i})| > 0$. Then for T_v, if $F_v = \emptyset$, then $|TM1[v]| \geq |\tilde{M}_v|$ where $\tilde{M}_v$ is the maximum 0-1 timed matching for T_v which includes at least one edge incident on v.*

Proof. We prove this lemma by contradiction. Assume that for each $u_i \in child(v)$, M_{u_i} is already computed and there is no 0-1 timed matching M'_{u_i}

Algorithm 2. computeFeasibleSet(v, $child(v)$, $\forall u_i \in child(v)\ M_{u_i}$)

Input: v, $child(v)$, $\forall u_i \in child(v)\ M_{u_i}$
Output: F_v

1: **for all** $u_i \in child(v)$ **do**
2:     $R_{M_{u_i}}(e_{vu_i}) := interSect(e_{vu_i}, M_{u_i})$

3: **for all** $u_i \in child(v)$ **do**
4:     **if** $|R_{M_{u_i}}(e_{vu_i})| = 0$ **then**
5:         $\tilde{\mathcal{R}}_0(v) := \tilde{\mathcal{R}}_0(v) \cup \{e_{vu_i}\}$

6: **if** $\tilde{\mathcal{R}}_0(v) \neq \emptyset$ **then**
7:     **if** $P(v)! = \phi$ **then**
8:         $\tilde{\mathcal{R}}_0^P(v) = \tilde{\mathcal{R}}_0(v) \cup \{e_{vP(v)}\}$
9:         **if** $|maxNonOverlap(\tilde{\mathcal{R}}_0(v))| < |maxNonOverlap(\tilde{\mathcal{R}}_0^P(v))|$ **then**
10:            $F_v := maxNonOverlap(\tilde{\mathcal{R}}_0^P(v)) \setminus \{e_{vP(v)}\}$
11:        **else**
12:            $F_v := maxNonOverlap(\tilde{\mathcal{R}}_0(v))$
13:    **else**
14:        $F_v := maxNonOverlap(\tilde{\mathcal{R}}_0(v))$
15: **else**
16:    $F_v := \emptyset$
17: $return(F_v)$

---

for T_{u_i} such that $|M'_{u_i}| = |M_{u_i}|$ and $|R_{M'_{u_i}}(e_{vu_i})| = 0$ but $|R_{M_{u_i}}(e_{vu_i})| > 0$. In such condition, we assume that there exists a 0-1 timed matching $\tilde{M}_v$ for T_v such that $\tilde{M}_v$ includes at least one edge incident on v and $|\tilde{M}_v| > |TM1[v]|$, when $F_v = \emptyset$. As $F_v = \emptyset$, then for all edges e_{vu_i} incident on v, $|R_{M_{u_i}}(e_{vu_i})| > 0$. As $TM1[v] = \bigcup_{\forall u_i \in child(v)} M_{u_i}$, $|\tilde{M}_v| > |TM1[v]|$ is possible if any of the following conditions hold.

- M_{u_i} is not the maximum 0-1 timed matching for T_{u_i}, but this contradicts our assumption about each already computed M_{u_i}.
- There is at least one edge e_{vu_i} for which $|\{e_{vu_i}\} \cup (M_{u_i} \setminus R_{M_{u_i}}(e_{vu_i}))| > |M_{u_i}|$ when $|R_{M_{u_i}}(e_{vu_i})| > 0$. This is impossible.
- There exists another 0-1 timed matching M'_{u_i} for T_{u_i} such that $|M'_{u_i}| = |M_{u_i}|$ and $|R_{M'_{u_i}}(e_{vu_i})| = 0$ but $|R_{M_{u_i}}(e_{vu_i})| > 0$. According to our assumption about each already computed M_{u_i}, this is also impossible. $\square$

Lemma 2. *Suppose that for each* $u_i \in child(v)$, M_{u_i} *is already computed and there is no 0-1 timed matching* M'_{u_i} *for* T_{u_i} *such that* $|M'_{u_i}| = |M_{u_i}|$ *and* $|R_{M'_{u_i}}(e_{vu_i})| = 0$ *but* $|R_{M_{u_i}}(e_{vu_i})| > 0$. *Then if* $F_v \neq \emptyset$, *Eq. 3 correctly computes* $TM2[v]$ *for* T_v.

Proof. We prove this lemma by contradiction. Assume that for each $u_i \in child(v)$, M_{u_i} is already computed and there is no 0-1 timed matching M'_{u_i} for T_{u_i} such that $|M'_{u_i}| = |M_{u_i}|$ and $|R_{M'_{u_i}}(e_{vu_i})| = 0$ but $|R_{M_{u_i}}(e_{vu_i})| > 0$. In such condition, we assume that there exists another 0-1 timed matching M_v^*

which includes edges incident on v, such that $|TM2[v]| < |M_v^*|$ when $F_v \neq \emptyset$. According to Eq. 3, $TM2[v]$ includes all edges in M_{u_i} where $u_i \in child(v)$. As each M_{u_i} is the maximum 0-1 timed matching for T_{u_i}, the cardinality of the edges in M_v^* which are not incident on v are smaller than or equal to such edges in $TM2[v]$. Thus $|TM2[v]| < |M_v^*|$ is possible in three cases:

- There exists another set of edges F_v^* incident on v, such that $|F_v^*| > |F_v|$ and $F_v^* \cup (\bigcup_{\forall u_i \in child(v)} M_{u_i})$ is also a 0-1 timed matching for T_v. But this contradicts the definition of F_v.
- There exists at least one edge $e_{vu_i} \in S_v$ ($S_v \subset M_v^*$ is the set of edges incident on v) for which $|R_{M_{u_i}}(e_{vu_i})| > 0$, but $|(M_{u_i} \setminus R_{M_{u_i}}(e_{vu_i})) \cup \{e_{vu_i}\}| > |M_{u_i}|$. This is impossible.
- There is at least one child u_i of v such that there exists another 0-1 timed matching M'_{u_i} for u_i such that $|M_{u_i}| = |M'_{u_i}|$, and $|R_{M'_{u_i}}(e_{vu_i})| = 0$ but $|R_{M_{u_i}}(e_{vu_i})| > 0$. This contradicts our assumption about computed M_{u_i}. $\square$

Lemma 3. *Algorithm 2 correctly computes F_v for T_v.*

Proof. Algorithm 1 invokes Algorithm 2 for computing F_v. Algorithm 1 processes each rooted temporal subtree according to the non-increasing order of the depth of its root node in $\mathcal{G}$, starting from a temporal subtree rooted at a node with the maximum depth (max_depth). Here max_depth is the maximum depth of a node in $\mathcal{G}$. We use induction on the height of the rooted temporal tree for which F_v is computed.

Base Case: *Height of the rooted temporal subtree is 0*: For the temporal subtrees with height 0 rooted on any leaf node x, the computed value F_x is $\emptyset^1$. As there are no edges in T_x, the computed F_x is correct and it satisfies the condition that $F_x \cup \{e_{xP(x)}\}$ is a 0-1 timed matching for $T_{P(x)}$.

Inductive Step: Let this lemma hold for the rooted temporal subtrees with height up to l. We have to show that this lemma holds for rooted temporal subtrees with height $l + 1$ also. Before processing a temporal subtree T_v rooted at v where the depth of v in $\mathcal{G}$ is $max\_depth - (l+1)$, M_{w_i} and F_{w_i} for every T_{w_i}, where the depth of w_i is greater than $max\_depth - (l + 1)$ in $\mathcal{G}$, are computed by Algorithms 1 and 2. Note that, the height of each T_{w_i} is at most l. As the height of each T_{u_i}, where $u_i \in child(v)$, is at most l, already computed F_{u_i} using Algorithm 2 for each T_{u_i} is correct.

Let the computed F_v for T_v by Algorithm 2 be incorrect. Let there exist another set of edges F'_v incident on v, such that $|F'_v| > |F_v|$ and $(\bigcup_{u_i \in child(v)} M_{u_i}) \cup F'_v$ is also a 0-1 timed matching for T_v. This is possible if at least one of the following cases is true.

1. The function $maxNonOverlap$, returns a non-maximum set of non-overlapping edges. But this is not true.

---

[1] This step is not shown explicitly in Algorithm 1.

2. Algorithm 2 only considers the edges e_{vu_i} for which $|R_{M_{u_i}}(e_{vu_i})| = 0$. Thus F'_v can exist if it includes some edge e_{vu_i} incident on v for which there exists another 0-1 timed matching M'_{u_i} such that $|M'_{u_i}| = |M_{u_i}|$ and $|R_{M_{u_i}}(e_{vu_i})| > 0$ but $|R_{M'_{u_i}}(e_{vu_i})| = 0$. As only the edges in F_{u_i} can be overlapping with edge e_{vu_i}, existence of such edge contradicts the fact that already computed F_{u_i} for T_{u_i} is correct.

Lines 7 to 13 of Algorithm 2 ensures that when $P(v) \neq \phi$, if possible, the algorithm returns F_v such that $F_v \cup \{e_{vP(v)}\}$ is a 0-1 timed matching for $T_{P(v)}$. Thus, F_v is correctly computed for T_v with height $l+1$. This proves the inductive step. $\qquad\square$

Theorem 3. *Algorithm 1 correctly computes the maximum 0-1 timed matching for a given rooted temporal tree $\mathcal{G}(\mathcal{V}, \mathcal{E})$.*

Proof. We prove this theorem by induction on the height of the rooted temporal subtree for which we are computing the maximum 0-1 timed matching. We prove that at each step the algorithm correctly computes M_v and if $P(v) \neq \phi$, then there is no other 0-1 timed matching M'_v for T_v, such that $|M'_v| = |M_v|$ and $|R_{M'_v}(e_{vP(v)}) = 0|$ but $|R_{M_v}(e_{vP(v)})| > 0$.

Base Case: *Height of the rooted temporal subtree is* 0: Algorithm 1, in Line 4, assigns $TM1$ and $TM2$ values for any temporal subtrees with height 0 rooted at any leaf node x to $\emptyset$. As T_x has no edges, computed M_x is correct and $|R_{M_x}(e_{xP(x)})| = 0$.

Inductive Step: Let this theorem hold for the rooted temporal subtrees with height up to l. We need to prove that it also holds for the rooted temporal subtrees with height $l + 1$. Algorithm 1 processes each rooted temporal subtree according to the non-increasing order of the depth of its root node in $\mathcal{G}$, starting from a temporal subtree rooted at a node with the maximum depth (max_depth). Thus, while processing T_v where the depth of v in $\mathcal{G}$ is $max\_depth - (l+1)$, each M_{w_i} is correctly computed for each T_{w_i} with height at most l, where the depth of w_i in $\mathcal{G}$ is at least $max\_depth - l$. It can be noted that, for each T_{u_i}, the depth of $u_i \in child(v)$ in $\mathcal{G}$ is $max\_depth - l$ and the height of each T_{u_i} is at most l. Hence each M_{u_i} is correct and there is no other 0-1 timed matching M'_{u_i} for T_{u_i} such that $|M'_{u_i}| = |M_{u_i}|$ and $|R_{M'_{u_i}}(e_{vu_i})| = 0$ but $|R_{M_{u_i}}(e_{vu_i})| > 0$.

In Line 8, Algorithm 1 computes $TM1[v]$ using Eq. 2. As $TM1[v]$ does not include any edge incident on v, the computed $TM1[v]$ is a 0-1 timed matching for T_v. As each M_{u_i} is correct, where $u_i \in child(v)$, the cardinality of $TM1[v]$ is the maximum among all such 0-1 timed matchings for T_v, which does not include any edge incident on v.

When $F_v = \emptyset$, Algorithm 1 returns $TM1[v]$ as M_v. Lemma 1 proves that when $F_v = \emptyset$, $|TM1[v]| \geq |\tilde{M}_v|$ where $\tilde{M}_v$ is the maximum 0-1 timed matching for T_v which includes at least one edge incident on v. Thus, $TM1[v]$ is M_v when $F_v = \emptyset$. As $TM1[v]$ does not include any edge incident on v, in this case $|R_{M_v}(e_{vP(v)})| = 0$. Thus, Algorithm 1 correctly computes M_v for T_v when $F_v = \emptyset$.

In Line 11, Algorithm 1 computes $TM2[v]$ using Eq. 3. Lemma 3 proves that the computed F_v by Algorithm 2 is correct. As for each $u_i \in child(v)$, M_{u_i} is already computed and there is no 0-1 timed matching M'_{u_i} for T_{u_i} such that $|M'_{u_i}| = |M_{u_i}|$ and $|R_{M'_{u_i}}(e_{vu_i})| = 0$ but $|R_{M_{u_i}}(e_{vu_i})| > 0$. Thus, from Lemma 2, the computed $TM2[v]$ is correct when $F_v \neq \emptyset$ and $TM2[v]$ is M_v.

It can be observed that, while including $e_{vP(v)}$ in a 0-1 timed matching for $T_{P(v)}$ only the edges in F_v can get removed. Thus, there can exist another 0-1 timed matching M'_v for T_v such that $|M_v| = |M'_v|$ and $|R_{M'_v}(e_{vP(v)})| = 0$ but $|R_{M_v}(e_{vP(v)})| > 0$ only when computed F_v is incorrect. This contradicts Lemma 3. This proves the inductive step. □

Theorem 4. *Time complexity of Algorithm 1 is* $O(n^3)$.

Proof. Algorithm 1 stores the nodes in the rooted temporal tree $\mathcal{G}$, in different lists according to their depth in $\mathcal{G}$. This can be done in $O(n)$ time. For any node v, to compute M_v when information about each M_{u_i} where $u_i \in child(v)$ is available, we need to compute F_v. Algorithm 1 calls Algorithm 2 to compute F_v. At Line 2, Algorithm 2 finds the number of overlapping edges in M_{u_i} with edge e_{vu_i} where $u_i \in child(v)$. As each edge is associated with one interval, function *interSect* does this in $O(n)$ time (number of edges in M_{u_i} is $O(n)$). Hence, the overall running time of this step is $O(n^2)$. The formation of $\tilde{\mathcal{R}}_0(v)$ at Lines 3 to 5 takes $O(n)$ time. Finding maximum non-overlapping set from $\tilde{\mathcal{R}}_0(v)$ and $\tilde{\mathcal{R}}_0^P(v)$ takes $O(n \log n)$ time using function *maxNonOverlap*. Thus the overall running time of Algorithm 2 is $O(n^2)$. Hence, the computation of M_v for any node v takes $O(n^2)$ time. Hence, the overall running time of the algorithm is $O(n^3)$. □

The following theorem follows from Theorems 3 and 4.

Theorem 5. *Algorithm 1 correctly computes the maximum 0-1 timed matching for a given rooted temporal tree* $\mathcal{G}(\mathcal{V}, \mathcal{E})$ *in* $O(n^3)$ *time, where* $n = |\mathcal{V}|$.

7   Conclusion

In this paper, we have defined 0-1 timed matching for a given temporal graph. We have shown that the problem of finding a maximum 0-1 timed matching is NP-Complete both for a bipartite temporal graph with a single time interval associated with each edge, and a rooted temporal tree with at most three time intervals associated with each edge. We have also given a $O(n^3)$ time algorithm for the problem for a rooted temporal tree with a single time interval associated with each edge, where n is the number of nodes.

References

1. Arkin, E.M., Silverberg, E.B.: Scheduling jobs with fixed start and end times. Discrete Appl. Math. **18**(1), 1–8 (1987)

2. Baste, J., Bui-Xuan, B.M., Roux, A.: Temporal matching. Theor. Comput. Sci. (2019)
3. Baswana, S., Gupta, M., Sen, S.: Fully dynamic maximal matching in O (log n) update time. In: Symposium on Foundations of Computer Science FOCS, pp. 383–392 (2011)
4. Berman, P., Fujito, T.: On approximation properties of the independent set problem for low degree graphs. Theor. Comput. Syst. **32**(2), 115–132 (1999)
5. Bernstein, A., Stein, C.: Fully dynamic matching in bipartite graphs. In: Halldórsson, M.M., Iwama, K., Kobayashi, N., Speckmann, B. (eds.) ICALP 2015. LNCS, vol. 9134, pp. 167–179. Springer, Heidelberg (2015). https://doi.org/10.1007/978-3-662-47672-7_14
6. Bhattacharya, S., Henzinger, M., Italiano, G.F.: Deterministic fully dynamic data structures for vertex cover and matching. In: ACM-SIAM Symposium on Discrete Algorithms SODA, pp. 785–804 (2015)
7. Bui-Xuan, B., Ferreira, A., Jarry, A.: Computing shortest, fastest, and foremost journeys in dynamic networks. Int. J. Found. Comput. Sci. **14**(2), 267–285 (2003)
8. Cheriyan, J.: Randomized õ(m(|v|)) algorithms for problems in matching theory. SIAM J. Comput. **26**(6), 1635–1669 (1997)
9. Edmonds, J.: Paths, trees, and flowers. Can. J. Math. **17**, 449–467 (1965)
10. Even, S., Kariv, O.: An o(n^2.5) algorithm for maximum matching in general graphs. In: Symposium on Foundations of Computer Science FOCS, pp. 100–112 (1975)
11. Ferreira, A.: On models and algorithms for dynamic communication networks: the case for evolving graphs. In: 4^e rencontres francophones sur les Aspects Algorithmiques des Telecommunications (ALGOTEL), pp. 155–161 (2002)
12. Hopcroft, J.E., Karp, R.M.: A n^5/2 algorithm for maximum matchings in bipartite graphs. In: Symposium on Switching and Automata Theory SWAT, pp. 122–125 (1971)
13. Huang, S., Fu, A.W., Liu, R.: Minimum spanning trees in temporal graphs. In: ACM SIGMOD International Conference on Management of Data, pp. 419–430 (2015)
14. Kostakos, V.: Temporal graphs. Phys. A **388**(6), 1007–1023 (2009)
15. Mandal, S., Gupta, A.: Approximation algorithms for permanent dominating set problem on dynamic networks. In: Negi, A., Bhatnagar, R., Parida, L. (eds.) ICDCIT 2018. LNCS, vol. 10722, pp. 265–279. Springer, Cham (2018). https://doi.org/10.1007/978-3-319-72344-0_22
16. Micali, S., Vazirani, V.V.: An o(sqrt(|v|) |e|) algorithm for finding maximum matching in general graphs. In: Symposium on Foundations of Computer Science FOCS, pp. 17–27 (1980)
17. Michail, O., Spirakis, P.G.: Traveling salesman problems in temporal graphs. Theoret. Comput. Sci. **634**, 1–23 (2016)
18. Mucha, M., Sankowski, P.: Maximum matchings in planar graphs via Gaussian elimination. Algorithmica **45**(1), 3–20 (2006)
19. Onak, K., Rubinfeld, R.: Maintaining a large matching and a small vertex cover. In: ACM Symposium on Theory of Computing STOC, pp. 457–464 (2010)

Arbitrary Pattern Formation by Opaque Fat Robots with Lights

Kaustav Bose[1](✉) , Ranendu Adhikary[1] , Manash Kumar Kundu[2] ,
and Buddhadeb Sau[1]

[1] Department of Mathematics, Jadavpur University, Kolkata, India
{kaustavbose.rs,ranenduadhikary.rs,manashkrkundu.rs,
buddhadeb.sau}@jadavpuruniversity.in
[2] Gayeshpur Government Polytechnic, Kalyani, India

Abstract. ARBITRARY PATTERN FORMATION is a widely studied problem in autonomous robot systems. The problem asks to design a distributed algorithm that moves a team of autonomous, anonymous and identical mobile robots to form any arbitrary pattern given as input. The majority of the existing literature investigates this problem for robots with unobstructed visibility. In a few recent works, the problem has been studied in the obstructed visibility model, where the view of a robot can be obstructed by the presence of other robots. However, in these works, the robots have been modelled as dimensionless points in the plane. In this paper, we have considered the problem in the more realistic setting where the robots have a physical extent. In particular, the robots are modelled as opaque disks. Furthermore, the robots operate under a fully asynchronous scheduler. They do not have access to any global coordinate system, but agree on the direction and orientation of one coordinate axis. Each robot is equipped with an externally visible light which can assume a constant number of predefined colors. In this setting, we have given a complete characterization of initial configurations from where any arbitrary pattern can be formed by a deterministic distributed algorithm.

Keywords: Distributed algorithm · Arbitrary Pattern Formation ·
Leader election · Opaque fat robots · Luminous robots · Asynchronous
scheduler

1 Introduction

ARBITRARY PATTERN FORMATION or $\mathcal{APF}$ is a fundamental coordination problem for distributed multi-robot systems. Given a team of autonomous mobile robots, the goal is to design a distributed algorithm that guides the robots to form any specific but arbitrary geometric pattern given to the robots as input. ARBITRARY PATTERN FORMATION is closely related to the LEADER ELECTION problem where a unique robot from the team is to be elected as the leader. In the traditional framework of theoretical studies, the robots are modelled as

© Springer Nature Switzerland AG 2020
M. Changat and S. Das (Eds.): CALDAM 2020, LNCS 12016, pp. 347–359, 2020.
https://doi.org/10.1007/978-3-030-39219-2_28

autonomous (there is no central control), *homogeneous* (they execute the same distributed algorithm), *anonymous* (they have no unique identifiers) and *identical* (they are indistinguishable by their appearance) computational entities that can freely move in the plane. Each robot is equipped with sensor capabilities to perceive the positions of other robots. The robots do not have access to any global coordinate system. The robots operate in LOOK-COMPUTE-MOVE *(LCM)* cycles: upon becoming active, a robot takes a snapshot of the positions of the other robots (LOOK), then computes a destination based on the snapshot (COMPUTE), and then moves towards the destination along a straight line (MOVE).

The ARBITRARY PATTERN FORMATION problem has been extensively studied in the literature in various settings (See [2,5,6,8–10,12,14–16] and references therein). Until recently, the problem had only been studied for robots with unobstructed visibility. In [13], the problem was first considered in the *opaque robots* or *obstructed visibility* model which assumes that the visibility of a robot can be obstructed by the presence of other robots. This is a more realistic model for robots equipped with camera sensors. They also assumed that the robots are equipped with persistent visible lights that can assume a constant number of predefined colors. This is known as the *luminous robot* model, introduced by Peleg [11], where the lights serve both as a medium of weak explicit communication and also as a form of memory. In [13], the robots are first brought to a configuration in which each robot can see all other robots, and then LEADER ELECTION is solved by a randomized algorithm. The first fully deterministic solutions for LEADER ELECTION and ARBITRARY PATTERN FORMATION were given in [4] for robots whose local coordinate systems agree on the direction and orientation of one coordinate axis. However, in both [4,13], the robots were modelled as dimensionless points in the plane. This assumption is obviously unrealistic, as real robots have a physical extent. In this work, we extend the results of [4] to the more realistic *opaque fat robots* model [1,7]. Furthermore, our algorithm also works for robots with *non-rigid* movements (a robot may stop before it reaches its computed destination), whereas the algorithm of [4] requires robots to have *rigid* movements (a robot reaches its computed destination without any interruption). Also, the total number of moves executed by the robots in our algorithm is asymptotically optimal. The contribution of this paper is summarized in Table 1.

Table 1. Comparison of this work with previous ones.

	Robots	Agreement in coordinate system	Scheduler	No. of colors used	Movement	Total no. of moves
[4]	Point	One-axis agreement	ASYNC	6	Rigid	$O(n^2)$
This paper	Fat	One-axis agreement	ASYNC	10	Non-rigid	$\Theta(n)$

2   Model and Definitions

In this section, we shall formally describe the robot model and also present the necessary definitions and notations that will be used in the rest of the paper.

Robots. We consider a set of $n \geq 3$ autonomous, anonymous, homogeneous and identical fat robots. Each robot is modelled as a disk of diameter equal to 1 unit. The robots do not have access to any global coordinate system, but their local coordinate systems agree on the direction and orientation of the X-axis. They also agree on the unit of length as the diameter of the robots are same and taken as 1 unit.

Lights. Each robot is equipped with an externally visible light which can assume a constant number of colors. Our algorithm will require in total ten colors, namely off, terminal, interior, failed, symmetry, ready, move, switch off, leader and done. Initially all robots have their lights set to off.

Visibility. The visibility range of a robot is unlimited, but can be obstructed by the presence of other robots. Formally, a point p in the plane is visible to a robot r_i if and only if there exists a point x_i on the boundary of r_i such that the line segment joining p and x_i does not contain any point of any other robot. This implies that a robot r_i can see another robot r_j if and only if there is at least one point on the boundary of r_j that is visible to r_i. Also, if r_i can see any portion of the boundary of r_j, then we assume that it can determine the position of (the center of) r_j.

Look-Compute-Move Cycles. The robots, when active, operate according to the so-called LOOK-COMPUTE-MOVE *(LCM)* cycle. In each cycle, a previously idle robot wakes up and executes the following steps. In LOOK, a robot takes the snapshot of the positions of the robots visible to it (represented in its own coordinate system), along with their respective colors. Then in COMPUTE, based on the perceived configuration, the robot performs computations according to a deterministic algorithm to decide a destination point and a color. Finally in MOVE, it sets its light to the decided color and moves towards the destination point. When a robot transitions from one LCM cycle to the next, all of its local memory (past computations and snapshots) are erased, except for the color of the light.

Scheduler. We assume that the robots are controlled by a fully asynchronous adversarial scheduler. The robots are activated independently and each robot executes its cycles independently. The amount of time spent in LOOK, COM-PUTE, MOVE and inactive states is finite but unbounded, unpredictable and not same for different robots. As a result, a robot can be seen while moving, and hence, computations can be made based on obsolete information about positions.

Movement. We assume that the robots have *non-rigid* movements. This means that a robot may stop before it reaches its destination. However, there is a fixed $\delta > 0$ so that each robot traverses at least the distance δ unless its destination is closer than δ. The value of δ, however, is not known to the robots. The existence of a fixed δ is necessary, because otherwise, a robot may stop after moving distances $\frac{1}{2}, \frac{1}{4}, \frac{1}{8}, \ldots$ and thus, not allowing any robot to traverse a distance of more than 1.

Definitions and Notations. We shall denote the set of robots by $\mathcal{R} = \{r_1, r_2, \ldots, r_n\}$, $n \geq 3$. When we say that a robot is at a point p on the plane, we shall mean that its center is at p. For any time t, the configuration of the robots at time t, denoted by $\mathbb{C}(t)$ or simply $\mathbb{C}$, is a sequence $(p_1(t), p_2(t), \ldots, p_n(t))$ of n points on the plane, where $p_i(t)$ is the position of (the center of) the robot r_i at t. At any time t, $r(t).light$ or simply $r.light$ will denote the color of the light of r at t. With respect to the local coordinate system of a robot, positive and negative directions of the X-axis will be referred to as *right* and *left* respectively, and the positive and negative directions of the Y-axis will be referred to as *up* and *down* respectively. Since the robots agree on the X-axis, they agree on horizontal and vertical. They also agree on left and right, but not on up and down. For a robot r, $\mathcal{L}_V(r)$ and $\mathcal{L}_H(r)$ are respectively the vertical and horizontal lines passing through the center of r. We denote by $\mathcal{H}_U^O(r)$ (resp. $\mathcal{H}_U^C(r)$) and $\mathcal{H}_B^O(r)$ (resp. $\mathcal{H}_B^C(r)$) the upper and bottom open (resp. closed) half-planes delimited by $\mathcal{L}_H(r)$ respectively. Similarly, $\mathcal{H}_L^O(r)$ (resp. $\mathcal{H}_L^C(r)$) and $\mathcal{H}_R^O(r)$ (resp. $\mathcal{H}_R^C(r)$) are the left and right open (resp. closed) half-planes delimited by $\mathcal{L}_V(r)$ respectively. For a configuration $\mathbb{C}$, a subset of robots that are on the same vertical line will be called a *batch*. Thus, any configuration $\mathbb{C}$ can be partitioned into batches $B_1, \ldots, B_k$, ordered from left to right. The vertical line passing through the centers of the robots of a batch will be called the *central axis* of that batch. When we say 'the distance between a batch B_i and a robot r (resp. another batch B_j)', we shall mean the horizontal distance between the central axis of B_i and the center of r (resp. central axis of B_j). A robot r belonging to batch B_i will be called *non-terminal* if it lies between two other robots of B_i, and otherwise it will be called *terminal*. Consider any batch B_j whose central axis is S and a horizontal line T. Let $\mathcal{H}_1$ and $\mathcal{H}_2$ be the closed half-planes delimited by T. For each $\mathcal{H}_i, i = 1, 2$, consider the distances of the robots on $S \cap \mathcal{H}_i$ from T arranged in increasing order. The string of real numbers thus obtained is denoted by λ_i. To make the lengths of the strings λ_1 and λ_2 equal, null elements Φ may be appended to the shorter string. Now the two strings are different if and only if the robots of B_j are not in symmetric positions with respect to T. In that case, $\mathcal{H}_i$ will be called the *dominant half with respect to T and B_j* if λ_i is the lexicographically smaller sequence (setting $x < \Phi$ for any $x \in \mathbb{R}$).

Problem Definition. Consider an initial configuration of n fat opaque robots in the Euclidean plane, all having their lights set to off. Each robot is given as input, a pattern $\mathbb{P}$, which is a list of n distinct elements from $\mathbb{R}_{\geq 0}^2 = \{(a, b) \in \mathbb{R}^2 | a, b \geq 0\}$. The ARBITRARY PATTERN FORMATION requires to design a distributed algorithm that guides the robots to a configuration that is similar to $\mathbb{P}$ with respect to translation, reflection, rotation and uniform scaling.

3   The Algorithm

The main result of the paper is Theorem 1. The proof of the 'only if' part is the same as in case for point robots, proved in [4]. The 'if' part will follow from the algorithm presented in this section.

Theorem 1. *For a set of opaque luminous fat robots with non-rigid movements and having one axis agreement, $\mathcal{APF}$ is deterministically solvable if and only if the initial configuration is not symmetric with respect to a line $\mathcal{K}$ which (1) is parallel to the agreed axis and (2) does not pass through the center of any robot.*

For the rest of the paper, we shall assume that the initial configuration $\mathbb{C}(0)$ does not admit the unsolvable symmetry stated in Theorem 1. Our algorithm works in two *stages*, namely *leader election* and *pattern formation from leader configuration*. The first stage is again divided into two phases, namely *Phase 1* and *Phase 2*. In the first stage, a single robot will be elected as the leader of the swarm. Since the robots do not have access to any global coordinate system, they do not agree on how the given pattern $\mathbb{P}$ would be realized in the plane. With the help of the elected leader, the robots can implicitly agree on a common coordinate system. Once an agreement on a common coordinate system is achieved, the robots will arrange themselves to form the given pattern in the agreed coordinate system in the second stage. Since the robots are oblivious, in each LCM cycle, a robot has to infer the current stage or phase from its local view. This is described in Algorithm 1.

Algorithm 1: Arbitrary Pattern Formation

    Input : The configuration of robots visible to me.
1 **Procedure** ARBITRARYPATTERNFORMATION()
2   **if** *there is a robot with light set to leader* **then**   // stage 2
3 | PATTERNFORMATIONFROMLEADERCONFIGURATION()
4   **else**   // stage 1
5   **if** *(the first batch has two robots with light set to terminal) and (the lights of all robots of the second batch are set to same color) and (the distance between the first and second batch is at least $\frac{n+3}{2}$ units)* **then**
6 | PHASE2()
7   **else if** *there is at least one robot with light set to failed, symmetry, ready, move or switch off* **then**
8 | PHASE2()
9   **else**
10 | PHASE1()

Due to space restrictions, we will not describe our algorithms here in much detail. For further details and formal proofs of correctness, the reader is referred to the full version of the paper [3].

3.1   Leader Election

In the leader election stage, a unique robot r_l will elect itself as leader by setting its light to `leader` (while the lights of all other robots should be set to `off`). We want the configuration to satisfy some additional properties as well, that will be useful in the second stage of the algorithm. In particular, we want (1) all the non-leader robots to lie inside $\mathcal{H}_R^O(r_l) \cap \mathcal{H}$ where $\mathcal{H} \in \{\mathcal{H}_U^O(r_l), \mathcal{H}_B^O(r_l)\}$, and (2) the distance of any non-leader robot from $\mathcal{L}_H(r_l)$ to be at least 2 units. We shall call this a *leader configuration*, and call r_l the *leader*.

3.2 Phase 1

Since the robots already have an agreement on left and right, if there is a unique leftmost robot, i.e., the first batch has only one robot, then that robot, say r, can identify this from its local view and elect itself as the leader. However, the robot r will not immediately change its light to `leader` as the additional conditions of a leader configuration might not be yet satisfied. So, it will start executing the procedure BECOMELEADER() to achieve these conditions. Only after these conditions are satisfied, r will change its light to `leader`. However, there might be more than one leftmost robots in the configuration. In the extreme case, all the robots may lie on the same vertical line, i.e., there may be only one batch. So if there are more than one leftmost robots, the aim of Phase 1 is to move the two terminal robots of the first batch leftwards by the same amount. We also want the distance between (the central axes of) the first batch and second batch in the new configuration to be at least $\frac{n+3}{2}$ units. Therefore, at the end of Phase 1, we shall either have a leader configuration or have at least two batches in the configuration with the first batch having exactly two robots and at least $\frac{n+3}{2}$ units to the left of the second batch. In the second case, the lights of the two robots of the first batch will be set to `terminal`, lights of all robots of the second batch will be set to either `interior` or `off`, and all other robots have lights set to `off`. A pseudocode description of the algorithm is given in Algorithm 2.

Algorithm 2: Phase1

```
 1  Procedure PHASE1()
 2  │  r ← myself
 3  │  if r.light = off then
 4  │  │  if I am in the first batch and I am the only robot in my batch then
 5  │  │  │  BECOMELEADER()
 6  │  │  else if I am in the first batch and I am not the only robot in my batch then
 7  │  │  │  if I am terminal then
 8  │  │  │  │  r.light ← terminal
 9  │  │  │  else
10  │  │  │  └  r.light ← interior
11  │  │  else if there is a robot with light interior on L_V(r) then
12  │  │  │  if I am not terminal then
13  │  │  │  │  r.light ← interior
14  │  │  │  else if (I am terminal) and (there is exactly one robot r' in H_L^O(r)) and
        │        (r'.light = terminal) then
15  │  │  │  │  r.light ← terminal
16  │  │  │  └  Move n+3/2 units to the left
17  │  else if r.light = terminal then
18  │  │  if there is a robot on L_V(r) with light interior then
19  │  │  │  Move n+3/2 units to the left
20  │  │  else if there is a robot on L_V(r) with light terminal then
21  │  │  │  d ← my horizontal distance from the leftmost robot in H_R^O(r)
22  │  │  │  if d < n+3/2 then
23  │  │  │  └  Move n+3/2 − d units to the left
24  │  │  else if there is a robot r' in H_L^O(r) with light terminal then
25  │  │  │  d ← my horizontal distance from r'
26  │  │  └  Move d units to the left
```

3.3   Phase 2

Assume that at the end of Phase 1, we have $k \geq 2$ batches and exactly two robots r_1^1 and r_2^1 in the first batch B_1 with light terminal that are at least $\frac{n+3}{2}$ units to the left of B_2. So now we are in Phase 2. Let $\mathcal{L}$ be the horizontal line passing through the mid-point of the line segment joining r_1^1 and r_2^1. Let $\mathcal{H}_1$ and $\mathcal{H}_2$ be the two open half-planes delimited by $\mathcal{L}$ such that $r_1^1 \in \mathcal{H}_1$ and $r_2^1 \in \mathcal{H}_2$. Our algorithm will achieve the following. Define $i > 1$ to be the smallest integer such that B_i is either (Case 1) asymmetric with respect to $\mathcal{L}$, or (Case 2) symmetric with respect to $\mathcal{L}$, but it has a robot lying on $\mathcal{L}$. In Case 1, a terminal robot from B_{i-1} will become the leader and in Case 2, the robot from B_i that lies on $\mathcal{L}$ will become the leader. From left to right, terminal robots of different batches will attempt to elect a leader either by electing itself as the leader or by asking a robot of the next batch to become the leader. In particular, when a batch B_j tries to elect leader, its terminal robots will check whether the next batch B_{j+1} is asymmetric or symmetric with respect to $\mathcal{L}$. In the first case, the terminal robot of B_j lying in the dominant half with respect to $\mathcal{L}$ and B_{j+1} will elect itself as the leader. In the later case, the terminal robots of B_j, using light, will communicate to the robots of B_{j+1} the fact that B_{j+1} is symmetric with respect to $\mathcal{L}$. If $\mathcal{L}$ passes through the center of a robot of B_{j+1}, then that robot will elect itself as the leader. Now there are three issues regarding the implementation of this strategy, which we shall discuss in the following three sections.

1. What happens when the robots of B_j are unable to see all the robots of B_{j+1}?
2. How will the robots ascertain $\mathcal{L}$ from their local view?
3. How will all the conditions of a leader configuration be achieved?

Coordinated Movement of a Batch
When the terminal robots of a batch B_j attempt to elect leader, they need to see all the robots of the next batch B_{j+1}. But since the robots are fat and opaque, a robot may not be able to see all the robots of the next batch. However, each robot of two consecutive batches will be able to see all robots of the other batch if the two batches are more than 1 unit distance apart. Recall that at the beginning of Phase 2, the robots of B_1 are at least $\frac{n+3}{2}$ units to the left of the robots of B_2. Therefore, when the robots of the first batch attempt to elect leader, they are able to see all robots of B_2. Now consider the case where the terminal robots of $B_j, j > 1$ are trying to elect leader. Therefore, the first $j - 1$ batches must have failed to elect leader. This implies that the first j batches are symmetric with respect to $\mathcal{L}$ and $\mathcal{L}$ does not pass through the center of a robot of the first j batches. After the terminal robots of B_{j-1} fail to elect leader, they will change their lights to failed and ask the next batch B_j to try to elect a leader. Then the robots of B_j will move left to position themselves exactly at a distance $1 + \frac{1}{n}$ units from the robots of B_{j-1}. It can be shown that B_j will have sufficient space to execute the movement and also, their horizontal distance from the robots of B_{j+1} will be at least 2 units after the movement. So, after the movements, the terminal robots of B_j can see all the robots of B_{j+1}.

However, since the scheduler is asynchronous and the movements are non-rigid, the robots of B_j can start moving at different times, move at different speeds and by different amounts. Therefore, we have to carefully coordinate the movements of the robots of a batch so that they do not get disbanded. We have to ensure that all the robots of the batch remain vertically aligned after their moves, and also the completion of the moves of the batch as a whole must be detectable. We will need two extra colors for this. When the robots of B_j find two terminal robots of B_{j-1} with light failed, they will not immediately move; they will first change their lights to ready. Having all robots of B_j with light set to ready will help these robots to identify their batchmates. On one hand, a robot that moves first will be able to identify the robots from its batch that are lagging behind and also detect when every one has completed their moves. On the other hand, a robot that has lagged behind will be able to remember that it has to move (from its own light ready) and determine how far it should move (from robots with light ready on its left) even if it can not see the terminal robots of batch B_{j-1}. Therefore, before moving, all robots of B_j must change their lights to ready. But they can not verify if all their batchmates have changed their lights as they can not see all the robots of their batch. But the robots of B_{j-1} are able to see all the robots of B_j, and thus can certify this. So when all the robots B_j have changed their lights to ready, the terminal robots of B_{j-1} will confirm this by turning their lights to move. Only after this, the robots of B_j will start moving. The robots will be able to detect that the movement of the batch has completed by checking that its distance from B_{j-1} is $1 + \frac{1}{n}$ and there are no robots with light ready on its right. When it detects that the movement of the batch has completed, it will try to elect leader if it is terminal, otherwise, it will change its light to off.

Electing Leader from Local View

When the terminal robots of a batch will attempt to elect leader, they will require the knowledge of $\mathcal{L}$. Therefore, as different batches try to elect leader from left to right, the knowledge of $\mathcal{L}$ also needs to be propagated along the way with the help of lights. Consider the terminal robots r_1^j and r_2^j of a batch $B_j, j \geq 1$, that are attempting to elect a leader. In order to do so, they need two things: (1) the knowledge of $\mathcal{L}$, and (2) a full view of the next batch B_{j+1}. First consider the case $j = 1$. The terminal robots of the first batch r_1^1 and r_2^1 (with lights set to terminal) obviously have the knowledge of $\mathcal{L}$ as it is the horizontal line passing through the mid-point of the line segment joining them. Also, since r_1^1 and r_2^1 are at least $\frac{n+3}{2}$ units apart from the robots of B_2, they can see all the robots of B_2. Now suppose that a batch $B_j, j > 1$, is attempting to elect leader. Then as discussed in the last section, the robots of B_j are horizontally exactly $1 + \frac{1}{n}$ units to the right of the robots of B_{j-1} and at least 2 units to the left of the robots of B_{j+1}. Therefore, r_1^j and r_2^j can see all the robots of both batches B_{j-1} and B_{j+1}. Now since B_j is attempting to elect leader, it implies that the first $j - 1$ batches have failed to break symmetry. Hence, the first j batches are symmetric with respect to $\mathcal{L}$. In particular, $\mathcal{L}$ passes through the mid-point of the line segment joining the terminal robots of B_{j-1}. Since r_1^j and r_2^j can see the terminal robots of B_{j-1} (having lights set to move), they can determine $\mathcal{L}$.

Now, for a batch $B_j, j \geq 1$, attempting to elect leader, there are three cases to consider. If the robots of B_{j+1} are asymmetric with respect to $\mathcal{L}$ (Case 1), the one of r_1^j and r_2^j which is in the dominant half will change its light to switch off and start executing BECOMELEADER() (described in the following section). If the robots of B_{j+1} are symmetric with respect to $\mathcal{L}$ and $\mathcal{L}$ passes through the center of a robot r' of B_{j+1} (Case 2), then r_1^j and r_2^j will change their lights to symmetry. When r' finds two robots on its left batch with light symmetry that are equidistant from it, it will change its light to switch off and start executing BECOMELEADER(). If the robots of B_{j+1} are symmetric with respect to $\mathcal{L}$ and $\mathcal{L}$ does not pass through the center of any robot of B_{j+1} (Case 3), then r_1^j and r_2^j will change their lights to failed. Then the robots of B_{j+1} execute movements as described in the previous section.

Executing BecomeLeader()

When a robot finds itself eligible to become leader, it sets its light to switch off and executes BECOMELEADER() in order to fulfill all the additional conditions of a leader configuration. A robot with light switch off will not do anything if it sees any robot with light other than off in its own batch or an adjacent batch, i.e., it will wait for those robots to turn their lights to off (See line 4 of Algorithm 3). A robot r that finds itself eligible to become leader, is either (Case 1) a terminal robot of a batch, or (Case 2) a middle robot of a batch. The first objective is to move vertically so that all robots are in $\mathcal{H} \in \{\mathcal{H}_U^O(r), \mathcal{H}_B^O(r)\}$ and at least 2 units away from $\mathcal{L}_H(r)$. In case 1, the robot has no obstruction to move vertically. But in case 2, it will have to move horizontally left first. We can show that it will have enough room to move and place itself at a position where there is no obstruction to move vertically. After the vertical movement, it will have to move horizontally so that all other robots are in $\mathcal{H}_R^O(r)$. But it will not try to do this in one go, as we have to also ensure that all other robots turn their lights to off. It will first move left to align itself with its nearest left batch, say B_j. From there it can see all robots of B_{j-1} and B_{j+1}, and it will wait until all robots of B_{j-1} and B_{j+1} turn their lights to off. Then it will move to align itself with B_{j-1} and so on. Eventually all the conditions of a leader configuration will be satisfied and it will change its light to leader.

Algorithm 3: Phase2

---

1 Procedure PHASE2()
2　　$r \leftarrow$ myself
3　　**if** $r.light \neq$ *switch off* **then**
4　　　　**if** *there is a robot with light* switch off *in my batch or an adjacent batch* **then**
5　　　　　　| $r.light \leftarrow$ off
6　　　　**else if** $r.light =$ off *or* interior **then**
7　　　　　　**if** *both terminal robots of my left batch have lights set to* failed *and the*
　　　　　　　　non-terminal robots (if any) have lights set to off **then**
8　　　　　　　　| $r.light \leftarrow$ ready
9　　　　　　**else if** *both terminal robots of my left batch have lights set to* symmetry *and*
　　　　　　　　the non-terminal robots (if any) have lights set to off **then**
10　　　　　　　　**if** *the two terminal robots of my left batch are equidistant from me* **then**
11　　　　　　　　　　| $r.light \leftarrow$ switch off

12　　　　**else if** $r.light =$ terminal **then**
13　　　　　　| ELECTLEADER()
14　　　　**else if** $r.light =$ failed **then**
15　　　　　　**if** *all robots of my right batch have their lights set to* ready **then**
16　　　　　　　　| $r.light \leftarrow$ move

17　　　　**else if** $r.light =$ ready **then**
18　　　　　　**if** *there is a robot* r' *in* $\mathcal{H}_L^O(r)$ *with light set to* ready **then**
19　　　　　　　　$d \leftarrow$ the horizontal distance of r' from me
20　　　　　　　　Move d units towards left
21　　　　　　**else if** *both terminal robots of my left batch have lights set to* move **then**
22　　　　　　　　$d \leftarrow$ the horizontal distance of my left batch from me
23　　　　　　　　**if** $d > 1 + \frac{1}{n}$ **then**
24　　　　　　　　　　| Move $d - 1 - \frac{1}{n}$ units towards left
25　　　　　　　　**else if** $d = 1 + \frac{1}{n}$ **then**
26　　　　　　　　　　**if** *there is no robot with light* ready *in* $\mathcal{H}_R^O(r)$ **then**
27　　　　　　　　　　　　**if** *I am terminal* **then**
28　　　　　　　　　　　　　　| ELECTLEADER()
29　　　　　　　　　　　　**else**
30　　　　　　　　　　　　　　| $r.light \leftarrow$ off

31　　**else**
32　　　　| BECOMELEADER()

33 Procedure ELECTLEADER()
34　　**if** *I am in the first batch* **then**
35　　　　| $\mathcal{L} \leftarrow$ the horizontal line passing through the mid-point of the line segment joining
　　　　　　me and the other robot (with light terminal) on $L_V(r)$
36　　**else**
37　　　　| $\mathcal{L} \leftarrow$ the horizontal line passing through the mid-point of the line segment joining
　　　　　　the terminal robots (with lights move) of my left batch
38　　**if** *my right batch is symmetric with respect to* $\mathcal{L}$ **then**
39　　　　**if** $\mathcal{L}$ *passes through the center of a robot of the right batch* **then**
40　　　　　　| $r.light \leftarrow$ symmetry
41　　　　**else**
42　　　　　　| $r.light \leftarrow$ failed
43　　**else if** *I am in the dominant half with respect to* $\mathcal{L}$ *and my right batch* **then**
44　　　　| $r.light \leftarrow$ switch off

3.4　Pattern Formation from Leader Configuration

In a leader configuration, the robots can reach an agreement on a common coordinate system. All non-leader robots in a leader configuration lie on one of the open half-planes delimited by the horizontal line passing through the leader

r_l. This half-plane will correspond to the positive direction of Y-axis or 'up'. Therefore, we have an agreement on 'up', 'down', 'left' and 'right'. Now the origin will be fixed at a point such that the coordinates of r_l are $(0, -2)$. Now the given pattern can be embedded on the plane with respect to the common coordinate system. Let us call these points the *target points*. Order these points as $t_0, t_1, \ldots, t_{n-1}$ from top to bottom, and from right to left in case multiple robots on the same horizontal line. Order the robots as $r_l = r_0, r_1, \ldots, r_{n-1}$ from bottom to up, and from left to right in case multiple robots on the same horizontal line. The non-leader robots will move sequentially according to this order and place themselves on $\mathcal{L}_H(r_l)$. Then sequentially $r_1, \ldots, r_{n-1}$ will move to the target points $t_0, \ldots, t_{n-2}$, and finally r_0 will move to t_{n-1}. Pseudocode of the algorithm is given in Algorithm 4.

Algorithm 4: Pattern Formation from Leader Configuration

Input   : The configuration of robots visible to me.
1 **Procedure** PATTERNFORMATIONFROMLEADERCONFIGURATION()
2 $\quad$ $r \leftarrow$ myself
3 $\quad$ $r_l \leftarrow$ the robot with light leader
4 $\quad$ if $r.light = off$ then
5 $\quad\quad$ if $(r_l \in \mathcal{H}_B^O(r))$ and (there is no robot in $\mathcal{H}_B^O(r) \cap \mathcal{H}_U^O(r_l)$) and ($r$ is leftmost on $\mathcal{L}_H(r)$) then
6 $\quad\quad\quad$ if there are no robots on $\mathcal{L}_H(r_l)$ other than r_l then
7 $\quad\quad\quad\quad$ if there is a robot with light done then
8 $\quad\quad\quad\quad\quad$ if I am at t_{n-2} then
9 $\quad\quad\quad\quad\quad$ | $r.light \leftarrow$ done
10 $\quad\quad\quad\quad$ else
11 $\quad\quad\quad\quad\quad$ Move to t_{n-2}
12 $\quad\quad\quad$ else
13 $\quad\quad\quad\quad$ Move to $(1, -2)$
14 $\quad\quad\quad$ else if there are i robots on $\mathcal{L}_H(r_l)$ other than r_l at $(1, -2), \ldots, (i, -2)$ then
15 $\quad\quad\quad\quad$ | Move to $(i + 1, -2)$
16 $\quad\quad\quad$ else if there are i robots on $\mathcal{L}_H(r_l)$ other than r_l at $(n - i, -2), \ldots, (n - 1, -2)$ then
17 $\quad\quad\quad\quad$ if I am at t_{n-i-2} then
18 $\quad\quad\quad\quad$ | $r.light \leftarrow$ done
19 $\quad\quad\quad$ else
20 $\quad\quad\quad\quad$ Move to t_{n-i-2}
21 $\quad\quad$ else if $r_l \in \mathcal{L}_H(r)$ and $\mathcal{H}_U^O(r)$ has no robots with light off then
22 $\quad\quad\quad$ if I am at $(i, -2)$ then
23 $\quad\quad\quad\quad$ Move to t_{i-1}
24 $\quad$ else if $r.light = leader$ then
25 $\quad\quad$ if there are no robots with light off then
26 $\quad\quad\quad$ if I am at t_{n-1} then
27 $\quad\quad\quad\quad$ | $r.light \leftarrow$ done
28 $\quad\quad\quad$ else
29 $\quad\quad\quad\quad$ Move to t_{n-1}

---

4 Conclusion

Using 4 extra colors, we have extended the results of [4] to the more realistic setting of fat robots with non-rigid movements and also improved the move complexity to $\Theta(n)$, which is asymptotically optimal. Techniques used in Phase 2 of our algorithm can be used to solve LEADER ELECTION without movement for luminous opaque point robots for any initial configuration where LEADER

ELECTION is solvable in full visibility model, except for the configuration where all robots are collinear. An interesting question is whether there is a no movement LEADER ELECTION algorithm for (luminous and opaque) fat robots. Another open question is whether it is possible to solve $\mathcal{APF}$ for opaque (point or fat) robots with only agreement in chirality.

Acknowledgements. The first two authors are supported by NBHM, DAE, Govt. of India and CSIR, Govt. of India, respectively. We would like to thank the anonymous reviewers for their valuable comments which helped us to improve the quality and presentation of the paper.

References

1. Agathangelou, C., Georgiou, C., Mavronicolas, M.: A distributed algorithm for gathering many fat mobile robots in the plane. In: Proceedings of the 2013 ACM symposium on Principles of distributed computing, pp. 250–259. ACM (2013)
2. Bose, K., Adhikary, R., Kundu, M.K., Sau, B.: Arbitrary pattern formation on infinite grid by asynchronous oblivious robots. In: Das, G.K., Mandal, P.S., Mukhopadhyaya, K., Nakano, S. (eds.) WALCOM 2019. LNCS, vol. 11355, pp. 354–366. Springer, Cham (2019). https://doi.org/10.1007/978-3-030-10564-8_28
3. Bose, K., Adhikary, R., Kundu, M.K., Sau, B.: Arbitrary pattern formation by opaque fat robots with lights. CoRR abs/1910.02706 (2019). http://arxiv.org/abs/1910.02706
4. Bose, K., Kundu, M.K., Adhikary, R., Sau, B.: Arbitrary pattern formation by asynchronous opaque robots with lights. In: Censor-Hillel, K., Flammini, M. (eds.) SIROCCO 2019. LNCS, vol. 11639, pp. 109–123. Springer, Cham (2019). https://doi.org/10.1007/978-3-030-24922-9_8
5. Bramas, Q., Tixeuil, S.: Arbitrary pattern formation with four robots. In: Izumi, T., Kuznetsov, P. (eds.) SSS 2018. LNCS, vol. 11201, pp. 333–348. Springer, Cham (2018). https://doi.org/10.1007/978-3-030-03232-6_22
6. Cicerone, S., Di Stefano, G., Navarra, A.: Asynchronous arbitrary pattern formation: the effects of a rigorous approach. Distrib. Comput., 1–42 (2018). https://doi.org/10.1007/s00446-018-0325-7
7. Czyzowicz, J., Gasieniec, L., Pelc, A.: Gathering few fat mobile robots in the plane. Theor. Comput. Sci. **410**(6–7), 481–499 (2009). https://doi.org/10.1007/11945529_25
8. Dieudonné, Y., Petit, F., Villain, V.: Leader election problem versus pattern formation problem. In: Lynch, N.A., Shvartsman, A.A. (eds.) DISC 2010. LNCS, vol. 6343, pp. 267–281. Springer, Heidelberg (2010). https://doi.org/10.1007/978-3-642-15763-9_26
9. Flocchini, P., Prencipe, G., Santoro, N., Widmayer, P.: Arbitrary pattern formation by asynchronous, anonymous, oblivious robots. Theor. Comput. Sci. **407**(1–3), 412–447 (2008). https://doi.org/10.1016/j.tcs.2008.07.026
10. Lukovszki, T., Meyer auf der Heide, F.: Fast collisionless pattern formation by anonymous, position-aware robots. In: Aguilera, M.K., Querzoni, L., Shapiro, M. (eds.) OPODIS 2014. LNCS, vol. 8878, pp. 248–262. Springer, Cham (2014). https://doi.org/10.1007/978-3-319-14472-6_17

11. Peleg, D.: Distributed coordination algorithms for mobile robot swarms: new directions and challenges. In: Pal, A., Kshemkalyani, A.D., Kumar, R., Gupta, A. (eds.) IWDC 2005. LNCS, vol. 3741, pp. 1–12. Springer, Heidelberg (2005). https://doi.org/10.1007/11603771_1

12. Suzuki, I., Yamashita, M.: Distributed anonymous mobile robots: formation of geometric patterns. SIAM J. Comput. **28**(4), 1347–1363 (1999). https://doi.org/10.1137/S009753979628292X

13. Vaidyanathan, R., Sharma, G., Trahan, J.L.: On fast pattern formation by autonomous robots. In: Izumi, T., Kuznetsov, P. (eds.) SSS 2018. LNCS, vol. 11201, pp. 203–220. Springer, Cham (2018). https://doi.org/10.1007/978-3-030-03232-6_14

14. Yamashita, M., Suzuki, I.: Characterizing geometric patterns formable by oblivious anonymous mobile robots. Theor. Comput. Sci. **411**(26–28), 2433–2453 (2010). https://doi.org/10.1016/j.tcs.2010.01.037

15. Yamauchi, Y., Yamashita, M.: Pattern formation by mobile robots with limited visibility. In: Moscibroda, T., Rescigno, A.A. (eds.) SIROCCO 2013. LNCS, vol. 8179, pp. 201–212. Springer, Cham (2013). https://doi.org/10.1007/978-3-319-03578-9_17

16. Yamauchi, Y., Yamashita, M.: Randomized pattern formation algorithm for asynchronous oblivious mobile robots. In: Kuhn, F. (ed.) DISC 2014. LNCS, vol. 8784, pp. 137–151. Springer, Heidelberg (2014). https://doi.org/10.1007/978-3-662-45174-8_10

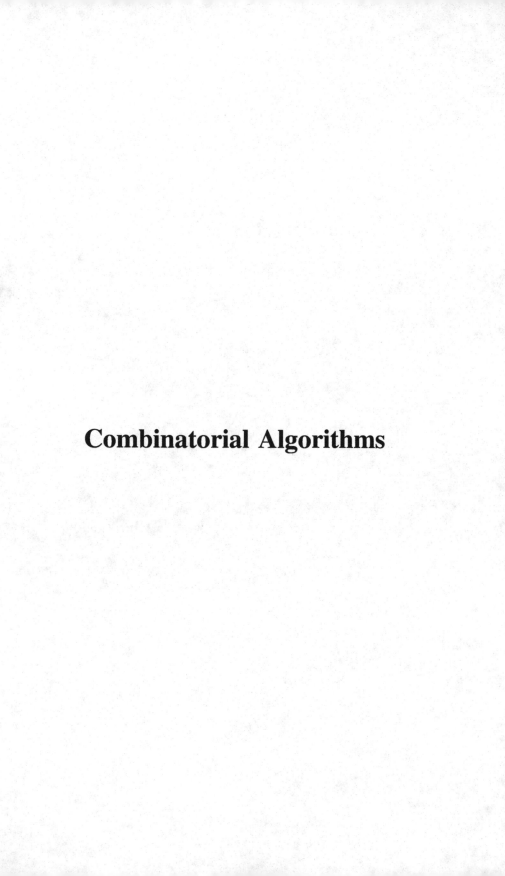

Combinatorial Algorithms

Greedy Universal Cycle Constructions for Weak Orders

Marsden Jacques[✉] and Dennis Wong[✉]

State University of New York, Incheon, Korea
`marsdenfernand.jacques@stonybrook.edu`, `cwong@uoguelph.ca`

Abstract. A weak order is a way to rank n objects where ties are allowed. In this paper, we extend the prefer-larger and the prefer-opposite algorithms for de Bruijn sequences to provide the first known greedy universal cycle constructions for weak orders.

Keywords: Universal cycle · Greedy algorithm · Weak orders · Prefer-larger · Prefer-opposite

1 Universal Cycles for Weak Orders

A *universal cycle* for a set $\mathbf{S}$ is a cyclic sequence of length $|\mathbf{S}|$ whose substrings of length n encode $|\mathbf{S}|$ distinct objects in $\mathbf{S}$. When $\mathbf{S}$ is the set of k-ary strings of length n, then a universal cycle for $\mathbf{S}$ is also known as a de Bruijn sequence. As an example, the cyclic sequence 002212011 is a universal cycle (also known as de Bruijn sequence) for the set of 3-ary strings of length 2; the 9 unique substrings of length 2 when considered cyclicly are:

$$00, 02, 22, 21, 12, 20, 01, 11, 10.$$

This sequence is known as the *prefer-larger de Bruijn sequence* and can be constructed using a simple greedy algorithm [6,14]. The *prefer-larger algorithm* starts with the sequence 0^n and applies a simple rule which can be summarized as follows:

> Greedily append the largest possible symbol such that the substrings of length n in the resulting sequence are distinct.

The greedy algorithm always terminates with a length $k^n + n - 1$ sequence that has the suffix 0^{n-1}. A de Bruijn sequence is thus obtained by removing the suffix 0^{n-1}.

Since the discovery of the prefer-larger de Bruijn sequence, similar algorithms have been proposed with different symbol insertion criteria [1–3,5,7,20]. One of the well-known greedy algorithm is the prefer-opposite algorithm. The *prefer-opposite algorithm* was first proposed by Alkahim in [1] for the construction of a binary de Bruijn sequence. The algorithm was generalized to construct a k-ary de Bruijn sequence by Alkahim and Sawada [15,16]. The prefer-opposite algorithm starts with the sequence 0^n and applies a simple rule which can be summarized as follows:

© Springer Nature Switzerland AG 2020
M. Changat and S. Das (Eds.): CALDAM 2020, LNCS 12016, pp. 363–370, 2020.
https://doi.org/10.1007/978-3-030-39219-2_29

Greedily append the value $t \bmod k$ such that t is the least possible value that is greater than the last symbol and the substrings of length n in the resulting sequence are distinct.

The greedy algorithm always terminates with a length $k^n + n - k + 1$ sequence that has the suffix $(k-1)^{n-1}(k-2)^{n-1} \cdots 0^{n-1}$ and missing the length n substrings t^n where $t \in \{1, 2, \ldots, k-1\}$. A de Bruijn sequence is thus obtained by inserting the symbols $t \in \{1, 2, \ldots, k-1\}$ after the length $n-1$ substrings t^{n-1} in the suffix $(k-1)^{n-1}(k-2)^{n-1} \cdots 0^{n-1}$ and also removing the suffix 0^{n-1}. As an example, the algorithm generates the cyclic sequence 001202211 when $n = 2$ and $k = 3$.

De Bruijn sequences are often used in education of discrete mathematics and theoretical computer science, including the *Concrete Mathematics* textbook by Graham, Knuth, and Patashnik [10] and *The Art of Computer Programming* by Knuth [13]. Historically, they have been referred to by many different names and have had many interesting applications, especially in the area of cryptography and communication systems. For example, Stein offered an interesting survey in the 'Memory Wheels' chapter in *Mathematics: The Man-Made Universe* [19]. An interesting problem in this research area is to generalize the idea of de Bruijn sequences to construct universal cycles for interesting combinatorial objects [12].

Table 1. The 13 possible ways to rank vanilla, chocolate and strawberry flavour ice cream.

First favorite(s)	Second favorite(s)	Third favorite(s)	Weak order	$W(n)$
Vanilla, Chocolate, Strawberry	–	–	$1 \equiv 2 \equiv 3$	000
Vanilla, Chocolate	Strawberry	–	$1 \equiv 2 \prec 3$	001
Vanilla, Strawberry	Chocolate	–	$1 \equiv 3 \prec 2$	010
Vanilla	Chocolate, Strawberry	–	$1 \prec 2 \equiv 3$	011
Chocolate, Strawberry	Vanilla	–	$2 \equiv 3 \prec 1$	100
Chocolate	Vanilla, Strawberry	–	$2 \prec 1 \equiv 3$	101
Strawberry	Vanilla, Chocolate	–	$3 \prec 1 \equiv 2$	110
Vanilla	Chocolate	Strawberry	$1 \prec 2 \prec 3$	012
Vanilla	Strawberry	Chocolate	$1 \prec 3 \prec 2$	021
Chocolate	Vanilla	Strawberry	$2 \prec 1 \prec 3$	102
Strawberry	Vanilla	Chocolate	$3 \prec 1 \prec 2$	120
Chocolate	Strawberry	Vanilla	$2 \prec 3 \prec 1$	201
Strawberry	Chocolate	Vanilla	$3 \prec 2 \prec 1$	210

In this paper, we extend the prefer-larger algorithm and the prefer-opposite algorithm to construct universal cycles for weak orders. An ordering on how n objects can be ranked when ties are allowed is known as a *weak order*. It can also be understood as a way to rank favourites. As an example, most people have a hard time picking a specific ice cream as their favourite. Table 1 lists out all 13 possible ways to rank vanilla, chocolate and strawberry flavour ice cream when ties are allowed.

The number of weak orders of order n is known as the ordered Bell number or Fubini number and the enumeration sequence is A000670 in the Online Encyclopedia of Integer Sequences [18]. The first six terms in this enumeration sequence are 1, 3, 13, 75, 541, and 4683 respectively.

Formally, a *weak order* for n objects is a binary relation $\preceq$ that is reflexive, anti-symmetric and transitive. We say $x \equiv y$ if $x \preceq y$ and $y \preceq x$, and we say $x \prec y$ if $x \preceq y$ but $y \not\preceq x$. A weak order over the set of n objects $\Sigma = \{1, 2, \cdots, n\}$ can be written as a permutation of the n objects with consecutive objects separated by either $\equiv$ or $\prec$ [13]. Let the *height* of an object be defined as the number of symbols that precedes it in the weak order. Let $W(n)$ denote the set of weak orders under the height notation. In [4], Diaconis and Graham introduced a notation to represent a weak order by the string $h_1 h_2 \cdots h_n$, where h_j denotes the height of the competitor j, and proved that universal cycles for weak orders exist under this notation. Table 1 lists out the weak orders (column 4) and their corresponding strings under the height notation (column 5) with the symbols 1, 2 and 3 representing vanilla, chocolate and strawberry flavour ice cream respectively. Note that the set of weak orders under the height notation is closed under rotation.

A more detailed proof on the existence of universal cycles for weak orders under the height notation is provided by Horan and Hurlbert in [11]. The problem of finding an efficient algorithm to construct a universal cycle for weak orders is listed as an open problem by Diaconis and Ruskey in [12] (Problem 2 of Problem 477) and has remained open.[1] There is, however, no known simple construction for such universal cycles for all n, albeit for an inefficient one. In this paper, we provide the first known greedy universal cycle constructions for weak orders in $W(n)$ by extending the prefer-larger and the prefer-opposite algorithms to $W(n)$. The following two simple greedy algorithms generate universal cycles for weak orders in $W(n)$ starting with the sequence 0^n:

Prefer-Larger Algorithm for $W(n)$: Greedily append the largest possible symbol such that the substrings of length n in the resulting sequence are distinct and in $W(n)$;

Prefer-Opposite Algorithm for $W(n)$: Greedily append the value $t \bmod n$ such that t is the least possible value that is greater than the last symbol and the substrings of length n in the resulting sequence are distinct and in $W(n)$.

As an example, the universal cycles generated by the prefer-larger algorithm and the prefer-opposite algorithm for $n = 3$ are 0001201102101 and 0001201011021 respectively. The 13 unique weak orders of length 3 for each sequence when considered cyclicly are:

---

[1] Sawada and Wong have recently discovered a new efficient algorithm to generate a universal cycle for weak orders for all n in [17] under another notation by applying the framework in [9].

Algorithm 1. The prefer-larger algorithm that generates a universal cycle for weak orders.

```
1: procedure PREFER-LARGER
2:     b₁b₂⋯bₙ ← 0ⁿ
3:     do
4:         Print(b₁)
5:         for z from n − 1 to 0 do
6:             if b₂b₃⋯bₙz ∈ W(n) and has not appeared before then
7:                 b₁b₂⋯bₙ ← b₂b₃⋯bₙz
8:                 break
9:     while b₁b₂⋯bₙ ≠ 0ⁿ
```

Algorithm 2. The prefer-opposite algorithm that generates a universal cycle for weak orders.

```
1: procedure PREFER-OPPOSITE
2:     b₁b₂⋯bₙ ← 0ⁿ
3:     do
4:         Print(b₁)
5:         for z from 1 to n − 1 do
6:             if b₂b₃⋯bₙ((bₙ + z) mod n) ∈ W(n) and has not appeared before then
7:                 b₁b₂⋯bₙ ← b₂b₃⋯bₙ((bₙ + z) mod n)
8:                 break
9:     while b₁b₂⋯bₙ ≠ 0ⁿ
```

Prefer-Larger Sequence: 000, 001, 012, 120, 201, 011, 110, 102, 021, 210, 101, 010, 100;

Prefer-Opposite Sequence: 000, 001, 012, 120, 201, 010, 101, 011, 110, 102, 021, 210, 100.

Pseudocode of the prefer-larger algorithm and the prefer-opposite algorithm for weak orders are given in Algorithms 1 and 2 respectively.

Theorem 1. *The prefer-larger algorithm generates a universal cycle for weak orders under the height notation for all n.*

Theorem 2. *The prefer-opposite algorithm generates a universal cycle for weak orders under the height notation for all n.*

The rest of the paper is outlined as follows. In Sect. 2, we prove Theorem 1. Then in Sect. 3, we prove Theorem 2. We conclude our paper in Sect. 4.

2   Proof of Theorem 1

Before we prove Theorem 1, we first prove the following lemmas for the prefer-larger universal cycle for weak orders.

Lemma 1. *The prefer-larger algorithm terminates after visiting the weak order* 10^{n-1}.

Proof. Assume the algorithm terminates after visiting some string $b_1 b_2 \cdots b_n \neq 10^{n-1}$. Let m be the number of unique weak orders in $W(n)$ that have the prefix $b_2 b_3 \cdots b_n$. As the algorithm terminates after visiting $b_1 b_2 \cdots b_n$, it follows by the greedy algorithm that all m weak orders with the prefix $b_2 b_3 \cdots b_n$ appear previously in the sequence. It further follows that there are $m + 1$ weak orders with the suffix $b_2 b_3 \cdots b_n$ appear in the sequence. However, observe that since weak orders in $W(n)$ are closed under rotation, the number of unique weak orders in $W(n)$ with the suffix $b_2 b_3 \cdots b_n$ is m. Thus by the pigeonhole principle, there are two duplicated weak orders in $W(n)$ with the suffix $b_2 b_3 \cdots b_n$ that appear in the sequence, a contradiction.

Lemma 2. *If* $b_1 b_2 \cdots b_n \in W(n)$ *does not appear in the prefer-larger sequence, then* $b_2 b_3 \cdots b_n 0$ *also does not appear in the prefer-larger sequence.*

Proof. Clearly the string $b_2 b_3 \cdots b_n 0$ does not appear in the sequence if it is not a weak order in $W(n)$. Now assume $b_2 b_3 \cdots b_n 0 \in W(n)$, we prove the lemma by contrapositive. Suppose the weak order $b_2 b_3 \cdots b_n 0 \neq 0^n$ is a substring in the sequence. Therefore by the greedy algorithm, all weak orders $b_2 b_3 \cdots b_n x \in W(n)$ with $x > 0$ appear before $b_2 b_3 \cdots b_n 0$ in the sequence. Since weak orders in $W(n)$ are closed under rotation, the number of weak orders in $W(n)$ with the suffix $b_2 b_3 \cdots b_n$ is the same as the number of weak orders with the prefix $b_2 b_3 \cdots b_n$. Furthremore, since each weak order before $b_2 b_3 \cdots b_n x$ in the sequence is unique and has the suffix $b_2 b_3 \cdots b_n$, it thus implies that all weak orders in $W(n)$ with the suffix $b_2 b_3 \cdots b_n$ also appear in the sequence, which includes the weak order $b_1 b_2 \cdots b_n$.

Lemma 3. *If* $b_1 b_2 \cdots b_n \in W(n)$ *with* b_1 *a unique but not a maximal symbol in* $b_1 b_2 \cdots b_n$, *then the non-existence of* $b_1 b_2 \cdots b_n$ *in the prefer-larger sequence implies* $b_2 b_3 \cdots b_n b_1$ *also does not appear in the prefer-larger sequence.*

Proof. Assume by contradiction that $b_2 b_3 \cdots b_n b_1$ exists in the sequence but $b_1 b_2 \cdots b_n$ does not exist in the sequence with b_1 a unique symbol but not a maximal symbol in $b_1 b_2 \cdots b_n$. Observe that since b_1 is unique and is not the maximal symbol in $b_1 b_2 \cdots b_n$, replacing b_1 with any other symbol creates a string that is not a weak order in $W(n)$. Therefore, the weak order that precedes $b_2 b_3 \cdots b_n b_1$ is $b_1 b_2 \cdots b_n$, a contradiction.

We now prove Theorem 1 using the lemmas we proved in this section.

Theorem 1. *The prefer-larger algorithm generates a universal cycle for weak orders under the height notation for all* n.

Proof. Since the greedy algorithm makes sure that there is no duplicated length n substring in the prefer-larger sequence, it suffices to show that each weak order in $W(n)$ appears as a substring in the sequence.

Assume by contradiction that there exists a weak order $b_1b_2 \cdots b_n \in W(n)$ that does not appear in the sequence and $b_1b_2 \cdots b_n \neq 0^n$. Let $b_r > 0$ be the first symbol in $b_1b_2 \cdots b_n$ which is not unique or is a maximal symbol in $b_1b_2 \cdots b_n$. Such a symbol always exists since $b_1b_2 \cdots b_n \neq 0^n$. Since weak orders in $W(n)$ are closed under rotation, the string $b_r b_{r+1} \cdots b_n b_1 b_2 \cdots b_{r-1}$ is a weak order in $W(n)$. By repeatedly applying Lemma 3, the weak order $b_r b_{r+1} \cdots b_n b_1 b_2 \cdots b_{r-1}$ also does not exist in the sequence. Furthermore by Lemma 2, the string $b_{r+1} b_{r+2} \cdots b_n b_1 b_2 \cdots b_{r-1} 0$ is a weak order in $W(n)$ and also does not exist in the sequence. Recursively applying the same argument implies that the weak order 10^{n-1} also does not exist in the sequence, a contradiction to Lemma 1. Therefore, each weak order in $W(n)$ appears as a substring in the prefer-larger sequence and thus the sequence is a universal cycle for $W(n)$.

3   Proof of Theorem 2

The proof of Theorem 2 is similar to the proof of Theorem 1. Before we prove Theorem 2, we first prove the following lemmas for the prefer-opposite universal cycle for weak orders.

Lemma 4. *The prefer-opposite algorithm terminates after visiting the weak order* 10^{n-1}.

Proof. The proof is similar to the proof of Lemma 1 and hence is omitted.

Lemma 5. *If* $b_1b_2 \cdots b_n \in W(n)$ *does not appear in the prefer-opposite sequence and* $b_n > 0$, *then* $b_2b_3 \cdots b_n(b_n - 1)$ *also does not appear in the prefer-opposite sequence.*

Proof. The proof is similar to the proof of Lemma 2. Clearly the string $b_2b_3 \cdots b_n(b_n - 1)$ does not appear in the sequence if it is not a weak order in $W(n)$. Now assume $b_2b_3 \cdots b_n(b_n - 1) \in W(n)$, we prove the lemma by contrapositive. Suppose the weak order $b_2b_3 \cdots b_n(b_n - 1) \neq 0^n$ is a substring in the sequence. Therefore by the greedy algorithm, all weak orders $b_2b_3 \cdots b_n x \in W(n)$ with $x \neq b_n - 1$ appear before $b_2b_3 \cdots b_n(b_n - 1)$ in the sequence. Since weak orders in $W(n)$ are closed under rotation, the number of weak orders in $W(n)$ with the suffix $b_2b_3 \cdots b_n$ is the same as the number of weak orders with the prefix $b_2b_3 \cdots b_n$. Furthremore, since each weak order before $b_2b_3 \cdots b_n x$ in the sequence is unique and has the suffix $b_2b_3 \cdots b_n$, it thus implies that all weak orders in $W(n)$ with the suffix $b_2b_3 \cdots b_n$ also appear in the sequence, which includes the weak order $b_1b_2 \cdots b_n$.

Lemma 6. *If* $b_1b_2 \cdots b_n \in W(n)$ *with* b_1 *a unique but not a maximal symbol in* $b_1b_2 \cdots b_n$, *then the non-existence of* $b_1b_2 \cdots b_n$ *in the prefer-opposite sequence implies* $b_2b_3 \cdots b_n b_1$ *also does not appear in the prefer-opposite sequence.*

Proof. The proof is similar to the proof of Lemma 3 and hence is omitted.

We now prove Theorem 2 using the lemmas we proved in this section.

Theorem 2. *The prefer-opposite algorithm generates a universal cycle for weak orders under the height notation for all n.*

Proof. Since the greedy algorithm makes sure that there is no duplicated length n substring in the prefer-opposite sequence, it suffices to show that each weak order in $W(n)$ appears as a substring in the sequence.

Assume by contradiction that there exists a weak order $b_1 b_2 \cdots b_n \in W(n)$ that does not appear in the sequence and $b_1 b_2 \cdots b_n \neq 0^n$. Let $b_r > 0$ be the first symbol in $b_1 b_2 \cdots b_n$ which is not unique or is a maximal symbol in $b_1 b_2 \cdots b_n$. Such a symbol always exists since $b_1 b_2 \cdots b_n \neq 0^n$. Since weak orders in $W(n)$ are closed under rotation, the string $b_r b_{r+1} \cdots b_n b_1 b_2 \cdots b_{r-1}$ is a weak order in $W(n)$. By repeatedly applying Lemma 6, the weak order $b_r b_{r+1} \cdots b_n b_1 b_2 \cdots b_{r-1}$ also does not exist in the sequence. Furthermore by Lemma 5, the string $b_{r+1} b_{r+2} \cdots b_n b_1 b_2 \cdots b_{r-1} (b_{r-1} - 1)$ is a weak order in $W(n)$ and also does not exist in the sequence, where $b_{r-1} - 1 < b_r$. Recursively applying the same argument implies that the weak order 10^{n-1} also does not exist in the sequence, a contradiction to Lemma 4. Therefore, each weak order in $W(n)$ appears as a substring in the prefer-opposite sequence and thus the sequence is a universal cycle for $W(n)$.

4   Conclusion

This paper extends the prefer-larger algorithm and the prefer-opposite algorithm to provide the first known greedy universal cycle constructions for weak orders. Future avenues of this research include investigating the following open problems.

1. The prefer-larger de Bruijn sequence is the k-ary complement of the lexicographically smallest de Bruijn sequence (also known as the GrandDaddy de Bruijn sequence and the Ford sequence), which can be constructed efficiently by a necklace concatenation approach in constant amortized time per symbol [8]. Naturally we would like to devise an efficient algorithm to generate the prefer-larger universal cycle for weak orders.
2. Another natural extension of this research is to generalize the greedy de Bruijn sequence constructions to generate universal cycles for other interesting combinatorial objects.

Acknowledgements. This research is supported by the MSIT (Ministry of Science and ICT), Korea, under the ICT Consilience Creative program (IITP-2019-2011-1-00783) supervised by the IITP (Institute for Information & communications Technology Planning & Evaluation).

The authors would like to thank Joe Sawada for his comments that greatly improve this paper. They would also like to thank Kyounga Woo for the fruitful discussions related to this research.

References

1. Alhakim, A.: A simple combinatorial algorithm for de Bruijn sequences. Am. Math. Mon. **117**(8), 728–732 (2010)

2. Alhakim, A.: Spans of preference functions for de Bruijn sequences. Discrete Appl. Math. **160**(7–8), 992–998 (2012)
3. Alhakim, A., Sala, E., Sawada, J.: Revisiting the prefer-same and prefer-opposite de Bruijn sequence constructions. Submitted manuscript (2019)
4. Diaconis, P., Graham, R.: Products of universal cycles. In: Demaine, E., Demaine, M., Rodgers, T. (eds.) A Lifetime of Puzzles, pp. 35–55. A K Peters (2008)
5. Eldert, C., Gray, H.J., Gurk, H.M., Rubinoff, M.: Shifting counters. AIEE Trans. **77**, 70–74 (1958)
6. Ford, L.R.: A cyclic arrangement of m-tuples. Report No. P-1071, RAND Corporation (1957)
7. Fredricksen, H.: A survey of full length nonlinear shift register cycle algorithms. SIAM Rev. **24**(2), 195–221 (1982)
8. Fredricksen, H., Maiorana, J.: Necklaces of beads in k colors and k-ary de Bruijn sequences. Discrete Math. **23**, 207–210 (1978)
9. Gabric, D., Sawada, J., Williams, A., Wong, D.: A successor rule framework for constructing k-ary de Bruijn sequences and universal cycles. IEEE Trans. Inf. Theory **66**, 679–687 (2019)
10. Graham, R.L., Knuth, D.E., Patashnik, O.: Concrete Mathematics: A Foundation for Computer Science, 2nd edn. Addison-Wesley Professional, Boston (1994)
11. Horan, V., Hurlbert, G.: Universal cycles for weak orders. SIAM J. Discrete Math. **27**(3), 1360–1371 (2013)
12. Jackson, B., Stevens, B., Hurlbert, G.: Research problems on Gray codes and universal cycles. Discrete Math. **309**(17), 5341–5348 (2009)
13. Knuth, D.E.: The Art of Computer Programming, Volume 4A, Combinatorial Algorithms. Addison-Wesley Professional, Boston (2011)
14. Martin, M.H.: A problem in arrangements. Bull. Am. Math. Soc. **40**, 859–864 (1934)
15. Mütze, T., Sawada, J., Williams, A.: The Combinatorial Object Server++. http://combos.org/index.html
16. Sawada, J.: Personal communication
17. Sawada, J., Wong, D.: An efficient universal cycle construction for weak orders. Submitted manuscript (2019)
18. Sloane, N.: The on-line encyclopedia of integer sequences. http://oeis.org. Sequence A000670
19. Stein, S.K.: Mathematics: The Man-Made Universe, 3rd edn. W. H. Freeman and Company, San Francisco (1994)
20. Wang, X., Wong, D., Zhang, W.: A simple greedy de Bruijn sequence construction. In: Proceedings of the 10th SEquences and Their Applications (SETA), Hong Kong (2018)

A New Model in Firefighting Theory

Rolf Klein[1], David Kübel[1(✉)], Elmar Langetepe[1], Jörg-Rüdiger Sack[2], and Barbara Schwarzwald[1]

[1] Department of Computer Science, Universität Bonn, 53115 Bonn, Germany
{rklein,dkuebel,schwarzwald}@uni-bonn.de,
elmar.langetepe@cs.uni-bonn.de
[2] School of Computer Science, Carleton University, Ottawa, ON K1S 5B6, Canada
sack@scs.carleton.ca

Abstract. Continuous and discrete models [1,5] for firefighting problems are well-studied in Theoretical Computer Science. We introduce a new, discrete, and more general framework based on a hexagonal cell graph to study firefighting problems in varied terrains. We present three different firefighting problems in the context of this model; for two of which, we provide efficient polynomial time algorithms and for the third, we show NP-completeness. We also discuss possible extensions of the model and their implications on the computational complexity.

Keywords: Cellular automaton · Combinatorial algorithms · Computational complexity · Discrete geometry · Fire spread models · Fire behaviour modeling · Firefighting · Forest fire simulation · Frontal propagation · Graph algorithms · Graph theory · NP-completeness · Undecidability

1 Introduction and Model Definition

Fighting multiple wildfires simultaneously or predicting their propagation involves many parameters one can neither foresee nor control. For the study of problems in this context, several models have been suggested and investigated in different communities.

In Theoretical Computer Science or Mathematics, models have been investigated, where fire spreads in the Euclidean plane or along edges of a graph; see e.g. [1,5]. Research in these models usually focuses on proving tight lower and upper bounds on what can be achieved with limited resources: In continuous models, researchers have been analysing the building speed of barriers which slow down or even stop the fire's expansion; in discrete models, the number of firefighters available to block/contain/extinguish the fire has been considered. Tight bounds are only available for simple cases in these models, e.g. [8]. For a survey, we refer to [4].

This work has been supported in part by DFG grant Kl 655/19 as part of a DACH project and by NSERC under grant no. RGPIN-2016-06253.

M. Changat and S. Das (Eds.): CALDAM 2020, LNCS 12016, pp. 371–383, 2020.
https://doi.org/10.1007/978-3-030-39219-2_30

In other communities, models have been developed to predict a fire's propagation in a given terrain. To make the forecast as realistic as possible, some models incorporate thermodynamic or chemical parameters as well as weather conditions including wind speed and direction. Some of the models are capable to distinguish between fires at different heights such as ground fires and crown fires. For a survey on theoretical and (semi-) empirical models, see [11].

We introduce a new model with the aim to develop a simple, theoretical framework for fire propagation forecast in large varied terrains and prove some initial results.

Definition 1 (basic hexagonal model). *Given a partition of the plane into hexagonal cells. The state of cell c at time t is given by two non-negative integers, $x(c,t)$ and $y(c,t)$. Cell c is called* burning *at time t if $x(c,t) = 0$ and $y(c,t) > 0$ hold;* alive *if $x(c,t) > 0$ and $y(c,t) > 0$; or* dead *if $y(c,t) = 0$ holds. At the transition from time t to $t+1$, the state of cell c changes as follows:*

- *If c is alive at time t, then $x(c,t+1) := \max\{x(c,t) - b, 0\}$, where b denotes the number of direct neighbours of c burning at time t.*
- *If c is burning at time t, then $y(c,t+1) := y(c,t) - 1$.*

Intuitively, x and y describe the (diminished) resistance against ignition and the (remaining) fuel of an individual cell at time t, respectively. Choosing suitable values for the cells, one can model natural properties of a given terrain: different types of ground and fuel; natural obstacles such as mountains or rivers. A cell of dry grassland might get small integers for both values such that it catches fire easily and burns down quickly. In contrast, we expect both values to be comparatively high for a moist forest such that the forest keeps burning for quite a while, once it caught fire. Figure 1 shows an example how a fire expands over time from a single source in a small lattice.

By this definition, our basic hexagonal model is a cellular automaton [12], whose cells can have state sets of different cardinality. We observe that in the basic hexagonal model a dead cell can never become alive or burning again. This is a major difference to cellular automata like *Conway's Game of Life* [6] or *Wolframs model* [13]. Another difference is that cells can die from overpopulation in Conway's Game of Life, for which there is no equivalent rule in our model.

Organization of the Paper. The rest of this paper is organized as follows. Section 2 discusses three problems in context of the basic hexagonal model and their algorithmic solutions. Variants of the basic model will be discussed in Sect. 3. We conclude with Sect. 4.

2   Results for the Basic Hexagonal Model

There is a multitude of interesting questions one can formulate within this model. In this paper, we will address the following three problems.

Suppose we restrict the hexagonal grid to a rectangular domain $\mathcal{R}$ consisting of n cells. Let us assume that initially all cells along the right boundary of $\mathcal{R}$ are

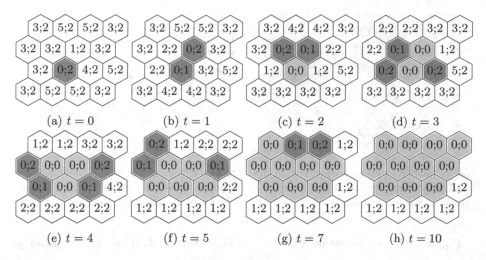

Fig. 1. Fire spreading in a hexagonal lattice. The x- and y-values are given in the cells as pairs of the form x; y at time t. The state of the cells is indicated by colours: Burning cells are red, alive cells are white, and dead cells are grey. At time $t = 10$, the propagation stops and several living cells remain. (Color figure online)

on fire and all cells along the left boundary of $\mathcal{R}$ represent a village that must be protected from the fire. To this end, we want to connect the upper to the lower boundary of $\mathcal{R}$ by a path π of cells that separates the village on the left and the fire approaching from the right; see Fig. 2a. To make π fire-resistant, we can fortify the cells on π by increasing their x-values.

In the *first version* of this problem, all cells on π will have their x-values raised by the same amount k. This corresponds to a fly-over by aircraft that douses each cell with the same amount of water. We want to compute the minimum k for which such a protecting path π exists. In Subsect. 2.1, we present a solution that runs in time $O(n \log n \log Y)$ where Y is the maximum sum of y-values of direct neighbours over all cells in $\mathcal{R}$. Our algorithm is based on a fast propagation routine that is interesting in its own right for simulation purposes.

In the *second version* of the above problem, firefighters can increase x-values of cells individually. Now we are interested in finding a separating path π for which the sum of these x-increments of cells on π is minimal. Although this appears to be a shortest-path problem, we have not been able to apply a classic graph algorithm like Dijkstra for reasons that will be explained in Subsect. 2.2. Our algorithm runs in time $O(n\sqrt{n} \log n)$, provided that all cells have identical y-values and each x-value is upper bounded by $2y + 1$.

In the *third version*, we no longer assume that cells along the right boundary of $\mathcal{R}$ are on fire, while cells along the left edge have to be protected. Instead, the cells of the village are given by a set $\mathcal{T}$ and the fire is allowed to start at cells of a set $\mathcal{F}$; see Fig. 2b. We now ask for a subset of $\mathcal{F}$ with m cells that will, when put on fire, burn all cells of the village to the ground. More precisely,

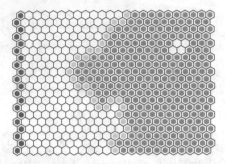
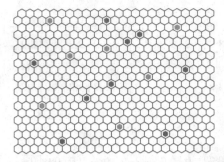

(a) A path π (grey cells) prevents a fire (red cells) from reaching the village (blue cells) at the left boundary of $\mathcal{R}$.

(b) Given two set of cells $\mathcal{F}$ (red) and $\mathcal{T}$ (blue). Can m cells out of $\mathcal{F}$ burn down *all* cells of $\mathcal{T}$, when put on fire?

Fig. 2. Problem variants on a rectangular finite domain $\mathcal{R}$. (Color figure online)

we are interested in answering the following decision problem: Are there m cells in $\mathcal{F}$ which, when put on fire, will eventually ignite all cells in $\mathcal{T}$? In Subsect. 2.3 we prove this problem to be NP-complete.

2.1 Homogeneous Fortification

For this problem, consider a rectangular domain $\mathcal{R}$ of our basic hexagonal model, in which all cells on the right boundary are on fire and all cells on the left boundary represent a village that must be protected. We call a path π connecting the lower to the upper boundary of $\mathcal{R}$ a *separating path*. We call π a *protecting path* for k if increasing the x-value of all cells along π by k ensures that the fire never ignites a cell of the village. The natural optimization problem is to find the minimum k for which a protecting path exists.

To solve this problem, we study the corresponding decision problem: Given $\mathcal{R}$ and k, does a protecting path π exist? For $k = 0$, it can be solved by simulating the fire propagation step-by-step, where all cells on the right boundary of $\mathcal{R}$ are initially ignited. Consider the map at the end of the simulation: All cells are either alive or dead but none burning; all dead cells form a connected component including the initially burning cells on the right boundary of $\mathcal{R}$. If no protecting path for $k = 0$ exists, at least one of the village cells will be dead. However, if a protecting path for $k = 0$ exists, then some of the alive cells form a connected component that includes all village cells on the left boundary of $\mathcal{R}$, see Fig. 2a. The *fire border* of this component which are alive cells with a direct dead neighbour, form a protecting path π for $k = 0$.

This approach can be extended to solve the decision problem for larger values of k: First increase the x-value of all cells by k, then run the simulation algorithm. If a cell of the village is dead at the end of the simulation, no protecting path for k exists. Otherwise, consider the fire border of the connected set of alive cells that includes the village and induces the separating path π. By construction, all cells of π stay alive, when their x-value is increased by k. This holds even if all cells right of π are burning or dead. This also holds when the x-values of all cells of $\mathcal{R} \setminus \pi$ remain untouched: Increasing the x-value of all burning cells is irrelevant for the survival of the village; increasing the x-value of any other alive cell is irrelevant as well, since only alive cells on π have burning or dead neighbours. Therefore, π is a protecting path in $\mathcal{R}$ for k.

To solve the optimization problem, we combine the decision-algorithm with binary search. It remains to give a sensible upper bound on k and an efficient algorithm for the simulation of fire propagation. Let Y be the maximum sum of y-values of all direct neighbours of a cell, over all cells in the grid. As every cell's x-value can only be decreased by at most the sum of its neighbours y, we know that $0 < k \leq Y$ holds. A brute force step-by-step simulation over time results in an algorithm with a worst-case running time of $O(n^2 \cdot x_{max})$, where x_{max} is the maximal value of x over all cells. However, given the state of all burning or dead cells at a time t, one can determine the next cell to ignite. This intuition gives rise to an $O(n \log n)$ Dijkstra-inspired algorithm independent of the cells values, which uses a priority queue for the retrieval of the next cell to ignite. Details and proofs can be found in the appendix of [9].

Theorem 1. *Let $\mathcal{R}$ be a rectangular domain, where all cells along the right boundary of $\mathcal{R}$ are on fire and all cells along the left boundary have to be protected.*

The minimum k for which a protecting path exists can be found in time $O(n \log n \log Y)$ where Y denotes the maximum sum of y-values of all direct neighbours of a cell, over all cells.

2.2 Selective Fortification

Similar to Subsect. 2.1, consider a rectangular domain $\mathcal{R}$ of our basic hexagonal model in which all cells on the right boundary are on fire and all cells on the left boundary represent a village that must be protected. We call a path π connecting the lower to the upper boundary a *separating path*. The path can be fortified to protect the village by individually increasing the x-value of each cell along π. For a given path π, we call the sum of those increments the *fortification cost* of π. The natural optimization problem is to find a separating path π that minimizes the fortification cost.

To begin with, observe that there is always a separating path with minimal fortification cost such that every cell along π has a direct, dead neighbour when the simulation ends: Any cell of π without a direct, dead neighbour can be excluded from π without increasing the fortification cost; all cells with a direct, dead neighbour that do not belong to π can be included to π since they do not require any fortification costs at all. Doing this, we obtain a separating path where all cells have dead neighbours. Therefore, we may restrict our search to separating paths to the right of which all cells are dead.

Moreover, this observation allows to compute the fortification costs of such a path: For a cell c of π, let Y_r be the sum of y-values of all neighbours of

c to the right of π. The fortification cost of c is the minimum k such that $x(c,0) + k = Y_r + 1$. The fortification cost of π is the sum of the fortification costs of all cells of π.

Finding a separating path π of cells is equivalent to finding a path π_b along corners and edges of the cells: Using the observation stated above, we may conclude that there is such a path π_b, where cells to the left belong to π and cells to the right are dead. This path π_b lies in the graph given by the corners and edges of the cells, which we call the border graph. Similar to the previous denotation, we call a path π_b in the border graph *separating* if it connects the upper to the lower boundary of $\mathcal{R}$. To distinguish left from right, we replace every edge $\{v, w\}$ in the border graph by two directed edges (v, w) and (w, v).

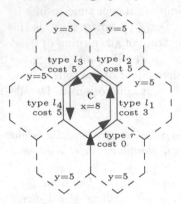

Fig. 3. A local-cost example.$\mathcal{R}$.

When transforming the optimization problem on the cells into a shortest-path problem on the border graph, it is not obvious how to assign the fortification cost of a cell to the adjacent edges; consider the example depicted in Fig. 3, where $y = 5$ for all cells and cell c has $x = 8$. The path π_b in question uses five edges of c, whose right neighbour cells are considered as burning. The crucial idea is to charge these edges for the fortification cost of c depending on their occurrence in π_b: The first directed edge of π_b along c gets cost 0 because no additional fortification is necessary to protect c from a single burning neighbour; the second edge gets cost 3; every further edge gets cost 5, since the x-value of c has to be increased by 5 for every additional burning neighbour. Unfortunately, this dynamic assignment of costs, where the cost of an edge depends on the previous edges of the path, rules out a direct solution via finding a shortest-path: A shortest path might visit the border of a cell several times; the edges along the same cell do not necessarily have to lie on the path in direct succession. Hence, the cost of an edge can be influenced by any previous edge in the path.

In general, the following two problems rule out a direct solution via shortest-path finding algorithms: (1) A shortest path π_b does not have to be simple and can have self-intersections, see Fig. 4a; (2) edges of π_b along the same cell c do not have to lie on π_b in direct succession, it can leave and *revisit* c multiple times, see Fig. 4b. In the following, we consider the problem for the case where the y-values are identical for all cells and $0 < x(c,0) \le 2y+1$ holds. This implies that each cell can always be ignited by three direct, burning neighbours. Based on these assumptions, we are able to prove that none of these problems occurs for a so-called shortest local-cost path.

Let e be an edge along a cell c of resistance $x_e := x(c,0)$. We say e is of type r if it comes after a right turn or is the very first edge of path π_b. We say e is of type l_k if it comes after the kth consecutive left turn of π_b along c. Thus, e is the $(k+1)^{th}$ consecutive edge along the same cell c on the left-hand side of π_b.

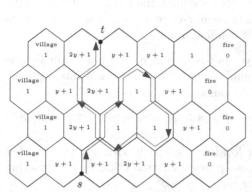

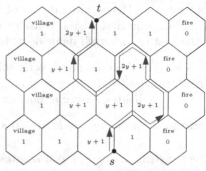

(a) A self-intersecting, shortest local-cost s-t-path with cost 0 and winding number 7.

(b) A simple s-t-path π_b with local-cost $4y$, winding number 1, and a re-visit. The corresponding separating path π has fortification cost $5y$.

Fig. 4. Two example domains which illustrate problems (1) and (2) for standard shortest-path finding algorithms. All cells have the same y-value and x-values as denoted in the cells.

Thus, we can define the local-cost of an edge e at cell c depending on x_e and its type:

$$c(e, \text{type}) = \begin{cases} \max(0, y + 1 - x_e), & \text{if type} = r \\ \min(y, 2y + 1 - x_e), & \text{if type} = l_1 \\ y, & \text{if type} \in \{l_2, l_3, l_4\}. \end{cases}$$

Definition 2 (Shortest local-cost path). *Let s, t be vertices in the border graph, where all cells have identical y-values and $0 < x(c, 0) < 2y(c, 0) + 1$ holds. Then, a shortest local-cost s-t-path is a path from s to t of minimum local edge cost as defined above.*

Moreover, we call the difference between the number of r-edges and l_k-edges of a path π_b in the border graph the *winding number* of π_b.

In the following, we prove that there is a shortest local-cost path π_b with winding number one that neither has (1) self-intersections, as shown in Fig. 4a, nor (2) revisits as shown in Fig. 4b: Lemmas 1 and 2 together prove (1), while Lemma 3 proves (2). Thus, the assigned local-cost of π_b are the true fortification costs of the corresponding separating path π.

Lemma 1. *Let s, l be vertices on the upper and lower boundary of the border graph of $\mathcal{R}$ and π_b be an s-t path. If π_b is simple, then it has winding number 1.*

Proof. As all turns in the regular grid have exactly the same angle, the winding number is a measure of the turn angle of the path. For any right turn, the total angular turn of the path decreases by $\frac{\pi}{3}$ and every left turn increases the turn angle by $\frac{\pi}{3}$. A simple s-t-path has a turn angle of 0 as first and last edge are both vertical. Hence, the number of left turns equals the number of right turns in π_b. Since the first edge is considered to be an r-edge, to assure that local-costs of the first cell are well defined, π_b has winding number 1.   □

Note that this does not directly solve our first problem, as it holds only for one direction: an s-t-path with winding number 1 might still contain intersections. The reverse holds because of our restrictions on the x- and y-values.

Lemma 2. *Any shortest local-cost s-t-path with winding number 1 is simple.*

Proof sketch. Assume π_b is a shortest local-cost s-t-path with winding number 1 and at least one intersection. Then, we prove a contradiction by constructing an intersection-free path π_b' with fewer costs.

Let π_b be given by the sequence $s, v_1, v_2, \ldots, v_i, \ldots, v_{n-1}, v_n, t$ of $n+2$ vertices in the border graph. Let $i < j$ be the smallest indices such that $v_i = v_j$ holds. By removing all vertices between v_i and v_{j+1}, we obtain a new path of which we show that its cost is strictly less than the cost of π_b.

While removed edges can no longer contribute to the cost of the path, removing them can change the type and hence the local cost of the edges from v up to the first unaffected r-edge after (v, v_{j+1}). A detailed proof, which shows that the local-cost of π_b' is strictly less than the local-cost of π_b, can be found in the appendix of [9].

All in all, repeated removal of loops results in a simple s-t-path with fewer costs than π_b which contradicts the assumption and completes the proof.   □

Lemma 3. *For any shortest local-cost s-t-path π with winding number 1, there exists a simple, shortest local-cost path without cell revisits that costs not more than π.*

Proof. Assume π_b is a shortest local-cost s-t-path with a winding number 1 where at least one cell on the left-hand of π_b is revisited. By Lemma 2, π_b is free of intersections. Then, we prove a contradiction by constructing a cell-revisit-free path π_b' with local-costs no more than that of π_b.

Let e be the first edge on the path along a revisited cell c. Let the edges of c be numbered counter clockwise from 0 to 5, where e is edge 0. Then, the situation at e can be restricted to the following two cases, also illustrated in Fig. 5.

1. The last edge along c on π_b is edge 2 and 1 does not lie on π_b.
2. The last edge along c on π_b is edge 3 and at least one of the edges 1 and 2 does not lie on π_b.

Note that π_b can neither include edge 4 nor 5. If it included edge 5, π_b would not be intersection free. If it included edge 4 but not 5, the edge preceding e on π_b would also lie along a revisited cell and we assumed e to be the first such edge on π_b.

We can construct π_b' from π_b by removing cell visits as follows. In case 1, we can replace all edges on π_b after edge 0 and before edge 2 by edge 1. This adds at

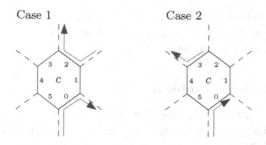

Fig. 5. Path π_b follows the blue edges upon its first visit of c and the red edges on its second visit. Cases are equivalent for rotation. (Color figure online)

most cost $2y$ (y for the new edge 1, and another y for the change of type edge 2 to l_2). In case 2, we replace all edges on π_b between edge 0 and edge 3 including these two edges by edge 4 and 5. This also adds at most cost $2y$ (y for edge 4 and 5 each). However, as π_b is free of intersection by Lemma 2, we know for both cases that the removed part includes enough l_k edges with $k > 1$ to counter the cost of adding the new edges to the path. □

Finally, we describe how to compute such a shortest local-cost path with a Dijkstra-inspired shortest-path algorithm tracking the winding number of the path: We use a priority queue of tuples (c, v, p, e, w), where c is the minimum known local-cost of a path to a vertex v via a predecessor p, where the last edge (p, v) on the path has edge type e and the whole path has winding number w. Due to regularity of the lattice, we know that: the vertex degree and hence the number of possible predecessors is constant; the number of edge types is constant. Moreover, w is limited by the size of our grid: A non-intersecting path with a very high winding number roughly forms a spiral, where the winding number corresponds to the number of spiralling rounds; the maximum size of such a spiral is limited by the width w and height h of the grid. Thus, it suffices to consider tuples with $|w| \leq 6 \min(w, h) = O(\sqrt{n})$.

Altogether, our priority queue contains at most $O(n\sqrt{n})$ many items, which results in a runtime of $O(n\sqrt{n} \log n)$ to find a simple, shortest local-cost path from a specific starting vertex s on the lower boundary to any vertex t on the upper boundary. To find the optimal separating path π, we have to compare shortest paths for all pairs of s and t. We can do this in a single run of the algorithm by initialising our priority queue with the outgoing edges of all possible s. We can terminate as soon as the minimal entry in our priority queue is a tuple where v is one of the vertices along the upper boundary of our grid and w is 1. A pseudocode description of this algorithm can be found in the appendix of [9].

Theorem 2. *Let $\mathcal{R}$ be a rectangular domain, where all cells have identical y-values and $0 < x(c, 0) \leq 2y(c, 0) + 1$ holds. All cells at the right boundary of $\mathcal{R}$ are on fire and all cells along the left boundary have to be protected.*

Then, we can compute a separating path of minimum fortification cost in time $O(n\sqrt{n} \log n)$.

For arbitrary values of x and y, it is still open whether the optimization problem can be solved in polynomial time. While the definition of local-costs can be adjusted, problems (1) and (2) remain.

2.3   An NP-Complete Problem

In this section, we no longer assume that cells along the right boundary of $\mathcal{R}$ are on fire, while cells along the left edge have to be protected. Instead, we consider two sets of cells $\mathcal{F}, \mathcal{T}$, where the fire is allowed to start from cells in $\mathcal{F}$ to burn cells in $\mathcal{T}$. We consider the following decision problem: Are there m cells in $\mathcal{F}$ which, when put on fire, will eventually ignite all cells in $\mathcal{T}$?

We prove this problem to be NP-complete by reduction from planar vertex cover. The planar vertex cover problem is as follows: Given a planar graph G, a vertex cover for G is a subset of vertices that contains at least one endpoint of every edge. This problem was proven to be NP-complete, even for planar graphs with maximum vertex degree three [7].

Given a planar graph G with n vertices, we have to show how to obtain an instance of the fire expansion problem in polynomial time. For simplicity, we assume that, instead of the basic hexagonal mode, we can use a basic grid model where the same rules apply only that a rectilinear grid is used instead of the hexagonal one. Consider the *rectilinear grid*, which is the infinite plane graph with a vertex at every positive integer coordinate and an edge between every pair of vertices at unit distance. In a first step, we compute a planar grid embedding $G_\square$ of G into the rectilinear grid such that the following holds: Disjoint vertices of G are mapped to disjoint integer coordinates; edges of G are mapped to rectilinear paths in the grid such that no two paths have a point in common, except, possibly, for the endpoints. The size of $G_\square$ is polynomial in n and can be computed in polynomial time; see [3]. In a second step, we scale $G_\square$ by a factor of two: a vertex at coordinates (a, b) is mapped to coordinates $(2a, 2b)$; edges of $G_\square$ are stretched accordingly. This introduces *buffer coordinates* between different edges in the embedding of $G_\square$. Finally, we place a cell at every integer coordinate spanned by $G_\square$ and set the (x, y)-values of the cell c as follows:

- If c corresponds to a vertex of $G_\square$ (vertex-cell) set the weights to $(n, 1)$;
- if c corresponds to an edge of $G_\square$ (edge-cell) set the weights to $(1, 1)$;
- set the weights of all remaining cells to $(0, 0)$.

Add all vertex-cells to $\mathcal{F}$ and all edge-cells incident to a vertex-cell to $\mathcal{T}$. The size of the resulting instance and the construction time are polynomial in n.

It remains to prove that this instance has m cells that ignite all cells in $\mathcal{T}$ iff G has a vertex cover of size m. On the one hand, if G has a vertex cover $\mathcal{C}$ of size m, the corresponding vertex-cells can be chosen to put on fire. Due to the choice of values, they will ignite all adjacent edge-cells. Since $\mathcal{C}$ is a vertex cover, all edge-cells will burn and therefore all cells in $\mathcal{T}$. On the other hand, assume there is a subset $\mathcal{S} \subset \mathcal{F}$ of size m that will eventually ignite all cells of $\mathcal{T}$. Due to the construction, all cells of $\mathcal{S}$ are vertex-cells, so only vertex-cells are put on fire. Observe that due to the choice of weights, a burning edge-cell can never ignite a neighbouring vertex-cell. Moreover, due to the buffer coordinates, any two edge-cells that belong to different edges of $G_\square$ are separated by at least one cell of weight $(0,0)$. Consequently, every edge-cell in $\mathcal{T}$ must be either ignited by its direct neighbouring vertex-cell or by a fire reaching it from the direct neighbouring edge-cell, emanating from a different vertex-cell. Since for every edge in G there is an edge-cell in $\mathcal{T}$, which is ignited via one of the adjacent vertex-cells of $\mathcal{S}$, the vertices of G corresponding to the vertex-cells in $\mathcal{S}$ constitute a vertex cover of size m for G.

Certainly the problem is in NP, since the subset-certificate of cells $\mathcal{S}$ can be verified in polynomial time via standard simulation. Thus, we obtain the following theorem.

Theorem 3. *Given an instance of the basic hexagonal model, $m \in \mathbb{N}$ and two finite sets of cells $\mathcal{F}, \mathcal{T}$. It is NP-complete to decide whether there is a subset $\mathcal{S} \subseteq \mathcal{F}$ with $|\mathcal{S}| \leq m$ such that putting all cells of $\mathcal{S}$ on fire will eventually ignite all cells of $\mathcal{T}$.*

Note that the hardness proof requires $\mathcal{T}$ to be possibly of size linear in $|\mathcal{F}|$. Restricting $\mathcal{T}$ to a single cell, we obtain a problem that might very well be easier to solve. We do not know, whether this restricted problem is still NP-hard and leave this open. Definitely, the restricted problem becomes undecidable in simple variants of the basic model, as the following section shows.

3 Variants of Basic Hexagonal Model

Our basic hexagonal model can be modified in many ways to model different circumstances, environments or other known problems.

Certainly, the basic model can also be defined for other types of lattices, like the rectilinear square lattice. Thus, the model covers grid versions of firefighting problems, e.g. [2], as a special case: Set $x(c,0) = 1$ and $y(c,0) = \infty$ for each cell c and model the blocking of c at time t via $x(c,t) = \infty$. In general, the values for x and y could be replaced by positive (not necessarily monotone) functions of the simulation time.

Another natural generalization is to stack several layers of cells on top of each other. For every cell, the x- and y-values could be defined for each layer individually. These extensions allow to model fire expansion in different heights, such as crown or ground fires.

Environmental factors can also be modelled by slightly adjusting the transition rules. For example, wind can be modelled by letting a burning cell decrease its neighbours' x-values by different amounts per round, depending on the direction in which the neighbour lies and in which direction the wind blows. Cooling down or regrowth of greenery can be modelled by having cells regain their x- or y-values if no neighbouring cells are burning. However, even given regrowth, seemingly similar models, like Conway's Game of Life, remain distinct.

Still, these variants can lead to surprisingly complex problems: With only three layers of cells in an infinite lattice with a constant description complexity, the question if putting fire to cell c will eventually ignite cell c' becomes undecidable. The problem remains undecidable, even with a single layer, if we allow cells to recover their initial x-values over time; see appendix of [9] for details.

Theorem 4. *In the version of our firefighting model, where regeneration or at least three layers of cells are allowed, there is no algorithm that can decide every instance of the following problem:*

Given a lattice with a finite description, a set of cells $\mathcal{F}$ and a single cell v. Each cell of $\mathcal{F}$ is set on fire at $t = 0$. Will cell v eventually catch fire?

4   Conclusion

In this paper, we present a new model for firefighting problems together with some solutions and hardness results. The basic hexagonal model is simple to understand and generalizes a discrete model that has been introduced before. It allows to incorporate additional parameters to model weather conditions or crown and ground fires. These extensions could be applied to single cells or the entire lattice.

Obvious questions are how to improve on the results and to widen their scopes. We did not address any *dynamic* aspects of firefighting, yet. How does the fire's propagation change, when single cells are fortified? Moreover, fortifying a path of cells takes time to refill an aircraft's water tanks and fly back and forth. Can this task be accomplished before the path is reached by the fire? Research on seemingly simple dynamic geometric problems [8,10] seem to indicate that one should not hope for provably optimal results in the basic hexagonal model, but strive for good approximations.

Acknowledgements. We thank all anonymous reviewers for their helpful comments and suggestions.

References

1. Bressan, A.: Differential inclusions and the control of forest fires. J. Differ. Equ. **243**(2), 179–207 (2007)
2. Develin, M., Hartke, S.G.: Fire containment in grids of dimension three and higher. Discrete Appl. Math. **155**(17), 2257–2268 (2007)
3. Eiglsperger, M., Fekete, S.P., Klau, G.W.: Orthogonal graph drawing. In: Drawing Graphs, Methods and Models, pp. 121–171 (1999)
4. Finbow, S., MacGillivray, G.: The firefighter problem: a survey of results, directions and questions. Australas. J. Comb. **43**, 57–78 (2009)
5. Fomin, F.V., Heggernes, P., van Leeuwen, E.J.: The firefighter problem on graph classes. Theor. Comput. Sci. **613**, 38–50 (2016)
6. Gardner, M.: Mathematical games: the fantastic combinations of john conway's new solitaire game "life". Sci. Am. **223**, 120–123 (1970)

7. Garey, M.R., Johnson, D.S.: The rectilinear Steiner tree problem in NP complete. SIAM J. Appl. Math. **32**, 826–834 (1977)
8. Kim, S.-S., Klein, R., Kübel, D., Langetepe, E., Schwarzwald, B.: Geometric firefighting in the half-plane. In: Friggstad, Z., Sack, J.-R., Salavatipour, M.R. (eds.) WADS 2019. LNCS, vol. 11646, pp. 481–494. Springer, Cham (2019). https://doi.org/10.1007/978-3-030-24766-9_35
9. Klein, R., Kübel, D., Langetepe, E., Sack, J.R., Schwarzwald, B.: A new model in firefighting theory. CoRR abs/1911.10341 (2019). https://arxiv.org/abs/1911.10341
10. Klein, R., Langetepe, E., Schwarzwald, B., Levcopoulos, C., Lingas, A.: On a fire fighter's problem. Int. J. Found. Comput. Sci. **30**(2), 231–246 (2019)
11. Pastor, E., Zárate, L., Planas, E., Arnaldos, J.: Mathematical models and calculation systems for the study of wildland fire behaviour. Prog. Energy Combust. Sci. **29**(2), 139–153 (2003)
12. Toffoli, T., Margolus, N.: Cellular Automata Machines: A New Environment for Modeling. MIT Press, Cambridge (1987)
13. Wolfram, S.: Statistical mechanics of cellular automata. Rev. Mod. Phys. **55**(3), 601 (1983)

An Algorithm for Strong Stability
in the Student-Project Allocation
Problem with Ties

Sofiat Olaosebikan$^{(\boxtimes)}$ ⓘ and David Manlove ⓘ

School of Computing Science, University of Glasgow, Glasgow, Scotland
s.olaosebikan.1@research.gla.ac.uk, David.Manlove@glasgow.ac.uk

Abstract. We study a variant of the *Student-Project Allocation problem with lecturer preferences over Students* where ties are allowed in the preference lists of students and lecturers (SPA-ST). We investigate the concept of *strong stability* in this context. Informally, a matching is *strongly stable* if there is no student and lecturer l such that if they decide to form a private arrangement outside of the matching via one of l's proposed projects, then neither party would be worse off and at least one of them would strictly improve. We describe the first polynomial-time algorithm to find a strongly stable matching or report that no such matching exists, given an instance of SPA-ST. Our algorithm runs in $O(m^2)$ time, where m is the total length of the students' preference lists.

1 Introduction

Matching problems, which generally involve the assignment of a set of agents to another set of agents based on preferences, have wide applications in many real-world settings, including, for example, allocating junior doctors to hospitals [25] and assigning students to projects [15]. In the context of assigning students to projects, each project is proposed by one lecturer and each student is required to provide a strictly-ordered preference list over the available projects that she finds acceptable. Also, lecturers may provide strictly-ordered preference lists over the students that find their projects acceptable, and/or over the projects that they propose. Typically, each project and lecturer have a specific capacity denoting the maximum number of students that they can accommodate. The goal is to find a *matching*, i.e., an assignment of students to projects that respects the stated preferences, such that each student is assigned at most one project, and the capacity constraints on projects and lecturers are not violated—the so-called *Student-Project Allocation problem* (SPA) [1,6,19].

Two major models of SPA exist in the literature: one permits preferences only from the students [15], while the other permits preferences from the students and lecturers [14,19]. In the latter case, three different variants have been

S. Olaosebikan was supported by a College of Science and Engineering Scholarship, University of Glasgow; whilst D. Manlove was supported by grant EP/P028306/1 from the Engineering and Physical Sciences Research Council.

M. Changat and S. Das (Eds.): CALDAM 2020, LNCS 12016, pp. 384–399, 2020.
https://doi.org/10.1007/978-3-030-39219-2_31

studied based on the nature of the lecturers' preference lists. These include SPA with lecturer preferences over (i) students [1], (ii) projects [12,21,22], and (iii) (student, project) pairs [2]. Outwith assigning students to projects, applications of each of these three variants can be seen in multi-cell networks where the goal is to find a stable association of users to channels at base-stations [3–5].

In this work, we will concern ourselves with variant (i), i.e., the *Student-Project Allocation problem with lecturer preferences over Students* (SPA-S). In this context, it has been argued in [25] that a natural property for a matching to satisfy is that of *stability*. Informally, a *stable matching* ensures that no student and lecturer would have an incentive to deviate from their current assignment. Abraham *et al.* [1] described two linear-time algorithms to find a stable matching in an instance of SPA-S where the preference lists are strictly ordered. In their paper, they also proposed an extension of SPA-S where the preference lists may include ties, known as the *Student-Project Allocation problem with lecturer preferences over Students with Ties* (SPA-ST) [23].

If we allow ties in the preference lists of students and lecturers, three stability definitions are possible, namely *weak stability*, *strong stability* and *super-stability* [8–10]. We give an informal definition in what follows. Suppose M is a matching in an instance of SPA-ST. Then M is (i) weakly stable, (ii) strongly stable, or (iii) super-stable, if there is no student and lecturer l such that if they decide to become assigned outside of M via one of l's proposed projects, respectively,

(i) both of them would strictly improve,
(ii) one of them would strictly improve and the other would not be worse off,
(iii) neither of them would be worse off.

Existing Results for SPA-ST. Manlove *et al.* [20] showed that every instance of SPA-ST admits a weakly stable matching, which could be of different sizes. Moreover, the problem of finding a maximum size weakly stable matching (MAX-SPA-ST) is NP-hard [11,20], even for the *Stable Marriage problem with Ties and Incomplete lists* (SMTI). Cooper and Manlove [7] described a $\frac{3}{2}$-approximation algorithm for MAX-SPA-ST. On the other hand, Irving *et al.* argued in [9] that super-stability is a natural and most robust solution concept to seek in cases where agents have incomplete information. Recently, Olaosebikan and Manlove [23] showed that if an instance of SPA-ST admits a super-stable matching M, then all weakly stable matchings in the instance are of the same size (equal to the size of M), and match exactly the same set of students. The main result of their paper was a polynomial-time algorithm to find a super-stable matching or report that no such matching exists, given an instance of SPA-ST. Their algorithm runs in $O(L)$ time, where L is the total length of all the preference lists.

Motivation for Strong Stability. It was motivated in [10] that weakly stable matching may be undermined by bribery or persuasion, in practical applications of the *Hospitals-Residents problem with Ties* (HRT). In what follows, we give a corresponding argument for an instance I of SPA-ST. Suppose that M is a weakly stable matching in I, and suppose that a student s_i prefers a project p_j (where p_j is offered by lecturer l_k) to her assigned project in M, say $p_{j'}$ (where $p_{j'}$ is offered by a lecturer different from l_k). Suppose further that p_j is full and l_k is indifferent between s_i and one of the worst student/s assigned to p_j in M, say $s_{i'}$. Clearly, the pair (s_i, p_j) does not constitute a blocking pair for the weakly stable matching M, as l_k would not improve by taking on s_i in the place of $s_{i'}$. However, s_i might be overly invested in p_j that she is ready to persuade or bribe l_k to reject $s_{i'}$ and accept her instead; l_k being indifferent between s_i and $s_{i'}$ may decide to accept s_i's proposal. We can reach a similar argument if the roles are reversed. However, if M is strongly stable, it cannot be potentially undermined by this type of (student, project) pair.

Henceforth, if a SPA-ST instance admits a strongly stable matching, we say that such instance is solvable. Unfortunately not every instance of SPA-ST is solvable. To see this, consider the case where there are two students, two projects and two lecturers, the capacity of each project and lecturer is 1, the students have exactly the same strictly-ordered preference list of length 2, and each of the lecturers preference list is a single tie of length 2 (any matching will be undermined by a student and lecturer that are not assigned together). However, it should be clear from the discussions above that in cases where a strongly stable matching exists, it should be preferred over a matching that is merely weakly stable. Previous results for strong stability in the literature include [8, 10, 13, 16, 18].

Our Contribution. We present the first polynomial-time algorithm to find a strongly stable matching or report that no such matching exists, given an instance of SPA-ST—thus solving an open problem given in [1, 23]. Our algorithm is student-oriented, which implies that if the given instance is solvable then our algorithm will output a solution in which each student has at least as good a project as she could obtain in any strongly stable matching. We note that our algorithm is a non-trivial extension of the strong stability algorithms for SMT (*Stable Marriage problem with Ties*), SMTI and HRT described in [8, 10, 18] (we discuss this further in [24, Sect. 4.3]).

The remainder of this paper is structured as follows. We give a formal definition of the SPA-S problem, the SPA-ST variant, and the three stability concepts in Sect. 2. We describe our algorithm for SPA-ST under strong stability in Sect. 3. Further, in Sect. 3, we illustrate an execution of our algorithm with respect to an instance of SPA-ST before moving on to present the algorithm's correctness and complexity results (all omitted proofs can be found in [24, Sect. 4.5]). Finally, we present some potential directions for future work in Sect. 4.

2   Preliminary Definitions

In this section, we give a formal definition of SPA-S as described in the literature [1, 23]. We also give a formal definition of SPA-ST as described in [23], which is a generalisation of SPA-S in which preference lists can include ties.

2.1   Formal Definition of SPA-S

An instance I of SPA-S involves a set $\mathcal{S} = \{s_1, s_2, \ldots, s_{n_1}\}$ of *students*, a set $\mathcal{P} = \{p_1, p_2, \ldots, p_{n_2}\}$ of *projects* and a set $\mathcal{L} = \{l_1, l_2, \ldots, l_{n_3}\}$ of *lecturers*. Each student s_i ranks a subset of $\mathcal{P}$ in strict order, which forms her preference list. We say that s_i finds p_j *acceptable* if p_j appears on s_i's preference list. We denote by A_i the set of projects that s_i finds acceptable.

Each lecturer $l_k \in \mathcal{L}$ offers a non-empty set of projects P_k, where $P_1, P_2, \ldots, P_{n_3}$ partitions $\mathcal{P}$, and l_k provides a preference list, denoted by $\mathcal{L}_k$, ranking in strict order of preference those students who find at least one project in P_k acceptable. Also l_k has a capacity $d_k \in \mathbb{Z}^+$, indicating the maximum number of students she is willing to supervise. Similarly each project $p_j \in \mathcal{P}$ has a capacity $c_j \in \mathbb{Z}^+$ indicating the maximum number of students that it can accommodate. We assume that for any lecturer l_k, $\max\{c_j : p_j \in P_k\} \leq d_k \leq \sum\{c_j : p_j \in P_k\}$ (i.e., the capacity of l_k is (i) at least the highest capacity of the projects offered by l_k, and (ii) at most the sum of the capacities of all the projects l_k is offering). We denote by $\mathcal{L}_k^j$, the *projected preference list* of lecturer l_k for p_j, which can be obtained from $\mathcal{L}_k$ by removing those students that do not find p_j acceptable (thereby retaining the order of the remaining students from $\mathcal{L}_k$).

Given a pair $(s_i, p_j) \in \mathcal{S} \times \mathcal{P}$, where p_j is offered by l_k, we refer to (s_i, p_j) as an *acceptable pair* if $p_j \in A_i$ and $s_i \in \mathcal{L}_k$. An *assignment* M is a collection of acceptable pairs in $\mathcal{S} \times \mathcal{P}$. If $(s_i, p_j) \in M$, we say that s_i *is assigned to* p_j, and p_j *is assigned* s_i. For convenience, if s_i is assigned in M to p_j, where p_j is offered by l_k, we may also say that s_i *is assigned to* l_k, and l_k *is assigned* s_i. For any project $p_j \in \mathcal{P}$, we denote by $M(p_j)$ the set of students assigned to p_j in M. Project p_j is *undersubscribed*, *full* or *oversubscribed* according as $|M(p_j)|$ is less than, equal to, or greater than c_j, respectively. Similarly, for any lecturer $l_k \in \mathcal{L}$, we denote by $M(l_k)$ the set of students assigned to l_k in M. Lecturer l_k is *undersubscribed*, *full* or *oversubscribed* according as $|M(l_k)|$ is less than, equal to, or greater than d_k, respectively. A *matching* M is an assignment such that $|M(s_i)| \leq 1$, $|M(p_j)| \leq c_j$ and $|M(l_k)| \leq d_k$. If s_i is assigned to some project in M, we let $M(s_i)$ denote that project; otherwise $M(s_i)$ is undefined.

2.2   Ties in the Preference Lists

We now give a formal definition, similar to the one given in [23], for the generalisation of SPA-S in which the preference lists can include ties. In the preference list of lecturer $l_k \in \mathcal{L}$, a set T of r students forms a *tie of length r* if l_k does not prefer s_i to $s_{i'}$ for any $s_i, s_{i'} \in T$ (i.e., l_k is *indifferent* between s_i and $s_{i'}$). A tie in a student's preference list is defined similarly. For convenience, in what follows we consider a non-tied entry in a preference list as a tie of length one. We denote by SPA-ST the generalisation of SPA-S in which the preference list of each student (respectively lecturer) comprises a strict ranking of ties, each comprising one or more projects (respectively students). An example SPA-ST instance I_1 is given in Fig. 1, which involves the set of students $\mathcal{S} = \{s_1, s_2, s_3\}$, the set of projects

Student preferences	Lecturer preferences	offers
s_1: $(p_1 \quad p_2)$	l_1: $s_3 \ (s_1 \quad s_2)$	p_1, p_2
s_2: $p_2 \quad p_3$	l_2: $(s_3 \quad s_2)$	p_3
s_3: $p_3 \quad p_1$		
	Project capacities: $c_1 = c_2 = c_3 = 1$	
	Lecturer capacities: $d_1 = 2, \ d_2 = 1$	

Fig. 1. An example SPA-ST instance I_1.

$\mathcal{P} = \{p_1, p_2, p_3\}$ and the set of lecturers $\mathcal{L} = \{l_1, l_2\}$. Ties in the preference lists are indicated by round brackets.

In the context of SPA-ST, we assume that all notation and terminology carries over from SPA-S with the exception of stability, which we now define. When ties appear in the preference lists, three types of stability arise, namely *weak stability, strong stability and super-stability* [8–10]. For our purpose in this paper, we only give a formal definition of strong stability in the context of SPA-ST. Henceforth, I is an instance of SPA-ST, (s_i, p_j) is an acceptable pair in I and l_k is the lecturer who offers p_j.

Definition 1 (Strong stability). We say that M is *strongly stable* in I if it admits no blocking pair, where a *blocking pair* for M is an acceptable pair $(s_i, p_j) \in (\mathcal{S} \times \mathcal{P}) \setminus M$ such that either (1a and 1b) or (2a and 2b) holds as follows:

(1a) either s_i is unassigned in M, or s_i prefers p_j to $M(s_i)$;
(1b) either (i), (ii), or (iii) holds as follows:
   (i) p_j is undersubscribed and l_k is undersubscribed;
   (ii) p_j is undersubscribed, l_k is full, and either $s_i \in M(l_k)$ or l_k prefers s_i to the worst student/s in $M(l_k)$ or is indifferent between them;
   (iii) p_j is full and l_k prefers s_i to the worst student/s in $M(p_j)$ or is indifferent between them.
(2a) s_i is indifferent between p_j and $M(s_i)$;
(2b) either (i), (ii), or (iii) holds as follows:
   (i) p_j is undersubscribed, l_k is undersubscribed and $s_i \notin M(l_k)$;
   (ii) p_j is undersubscribed, l_k is full, $s_i \notin M(l_k)$, and l_k prefers s_i to the worst student/s in $M(l_k)$;
   (iii) p_j is full and l_k prefers s_i to the worst student/s in $M(p_j)$.

Some intuition for the strong stability definition is given in [24, Sect. 3]. In the remainder of this paper, any usage of the term *blocking pair* refers to the version of this term for strong stability as defined above.

3   An Algorithm for SPA-ST under strong stability

In this section we present our algorithm for SPA-ST under strong stability, which we will refer to as `Algorithm SPA-ST-strong`. In Sect. 3.1, we give some definitions relating to the algorithm. In Sect. 3.2, we give a description of our algorithm and present it in pseudocode form. We illustrate an execution of our algorithm with respect to a SPA-ST instance in Sect. 3.3. Finally, we present the algorithm's correctness and complexity results in Sect. 3.4.

3.1 Definitions Relating to the Algorithm

Given a pair $(s_i, p_j) \in M$, for some strongly stable matching M in I, we call (s_i, p_j) a *strongly stable pair*. During the execution of the algorithm, students become *provisionally assigned* to projects (and implicitly to lecturers), and it is possible for a project (and lecturer) to be provisionally assigned a number of students that exceeds its capacity. We describe a project (respectively lecturer) as *replete* if at any time during the execution of the algorithm it has been full or oversubscribed. We say that a project (respectively lecturer) is *non-replete* if it is not replete.

The *provisional assignment graph* is an undirected bipartite graph $G = (S \cup P, E)$, with $S \subseteq \mathcal{S}$ and $P \subseteq \mathcal{P}$ such that there is an edge $(s_i, p_j) \in E$ if and only if s_i is provisionally assigned to p_j. During the execution of the algorithm, it is possible for a student to be adjacent to more than one project in G. Thus, we denote by $G(s_i)$ the set of projects that are adjacent to s_i in G. Given a project $p_j \in P$, we denote by $G(p_j)$ the set of students who are provisionally assigned to p_j in G and we let $d_G(p_j) = |G(p_j)|$. Similarly, we denote by $G(l_k)$ the set of students who are provisionally assigned to a project offered by l_k in G and we let $d_G(l_k) = |G(l_k)|$.

As stated earlier, for a project p_j, it is possible that $d_G(p_j) > c_j$ at some point during the algorithm's execution. Thus, we denote by $q_j = \min\{c_j, d_G(p_j)\}$ the *quota of p_j in G*, which is the minimum between p_j's capacity and the number of students who are provisionally assigned to p_j in G. Similarly, for a lecturer l_k, it is possible that $d_G(l_k) > d_k$ at some point during the algorithm's execution. At this point, we denote by $\alpha_k = \sum\{q_j : p_j \in P_k \cap P\}$ the total quota of projects offered by l_k that is provisionally assigned to students in G and we denote by $q_k = \min\{d_k, d_G(l_k), \alpha_k\}$ the *quota of l_k in G*.

The algorithm proceeds by deleting from the preference lists certain (s_i, p_j) pairs that are not strongly stable. By the term *delete* (s_i, p_j), we mean the removal of p_j from s_i's preference list and the removal of s_i from $\mathcal{L}_k^j$ (the projected preference list of lecturer l_k for p_j); in addition, if $(s_i, p_j) \in E$ we delete the edge from G. By the *head* and *tail* of a preference list at a given point we mean the first and last tie respectively on that list after any deletions might have occurred (recalling that a tie can be of length 1). Given a project p_j, we say that a student s_i is *dominated in* $\mathcal{L}_k^j$ if s_i is worse than at least c_j students who are provisionally assigned to p_j. The concept of a student becoming dominated in a lecturer's preference list is defined in a slightly different manner.

Definition 2 (Dominated in $\mathcal{L}_k$). At a given point during the algorithm's execution, let α_k and $d_G(l_k)$ be as defined above. We say that a student s_i is *dominated in* $\mathcal{L}_k$ if $\min\{d_G(l_k), \alpha_k\} \geq d_k$, and s_i is worse than at least d_k students who are provisionally assigned in G to a project offered by l_k.

Definition 3 (Lower rank edge). We define an edge $(s_i, p_j) \in E$ as a *lower rank edge* if s_i is in the tail of $\mathcal{L}_k$ and $\min\{d_G(l_k), \alpha_k\} > d_k$.

Definition 4 (Bound). Given an edge $(s_i, p_j) \in E$, we say that s_i is *bound to* p_j if (i) p_j is not oversubscribed or s_i is not in the tail of $\mathcal{L}_k^j$ (or both), and (ii) (s_i, p_j) is not a lower rank edge or s_i is not in the tail of $\mathcal{L}_k$ (or both). If s_i is bound to p_j, we may also say that (s_i, p_j) is a *bound edge*. Otherwise, we refer to it as an *unbound edge*.[1]

We form a *reduced assignment graph* $G_r = (S_r, P_r, E_r)$ from a provisional assignment graph G as follows. For each edge $(s_i, p_j) \in E$ such that s_i is bound to p_j, we remove the edge (s_i, p_j) from G_r and we reduce the quota of p_j in G_r (and implicitly $l_k{}^2$) by one. Further, we remove all other unbound edges incident to s_i in G_r. Each isolated student vertex is then removed from G_r. Finally, if the quota of any project is reduced to 0, or p_j becomes an isolated vertex, then p_j is removed from G_r. For each surviving p_j in G_r, we denote by q_j^* the *revised quota of* p_j, where q_j^* is the difference between p_j's quota in G (i.e., q_j) and the number of students that are bound to p_j. Similarly, we denote by q_k^* the *revised quota of* l_k in G_r, where q_k^* is the difference between l_k's quota in G (i.e., q_k) and the number of students that are bound to a project offered by l_k. Further, for each l_k who offers at least one project in G_r, we let $n = \sum\{q_j^* : p_j \in P_k \cap P_r\} - q_k^*$, where n is the difference between the total revised quota of projects in G_r that are offered by l_k and the revised quota of l_k in G_r. Now, if $n \leq 0$, we do nothing; otherwise, we extend G_r as follows. We add n dummy student vertices to S_r. For each of these dummy vertices, say s_{d_i}, and for each project $p_j \in P_k \cap P_r$ that is adjacent to a student vertex in S_r via a lower rank edge, we add the edge (s_{d_i}, p_j) to E_r.[3]

Given a set $X \subseteq S_r$ of students, define $\mathcal{N}(X)$, the *neighbourhood of* X, to be the set of project vertices adjacent in G_r to a student in X. If for all subsets X of S_r, each student in X can be assigned to one project in $\mathcal{N}(X)$, without exceeding the revised quota of each project in $\mathcal{N}(X)$ (i.e., $|X| \leq \sum\{q_j^* : p_j \in \mathcal{N}(X)\}$ for all $X \subseteq S_r$); then we say that G_r admits a *perfect matching* that saturates S_r.

Definition 5 (Critical set). It is well known in the literature [17] that if G_r does not admit a perfect matching that saturates S_r, then there must exist a *deficient* subset $Z \subseteq S_r$ such that $|Z| > \sum\{q_j^* : p_j \in \mathcal{N}(Z)\}$. To be precise, the *deficiency* of Z is defined by $\delta(Z) = |Z| - \sum\{q_j^* : p_j \in \mathcal{N}(Z)\}$. The *deficiency* of G_r, denoted $\delta(G_r)$, is the maximum deficiency taken over all subsets of S_r.

---

[1] An edge $(s_i, p_j) \in E$ can change state from *bound* to *unbound*, but not vice versa.

[2] If s_i is bound to more than one projects offered by l_k, for all the bound edges involving s_i and these projects that we remove from G_r, we only reduce l_k's quota in G_r by one.

[3] An intuition as to why we add dummy students to G_r is as follows. Given a lecturer l_k whose project is provisionally assigned to a student in G_r. If $q_k^* < \sum\{q_j^* : p_j \in P_k \cap P_r\}$, then we need n dummy students to offset the difference between $\sum\{q_j^* : p_j \in P_k \cap P_r\}$ and q_k^*, so that we do not oversubscribe l_k in any maximum matching obtained from G_r.

Thus, if $\delta(Z) = \delta(G_r)$, we say that Z is a *maximally deficient* subset of S_r, and we refer to Z as a *critical set*.

We denote by P_R the set of replete projects in G and we denote by P_R^* a subset of projects in P_R which is obtained as follows. For each project $p_j \in P_R$, let l_k be the lecturer who offers p_j. For each student s_i such that (s_i, p_j) has been deleted, we add p_j to P_R^* if (i) and (ii) holds as follows:

(i) either s_i is unassigned in G, or $(s_i, p_{j'}) \in G$ where s_i prefers p_j to $p_{j'}$, or $(s_i, p_{j'}) \in G$ and s_i is indifferent between p_j and $p_{j'}$ where $p_{j'} \notin P_k$;
(ii) either l_k is undersubscribed in G, or l_k is full in G and either $s_i \in G(l_k)$ or l_k prefers s_i to some student assigned to l_k in G.

Definition 6 (Feasible matching). A *feasible matching* in the final provisional assignment graph G is a matching M obtained as follows:

1. Let G^* be the subgraph of G induced by the students who are adjacent to a project in P_R^*. First, find a maximum matching M^* in G^*;
2. Using M^* as an initial solution, find a maximum matching M in G.

3.2 Description of the Algorithm

Algorithm SPA-ST-strong, described in Algorithm 1, begins by initialising an empty bipartite graph G which will contain the provisional assignments of students to projects (and implicitly to lecturers). We remark that such assignments (i.e., edges in G) can subsequently be broken during the algorithm's execution.

The while loop of the algorithm involves each student s_i who is not adjacent to any project in G and who has a non-empty list applying in turn to each project p_j at the head of her list. Immediately, s_i becomes provisionally assigned to p_j in G (and to l_k). If, by gaining a new provisional assignee, project p_j becomes full or oversubscribed then we set p_j as replete. Further, for each student s_t in $\mathcal{L}_k^j$, such that s_t is dominated in $\mathcal{L}_k^j$, we delete the pair (s_t, p_j). As we will prove later, such pairs cannot belong to any strongly stable matching. Similarly, if by gaining a new provisional assignee, l_k becomes full or oversubscribed then we set l_k as replete. For each student s_t in $\mathcal{L}_k$, such that s_t is dominated in $\mathcal{L}_k$ and for each project $p_u \in P_k$ that s_t finds acceptable, we delete the pair (s_t, p_u). This continues until every student is provisionally assigned to one or more projects or has an empty list. At the point where the while loop terminates, we form the reduced assignment graph G_r and we find the critical set Z of students in G_r (we describe how to find Z on Page 9). As we will see later, no project $p_j \in \mathcal{N}(Z)$ can be assigned to any student in the tail of $\mathcal{L}_k^j$ in any strongly stable matching, so all such pairs are deleted.

At the termination of the inner repeat-until loop in line 21, i.e., when Z is empty, if some project p_j that is replete ends up undersubscribed, we carry out some certain deletions[4]. We let s_r be any one of the most preferred students

---

[4] This type of deletion was also carried out in Algorithm SPA-ST-super for superstability [23].

(according to $\mathcal{L}_k^j$) who was provisionally assigned to p_j during some iteration of the algorithm but is not assigned to p_j at this point (for convenience, we henceforth refer to such s_r as the most preferred student rejected from p_j according to $\mathcal{L}_k^j$). If the students at the tail of $\mathcal{L}_k$ (recalling that the tail of $\mathcal{L}_k$ is the least-preferred tie in $\mathcal{L}_k$ after any deletions might have occurred) are no better than s_r, it turns out that none of these students s_t can be assigned to any project offered by l_k in any strongly stable matching – such pairs (s_t, p_u), for each project $p_u \in P_k$ that s_t finds acceptable, are deleted. The repeat-until loop is then potentially reactivated, and the entire process continues until every student is provisionally assigned to a project or has an empty list.

At the termination of the outer repeat-until loop in line 30, if a student is adjacent in G to a project p_j via a bound edge, then we may potentially carry out extra deletions. First, we let l_k be the lecturer who offers p_j and we let U be the set of projects that are adjacent to s_i in G via an unbound edge. For each project $p_u \in U \setminus P_k$, it turns out that the pair (s_i, p_u) cannot belong to any strongly stable matching, thus we delete all such pairs. Finally, we let M be any feasible matching in the provisional assignment graph G. If M is strongly stable relative to the given instance I then M is output as a strongly stable matching in I. Otherwise, the algorithm reports that no strongly stable matching exists in I. We present Algorithm SPA-ST-strong in pseudocode form in Algorithm 1.

Finding the Critical Set. Consider the reduced assignment graph $G_r = (S_r, P_r, E_r)$ formed from G at a given point during the algorithm's execution (at line 15). To find the critical set of students in G_r, first we need to construct a maximum matching M_r in G_r, with respect to the revised quota q_j^*, for each $p_j \in P_r$. In this context, a *matching* $M_r \subseteq E_r$ is such that $|M_r(s_i)| \leq 1$ for all $s_i \in S_r$, and $|M_r(p_j)| \leq q_j^*$ for all $p_j \in P_r$. We describe how to construct M_r as follows:

1. Let G_r' be the subgraph of G_r induced by the dummy students adjacent to a project in G_r. First, find a maximum matching M_r' in G_r'.
2. Using M_r' as an initial solution, find a maximum matching M_r in G_r.[5]

Given a maximum matching M_r in the reduced assignment graph G_r, the critical set Z consists of the set U of unassigned students together with the set U' of students reachable from a student in U via an alternating path (see [24, Lemma 1] for a proof).

3.3   Example Algorithm Execution

In this section, we illustrate an execution of Algorithm SPA-ST-strong with respect to the SPA-ST instance I_3 shown in Fig. 2 (Page 10), which involves the set of students $\mathcal{S} = \{s_i : 1 \leq i \leq 8\}$, the set of projects $\mathcal{P} = \{p_j : 1 \leq j \leq 6\}$ and

---

[5] By making sure that all the dummy students are matched in step 1, we are guaranteed that no lecturer is oversubscribed with non-dummy students in G_r.

Algorithm 1. Algorithm SPA-ST-strong

Input: SPA-ST instance I

Output: a strongly stable matching in I or "no strongly stable matching exists in I"

```
 1: G ← ∅
 2: repeat
 3:     repeat
 4:         while some student sᵢ is unassigned and has a non-empty list do
 5:             for each project pⱼ at the head of sᵢ's list do
 6:                 lₖ ← lecturer who offers pⱼ
 7:                 add the edge (sᵢ, pⱼ) to G
 8:                 if pⱼ is full or oversubscribed then
 9:                     for each student sₜ dominated in Lₖʲ do
10:                         delete (sₜ, pⱼ)
11:                 if lₖ is full or oversubscribed then
12:                     for each student sₜ dominated in Lₖ do
13:                         for each project pᵤ ∈ Pₖ ∩ Aₜ do
14:                             delete (sₜ, pᵤ)
15:         form the reduced assignment graph Gᵣ
16:         find the critical set Z of students
17:         for each project pᵤ ∈ N(Z) do
18:             lₖ ← lecturer who offers pᵤ
19:             for each student sₜ at the tail of Lₖᵘ do
20:                 delete (sₜ, pᵤ)
21:     until Z is empty
22:     for each pⱼ ∈ P do
23:         if pⱼ is replete and pⱼ is undersubscribed then
24:             lₖ ← lecturer who offers pⱼ
25:             sᵣ ← most preferred student rejected from pⱼ in Lₖʲ {any if > 1}
26:             if the students at the tail of Lₖ are no better than sᵣ then
27:                 for each student sₜ at the tail of Lₖ do
28:                     for each project pᵤ ∈ Pₖ ∩ Aₜ do
29:                         delete (sₜ, pᵤ)
30: until every unassigned student has an empty list
31: for each student sᵢ in G do
32:     if sᵢ is adjacent in G to a project pⱼ via a bound edge then
33:         lₖ ← lecturer who offers pⱼ
34:         U ← unbound projects adjacent to sᵢ in G
35:         for each pᵤ ∈ U \ Pₖ do
36:             delete (sᵢ, pᵤ)
37: M ← a feasible matching in G
38: if M is a strongly stable matching in I then
39:     return M
40: else
41:     return "no strongly stable matching exists in I"
```

the set of lecturers $\mathcal{L} = \{l_k : 1 \leq k \leq 3\}$. The algorithm starts by initialising the bipartite graph $G = \{\}$, which will contain the provisional assignment of students to projects. We assume that the students become provisionally assigned to each

Student preferences	Lecturer preferences	offers
s_1: p_1 p_6	$\{3\}$ l_1: s_8 s_7 $(s_1$ s_2 $s_3)$ $(s_4$ $s_5)$ s_6	p_1, p_2
s_2: p_1 p_2	$\{2\}$ l_2: s_6 s_5 $(s_7$ $s_3)$	p_3, p_4
s_3: $(p_1$ $p_4)$	$\{3\}$ l_3: $(s_1$ $s_4)$ s_8	p_5, p_6
s_4: p_2 $(p_5$ $p_6)$		
s_5: $(p_2$ $p_3)$		
s_6: $(p_2$ $p_4)$	Project capacities: $c_1 = c_2 = c_6 = 2$, $c_3 = c_4 = c_5 = 1$	
s_7: p_3 p_1	Lecturer capacities: $d_1 = d_3 = 3$, $d_2 = 2$	
s_8: p_5 p_1		

Fig. 2. An instance I_3 of SPA-ST.

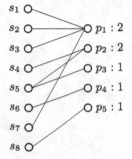 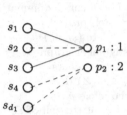 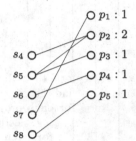

(a) The provisional assignment graph $G^{(1)}$ at the end of the **while** loop, with the quota of each project labelled beside it.

(b) The reduced assignment graph $G_r^{(1)}$, with the revised quota of each project labelled beside it. The collection of the dashed edges is the maximum matching $M_r^{(1)}$.

(c) The provisional assignment graph $G^{(1)}$ at the termination of iteration (1).

Fig. 3. Iteration (1).

project at the head of their list in subscript order. Figures 3, 4 and 5 illustrate how this execution of Algorithm SPA-ST-strong proceeds with respect to I_3.

Iteration 1: At the termination of the **while** loop during the first iteration of the inner **repeat-until** loop, every student, except s_3, s_6 and s_7, is provisionally assigned to every project in the first tie on their preference list. Edge $(s_3, p_4) \notin G^{(1)}$ because (s_3, p_4) was deleted as a result of s_6 becoming provisionally assigned to p_4, causing s_3 to be dominated in $\mathcal{L}_2^4$. Also, edge $(s_6, p_2) \notin G^{(1)}$ because (s_6, p_2) was deleted as a result of s_4 becoming provisionally assigned to p_2, causing s_6 to be dominated in $\mathcal{L}_1$ (at that point in the algorithm, $\min\{d_G(l_1), \alpha_1\} = \min\{4, 3\} = 3 = d_1$ and s_6 is worse than at least d_1 students who are provisionally assigned to l_1). Finally, edge $(s_7, p_3) \notin G^{(1)}$ because (s_7, p_3) was deleted as a result of s_5 becoming provisionally assigned to p_5, causing s_7 to be dominated in $\mathcal{L}_2^3$.

To form $G_r^{(1)}$, the bound edges $(s_5, p_3), (s_6, p_4), (s_7, p_1)$ and (s_8, p_5) are removed from the graph. We can verify that edges (s_4, p_2) and (s_5, p_2) are unbound, since they are lower rank edges for l_1. Also, since p_1 is oversubscribed, and each of s_1, s_2 and s_3 is at the tail of $\mathcal{L}_1^1$, edges (s_1, p_1), (s_2, p_1) and (s_3, p_1) are unbound. Further, the revised quota of l_1 in $G_r^{(1)}$ is 2, and the total revised quota of projects offered by l_1 (i.e., p_1 and p_2) is 3. Thus, we add one dummy student vertex s_{d_1} to G_r^1, and we add an edge between s_{d_1} and p_2 (since p_2 is the only project in $G_r^{(1)}$ adjacent to a student in the tail of $\mathcal{L}_1$ via a lower rank edge). With respect to the maximum matching $M_r^{(1)}$, it is clear that the critical set $Z^{(1)} = \{s_1, s_2, s_3\}$, thus we delete the edges (s_1, p_1), (s_2, p_1) and (s_3, p_1) from $G^{(1)}$; and the inner `repeat-until` loop is reactivated.

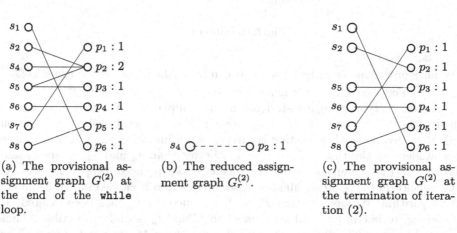

(a) The provisional assignment graph $G^{(2)}$ at the end of the `while` loop.

(b) The reduced assignment graph $G_r^{(2)}$.

(c) The provisional assignment graph $G^{(2)}$ at the termination of iteration (2).

Fig. 4. Iteration (2).

Iteartion 2: At the beginning of this iteration, each of s_1 and s_2 is unassigned and has a non-empty list; thus we add edges (s_1, p_6) and (s_2, p_2) to the provisional assignment graph obtained at the termination of iteration (1) to form $G_r^{(2)}$. It can be verified that every edge in $G_r^{(2)}$, except (s_4, p_2) and (s_5, p_2), is a bound edge. Clearly, the critical set $Z^{(2)} = \emptyset$, thus the inner `repeat-until` loop terminates. At this point, project p_1, which was replete during iteration (1), is undersubscribed in iteration (2). Moreover, the students at the tail of $\mathcal{L}_1$ (i.e., s_4 and s_5) are no better than s_3, where s_3 is one of the most preferred students rejected from p_1 according to $\mathcal{L}_1^1$; thus we delete edges (s_4, p_2) and (s_5, p_2). The outer `repeat-until` loop is then reactivated (since s_4 is unassigned and has a non-empty list).

Iteration 3: At the beginning of this iteration, the only student that is unassigned and has a non-empty list is s_4; thus we add edges (s_4, p_5) and (s_4, p_6) to the provisional assignment graph obtained at the termination of iteration (2) to form $G_r^{(3)}$. The provisional assignment of s_4 to p_5 led to p_5 becoming oversubscribed; thus (s_8, p_5) is deleted (since s_8 is dominated on $\mathcal{L}_3^5$). Further, s_8

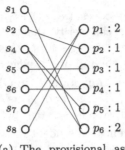

(a) The provisional assignment graph $G^{(3)}$ at the end of the **while** loop.

Fig. 5. Iteration (3).

becomes provisionally assigned to p_1. It can be verified that all the edges in $G_r^{(3)}$ are bound edges. Moreover, the reduced assignment graph $G_r^{(3)} = \emptyset$.

Again, every unassigned students has an empty list. We also have that a project p_2, which was replete in iteration (2), is undersubscribed in iteration (3). However, no further deletion is carried out in line 29 of the algorithm, since the student at the tail of $\mathcal{L}_1$ (i.e., s_2) is better than s_4 and s_5, where s_4 and s_5 are the most preferred students rejected from p_2 according to $\mathcal{L}_1^2$. Hence, the **repeat-until** loop terminates. Also, no deletion is carried out in line 36 of the algorithm. We observe that $P_R^* = \{p_5\}$, since (s_8, p_5) has been deleted, s_8 prefers p_5 to her provisional assignment in G and l_3 is undersubscribed. Thus we need to ensure p_5 fills up in the feasible matching M constructed from G, so as to avoid (s_8, p_5) from blocking M. Finally, the algorithm outputs the feasible matching $M = \{(s_1, p_6), (s_2, p_2), (s_4, p_5), (s_5, p_3), (s_6, p_4), (s_7, p_1), (s_8, p_1)\}$ as a strongly stable matching in I_3.

3.4 Correctness of the algorithm

The correctness and complexity of **Algorithm SPA-ST-strong** is established via a sequence of lemmas, namely Lemmas 4–14 in [24, Sect. 4.5]. These are omitted here for space reasons, but may be summarised as follows:

1. no strongly stable pair is deleted during the execution of the algorithm;
2. no strongly stable matching exists if some:
   - (a) non-replete lecturer l_k has fewer assignees in the feasible matching M than provisional assignees in the final assignment graph G, or
   - (b) replete lecturer is not full in M, or
   - (c) student is bound to two or more projects that are offered by different lecturers, or
   - (d) pair (s_i, p_j) was deleted where p_j is offered by l_k, each of p_j and l_k is undersubscribed in M, and for any $p_{j'} \in P_k$ such that s_i is indifferent between p_j and $p_{j'}$, $(s_i, p_{j'}) \notin M$;

3. if the algorithm outputs "no strongly stable matching exists" then at least one of the properties in (2) above must hold;
4. Algorithm SPA-ST-strong may be implemented to run in $O(m^2)$ time, where m is the total length of the students' preference lists.

The following theorem collects together Lemmas 4–14 in [24] and establishes the correctness and complexity of Algorithm SPA-ST-strong.

Theorem 1. *For a given instance I of* SPA-ST, *Algorithm SPA-ST-strong determines in $O(m^2)$ time whether or not a strongly stable matching exists in I. If such a matching does exist, all possible executions of the algorithm find one in which each assigned student is assigned at least as good a project as she could obtain in any strongly stable matching, and each unassigned student is unassigned in every strongly stable matchings.*

Given the optimality property established by Theorem 1, we define the strongly stable matching found by Algorithm SPA-ST-strong to be *student-optimal*. For example, in the SPA-ST instance illustrated in Fig. 1, it may be verified that the student-optimal strongly stable matching is $\{(s_1, p_1), (s_2, p_2), (s_3, p_3)\}$.

4   Conclusion

We leave open the formulation of a lecturer-oriented counterpart to Algorithm SPA-ST-strong. From an experimental perspective, an interesting direction would be to carry out an empirical analysis of Algorithm SPA-ST-strong, to investigate how various parameters (e.g., the density and position of ties in the preference lists, the length of the preference lists, or the popularity of some projects) affect the existence of a strongly stable matching, based on randomly generated and/or real instances of SPA-ST.

Acknowledgement. The authors would like to convey their sincere gratitude to Adam Kunysz for valuable discussions concerning Algorithm SPA-ST-strong. They would also like to thank the anonymous reviewers for their helpful suggestions.

References

1. Abraham, D.J., Irving, R.W., Manlove, D.F.: Two algorithms for the Student-Project allocation problem. J. Discrete Algorithms **5**(1), 79–91 (2007)
2. Abu El-Atta, A.H., Moussa, M.I.: Student project allocation with preference lists over (student, project) pairs. In: Proceedings of ICCEE 2009, pp. 375–379 (2009)
3. Baidas, M., Bahbahani, Z., Alsusa, E.: User association and channel assignment in downlink multi-cell NOMA networks: a matching-theoretic approach. EURASIP J. Wirel. Commun. Netw. **2019**, 220 (2019)
4. Baidas, M., Bahbahani, M., Alsusa, E., Hamdi, K., Ding, Z.: D2D group association and channel assignment in uplink multi-cell NOMA networks: a matching theoretic approach. IEEE Trans. Commun. **67**, 8771–8785 (2019)

5. Baidas, M., Bahbahani, Z., El-Sharkawi, N., Shehada, H., Alsusa, E.: Joint relay selection and max-min energy-efficient power allocation in downlink multicell NOMA networks: a matching-theoretic approach. Trans. Emerg. Telecommun. Technol. **30**, 5 (2019)
6. Chiarandini, M., Fagerberg, R., Gualandi, S.: Handling preferences in student-project allocation. Ann. Oper. Res. **275**(1), 39–78 (2019)
7. Cooper, F., Manlove, D.: A 3/2-Approximation algorithm for the student-project allocation problem. In: Proceedings of SEA 2018. Leibniz International Proceedings in Informatics (LIPIcs), vol. 103, pp. 8:1–8:13 (2018)
8. Irving, R.W.: Stable marriage and indifference. Discrete Appl. Math. **48**, 261–272 (1994)
9. Irving, R.W., Manlove, D.F., Scott, S.: The hospitals/residents problem with ties. SWAT 2000. LNCS, vol. 1851, pp. 259–271. Springer, Heidelberg (2000). https://doi.org/10.1007/3-540-44985-X_24
10. Irving, R.W., Manlove, D.F., Scott, S.: Strong stability in the hospitals/residents problem. In: Alt, H., Habib, M. (eds.) STACS 2003. LNCS, vol. 2607, pp. 439–450. Springer, Heidelberg (2003). https://doi.org/10.1007/3-540-36494-3_39
11. Iwama, K., Miyazaki, S., Morita, Y., Manlove, D.: Stable marriage with incomplete lists and ties. In: Wiedermann, J., van Emde Boas, P., Nielsen, M. (eds.) ICALP 1999. LNCS, vol. 1644, pp. 443–452. Springer, Heidelberg (1999). https://doi.org/10.1007/3-540-48523-6_41
12. Iwama, K., Miyazaki, S., Yanagisawa, H.: Improved approximation bounds for the student-project allocation problem with preferences over projects. J. Discrete Algorithms **13**, 59–66 (2012)
13. Kavitha, T., Mehlhorn, K., Michail, D., Paluch, K.: Strongly stable matchings in time $O(nm)$ and extension to the hospitals-residents problem. In: Diekert, V., Habib, M. (eds.) STACS 2004. LNCS, vol. 2996, pp. 222–233. Springer, Heidelberg (2004). https://doi.org/10.1007/978-3-540-24749-4_20
14. Kazakov, D.: Co-ordination of student-project allocation. Manuscript, University of York, Department of Computer Science (2001). http://www-users.cs.york.ac.uk/kazakov/papers/proj.pdf. Accessed 25 Nov 2019
15. Kwanashie, A., Irving, R.W., Manlove, D.F., Sng, C.T.S.: Profile-based optimal matchings in the Student/project Allocation problem. In: Kratochvíl, J., Miller, M., Froncek, D. (eds.) IWOCA 2014. LNCS, vol. 8986, pp. 213–225. Springer, Cham (2015). https://doi.org/10.1007/978-3-319-19315-1_19
16. Kunysz, A.: An algorithm for the maximum weight strongly stable matching problem. In: Proceedings of ISAAC 2018. LIPIcs, vol. 123, pp. 42:1–42:13 (2018)
17. Liu, C.L.: Introduction to Combinatorial Mathematics. McGraw-Hill, New York (1968)
18. Manlove, D.: Stable marriage with ties and unacceptable partners. Technical report TR-1999-29, University of Glasgow, Department of Computing Science, January 1999
19. Manlove, D.: Algorithmics of Matching Under Preferences. World Scientific, Singapore (2013)
20. Manlove, D., Irving, R.W., Iwama, K., Miyazaki, S., Morita, Y.: Hard variants of stable marriage. Theor. Comput. Sci. **276**(1–2), 261–279 (2002)
21. Manlove, D., Milne, D., Olaosebikan, S.: An integer programming approach to the student-project allocation problem with preferences over projects. In: Lee, J., Rinaldi, G., Mahjoub, A.R. (eds.) ISCO 2018. LNCS, vol. 10856, pp. 313–325. Springer, Cham (2018). https://doi.org/10.1007/978-3-319-96151-4_27

22. Manlove, D., O'Malley, G.: Student project allocation with preferences over projects. J. Discrete Algorithms **6**, 553–560 (2008)
23. Olaosebikan, S., Manlove, D.: Super-stability in the student-project allocation problem with ties. In: Kim, D., Uma, R.N., Zelikovsky, A. (eds.) COCOA 2018. LNCS, vol. 11346, pp. 357–371. Springer, Cham (2018). https://doi.org/10.1007/978-3-030-04651-4_24
24. Olaosebikan, S., Manlove, D.F.: An Algorithm for Strong Stability in the Student-Project Allocation problem with Ties. CoRR, abs/1911.10262 (2019). http://arxiv.org/abs/1911.10262
25. Roth, A.E.: The evolution of the labor market for medical interns and residents: a case study in game theory. J. Polit. Econ. **92**(6), 991–1016 (1984)

Computational Complexity

Overlaying a Hypergraph with a Graph with Bounded Maximum Degree

Frédéric Havet[1,2] ⬨, Dorian Mazauric[2] ⬨, Viet-Ha Nguyen[1,2(✉)], and Rémi Watrigant[3] ⬨

[1] CNRS, I3S, Sophia Antipolis, France
[2] Inria, Université Côte d'Azur, Sophia Antipolis, France
thi-viet-ha.nguyen@inria.fr
[3] Université de Lyon, Lyon, France

Abstract. Let G and H be respectively a graph and a hypergraph defined on a same set of vertices, and let F be a fixed graph. We say that G *F-overlays* a hyperedge S of H if F is a spanning subgraph of the subgraph of G induced by S, and that it *F-overlays* H if it F-overlays every hyperedge of H. Motivated by structural biology, we study the computational complexity of two problems. The first problem, $(\Delta \leq k)$ F-OVERLAY, consists in deciding whether there is a graph with maximum degree at most k that F-overlays a given hypergraph H. It is a particular case of the second problem MAX $(\Delta \leq k)$ F-OVERLAY, which takes a hypergraph H and an integer s as input, and consists in deciding whether there is a graph with maximum degree at most k that F-overlays at least s hyperedges of H.

We give a complete polynomial/$\mathcal{NP}$-complete dichotomy for the MAX $(\Delta \leq k)$-F-OVERLAY problems depending on the pairs (F, k), and establish the complexity of $(\Delta \leq k)$ F-OVERLAY for many pairs (F, k).

1 Introduction

A major problem in structural biology is the characterization of low resolution structures of macro-molecular assemblies [5,20]. To attack this very difficult question, one has to determine the plausible contacts between the subunits (e.g. proteins) of an assembly, given the lists of subunits involved in all the complexes. We assume that the composition, in terms of individual subunits, of selected complexes is known. Indeed, a given assembly can be chemically split into complexes by manipulating chemical conditions. This problem can be conveniently modeled by graphs and hypergraphs. We consider the hypergraph H whose vertices represent the subunits and whose hyperedges are the complexes. We are then looking for a graph G with the same vertex set as H whose edges represent the contacts between subunits, and satisfying (i) some local properties for every complex (*i.e.* hyperedge), and (ii) some other global properties.

We first focus on the local properties. They are usually modeled by a (possibly infinite) family $\mathcal{F}$ of admissible graphs to which each complex must belong: to this end, we define the notion of *enforcement* of a hyperedge and a hypergraph.

© Springer Nature Switzerland AG 2020
M. Changat and S. Das (Eds.): CALDAM 2020, LNCS 12016, pp. 403–414, 2020.
https://doi.org/10.1007/978-3-030-39219-2_32

A graph G $\mathcal{F}$-*enforces* a hyperedge $S \in E(H)$ if the subgraph $G[S]$ of G induced by S belongs to $\mathcal{F}$, and it $\mathcal{F}$-*enforces* H if it $\mathcal{F}$-enforces all hyperedges of H. Very often, the considered family $\mathcal{F}$ is closed on taking edge supergraphs [1,8]: if $F \in \mathcal{F}$, then every graph obtained from G by adding edges is also in $\mathcal{F}$. Such a family is completely defined by its set $\mathcal{M} = \mathcal{M}(\mathcal{F})$ of minimal graphs that are the elements of $\mathcal{F}$ which are not edge supergraphs of any other. In this case, a graph G $\mathcal{F}$-enforcing S is such that there is an element of $\mathcal{M}$ which is a spanning subgraph of $G[S]$. This leads to the following notion of *overlayment* when considering minimal graph families.

Definition 1. A graph G $\mathcal{F}$-*overlays* a hyperedge S if there exists $F \in \mathcal{F}$ such that F is a spanning subgraph of $G[S]$, and it $\mathcal{F}$-*overlays* H if it $\mathcal{F}$-overlays every hyperedge of H.

As said previously, the graph sought will also have to satisfy some global constraints. Since in a macro-molecular assembly the number of contacts is small, the first natural idea is to look for a graph G with the minimum number of edges. This leads to the MIN-$\mathcal{F}$-OVERLAY problem: given a hypergraph H and an integer m, decide if there exists a graph G $\mathcal{F}$-overlaying H such that $|E(G)| \leq m$.

A typical example of a family $\mathcal{F}$ is the set of all connected graphs, in which case $\mathcal{M}(\mathcal{F})$ is the set of all trees. Agarwal et al. [1] focused on MIN-$\mathcal{M}(\mathcal{F})$-OVERLAY for this particular family in the aforementioned context of structural biology. However, this problem was previously studied by several communities in other domains, as pointed out by Chen *et al.* [6]. Indeed, it is also known as SUBSET INTERCONNECTION DESIGN, MINIMUM TOPIC-CONNECTED OVERLAY or INTERCONNECTION GRAPH PROBLEM, and was considered (among others) in the design of vacuum systems [10,11], scalable overlay networks [7,18], and reconfigurable interconnection networks [12,13]. Some variants have also been considered in the contexts of inferring a most likely social network [2], determining winners of combinatorial auctions [9], as well as drawing hypergraphs [4,14].

Cohen et al. [8] presented a dichotomy regarding the polynomial vs. $\mathcal{NP}$-hard status of the problem MIN-$\mathcal{F}$-OVERLAY with respect to the considered family $\mathcal{F}$. Roughly speaking, they showed that the easy cases one can think of (e.g. when edgeless graphs of the right sizes are in $\mathcal{F}$, or if $\mathcal{F}$ contains only cliques) are the only families giving rise to a polynomial-time solvable problem: all others are $\mathcal{NP}$-complete. They also considered the FPT/W[1]-hard dichotomy for several families $\mathcal{F}$.

In this paper, we consider the variant in which the additional constraint is that G must have a bounded maximum degree: this constraint is motivated by the context of structural biology, since a subunit (e.g. a protein) cannot be connected to many other subunits. This yields the following problem for any family $\mathcal{F}$ of graphs and an integer k.

$(\Delta \leq k)$-$\mathcal{F}$-OVERLAY

Input: A hypergraph H.

Question: Does there exist a graph G $\mathcal{F}$-overlaying H such that $\Delta(G) \leq k$?

We denote by $over_{\mathcal{F}}(H, G)$ the number of hyperedges of H that are $\mathcal{F}$-overlaid by G. A natural generalization is to find $over_{\mathcal{F}}(H, k)$, the maximum number of hyperedges $\mathcal{F}$-overlaid by a graph with maximum degree at most k.

MAX $(\Delta \leq k)$-$\mathcal{F}$-OVERLAY

Input: A hypergraph H and a positive integer s.

Question: Does there exist a graph G such that $\Delta(G) \leq k$ and $over_{\mathcal{F}}(H, G) \geq s$?

Observe that there is an obvious reduction from $(\Delta \leq k)$-$\mathcal{F}$-OVERLAY to MAX $(\Delta \leq k)$-$\mathcal{F}$-OVERLAY (by setting $s = |E(H)|$) (Fig. 1).

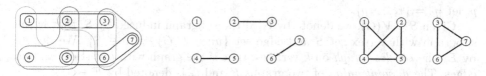

Fig. 1. Example of $(\Delta \leq k)$-$\mathcal{F}$-OVERLAY and MAX $(\Delta \leq k)$-$\mathcal{F}$-OVERLAY. In the figure, an instance H (left), a graph G with $\Delta(G) \leq 1$ that O_3-overlays H (with O_3 being the graph with three vertices and one edge) (center), and a solution to MAX $(\Delta \leq 3)$-C_3-OVERLAY (with C_3 being the cycle on three vertices) (right).

In this paper, we mainly consider the case when the family $\mathcal{F}$ contains a unique graph F. We abbreviate $(\Delta \leq k)$-$\{F\}$-OVERLAY and MAX $(\Delta \leq k)$-$\{F\}$-OVERLAY as $(\Delta \leq k)$-F-OVERLAY and MAX $(\Delta \leq k)$-F-OVERLAY, respectively. By definition those two problems really make sense only for $|F|$-uniform hypergraphs *i.e.* hypergraphs whose hyperedges are of size $|F|$. Therefore, we always assume the hypergraph to be $|F|$-uniform.

If F is a graph with maximum degree greater than k, then solving $(\Delta \leq k)$-F-OVERLAY or MAX $(\Delta \leq k)$-F-OVERLAY is trivial as the answer is always 'No'. So we only study the problems when $\Delta(F) \leq k$.

If F is an empty graph, then MAX $(\Delta \leq k)$-F-OVERLAY is also trivial, because for any hypergraph H, the empty graph on $V(H)$ vertices F-overlays H. Hence the first natural interesting cases are the graphs with one edge. For every integer $p \geq 2$, we denote by O_p the graph with p vertices and one edge. In Sect. 2, we prove the following dichotomy theorem.

Theorem 1. *Let $k \geq 1$ and $p \geq 2$ be integers. If $p = 2$ or if $k = 1$ and $p = 3$, then* MAX $(\Delta \leq k)$-O_p-OVERLAY *and* $(\Delta \leq k)$-O_p-OVERLAY *are polynomial-time solvable. Otherwise, they are $\mathcal{NP}$-complete.*

Then, in Sect. 3, we give a complete polynomial/$\mathcal{NP}$-complete dichotomy for the MAX $(\Delta \leq k)$-F-OVERLAY problems.

Theorem 2. MAX $(\Delta \leq k)$-F-OVERLAY *is polynomial-time solvable if either* $\Delta(F) > k$, *or* F *is an empty graph, or* $F = O_2$, *or* $k = 1$ *and* $F = O_3$. *Otherwise it is* $\mathcal{NP}$-*complete.*

In Sect. 4, we investigate the complexity of $(\Delta \leq k)$-F-OVERLAY problems. We believe that each such problem is either polynomial-time solvable or $\mathcal{NP}$-complete. However the dichotomy seems to be more complicated than the one for MAX $(\Delta \leq k)$-F-OVERLAY. We exhibit several pairs (F, k) such that $(\Delta \leq k)$-F-OVERLAY is polynomial-time solvable, while MAX $(\Delta \leq k)$-F-OVERLAY is $\mathcal{NP}$-complete. This is in particular the case when F is a complete graph (Proposition 3), F is connected k-regular (Proposition 4), F is a path and $k = 2$ (Theorem 8), and when F is the cycle on 4 vertices and $k \leq 3$ (Theorem 7).

Due to space constraints, some proofs (marked with a $\star$) were omitted.

Most notations of this paper are standard. We now recall some of them, and we refer the reader to [3] for any undefined terminology. For a positive integer p, let $[p] = \{1, \ldots, p\}$.

Given $S \subseteq V(G)$, we denote by $G[S]$ the subgraph induced by S, that is the subgraph with vertex set S and edge set $\{uv \in E(G) \mid u, v \in S\}$. We denote by E_k the *edgeless graph* on k vertices, that is the graph with k vertices and no edges. The *disjoint union* of two graphs F and G is denoted by $F + G$.

Let H be a hypergraph. Two hyperedges are *adjacent* if their intersection has size at least 2. A hypergraph is *neat* if any two distinct hyperedges intersect in at most one vertex. In other words, a hypergraph is neat if there is no pair of adjacent hyperedges. We denote by $K(H)$, the graph obtained by replacing each hyperedge by a complete graph. In other words, $V(K(H)) = V(H)$ and $E(K(H)) = \{xy \mid \exists S \in E(H), \{x, y\} \subseteq S\}$. The *edge-weight function induced by* H on $K(H)$, denoted by w_H, is defined by $w_H(e) = |\{S \in E(H) \mid e \subseteq S\}|$. In words, $w_H(e)$ is the number of hyperedges of H containing e. A hypergraph H is *connected* if $K(H)$ is connected, and the *connected components* of a hypergraph H are the connected components of $K(H)$. Finally, a graph G $\mathcal{F}$-overlaying H with maximum degree at most k is called an $(\mathcal{F}, H, k)$-*graph*.

2   The Graphs with One Edge

In this section, we establish Theorem 1. Let $p \geq 2$, and H be a p-uniform hypergraph. Consider the edge-weighted graph $(K(H), w_H)$. For every matching M of this graph, let $G_M = (V(H), M)$. Every hyperedge O_p-overlaid by G_M contains at least one edge of M and at most $\lfloor \frac{p}{2} \rfloor$ edges of M. We thus have the following:

Observation 3. *For every matching M of $K(H)$, we have:*

$$\frac{1}{\lfloor \frac{p}{2} \rfloor} w_H(M) \leq over_{O_p}(H, G_M) \leq w_H(M), \tag{1}$$

where $w_H(M) = \sum_{e \in M} w_H(e)$.

Consider first the case when $p = 2$. Let H be a 2-uniform hypergraph. Every hyperedge is an edge, so $K(H) = H$. Moreover, a (hyper)edge of H is O_2-overlaid by G if and only if it is in $E(G)$. Hence MAX $(\Delta \leq k)$-O_2-OVERLAY is equivalent to finding a maximum k-matching (that is a subgraph with maximum degree at most k) in $K(H)$. This problem is polynomial-time solvable, see [19, Chap. 31], hence:

Proposition 1. MAX $(\Delta \leq k)$-O_2-OVERLAY *is polynomial-time solvable for all positive integer k.*

If $p = 3$, Inequalities (1) are equivalent to $over_{O_3}(H, G_M) = w_H(M)$. Since the edge set of a graph with maximum degree 1 is a matching, MAX $(\Delta \leq 1)$-O_3-OVERLAY is equivalent to finding a maximum-weight matching in the edge-weighted graph $(K(H), w_H)$. This can be done in polynomial-time, see [15, Chap. 14].

Proposition 2. MAX $(\Delta \leq 1)$-O_3-OVERLAY *is polynomial-time solvable.*

We shall now prove that if $p \geq 4$, or $p = 3$ and $k \geq 2$, then MAX $(\Delta \leq k)$-O_p-OVERLAY is $\mathcal{NP}$-complete. We prove it by a double induction on k and p. Theorems 4 and 5 first prove the base cases of the induction and Lemma 1 corresponds to the inductive steps.

Theorem 4 $(\star)$. $(\Delta \leq 1)$-O_4-OVERLAY *is $\mathcal{NP}$-complete.*

Theorem 5 $(\star)$. $(\Delta \leq 2)$-O_3-OVERLAY *is $\mathcal{NP}$-complete.*

Lemma 1 $(\star)$. *If $(\Delta \leq k)$-O_p-OVERLAY is $\mathcal{NP}$-complete, then $(\Delta \leq k)$-O_{p+1}-OVERLAY and $(\Delta \leq k + 1)$-O_p-OVERLAY are $\mathcal{NP}$-complete.*

Propositions 1 and 2, Theorems 4 and 5, and Lemma 1 imply Theorem 1.

3   Complexity of MAX $(\Delta \leq k)$-F-OVERLAY

The aim of this section is to establish Theorem 2 that gives the polynomial/$\mathcal{NP}$-complete dichotomy for the MAX $(\Delta \leq k)$-F-OVERLAY problems.

As noticed in the introduction, if $\Delta(F) > k$ or F is an empty graph then MAX $(\Delta \leq k)$-$\mathcal{F}$-OVERLAY is trivially polynomial-time solvable. Moreover, by Propositions 1 and 2, MAX $(\Delta \leq 1)$-O_3-OVERLAY as well as MAX $(\Delta \leq k)$-O_2-OVERLAY (for all positive integers k) are also polynomial-time solvable.

We shall now prove that if we are not in one of the above cases, then MAX $(\Delta \leq k)$-F-OVERLAY is $\mathcal{NP}$-complete. We first establish the $\mathcal{NP}$-completeness when F has no isolated vertices.

Theorem 6. *Let F be a graph on at least three vertices with no isolated vertices. If $k \geq \Delta(F)$, then MAX $(\Delta \leq k)$-F-OVERLAY is $\mathcal{NP}$-complete on neat hypergraphs.*

Proof. Assume $k \geq \Delta(F)$. Let $n = |F|$, $a_1, \ldots, a_n$ be an ordering of the vertices of F such that $\delta(F) = d(a_1) \leq d(a_2) \leq \cdots \leq d(a_n) = \Delta(F)$.

Let $\gamma = \lfloor k/\delta(F) \rfloor - 1$, $\beta = k - \gamma\delta(F)$. Observe that $\delta(F) \leq \beta \leq 2\delta(F) - 1$.

We shall give a reduction from INDEPENDENT SET which is a well-known $\mathcal{NP}$-complete problem even for cubic graphs (see [16].) We distinguish two cases depending on whether $d(a_2) > \beta$ or not. The two reductions are very similar.

<u>Case 1</u>: $d(a_2) > \beta$. Set $\gamma_1 = \gamma_2 = \lfloor (k - d(a_2))/\delta(F) \rfloor$ and $\gamma_3 = \lfloor (k - d(a_3))/\delta(F) \rfloor$.

Let Γ be a cubic graph. For each vertex $v \in V(\Gamma)$, let $(e_1(v), e_2(v), e_3(v))$ be an ordering of the edges incident to v. We shall construct the neat hypergraph $H = H(\Gamma)$ as follows.

- For each vertex $v \in \Gamma$, we create a hyperedge $S_v = \{a_1^v, \ldots, a_n^v\}$. Then, for $1 \leq i \leq 3$, we add γ_i a_i^v-*leaves*, that are hyperedges containing a_i^v and $n - 1$ new vertices.
- For each edge $e = uv \in \Gamma$, let i and j be the indices such that $e = e_i(u) = e_j(v)$. We create a new vertex z_e and hyperedges S_u^e (S_v^e) containing z_e, a_i^u (a_j^v), and $n - 2$ new vertices, respectively. Then, we add γ z_e-*leaves*, that are hyperedges containing z_e and $n - 1$ new vertices.

We shall prove that $over_F(H, k) = (\gamma_1 + \gamma_2 + \gamma_3)|V(\Gamma)| + (\gamma + 1)|E(\Gamma)| + \alpha(\Gamma)$, where $\alpha(\Gamma)$ denotes the cardinality of a maximum independent set in Γ.

The following claim shows that there are optimal solutions with specific structure. This leads to the inequality:

$$over_F(H, k) \leq (\gamma_1 + \gamma_2 + \gamma_3)|V(\Gamma)| + (\gamma + 1)|E(\Gamma)| + \alpha(\Gamma)$$

Claim 1 ($\star$). *There is a graph G with $\Delta(G) \leq k$ that F-overlays $over_F(H, k)$ hyperedges of H such that:*

(a) *each x-leaf L is F-overlaid and x is incident to $\delta(F)$ edges in $G[L]$ (with $x = a_i^v$ or $x = z_e$).*
(b) *for each edge $e = uv \in E(\Gamma)$, exactly one of the two hyperedges S_u^e and S_v^e is F-overlaid. Moreover if S_u^e (S_v^e) is F-overlaid, then a_i^u (a_j^v) is incident to $d(a_2)$ edges in S_u^e (S_v^e), respectively.*
(c) *the set of vertices v such that S_v is F-overlaid is an independent set in Γ.*

Conversely, consider W a maximum independent set of Γ.

Let G be the graph with vertex $V(H)$ which is the union of the following subgraphs:

- for each x-leaf L, we add a copy of F on L in which x has degree $\delta(F)$;
- for each vertex $v \in W$, we add a copy of F on S_v in which a_i^v has degree $d(a_i)$ for all $1 \leq i \leq n$.
- for each edge $e \in E(\Gamma)$, we choose an endvertex u of e such that $u \notin W$, and add a copy of F in which z_e has degree $d(a_1)$ and a_i^u has degree $d(a_2)$ (with i the index such that $e_i(u) = e$).

It is simple matter to check that $\Delta(G) \leq k$ and that G F-overlays $(\gamma_1 + \gamma_2 + \gamma_3)|V(\Gamma)| + (\gamma + 1)|E(\Gamma)| + \alpha(\Gamma)$ hyperedges of H. Thus $over_F(H, k) \geq (\gamma_1 + \gamma_2 + \gamma_3)|V(\Gamma)| + (\gamma + 1)|E(\Gamma)| + \alpha(\Gamma)$.

<u>Case 2</u>: $d(a_2) \leq \beta$. The proof is very similar to Case 1. The main difference is the definition of the γ_i. In this case, we set $\gamma_i = \lfloor (k - d(a_i))/\delta(F) \rfloor$ for $1 \leq i \leq 3$, and we can adapt the proof of Claim 1.

Conversely, if we have W a maximum independent set of Γ, then we construct graph G, union of the subgraphs as Case 1 except the subgraphs for hyperedges $S_u^{e_1}(v)$, that we add a copy of F in which $d(z_e) = d(a_2)$ and $d(a_1^u) = d(a_1)$.

We then establish the following lemma, which allows to derive the $\mathcal{NP}$-completeness of MAX $(\Delta \leq k)$-F-OVERLAY when F has isolated vertices.

Lemma 2 (⋆). *Let k be a positive integer, let F be a graph with $\delta(F) \geq 1$, and let q be a non-negative integer. If MAX $(\Delta \leq k)$-$(F + E_q)$-OVERLAY is $\mathcal{NP}$-complete, then MAX $(\Delta \leq k)$-$(F + E_{q+1})$-OVERLAY is also $\mathcal{NP}$-complete.*

Now we can prove Theorem 2. As explained in the beginning of the section, it suffices to prove that MAX $(\Delta \leq k)$-F-OVERLAY remains $\mathcal{NP}$-complete when $\Delta(F) \leq k$, $F \neq E_{|F|}$, $|F| \geq 3$ and $(F, k) \neq (O_3, 1)$. Assume that the above conditions are satisfied. Let F' be the graph induced by the non-isolated vertices of F. Then $F = F' + E_q$ with $q = |F| - |F'|$. If $|F'| = 2$, then $F = O_{|F|}$, and we have the result by Theorem 1. If $|F'| \geq 3$, then the result follows from Theorem 6, Lemma 2, and an immediate induction.

4   Complexity of $(\Delta \leq K)$-$\mathcal{F}$-OVERLAY

4.1   Regular Graphs

Proposition 3. *For every complete graph K and every positive integer k, $(\Delta \leq k)$-K-OVERLAY is polynomial-time solvable.*

Proof. Observe that a $|V(K)|$-uniform hypergraph H is a positive instance of $(\Delta \leq k)$-K-OVERLAY if and only if $K(H)$ is a (K, H, k)-graph.

Proposition 4. *For every connected k-regular graph F, $(\Delta \leq k)$-F-OVERLAY is polynomial-time solvable.*

Proof. One easily sees that a $|V(F)|$-uniform hypergraph H admits an (F, H, k)-graph if and only if the hyperedges of H are pairwise non-intersecting.

Let C_4 denote the cycle on 4 vertices. Proposition 4 implies that $(\Delta \leq 2)$-C_4-OVERLAY is polynomial-time solvable. We now show that $(\Delta \leq 3)$-C_4-OVERLAY is also polynomial-time solvable.

Theorem 7. *$(\Delta \leq 3)$-C_4-OVERLAY is polynomial-time solvable.*

Proof. Let H be a 4-uniform hypergraph.

Let us describe an algorithm to decide whether there is a $(C_4, H, 3)$-graph. It is sufficient to do it when H is connected since the disjoint union of the $(C_4, K, 3)$-graphs for connected components K of H is a $(C_4, H, 3)$-graph.

Observe first that if two hyperedges of H intersect in exactly one vertex u, then no such graph exists, since u must have degree 2 in each of the hyperedges if they are C_4-overlaid, and thus degree 4 in total. Therefore if there are two such hyperedges, we return 'No'. At this point we may assume that $|E(H)| \geq 2$ for otherwise we return 'Yes'.

From now on we may assume that two hyperedges either do not intersect, or are adjacent (intersect on at least two vertices).

Claim 2. *If two hyperedges S_1 and S_2 intersect on three vertices and there is a $(C_4, H, 3)$-graph G, then $|V(H)| \leq 6$.*

Proof of claim:. Assume $S_1 = \{a_1, b, c, d\}$ and $S_2 = \{a_2, b, c, d\}$. Let G be a $(C_4, H, 3)$-graph. In G, a_1 and a_2 have the same two neighbours in $\{b, c, d\}$ and the third vertex of $\{b, c, d\}$ is also adjacent to those two. Consider a hyperedge S_3 intersecting $S_1 \cup S_2$. Since it is C_4-overlaid by G, at least two edges connect $S_3 \cap (S_1 \cup S_2)$ to $S_3 \setminus (S_1 \cup S_2)$. The endvertices of those edges in $S_1 \cup S_2$ must have degree 2 in $G[S_1 \cup S_2]$. Hence, without loss of generality, either $S_3 = \{a_1, a_2, b, e\}$, or $S_3 = \{a_1, b, c, e\}$ for some vertex e not in $S_1 \cup S_2$. Now no hyperedge can both intersect $S_1 \cup S_2 \cup S_3$ and contain a vertex not in $S_1 \cup S_2 \cup S_3$, for such a hyperedge must contain either the vertices c, e or a_2, e which are at distance 3 in $G[S_1 \cup S_2 \cup S_3]$. (However there can be more hyperedges contained in $S_1 \cup S_2 \cup S_3$.) Hence $|V(H)| \leq 6$.   ◁

In view of Claim 2, if there are two hyperedges with three vertices in common, either we return 'No' if $|V(H)| > 6$, or we check all possibilities (or follow the proof of the above claim) to return the correct answer otherwise. Henceforth, we may assume that any two adjacent hyperedges intersect in exactly two vertices.

Let S_1 and S_2 be two adjacent hyperedges, say $S_1 = \{a, b, c, d\}$ and $S_2 = \{c, d, e, f\}$. Note that every $(C_4, H, 3)$-graph contains the edges ab, cd and ef, and that $N(c) \cup N(d) = S_1 \cup S_2$.

Claim 3. *If there is another hyperedge than S_1 and S_2 containing c or d, and there is a $(C_4, H, 3)$-graph G, then $|V(H)| \leq 8$.*

Proof of claim:. Without loss of generality, we may assume that G contains the cycle (a, b, d, f, e, c, a) and the edge cd. Hence the only possible hyperedges containing c or d and a vertex not in $S_1 \cup S_2$ are $S_3 = \{a, c, e, g\}$ for some $g \notin S_1 \cup S_2$ and $S_4 = \{b, d, f, h\}$ for some $h \notin S_1 \cup S_2$.

If H contains both S_3 and S_4, then G contains the edges ag, eg, bh and hf. If G contains also gh, then $G[S_1 \cup S_2 \cup S_3 \cup S_4]$ is 3-regular, so $G = G[S_1 \cup S_2 \cup S_3 \cup S_4]$. If G does not contain gh, then the only vertices of degree 2 in $G[S_1 \cup S_2 \cup S_3 \cup S_4]$ are g and h, and they are at distance at least 3 in this graph. Thus every hyperedge intersecting $S_1 \cup S_2 \cup S_3 \cup S_4$ is contained in this set, so $|V(H)| = 8$.

Assume now that G contains only one of S_3, S_4. Without loss of generality, we may assume that this is S_3. Hence G also contains the edges ag and eg. If $V(G) \neq S_1 \cup S_2 \cup S_3$, then there is a hyperedge S that intersects $S_1 \cup S_2 \cup S_3$ and that is not contained in $S_1 \cup S_2 \cup S_3$. It does not contain c and d. Hence it must contain one of the vertices a or e, because it intersects each S_i along an edge of G or not at all. Without loss of generality, $a \in S$. Hence $S = \{a, b, i, g\}$ for some vertex i not in $S_1 \cup S_2 \cup S_3$, and G contains the edges bi and ig. Now, as previously, either i and f are adjacent and $G = G[S_1 \cup S_2 \cup S_3 \cup S]$ or they are not adjacent, and every hyperedge intersecting $S_1 \cup S_2 \cup S_3 \cup S$ is contained in this set. In both cases, $|V(H)| = 8$.   ◁

We now summarize the algorithm: if $|V(G)| \leq 8$, then we solve the instance by brute force. Otherwise, for every pair of hyperedges S_1, S_2, if their intersection is of size 1 or 3, we answer 'No'. In the remaining cases, if S_1 and S_2 have non-empty intersection, then, they must intersect on two vertices c and d, and these vertices do not belong to any other hyperedges but S_1 and S_2.

In this case, let H' be the hypergraph with vertex set $V(H) \backslash \{c, d\}$ and hyperedge set $(E(H) \cup \{\{a, b, e, f\}\}) \backslash \{S_1, S_2\}$. It is simple matter to check that there is a $(C_4, H, 3)$-graph if and only if there is a $(C_4, H', 3)$-graph. Consequently, we recursively apply the algorithm on H'.

Clearly, the above-described algorithm runs in polynomial time.

4.2   Paths

Let $\mathcal{P}$ be the set of all paths. We have the following:

Theorem 8. $(\Delta \leq 2)$-$\mathcal{P}$-OVERLAY *is linear-time solvable.*

Proof. Clearly, if H is not connected, it suffices to solve the problem on each of the components and to return 'No' if the answer is negative for at least one of the components, and 'Yes' otherwise. Henceforth, we shall now assume that H is connected. In such a case, a $(\mathcal{P}, H, 2)$-graph is either a path or a cycle. However, if H is $\mathcal{P}$-overlaid by a path P, then it is also $\mathcal{P}$-overlaid by the cycle obtained from P by adding an edge between its two endvertices. Thus, we focus on the case where G is a cycle.

Let S be a family of sets. The *intersection graph* of a set S is the graph $IG(S)$ whose vertices are the sets of S, and in which two vertices are adjacent if the corresponding sets in S intersect.

The *intersection graph* of a hypergraph H, denoted by $IG(H)$, is the intersection graph of its hyperedge set. We define two functions l_H and s_H as follows:

$$l_H(S) = |S| - 1 \text{ for all } S \in E(H) \quad \text{and} \quad s_H(S, S') = |S \cap S'| - 1 \text{ for all } S, S' \in E(H).$$

Let $\mathbb{C}_\ell$ be the circle of circumference ℓ. We identify the points of $\mathbb{C}_\ell$ with the integer numbers (points) of the segment $[0, \ell]$, (with 0 identified with ℓ). A *circular-arc graph* is the intersection graph of a set of arcs on $\mathbb{C}_\ell$. A set $\mathcal{A}$ of arcs such that $IG(\mathcal{A}) = G$ is called an *arc representation* of G. We denote by

A_v the arc corresponding to v in $\mathcal{A}$. Let G be a graph and let $l : V(G) \to \mathbb{N}$ and $s : E(G) \to \mathbb{N}$ be two functions. An arc representation $\mathcal{A}$ of G is *l-respecting* if A_v has length $l(v)$ for any $v \in V(G)$, *s-respecting* if $A_v \cap A_u$ has length $s(u,v)$ for all $uv \in E(G)$, and (l,s)-*respecting* if it is both l-respecting and s-respecting. One can easily adapt the algorithm given by Köbler et al. [17] for (l,s)-respecting interval representations to decide in linear time whether a graph admits an (l,s)-respecting arc representation in $\mathbb{C}_n$ for every integer n.

Claim 4. *Let H be a connected hypergraph on n vertices. There is a cycle $\mathcal{P}$-overlaying H if and only if $IG(H)$ admits an (l_H, s_H)-respecting arc representation into $\mathbb{C}_n$.*

Proof of claim:. Assume that H is $\mathcal{P}$-overlaid by a cycle $C = (v_0, v_1, \ldots, v_{n-1}, v_0)$. There is a canonical embedding of C to $\mathbb{C}_n$ in which every vertex v_i is mapped to i and every edge $v_i v_{i+1}$ to the circular arc $[i, i+1]$. For every hyperedge $S \in E(H)$, $P[S]$ is a subpath, which is mapped to the circular arc A_S of $\mathbb{C}_n$ that is the union of the circular arcs to which its edges are mapped. Clearly, $\mathcal{A} = \{A_S \mid S \in E(H)\}$ is an (l_H, s_H)-respecting interval representation of $IG(H)$.

Conversely, assume that $IG(H)$ admits an (l_H, s_H)-respecting interval representation $\mathcal{A} = \{A_S \mid S \in E(H)\}$ into $\mathbb{C}_n$. Let S_0 be a hyperedge of minimum size. Free to rotate all intervals, we may assume that A_{S_0} is $[1, |S_0|]$. Now since $\mathcal{A}$ is (l_H, s_H)-respecting and H is connected, we deduce that the extremities of A_S are integers for all $S \in E(H)$. Let v_1 be a vertex of H that belongs to the hyperedges whose corresponding arcs of $\mathcal{A}$ contain 1. Then for all $i = 2$ to $n = |V(H)|$, denote by v_i an arbitrary vertex not in $\{v_1, \ldots, v_{i-1}\}$ that belongs to the hyperedges whose corresponding arcs of $\mathcal{A}$ contain i. Such a vertex exists because $\mathcal{A}$ is (l_H, s_H)-respecting. Observe that such a construction yields $S = \{v_i \mid i \in A_S\}$ for all $S \in E(H)$. Furthermore, the cycle $C = (v_1, \ldots, v_n, v_1)$ $\mathcal{P}$-overlays H. Indeed, for each $S \in E(H)$, $C[S]$ is the subpath corresponding to A_S, that is $V(C[S]) = \{v_i \mid i \in A_S\}$ and $E(C[S]) = \{v_i v_{i+1} \mid [i, i+1] \subseteq A_S\}$.

The algorithm to solve $(\Delta \le 2)$-$\mathcal{P}$-OVERLAY for a connected hypergraph H in linear time is thus the following:

1. Construct the intersection graph $IG(H)$ and compute the associated functions l_H and s_H.
2. Check whether graph $IG(H)$ has an (l_H, s_H)-respecting interval representation. If it is the case, return 'Yes'. If not return 'No'.

Remark 1. We can also detect in polynomial time whether a connected hypergraph H is $\mathcal{P}$-overlaid by a path. Indeed, similarly to Claim 4, one can show that there is a path $\mathcal{P}$-overlaying H if and only if $IG(H)$ admits an (l_H, s_H)-respecting interval representation.

5 Further Research

Theorem 2 characterizes the complexity of MAX $(\Delta \leq k)$-$\mathcal{F}$-OVERLAY when $\mathcal{F}$ contains a unique graph. It would be nice to extend this characterization to families $\mathcal{F}$ of arbitrary size.

Problem 1. Characterize the pairs $(\mathcal{F}, k)$ for which MAX $(\Delta \leq k)$-$\mathcal{F}$-OVERLAY is polynomial-time solvable and those for which it is $\mathcal{NP}$-complete.

Theorem 1 and the results obtained in Sect. 4 give a first view of the complexity of $(\Delta \leq k)$-F-OVERLAY. A natural problem is to close the dichotomy:

Problem 2. Characterize the pairs (F, k) for which $(\Delta \leq k)$-F-OVERLAY is polynomial-time solvable and those for which it is $\mathcal{NP}$-complete.

It would be interesting to consider the complexity of this problem when F is k-regular but non-connected, and when F is a cycle. In order to attack Problem 2, it would be helpful to prove the following conjecture.

Conjecture 1. If $(\Delta \leq k)$-F-OVERLAY is $\mathcal{NP}$-complete, then $(\Delta \leq k + 1)$-F-OVERLAY is also $\mathcal{NP}$-complete.

Furthermore, for each pair $(\mathcal{F}, k)$ such that MAX $(\Delta \leq k)$-$\mathcal{F}$-OVERLAY is $\mathcal{NP}$-complete and $(\Delta \leq k)$-$\mathcal{F}$-OVERLAY is polynomial-time solvable, it is natural to consider the parameterized complexity of MAX $(\Delta \leq k)$-$\mathcal{F}$-OVERLAY when parameterized by $|E(H)| - s$, because $(\Delta \leq k)$-$\mathcal{F}$-OVERLAY is the case $s = 0$.

Finally, it would be interesting to obtain approximation algorithms for MAX $(\Delta \leq k)$-$\mathcal{F}$-OVERLAY when this problem is $\mathcal{NP}$-complete.

References

1. Agarwal, D., Caillouet, C., Coudert, D., Cazals, F.: Unveiling contacts within macro-molecular assemblies by solving minimum weight connectivity inference problems. Mol. Cell. Proteomics **14**, 2274–2284 (2015)
2. Angluin, D., Aspnes, J., Reyzin, L.: Inferring social networks from outbreaks. In: Hutter, M., Stephan, F., Vovk, V., Zeugmann, T. (eds.) ALT 2010. LNCS (LNAI), vol. 6331, pp. 104–118. Springer, Heidelberg (2010). https://doi.org/10.1007/978-3-642-16108-7_12
3. Bondy, A.J., Murty, U.S.R.: Graph Theory. Graduate Texts in Mathematics, vol. 244. Springer, New York (2008)
4. Brandes, U., Cornelsen, S., Pampel, B., Sallaberry, A.: Blocks of hypergraphs. In: Iliopoulos, C.S., Smyth, W.F. (eds.) IWOCA 2010. LNCS, vol. 6460, pp. 201–211. Springer, Heidelberg (2011). https://doi.org/10.1007/978-3-642-19222-7_21
5. Burnley, R., Damoc, E., Denisov, E., Makarov, A., Heck, A.: High-sensitivity orbitrap mass analysis of intact macromolecular assemblies. Nat. Methods **9**, 10 (2012)
6. Chen, J., Komusiewicz, C., Niedermeier, R., Sorge, M., Suchý, O., Weller, M.: Polynomial-time data reduction for the subset interconnection design problem. SIAM J. Discret. Math. **29**(1), 1–25 (2015)

7. Chockler, G., Melamed, R., Tock, Y., Vitenberg, R.: Constructing scalable overlays for pub-sub with many topics. In: PODC 2007, pp. 109–118. ACM, New York (2007)
8. Cohen, N., Havet, F., Mazauric, D., Sau, I., Watrigant, R.: Complexity dichotomies for the minimum F-overlay problem. J. Discret. Algorithms **52**, 133–142 (2019)
9. Conitzer, V., Derryberry, J., Sandholm, T.: Combinatorial auctions with structured item graphs. In: 16th Conference on Innovative Applications of Artificial Intelligence (AAAI 2004), pp. 212–218 (2004)
10. Du, D.Z., Kelley, D.F.: On complexity of subset interconnection designs. J. Global Optim. **6**(2), 193–205 (1995)
11. Du, D.-Z., Miller, Z.: Matroids and subset interconnection design. SIAM J. Discret. Math. **1**(4), 416–424 (1988)
12. Fan, H., Hundt, C., Wu, Y.-L., Ernst, J.: Algorithms and implementation for interconnection graph problem. In: Yang, B., Du, D.-Z., Wang, C.A. (eds.) COCOA 2008. LNCS, vol. 5165, pp. 201–210. Springer, Heidelberg (2008). https://doi.org/10.1007/978-3-540-85097-7_19
13. Fan, H., Wu, Y.: Interconnection graph problem. In: Proceedings of the 2008 International Conference on Foundations of Computer Science, FCS 2008, pp. 51–55 (2008)
14. Johnson, D.S., Pollak, H.O.: Hypergraph planarity and the complexity of drawing venn diagrams. J. Graph Theory **11**(3), 309–325 (1987)
15. Jungnickel, D.: Graphs, Networks and Algorithms. Algorithmsand Computation in Mathematics, vol. 5. Springer, Heidelberg (2013)
16. Karp, R.M.: Reducibility among Combinatorial Problems. Springer, New York (1972)
17. Köbler, J., Kuhnert, S., Watanabe, O.: Interval graph representation with given interval and intersection lengths. J. Discret. Algorithms **34**, 108–117 (2015)
18. Onus, M., Richa, A.W.: Minimum maximum-degree publish-subscribe overlay network design. IEEE/ACM Trans. Networking **19**(5), 1331–1343 (2011)
19. Schrijver, A.: Combinatorial Optimization. Polyhedra and Efficiency. Algorithms and Combinatorics, vol. 24. Springer-Verlag, Berlin (2003)
20. Sharon, M., Robinson, C.V.: The role of mass spectrometry in structure elucidation of dynamic protein complexes. Annu. Rev. Biochem. **76**(1), 167–193 (2007). PMID: 17328674

Parameterized Algorithms for Directed Modular Width

Raphael Steiner🄳 and Sebastian Wiederrecht[✉]🄳

Technische Universität Berlin, Berlin, Germany
`sebastian.wiederrecht@tu-berlin.de`

Abstract. Many well-known NP-hard algorithmic problems on directed graphs resist efficient parameterizations with most known width measures for directed graphs, such as directed treewidth, DAG-width, Kelly-width and many others. While these focus on measuring how close a digraph is to an oriented tree resp. a directed acyclic graph, in this paper, we investigate *directed modular width* as a parameter, which is closer to the concept of clique-width. We investigate applications of modular decompositions of directed graphs to a wide range of algorithmic problems and derive FPT algorithms for several well-known digraph-specific NP-hard problems, namely *minimum (weight) directed feedback vertex set*, *minimum (weight) directed dominating set*, *digraph colouring*, *directed Hamiltonian path/cycle*, *partitioning into paths*, *(capacitated) vertex-disjoint directed paths*, and the *directed subgraph homeomorphism problem*. The latter yields a polynomial-time algorithm for detecting topological minors in digraphs of bounded directed modular width. Finally we illustrate that other structural digraph parameters, such as *directed pathwidth* and *cycle-rank* can be computed efficiently using directed modular width as a parameter.

Keywords: Parameterized complexity · Fixed-parameter-tractability · Width measures · Modular decomposition · Integer Linear Programming

1 Introduction

Width measures for graphs have become a fundamental pillar in both structural graph theory and parameterized complexity. From an algorithmic point of view a *good* width measure should provide three things: (1) graph classes of bounded width should have a reasonably rich structure, (2) a large number of different problems should be efficiently solvable on those classes, and at last (3) a decomposition witnessing the width of a graph should be computable in a reasonable amount of time. For undirected graphs width measures have been extremely successful. For sparse graph classes the notion of *treewidth* [33] has lead to a plethora

---

A full version of this article is available at https://arxiv.org/abs/1905.13203.

R. Steiner—Supported by DFG-GRK 2434.

S. Wiederrecht—Supported by the ERC consolidator grant DISTRUCT-648527.

M. Changat and S. Das (Eds.): CALDAM 2020, LNCS 12016, pp. 415–426, 2020.
https://doi.org/10.1007/978-3-030-39219-2_33

of different algorithmic approaches to solve otherwise untractable problems [4]. Among the methods utilising treewidth there are also so called *algorithmic meta theorems* like Courcelle's Theorem [6] stating that any MSO_2-definable problem can be solved on graphs of bounded treewidth in linear time. For dense graphs the approach of *clique-width* [9] still leads to the tractability of several otherwise hard problems on classes of bounded width [14]. In fact, in the case of clique-width still a powerful algorithmic meta theorem exists [7,10].

In harsh contrast to the immense success of undirected graph width measures for algorithmic applications, the setting of *digraphs* has resisted all approaches that tried to replicate the algorithmic power of treewidth. The most popular among the treewidth-like measures are *directed treewidth* [27], *DAG-width* [3], and *Kelly-width* [26]. For most of these concepts there exist some (XP) algorithms for a couple of specialized problems, usually involving some kind of routing, however the spectrum of algorithmic applications does not even come close to that of undirected treewidth. Ganian et al. [20] introduce a concept to capture the algorithmic abilities of a directed width parameter and reach the conclusion that no directed width measure can be 'algorithmically useful' and have 'nice structural properties' at the same time. So with all of the above mentioned directed analogues of treewidth being nicely behaved with respect to subdigraphs and butterfly minors, one cannot expect much more from the algorithmic side of these width measures.

The authors of [20] conclude their work by suggesting that the directed version of *clique-width* [9] would be a more suitable parameter, especially since the afore mentioned algorithmic meta theorem carries over to the digraph setting. However, it is not entirely clear how a directed clique-width expression of minimum width can be approximated in reasonable time. The current technique is to compute a bi-rank-width decomposition of optimal width, say k, and then use this decomposition to obtain a directed clique-width expression of width at most $2^{k+1} - 1$ in FPT-time [8]. There are elementary problems that still are $W[1]$-hard on graphs of bounded clique-width like the *Hamiltonian cycle* problem [16].

This naturally motivates finding a digraph width measure that does not struggle with more global decision problems like Hamiltonian cycles, while at the same time offers a broad variety of different problems that become efficiently computable in digraphs of bounded width and uses a decomposition concept that can be computed, or at least approximated within a reasonable span in FPT-time. In the case of undirected graphs clique width has similar problems, which led Gajarsky et al. to consider the more restrictive parameter of *modular width* [19]. For undirected graphs modular width fills the very specific niche as it covers dense undirected graphs, allows for FPT-time algorithms for a broad spectrum of problems and an optimal decomposition can be computed efficiently [28].

Contribution. The main contribution of this paper is to consider the directed version of modular width as a structural parameter for directed graphs in terms of its usefulness in parameterized algorithms. While *directed modular width* is more restrictive than the digraph parameters discussed above, the advantage is a wealth of otherwise untractable problems that turn out to admit FPT algorithms when parameterized with the directed modular width of the input digraph. Similar to the undirected case a directed modular decomposition of optimal width can be computed in polynomial time [29]. Besides other classical hard problems we give FPT algorithms for the problems *digraph colouring, Hamiltonian cycle, partitioning into paths, capacitated k-disjoint paths*, the *directed subgraph homeomorphism problem*, and for computing the *directed pathwidth* and the *cycle rank*. We obtain polynomial-time algorithms for finding topological minors in digraphs of bounded modular width. The dynamic programming approach we take utilizes the recursive nature of a directed modular decomposition, however, combining the dynamic programming tables of the children of some node in our recursion tree turns out to be a non-trivial problem on its own.

The paper is organized as follows. In the next section we give a short introduction to the tools we will use to construct our algorithms while in Sect. 3 we give a very high level overview on the strategy used in (almost) all of these algorithms. In Sect. 4 we show how to use this strategy to obtain a parameterized algorithm for a generalized version of the disjoint path problem which then can be applied to solve the directed subgraph homeomorphism problem. In Sect. 5 we present algorithms to compute the cycle rank and the directed pathwidth of a given digraph. For these problems we do not use the approach involving integer programming, instead we exploit the fact that these parameters can be equivalently formulated as vertex orderings. We conclude with an overview over other algorithmic results we obtained for directed modular width in a compact table. For detailed proofs of the theorems and discussions on the individual problems the reader is referred to the arXiv-version[1] of this paper.

2   Preliminaries

Digraphs in this paper are considered loopless and without parallel edges, but may have pairs of anti-parallel directed edges (called *digons*). Given a vertex x in a digraph D, we denote by $N_D^+(x) = \{y \in V(D) | (x,y) \in E(D)\}$ and $N_D^-(x) = \{y \in V(D) | (y,x) \in E(D)\}$ the out- and in-neighbourhood of x in D.

Directed Modular Width. Modules in digraphs are sets of vertices with the same relations to vertices outside the set Directed modular width measures the ability to repeatedly decompose a digraph into a bounded number of modules.

Definition 1. *Let D be a digraph. A non-empty subset $M \subseteq V(D)$ of vertices is called a* module, *if all the vertices in M have the same sets of out-neighbours and the same sets of in-neighbours outside the module. Formally, we have $N_D^+(u_1) \setminus M = N_D^+(u_2) \setminus M$ and $N_D^-(u_1) \setminus M = N_D^-(u_2) \setminus M$ for all $u_1, u_2 \in M$.*

Definition 2 (Directed Modular Width). *Let $k \in \mathbb{N}_0$, and let D be a digraph. D has* directed modular width *at most k if one of the following holds:*

- $|V(D)| \leq k$, *or*

---

[1] https://arxiv.org/abs/1905.13203.

- *There exists a partition of $V(D)$ into $\ell \in \{2, \ldots, k\}$ modules $M_1, \ldots, M_\ell$ such that for every i, $D[M_i]$ has directed modular width at most k.*

The least $k \geq 1$ for which a digraph D has directed modular width at most k is now defined to be the directed modular width, *denoted by $\mathrm{dmw}(D)$, of D.*

Fact 1 *Let D be a digraph, and let D' be an induced subdigraph of D. Then*

$$\mathrm{dmw}(D') \leq \mathrm{dmw}(D).$$

Given a digraph D and a partition $M_1, \ldots, M_\ell$ of $V(D)$ into modules, we will frequently use D_M to denote the *module-digraph* of D corresponding to the module-decomposition $M_1, \ldots, M_\ell$: D_M is obtained from D by identifying $M_i, i \in [\ell]$ each into a single vertex $v_i \in V(D_M)$ and deleting parallel directed edges afterwards. By the definition of a module, for every edge $(v_i, v_j) \in E(D_M)$ we have that $(u, w) \in E(D)$ for all $u \in M_i, w \in M_j$.

In the undirected case, clique-width is a lower bound on the modular width. The same is true in the case of *directed clique-width*, denoted by $\mathrm{dcw}(D)$, and directed modular width. A full definition of this parameter is given in [1].

Theorem 1. *Let D be a directed graph, then $\mathrm{dcw}(D) \leq \mathrm{dmw}(D)$.*

Computing a Non-Trivial Module-Decomposition. An important tool for all algorithms proposed in this paper is the ability to find a non-trivial decomposition of the vertex set of a given digraph into modules, in polynomial time. In fact, this task can be executed in a much stronger form. In [29], it was shown that a so-called *canonical module-decomposition* of a given digraph can be obtained in linear time. For us, the following weaker form of this result will be sufficient.

Theorem 2 *([29]).* *There is an algorithm that, given a digraph D on at least two vertices as input, returns a decomposition of $V(D)$ into $\ell \in \{2, \ldots, \mathrm{dmw}(D)\}$ modules. The running time is $\mathcal{O}(n + m)$, where $n := |V(D)|$ and $m := |E(D)|$.*

Integer Linear Programming with Bounded Number of Variables. The following strong theoretical result on the fixed-parameter tractability of Integer Linear Programs is an important part of most of our algorithms.

Theorem 3 *([15], Theorem 12).* *There exists an algorithm that, given as input a matrix $A \in \mathbb{Z}^{n \times p}$, vectors $c \in \mathbb{Z}^p, b \in \mathbb{Z}^n$, and some $U_1, U_2 \in \mathbb{Z}_+$, tests feasibility and if applicable outputs an optimal solution of the ILP*

$$\min c^T x \tag{1}$$

$$\text{subj. to } Ax \geq b, x \in \mathbb{Z}^p \tag{2}$$

in time $\mathcal{O}(p^{2.5p+o(p)} L \log(U_1 U_2))$ where L denotes the coding length of the input (A, b, c). Here we assume that the optimal value of the program lies within $[-U_1, U_1]$ and that U_2 is an upper bound on the largest absolute value any entry in an optimal solution vector can take.

3   Strategy

Our FPT algorithms are based on a common general strategy, which shall be outlined in the following.

- In most cases, we consider a well-chosen generalisation of the original problem we want to solve. This often involves additional inputs, such as integer weights or capacities on the vertices. In many cases, this is a crucial step to enable the recursion.
- We derive auxiliary theoretical results, that deal with a given module-decomposition of a digraph and describe how the studied parameters or objects which shall be computed on the whole digraph interact with corresponding objects on the digraphs induced by the modules and the module-digraph. These results are at the core of the construction of these algorithms.
- We describe an algorithm that, given solutions to the considered problem on the modules and the module-digraph, constructs a solution to the problem for the whole digraph in polynomial time.
- To solve the problem on the module-digraph, we make use of the fact that for bounded directed modular width, we can bound the number of vertices of the module-digraph by a constant. We then reformulate the problem on the module-digraph as an Integer Linear Program, which has bounded number of variables. However, the additional inputs such as weights on the vertices may still have polynomial size. We then make use of Theorem 3 to solve the problem on the module-digraph in FPT-time.
- Now we recurse until we end up solving the problem on digraphs consisting of single vertices. Because in each step, we further decompose an induced subdigraph into modules, the size of the recursion-tree is linear in the number of vertices of the input digraph. Because the module-decompositions can be computed in polynomial time in each step, we manage to prove an upper bound on the run-time of the form $\mathcal{O}(f(\omega)p(n)q(\log \tau))$, where f is some function, p and q are polynomials, ω denotes the directed modular width of the input digraph, n the number of vertices of the input digraph, and τ bounds the additional information carried in the input (for instance an upper bound on the sum of the weights distributed on the digraph).

4   Disjoint Paths and the Directed Subgraph Homeomorphism Problem

Instead of stating the actual problem we will describe a generalized version which is specifically designed for the needs of our strategy. For an overview on the r-Directed Disjoint Paths Problem, we refer to [18,31]. We list some known results.

- The 1-VDDP is polynomial-time solvable for arbitrary digraphs[2].

---

[2] Reachability in directed graphs is a simple special case of the max-flow problem and can be solved using one of the well-known polynomial algorithms for this task (see for instance [12]).

- For general digraphs, the existence version of r-VDDP is NP-complete, for any fixed $r \geq 2$ ([17]).
- There is an FPT algorithm solving r-VDPP with respect to parameter r for planar digraphs [11].
- The r-VDDP is polynomial-time solvable on DAGs ([17]), for any fixed $r \geq 1$. More generally, for any fixed $r \geq 1$, there exists an XP-algorithm for r-VDDP when parametrized with the directed treewidth $\mathrm{dtw}(D)$ of the digraph [27]. No FPT algorithms with respect to parameters r and $\mathrm{dtw}(D)$ are known.

Given a digraph D and vertices $s, t \in V(D)$, we use the term s-t-*path* to refer to a usual directed path starting in s and ending in t in the case of $s \neq t$, and to a directed cycle with a designated 'beginning' and 'end' at the vertex $s = t$ otherwise. For the generalized problem, we allow that in the given pairs $(s_1, t_1), \ldots, (s_r, t_r)$, vertices coincide, i.e., $s_i = t_j$ or $s_i = s_j$ for some $i, j \in [r]$. Our goal is to find an s_i-t_i-path P_i in D for every $i \in [r]$. However, we do not require the paths to be disjoint any more, they are allowed to share vertices. Instead, we bound the number of times a vertex can be traversed by the paths in total by a respective non-negative integer-capacity. For an s-t-path with $s = t$, we count two traversals of the vertex $s = t$ (at the beginning and at the end). The last important alteration of the original problem is that in addition to testing whether a collection of s_i-t_i-paths as required exists, in case it does, we want to find one that minimizes the *sum of the sizes* of the paths. In the following, the *size* of an s-t-path P shall be defined as $|P| := |E(P)| + 1$. $|P|$ is defined such that (independent of whether $s = t$ or $s \neq t$), it counts the total number of traversals of vertices by P.

For the purpose of applying the method from Theorem 3, we use a parameter $\tau \in \mathbb{N}$ as additional input, which bounds the sum of the vertex-capacities. To formulate our generalized problem properly, we need the following terminology. Given a list $\mathcal{S} = [(s_1, t_1), \ldots, (s_r, t_r)]$ of vertex-pairs equipped with vertex-capacities $w : V(D) \to \mathbb{N}_0$ of a digraph D, let us say that a collection $P_1, \ldots, P_r$ of s_i-t_i-paths in D is *compatible* with $(w, \mathcal{S})$, if any vertex $z \in V(D)$ is traversed at most $w(z)$ times in total by the collection. Furthermore, let us define $W(D, w, \mathcal{S})$ to be the minimum of $W(P_1, \ldots, P_r) := \sum_{i=1}^{r} |P_i|$ over all compatible collections $\mathcal{P} = (P_1, \ldots, P_r)$ of s_i-t_i-paths in D. If such a collection of s_i-t_i-paths does not exist, by convention, we put $W(D, w, \mathcal{S}) = \infty$. A feasible collection $\mathcal{P}$ for the pair $(w, \mathcal{S})$ is called *optimal*, if it attains the minimum.

r-Vertex-Disjoint Directed Paths (Capacity Version)
short: r-VDDP-C

Input A digraph D, a list $\mathcal{S} = [(s_1, t_1), \ldots, (s_r, t_r)]$ of not necessarily disjoint pairs of not necessarily distinct vertices in D, non-negative capacities $w : V(D) \to \mathbb{N}_0$ on the vertices and a threshold $\tau \in \mathbb{N}$ such that $\sum_{z \in V(D)} w(z) \leq \tau$.

Task Does there exist a feasible collection $P_1, \ldots, P_r$ compatible with $(w, \mathcal{S})$? If so, output the value of $W(D, w, \mathcal{S})$ and a corresponding optimal collection.

Setting $w(v) := 1, v \in V(D)$ we see that the original r-VDDP forms a special case of r-VDDP-C, where we can take $\tau = n$. Our main result is the following.

Theorem 4. *There exists an algorithm solving the r-VDDP-C which runs in time $\mathcal{O}\left(n^3 + f(r,\omega)(n^2 + n \log \tau)\right)$, where $n := |V(D)|$ and $\omega := \mathrm{dmw}(D)$. Furthermore, $f(r,\omega) = 2^{\mathcal{O}(r \log r \cdot 2^\omega \omega)}$. As a consequence, the r-VDDP on D can be solved in time $\mathcal{O}(n^3 + f(r,\omega)n^2)$.*

Proof (High-Level-Sketch). A first step towards an algorithm will be to observe that we can restrict our attention to *reduced paths*, these are s-t-paths P such that there is no s-t-path P' with $V(P') \subsetneq V(P)$. Given a module-decomposition $M_1, \ldots, M_\ell$, such a path is either completely contained in a module M_j of D, or it uses at most one vertex of each module, except maybe for the endpoints, which may be contained in the same module.

Suppose that s_i, t_i lie in the same module for $1 \leq i \leq r'$ and in different modules for all $i > r'$. For $i \leq r'$, we have to make a decision whether to route the path P_i connecting s_i, t_i within the module or to leave the module and reenter eventually. Let $B \subseteq [r']$ be the set of indices for which we choose the second option. It is now possible to split the pairs s_i, t_i into two parts, namely the ones with $i \in B \cup \{r'+1, \ldots, r\}$, which are routed between the modules, and the ones with $i \in [r'] \setminus B$, which are routed within their module. The first part of this problem is taken care of by solving an equivalent routing problem in the module-digraph D_M using an integer-program with a bounded number of variables (see Theorem 3), the second part splits into individual routing problems within the respective modules. For each of those modular routing problems we have to solve, we need to correspondingly adapt the weight functions. The basic idea is to solve the problem for all possible sets $B \subseteq [r']$ and to choose the best collection of paths in the end. However, a central problem here is to control the complexity of the dynamic programming tree and therefore to control the different possible weight functions which can arise in the subproblems. This problem is overcome by computing in each step not only the optimal collection for the problem with a fixed given weight function w, but instead for many different well-chosen weight functions w_A on the vertices, determined by an index-set $A \subseteq [r]$, which include via $w_{[r]} = w$ the original weight function we are interested in. We can then show that given optimal solutions to this more general problem for all digraphs $D[M_j], 1 \leq j \leq r$, it is possible to recover the optimal path-collections for D and all weightings $w_A, A \subseteq [r]$.

Consider a pair of an *input digraph* D and a *pattern digraph* $\mathcal{H}$. The pattern digraph $\mathcal{H}$ (exceptionally in this paper) is allowed to admit loops, as well as multiple parallel or anti-parallel edges. A *homeomorphism* from $\mathcal{H}$ into D maps vertices of $\mathcal{H}$ to distinct vertices in D and directed edges in $\mathcal{H}$ to directed paths in D connecting the images of the corresponding end vertices, such that the paths only intersect at common endpoints. For a loop, this means that its image forms a directed cycle passing through the image of the incident vertex in $\mathcal{H}$. The *directed subgraph homeomorphism problem* ($\mathcal{H}$-DSHP) is to decide whether a given digraph D contains a homeomorphic image of a pattern digraph $\mathcal{H}$ using specified vertices. This can be seen as a generalized path finding problem. In fact, the r-VDDP is the special case of this problem where the pattern digraph $\mathcal{H}$ forms an oriented matching consisting of r disjoint edges. To keep control of the complexity of this problem, one usually regards the pattern digraph $\mathcal{H}$ as part of the problem description rather than as part of the input.

The authors of [17] show that $\mathcal{H}$-DSHP is in P for pattern digraphs $\mathcal{H}$ which admit a dominating source or a dominating sink, and NP-complete in every other case. They established polynomial-time algorithms to solve the $\mathcal{H}$-DSHP on DAGs for every fixed pattern digraph $\mathcal{H}$. Using a suitable weight function w on the vertices it is possible to formulate the $\mathcal{H}$-DSHP as a subproblem of the r-VDDP-C. As a Corollary of Theorem 4, we obtain the following.

Theorem 5. *Let $\mathcal{H}$ be a pattern digraph with vertex set $h_1, \ldots, h_r$, and let $m := |E(H)|$. There exists an algorithm that, given as input a digraph D equipped with a list $s_1, s_2, \ldots, s_r$ of pairwise distinct vertices in D, solves the $\mathcal{H}$-DSHP with this instance in time*

$$\mathcal{O}(n^3 + f(m, \omega)n^2),$$

where $n := |V(D)|$, $\omega := \mathrm{dmw}(D)$, and $f(m, \omega) = 2^{\mathcal{O}(m \log m \cdot 2^{\omega} \omega)}$.

If a digraph D contains a homeomorphic image (also called subdivision) of a digraph $\mathcal{H}$, $\mathcal{H}$ is called a *topological minor* of D. For undirected graphs, it has been proven that the detection of topological minors is fixed parameter-tractable [24] when parametrizing with the size of the minor. However, for directed graphs, there exist instances $\mathcal{H}$ for which it is NP-complete to test whether a given digraph D contains a subdivision of $\mathcal{H}$ (see Theorem 33 in [2] for an example). We therefore believe that the following result is a relevant contribution to minor-testing in directed graphs.

Theorem 6. *Let $\mathcal{H}$ be a multi-digraph (possibly containing loops, and multiple parallel and anti-parallel edges). There exists an algorithm that decides whether $\mathcal{H}$ is a topological minor of a given digraph D, and if so, returns a subdivision of $\mathcal{H}$ which is a subdigraph of D. This algorithm runs in time*

$$\mathcal{O}(f(m, \omega)n^{r+3}).$$

Here, we have $m := |E(H)|, r := |V(H)|, n := |V(D)|$ and $\omega := \mathrm{dmw}(D)$. We furthermore have $f(m, \omega) = 2^{\mathcal{O}(m \log m \cdot 2^{\omega} \omega)}$.

5   Other Directed Width Measures

Although most directed width measures are not very powerful in the algorithmic context, some of them like *directed pathwidth* or the *cycle-rank* which have high theoretical importance, find specialized applications and there are no FPT algorithms known to compute their corresponding optimal decompositions.

A *directed path decomposition* for a digraph D is a tuple (P, β) where P is a directed path and $\beta \colon V(P) \to 2^{V(D)}$ a function of *bags* such that $\bigcup_{p \in V(P)} \beta(p) = V(D)$. Moreover for every $v \in V(D)$ the vertices $p \in V(P)$ with $v \in \beta(p)$ induce a subpath of P and for every edge $(p, t) \in E(P)$ every edge with tail in $\bigcup_{t' \in V(tP)} \beta(t')$ and head in $\bigcup_{p' \in Pp} \beta(p')$ contains a vertex of $\beta(p) \cap \beta(t)$. Here we denote the maximal subpath of P ending in p by Pp while the maximal subpath starting in t is denoted by tP. The *width* of (P, β) is $\max_{p \in V(P)} |\beta(p)| - 1$ and the directed pathwidth of D is the minimum width over all directed path decompositions for D.

We want to produce an optimal directed path decomposition for D from partial solutions obtained for its modules, so we need to be able to merge these partial solutions. To do this we differ between two kinds of vertices: those which appear in exactly one bag and need no further care, and those that are contained in at least two bags. Since the size of the module digraph is bounded we are able to consider all possible directed paths of length at most $\mathrm{dmw}(D) - 1$ together with all possible bag functions β for these paths. So the key to our algorithm is to show that there always exists an optimal directed path decomposition for D such that for every module M, either all vertices of M need to be contained in a single bag, or we may use an optimal decomposition of M independently of the rest of D. This yields the following.

Theorem 7. *There exists an algorithm that given a digraph D as input, outputs a directed path decomposition for D of minimum width. The algorithm runs in time $\mathcal{O}\left(\omega n^3 + \omega^3 2^{\omega^2} n\right)$, where $n := |V(D)|$ and $\omega := \mathrm{dmw}(D)$.*

Proof (High-Level-Sketch). Let M be a module and v_M the corresponding vertex in the module digraph. To realize the above ideas we first have to show that we can always find an optimal directed path decomposition (P, β) for the module digraph such that P has at most $\mathrm{dmw}(D)$ vertices. Now suppose with $\beta'(p) := \beta(p) \setminus \{v_M\}$, (P, β') is still a directed path decomposition. Then v_M is not needed to block an edge and thus we may incorporate an optimal directed path decomposition of M into our solution. Otherwise v_M is essential as a vertex in some separator defined by the decomposition. In this case we do not have to bother decomposing M, since each vertex of M has to be part of said separator. We show that every v_M belongs to one of the two categories and that we can always find an optimal directed path decomposition decomposing D in this fashion.

In a similar way, the *cycle-rank* of D can be expressed as an ordering of its vertices, see [13,23] for the definitions.

Theorem 8. *There exists an algorithm that, given a digraph D as input, outputs an ordering σ for $V(D)$ such that $\mathrm{rank}(\sigma) = \mathrm{cr}(D)$. The algorithm runs in time $\mathcal{O}\left(n^3 + \omega^3 \omega! n\right)$, where $n := |V(D)|$ and $\omega := \mathrm{dmw}(D)$.*

Neither of the two algorithms makes use of the integer programming approach. The factor $\omega!$ in Theorem 8 arises from the enumeration of all possible orderings of the vertices of the module digraph, while in Theorem 7 the factor 2^{ω^2} expresses the enumeration of all possible directed paths of length at most ω.

6   Conclusion

By applying the strategy from Sect. 3 and by revisiting some ideas from [19] we are able to construct parameterized algorithms for several classical problems. Since the main ideas do not differ to much, we present them here in form of a table, moreover we give additional information on known hardness results for these problems with regards to the width parameters directed treewidth (dtw) and directed clique-width (dcw). If hardness results for very restricted classes of digraphs exist, or a problem is known to be parameterizable by the solution-size, we will also mention those.

Problem	dmw[a]	dtw	dcw	General complexity
Directed Feedback Vertex Set	FPT	-	FPT [10]	NP-hard for planar digraphs with Δ^+, $\Delta^- \leq 3$ [22], FPT in general [5]
Directed Dominating Set	FPT	NP-h [21]	FPT [1]	-
Dichromatic Number	FPT	NP-h [30]	-	NP-hard even for bounded directed tree-width and bounded degeneracy [30]
Partitioning into Directed Paths	FPT	-	-	NP-hard even for planar digraphs and digraphs of bounded degree [32]
Directed Hamiltonian Cycle	FPT	XP [27]	W[1]-h [16], XP [14]	-

[a]All results in this column are due to the work presented in this report.

To summarize we have initiated the study of the *directed modular width* in parameterized algorithmics, which is a natural structural parameter for digraphs. We have seen that by combining dynamical programming with the strong tool of bounded-variable ILP solving, one can obtain FPT algorithms for many intrinsically hard problems on directed graphs, which are intractable or unsolved for classes of bounded directed tree-width, DAG-width or clique-width. Moreover, the recursive nature of module-decompositions allows us to find fast FPT algorithms for generalisations of these problems, such as r-VDDP-C for r-VDDP.

Our results show that the directed modular width covers a nice niche in the landscape of directed width measures, as it can be computed efficiently, is small on dense but structured networks and avoids the algorithmic price of generality paid by most other width measures for directed graphs. Directed modular width furthermore generalizes directed co-graphs (digraphs of modular width 2), which have received attention in terms of algorithmics previously (cf. [25]).

References

1. Bang-Jensen, J., Gutin, G.: Classes of Directed Graphs. Springer, Cham (2018). https://doi.org/10.1007/978-3-319-71840-8
2. Bang-Jensen, J., Havet, F., Trotignon, N.: Finding an induced subdivision of a digraph. Theor. Comput. Sci. **443**, 10–24 (2012)
3. Berwanger, D., Dawar, A., Hunter, P., Kreutzer, S.: DAG-width and parity games. In: Durand, B., Thomas, W. (eds.) STACS 2006. LNCS, vol. 3884, pp. 524–536. Springer, Heidelberg (2006). https://doi.org/10.1007/11672142_43
4. Bodlaender, H.L.: Treewidth: algorithmic techniques and results. In: Prívara, I., Ružička, P. (eds.) MFCS 1997. LNCS, vol. 1295, pp. 19–36. Springer, Heidelberg (1997). https://doi.org/10.1007/BFb0029946
5. Chen, J., Liu, Y., Lu, S., O'Sullivan, B., Razgon, I.: A fixed-parameter algorithm for the directed feedback vertex set problem. J. ACM **55**(5), 21 (2008)
6. Courcelle, B.: The monadic second-order logic of graphs. i. recognizable sets of finite graphs. Inf. Comput. **85**(1), 12–75 (1990)
7. Courcelle, B.: The expression of graph properties and graph transformations in monadic second-order logic. In: Handbook of Graph Grammars and Computing By Graph Transformation: Volume 1: Foundations, pp. 313–400. World Scientific (1997)
8. Courcelle, B., Engelfriet, J.: Graph Structure and Monadic Second-Order Logic: A Language-Theoretic Approach, vol. 138. Cambridge University Press, New York (2012)
9. Courcelle, B., Engelfriet, J., Rozenberg, G.: Handle-rewriting hypergraph grammars. J. Comput. Syst. Sci. **46**(2), 218–270 (1993)
10. Courcelle, B., Makowsky, J.A., Rotics, U.: Linear time solvable optimization problems on graphs of bounded clique-width. Theory Comput. Syst. **33**(2), 125–150 (2000)
11. Cygan, M., Marx, D., Pilipczuk, M.: The planar directed k-vertex-disjoint paths problem is fixed-parameter tractable. In: 2013 IEEE 54th Annual Symposium on Foundations of Computer Science, pp. 197–206 (2013)
12. Edmonds, J., Karp, R.M.: Theoretical improvements in algorithmic efficiency for network flow problems. J. ACM **19**, 248–264 (1972)
13. Eggan, L.C., et al.: Transition graphs and the star-height of regular events. The Mich. Math. J. **10**(4), 385–397 (1963)
14. Espelage, W., Gurski, F., Wanke, E.: How to solve NP-hard graph problems on clique-width bounded graphs in polynomial time. In: Brandstädt, A., Le, V.B. (eds.) WG 2001. LNCS, vol. 2204, pp. 117–128. Springer, Heidelberg (2001). https://doi.org/10.1007/3-540-45477-2_12
15. Fellows, M.R., Lokshtanov, D., Misra, N., Rosamond, F.A., Saurabh, S.: Graph layout problems parameterized by vertex cover. In: Hong, S.-H., Nagamochi, H., Fukunaga, T. (eds.) ISAAC 2008. LNCS, vol. 5369, pp. 294–305. Springer, Heidelberg (2008). https://doi.org/10.1007/978-3-540-92182-0_28
16. Fomin, F.V., Golovach, P.A., Lokshtanov, D., Saurabh, S.: Intractability of clique-width parameterizations. SIAM J. Comput. **39**(5), 1941–1956 (2010)
17. Fortune, S., Hopcroft, J., Wyllie, J.: The directed subgraph homeomorphism problem. Theor. Comput. Sci. **10**, 111–121 (1980)
18. Frank, A.: Packing paths, circuits and cuts-a survey. In: Paths, Flows, and VLSI-Layout, pp. 47–100. Springer-Verlag, Berlin (1990)

19. Gajarský, J., Lampis, M., Ordyniak, S.: Parameterized algorithms for modular-width. In: Gutin, G., Szeider, S. (eds.) IPEC 2013. LNCS, vol. 8246, pp. 163–176. Springer, Cham (2013). https://doi.org/10.1007/978-3-319-03898-8_15

20. Ganian, R., et al.: Are there any good digraph width measures? J. Comb. Theory Ser. B **116**, 250–286 (2016)

21. Ganian, R., Hliněný, P., Kneis, J., Langer, A., Obdržálek, J., Rossmanith, P.: Digraph width measures in parameterized algorithmics. Discrete Appl. Math. **168**, 88–107 (2014)

22. Garey, M.R., Johnson, D.S.: Computers and Intractability; A Guide to the Theory of NP-Completeness. W. H. Freeman & Co., New York (1990)

23. Giannopoulou, A.C., Hunter, P., Thilikos, D.M.: Lifo-search: a min-max theorem and a searching game for cycle-rank and tree-depth. Discrete Appl. Math. **160**(15), 2089–2097 (2012)

24. Grohe, M., Kawarabayashi, K.I., Marx, D., Wollan, P.: Finding topological subgraphs is fixed-parameter tractable. In: Proceedings of the Forty-third Annual ACM Symposium on Theory of Computing, STOC 2011, New York, NY, USA, pp. 479–488 (2011)

25. Gurski, F., Komander, D., Rehs, C.: Computing digraph width measures on directed co-graphs. In: Gąsieniec, L.A., Jansson, J., Levcopoulos, C. (eds.) FCT 2019. LNCS, vol. 11651, pp. 292–305. Springer, Cham (2019). https://doi.org/10.1007/978-3-030-25027-0_20

26. Hunter, P., Kreutzer, S.: Digraph measures: kelly decompositions, games, and orderings. Theor. Comput. Sci. **399**(3), 206–219 (2008)

27. Johnson, T., Robertson, N., Seymour, P., Thomas, R.: Directed tree-width. J. Comb. Theory Ser. B **82**(1), 138–154 (2001)

28. McConnell, R.M., Spinrad, J.P.: Modular decomposition and transitive orientation. Discrete Math. **201**(1–3), 189–241 (1999)

29. McConnella, R.M., de Montgolfier, F.: Linear time modular decomposition of directed graphs. Discrete Appl. Math. **145**, 198–209 (2005)

30. Millani, M.G., Steiner, R., Wiederrecht, S.: Colouring non-even digraphs. Technical report (2019, submitted). arXiv Preprint, arXiv:1903.02872

31. N. Robertson, P.D.S.: An outline of a disjoint paths algorithm. In: Paths, Flows, and VLSI-Layout, pp. 267–292. Springer-Verlag, Berlin (1990)

32. Plesník, J.: The NP-completeness of the Hamiltonian cycle problem in planar digraphs with degree bound two. Inform. Process. Lett. **8**(4), 199–201 (1979)

33. Robertson, N., Seymour, P.D.: Graph minors. ii. algorithmic aspects of tree-width. J. Algorithms **7**(3), 309–322 (1986)

On the Parameterized Complexity of Spanning Trees with Small Vertex Covers

Chamanvir Kaur and Neeldhara Misra[(✉)]

Indian Institute of Technology, Gandhinagar, Gujarat, India
{chamanvir.kaur,neeldhara.m}@iitgn.ac.in

Abstract. We consider the minimum power spanning tree (MPST) problem with general and unit demands from a parameterized perspective. The case of unit demands is equivalent to the problem of finding a spanning tree with the smallest possible vertex cover (MCST). We show that MPST is W[1]-hard when parameterized by the vertex cover of the input graph, and is W[2]-hard when parameterized by the solution size—the latter holds even in the case of unit demands. For the special case of unit demands, however, we demonstrate an FPT algorithm when parameterized by treewidth. In the context of kernelization, we show that even MCST is unlikely to admit a polynomial kernel under standard complexity-theoretic assumptions when parameterized by the vertex cover of the input graph.

Keywords: Vertex cover · Spanning trees · Treewidth

1 Introduction

A spanning tree is a minimally connected subgraph of a graph that spans all of its vertices. The problem of finding spanning trees of minimum weight is a fundamental algorithmic question and has received much attention. Variations of this question have also attracted a lot of interest, wherein one is interested in spanning trees with special structural properties, such as having bounded diameter [10], many leaves [4], or minimum poise (the sum of the diameter and the maximum degree [11]).

Our focus, in this work, is on the problem of finding spanning trees with small vertex covers. This is a special case of a more general problem which we also address, namely the minimum power spanning tree problem (MPST). Here, we are given an edge-weighted graph $G = (V, E)$, where the weights can be thought of as "demands", and the objective is to find a spanning tree and to assign "power" values to vertices such that all the edges of the spanning tree are covered. An edge is *covered* if the assigned power in one of its extremities is at least its demand. The goal is to minimize the sum of powers over all vertices.

This work is supported by the Science and Engineering Research Board (SERB), India.

M. Changat and S. Das (Eds.): CALDAM 2020, LNCS 12016, pp. 427–438, 2020.
https://doi.org/10.1007/978-3-030-39219-2_34

Observe that if all edges have unit demands, then this question is equivalent to finding a spanning tree with the smallest possible vertex cover. As an example, consider the complete graph, which has two spanning trees that are extreme from this perspective: the first is a star, which has a vertex cover of size one and the other is a Hamiltonian path, where we need roughly half the vertices of the graph to cover all the edges involved.

The minimum power variation of various optimization problems—notably, vertex cover (power vertex cover), Steiner trees (minimum power Steiner trees), and cut problems—are widely studied for their suitability in application scenarios. As a consequence of being more general, these problems usually model more sophisticated scenarios. Several applications arise in the context of connectivity questions in the domain of wireless networks and placements of sensors and cameras on road and home networks.

Our Contributions. In this contribution, we explore the parameterized complexity of the MPST problem and special cases. Our main result is that MPST is W[1]-hard when parameterized by the vertex cover of the input graph (Theorem 2), even when the weights are polynomially bounded in the size of the input. In fact, assuming ETH (Exponential Time Hypothesis), this also implies that there is no algorithm with a running time of $n^{o(\ell)}$, where ℓ denotes the size of the vertex cover.

Motivated by this intractibility, we turn to the special case with unit demands, which we refer to as the Minimum Cover Spanning Tree problem (MCST). We show that MCST is W[2]-hard when parameterized by the solution size (Theorem 1). On the other hand, we show an FPT algorithm when parameterized by the treewidth of the input graph (Theorem 3). In the context of kernelization, we show that even MCST is unlikely to admit a polynomial kernel under standard complexity-theoretic assumptions when parameterized by the vertex cover of the input graph (Corollary 3).

Related Work. The MPST problem, in the form that we propose and study it here, was considered by Angel et al. [2], who establish the hardness of approximation for this problem by a reduction from Dominating Set. In fact, our FPT reduction showing that MCST is W[2]-hard is in similar spirit. For a treatment of MCST from the perspective of approximation algorithms and on special classes of graphs, see [8].

Our result showing the hardness of MPST when parameterized by the vertex cover uses some ideas from the reduction employed by Angel et al. [3] to demonstrate that the Power Vertex Cover problem is also similarly intractable when parameterized by treewidth. The Power Vertex Cover problem is the natural "power"-based analog of the traditional vertex cover problem, where we seek an assignment of power values to the vertices, minimizing the total power assigned, so that the demand of every edge is met.

Closely related to MPST and MCST is the minimum power analog of the Steiner Tree problem, which has also been studied quite extensively (see, for instance, [1,9]). However, we remark that in these treatments, the demand of an edge is met only if both of its endpoints receive a power value that is at least as much as its demand. For recent developments in approximation algorithms for power covering problems, see [5].

2   Preliminaries

We use $[n]$ to denote the set $\{1, 2, \ldots, n\}$. We follow standard notation and terminology from parameterized complexity [6] and graph theory [7]. We recall some of the definitions that will be relevant to our discussions.

The *neighborhood* of a vertex is denoted $N(v)$ and consists of all vertices u adjacent to v. The *closed neighborhood* of a vertex is denoted $N[v]$ and is defined as $N(v) \cup \{v\}$. A *tree* is a connected and acyclic graph. Given a graph G, a *spanning tree* T is a subgraph of G such that $V(T) = V(G)$ and T is a tree. For any two vertices u and v, we let $d(u, v)$ denote the length of the shortest path between the vertices. For $S \subseteq V$, $G[S]$ denotes the graph *induced* by S in G. The vertex set of $G[S]$ is S, and the edge set is $\{(u, v) \mid u \in S, v \in S$ and $(u, v) \in E\}$. We say that a vertex v is *global* to a set S of vertices if v is adjacent to every vertex in S. We now turn to the description of some of the problems that will be considered in this contribution.

The MINIMUM POWER SPANNING TREE (MPST) problem is the following.

MINIMUM POWER SPANNING TREE (MPST)
Input: A graph G, a demand function $w : E \to \mathbb{R}^+$, and $k \in \mathbb{Z}^+$.
Question: Does G admit a spanning tree T and an assignment of power values $\rho : V(G) \to \mathbb{R}^+$ such that $\sum_{v \in V(G)} \rho(v) \leqslant k$ and for every edge $e \in E(T)$ with endpoints u and v, $\max(\rho(u), \rho(v)) \geqslant w(e)$?

We now consider this problem in the context of unit demands. Noting that this is equivalent to finding a spanning tree with the smallest vertex cover, we refer to this as MINIMUM COVER SPANNING TREE (MCST) problem.

MINIMUM COVER SPANNING TREE (MCST)
Input: A graph G and a positive integer k.
Question: Does G admit a spanning tree T with a vertex cover of size at most k?

The problems RBDS and MCIS (defined below) are known to be W[2]-hard and W[1]-hard, respectively, when parameterized by the solution size [6].

RED-BLUE DOMINATING SET (RBDS)
Input: A bipartite graph $G = (R \cup B, E)$ and a positive integer k.
Question: Does there exist a subset $S \subseteq R$ of size at most k such that for every $v \in B$, $|N(v) \cap S| \geqslant 1$?

MULTI-COLORED INDEPENDENT SET (MCIS)
Input: A $G = (V, E)$ and a partition of $V = (V_1, \ldots, V_k)$ into k parts.
Question: Does there exist a subset $S \subseteq V$ such that S is independent
in G and for every $i \in [k]$, $|V_i \cap S| = 1$?

We now turn to the notion of the treewidth of a graph.

Definition 1. *Let G be a graph. A* tree-decomposition *of G is a pair* $\mathbb{T} = (T, (B_t)_{t \in V(T)})$, *where T is a rooted tree, and for all* $t \in V(T)$, $B_t \subseteq V(G)$ *such that*

▷ $\bigcup_{t \in V(T)} B_t = V(G)$,
▷ *for every edge* $xy \in E(G)$ *there is a* $t \in V(T)$ *such that* $\{x, y\} \subseteq B_t$, *and*
▷ *for every vertex* $v \in V(G)$ *the subgraph of* T *induced by the set* $\{t \mid v \in B_t\}$ *is connected.*

The width *of a tree decomposition is* $\max_{t \in V(T)} |B_t| - 1$ *and the* treewidth *of G is the minimum width over all tree decompositions of G and is denoted by* **tw**(G).

For completeness, we also define here the notion of a *nice tree decomposition with introduce edge nodes*, as this is what we will work with in due course. We note that a given tree decomposition can be modified in linear time to fulfill the above constraints; moreover, the number of nodes in such a tree decomposition of width w is $O(w \cdot n)$ [12].

Definition 2. *A tree decomposition* $\mathbb{T} = (T, (B_\alpha)_{\alpha \in V(T)})$ *is a* nice tree decomposition with introduce edge nodes *if the following conditions hold.*

1. *The tree* T *is rooted and binary.*
2. *For all edges in* $E(G)$ *there is exactly one* introduce edge node *in* T, *where an introduce edge node is a node* α *in the tree decomposition* T *of G labeled with an edge* $\{u, v\} \in E(G)$ *with* $u, v \in B_\alpha$ *that has exactly one child node* α'; *furthermore* $B_\alpha = B_{\alpha'}$.
3. *Each node* $\alpha \in V(T)$ *is of one of the following types:*
    ▷ introduce edge node: *as described above;*
    ▷ leaf node: α *is a leaf of* T *and* $B_\alpha = \emptyset$;
    ▷ introduce vertex node: α *is an inner node of* T *with exactly one child node* $\beta \in V(T)$; *furthermore* $B_\beta \subseteq B_\alpha$ *and* $|B_\alpha \backslash B_\beta| = 1$;
    ▷ forget node: α *is an inner node of* T *with exactly one child node* $\beta \in V(T)$; *furthermore* $B_\alpha \subseteq B_\beta$ *and* $|B_\beta \backslash B_\alpha| = 1$;
    ▷ join node: α *is an inner node of* T *with exactly two child nodes* $\beta, \gamma \in V(T)$; *furthermore* $B_\alpha = B_\beta = B_\gamma$.

3 The Standard Parameter

In this section, we show that MCST is W[2]-hard with respect to the standard parameter (i.e, the solution size). In particular, we describe an FPT reduction from RBDS to MCST.

Theorem 1. MCST *is* W[2]-*hard when parameterized by the solution size.*

Proof. Let (G, k) be an instance of RBDS where $G = (R \cup B, E)$. Without loss of generality, we assume that every vertex in B has at least one neighbor in R, because if this is not the case, we may return a trivial No instance. We now describe the transformed instance of MCST, which we denote by (H, k'). The graph H is obtained from G by adding a global vertex to R, which in turn has $(k + 2)$ new neighbors of degree one, which we refer to as *guards*. In particular, we have:

$$V(H) = V(G) \cup \{g, g_1, \ldots, g_{k+2}\},$$

and:

$$E(H) = E(G) \cup \{(g, g_1), (g, g_2), \ldots, (g, g_{k+2})\} \cup \{(g, v) \mid v \in R\}.$$

We let $k' = k + 1$. This completes the description of the reduction, and we now turn to a proof of the equivalence of the instances.

The Forward Direction. Let $S \subseteq R$ be a dominating set. For a vertex $v \in B$, let $p_v \in R$ denote a vertex from S such that $p_v \in N(v)$. In the event that there are multiple vertices in S that are adjacent to v, the choice of p_v is arbitrary. Consider the spanning tree T given by the following edges:

$$E(T) = \{(g, g_1), (g, g_2), \ldots, (g, g_{k+2})\} \cup \{(g, v) \mid v \in R\} \cup \{(v, p_v) \mid v \in B\}.$$

It is easy to verify that $\{g\} \cup S$ is a vertex cover of size $(k + 1)$ for the edges of T. Indeed, it is evident that any edge in T that is not incident to g is incident to a vertex from S and this concludes the argument in the forward direction.

The Reverse Direction. Let T be a spanning tree of H with a vertex cover S of size $(k + 1)$. Observe that the edge (g, g_i) belongs to T for any $1 \leqslant i \leqslant k + 2$: if not, the vertex g_i would be isolated in T. Therefore, we also conclude that $g \in S$, since it would otherwise be too expensive for S to account for covering the edges incident to all the guards.

Let $S' := S \cap (R \cup B)$. By the argument above, we have that $|S'| \leqslant k$. For any $v \in B \cap S'$, let p_v denote an arbitrarily chosen neighbor of v in R. Now consider the set given by:

$$S^\star := (S \cap R) \cup \{p_v \mid v \in S \cap B\}.$$

It is easy to see that $|S^\star| \leqslant k$ and that $S^\star \subseteq R$ dominates all vertices in B. Indeed, consider any $v \in B$. If $v \in S \cap B$, then $p_v \in S^\star$ dominates v. If not, then observe that for some vertex $u \in N(v)$, the edge (u, v) must belong to T (if not, v is isolated in T, a contradiction). By the case we are in, $v \notin S$, so to cover the edge (u, v), we must have that $u \in S$. Further, since G is bipartite, $u \in R$, thus $u \in (S \cap R)$. This implies, by construction, that $u \in S^\star$, and u dominates v, as required. This concludes the proof. $\square$

4   Vertex Cover and Treewidth

We begin by establishing that MPST is W[1]-hard when parameterized even by the vertex cover of the input graph (and therefore also its treewidth). Thereafter, we show a FPT algorithm for the special case of unit demands, i.e, the MCST problem, when parameterized by treewidth. We remark that the ideas in the reduction demonstrating the hardness parameterized by vertex cover are inspired by the construction used for showing the hardness of the power vertex cover problem when parameterized by treewidth [3].

Theorem 2. MPST *is* W[1]-*hard when parameterized by the vertex cover of the input graph.*

Proof. We reduce from MCIS. Let (G, k) be an instance of MCIS where $G = (V, E)$ and further, let $V = (V_1, \ldots, V_k)$ denote the partition of the vertex set V. We assume, without loss of generality, that $|V_i| = n$ for all $i \in [k]$. Specifically, we denote the vertices of V_i by $\{v_1^i, \ldots, v_n^i\}$. We are now ready to describe the transformed instance of MCST, which we denote by (H, w, k'). The construction is a combination of *choice gadgets* and *checker gadgets*.

Choice Gadgets. For each $1 \leqslant i \leqslant k$, introduce two vertices x_i and y_i, and further, introduce n vertices $z_1^i, \ldots, z_n^i$. We add the edges (x_i, z_j^i) and (y_i, z_j^i) for all $j \in [n]$. The weight function is defined as follows, for all $1 \leqslant j \leqslant n$.

$$w(u, z_j^i) = \begin{cases} j & \text{if } u = x_i, \\ n - j + 1 & \text{if } u = y_i. \end{cases}$$

In other words, for all $1 \leqslant i \leqslant k$ and $1 \leqslant j \leqslant n$, the vertex z_j^i is adjacent to x_i with an edge of weight j and to y_i with an edge of weight $(n-j+1)$. We refer to x_i and y_i as *anchors* and their neighbors in the choice gadget as *guards*.

Checker Gadgets. For each edge $e = (v_p^a, v_q^b) \in E(G)$, we introduce a vertex c_e, which is adjacent to the vertices x_a, y_a, x_b, y_b in the choice gadgets. The edge weights are given by:

$$w(c_e, u) = \begin{cases} p + 1 & \text{if } u = x_a, \\ n - p + 1 & \text{if } u = y_a, \\ q + 1 & \text{if } u = x_b, \\ n - q + 1 & \text{if } u = y_b. \end{cases}$$

Our transformed instance comprises of all the k choice gadgets and m checker gadgets. Further, we add a vertex g universal to all the anchor vertices and introduce b vertices $\{g_1, \ldots, g_b\}$ that are adjacent only to g, where $b = nk + 2$. All edges incident to g have a weight of one. This completes the description of the graph H. We refer the reader to Fig. 1 for a schematic depiction of the graph H.

The target total power, k', is set to $nk+1$. Observe that the anchor vertices along with the vertex g form a vertex cover of size $(2k+1)$ for H—indeed, all edges not incident to the universal vertices are either edges between anchors and guards, or between anchors and the vertices in the checker gadgets representing the edges of G. We now turn to a proof of equivalence.

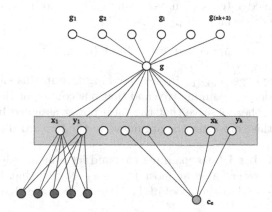

Fig. 1. An overview of the reduction used in the proof of Theorem 2. The anchor vertices are placed in the blue box, and along with the vertex g, form a vertex cover of size $2k+1$. The red vertices are the guards while the green vertex is an example of a vertex from a checker gadget. (Color figure online)

Forward Direction. Let $S \subseteq V$ be a multicolored independent set of G. Further, let $\tau : [k] \rightarrow [n]$ denote the choice of S from the parts of V. Specifically:

$$S \cap V_i := \{v^i_{\tau(i)}\}.$$

We now describe our spanning tree T and a power assignment ρ of total cost at most $k' = nk+1$. First, we choose all edges incident on g. Next, from the choice gadget containing vertices x_i and y_i, we pick the following edges:

$$\{(x_i, z^i_1), \ldots, (x_i, z^i_t), (y_i, z^i_{t+1}), \ldots, (y_i, z^i_n)\},$$

where $t := \tau(i)$. Before describing how we will connect c_e to the structure developed so far, it will be useful to describe the power value assignment ρ. We have the following, where $i \in [k]$:

$$\rho(u) = \begin{cases} 1 & \text{if } u = g, \\ \tau(i) & \text{if } u = x_i, \\ n - \tau(i) & \text{if } u = y_i, \\ 0 & \text{for all other vertices.} \end{cases}$$

We now consider again the checker gadgets. Let $e = (v_p^a, v_q^b) \in E(G)$. Consider the case when $v_p^a \in S$ (the other scenarios are symmetric). Then $v_q^b \notin S$. Consider $\rho(x_b)$. We know that $\rho(x_b) \neq q$. If $\rho(x_b) < q$, then:

$$\rho(y_b) = n - \rho(x_b) > n - q \geqslant n - q + 1 = w(c_e, y_b),$$

and we choose the edge (c_e, y_b) in our spanning tree, noting that its demand is met by ρ. On the other hand, if $\rho(x_b) > q$, then:

$$\rho(x_b) > q \geqslant q + 1 = w(c_e, x_b),$$

and we choose the edge (c_e, y_b) in our spanning tree in this case. It is easy to verify that T is indeed a spanning tree and that ρ accounts for the demands of all the edges in T. Further, observe that the total power assigned by ρ is $(nk + 1)$, as required, and this completes the argument in the forward direction.

Reverse Direction. Let T be a spanning tree and let ρ be a assignment of power values with total power value at most $(nk + 1)$. Notice that by an argument similar to the one used in the proof of Theorem 1, we have that:

$$\sum_{v \in V(H) \setminus \{g, g_1, \ldots, g_b\}} \rho(v) \leqslant nk. \tag{1}$$

This is to say that any valid assignment of power values must assign a power value of one or more to the vertex g. Therefore, we are left with nk "units of power" to distribute amongst the k choice gadgets and the m checker gadgets. Our next claim is that if $C_i \subseteq V(H)$ denotes the set of vertices used in the ith choice gadget, then:

$$\sum_{v \in C_i} \rho(v) \geqslant n. \tag{2}$$

Indeed, suppose not. Fix $i \in [k]$ and suppose $\rho(x_i) = p$ and $\rho(y_i) = q$. If $p + q \geqslant n$, then there is nothing to prove, therefore, assume that $p + q < n$. This implies that for vertices z_j^i, where $p < j < n - q + 1$, $\rho(z_j^i) \geqslant 1$, since the assignment to the vertices x_i and y_i will not be enough to meet the demands of either of the edges incident to z_j^i. This implies that the total power assigned by ρ in this gadget is:

$$p + q + ((n - q + 1) - p) - 1 = n,$$

and the claim follows. The inequalities (1) and (2) imply that every choice gadget is assigned a total power of n according to ρ, which also gives us that $\rho(c_e) = 0$ for all $e \in E$.

Now, we choose our independent set as follows. Let $\tau(i)$ denote $\rho(x_i)$. Then from V_i we choose the vertex $v_{\tau(i)}^i$. We claim that these chosen vertices must be independent in G. Indeed, suppose not, and in particular, suppose the vertices chosen from parts $a, b \in [k]$ are adjacent in G. Let the chosen vertices be v_p^a and

v_q^b. Note that the power assignments made by ρ on the relevant choice gadget can be inferred to be:

$$\rho(u) = \begin{cases} p & \text{if } u = x_a, \\ q & \text{if } u = x_b, \\ p' & \text{if } u = y_a, \\ q' & \text{if } u = y_b. \end{cases}$$

Here, we know that $p' \leqslant n - p$ and $q' \leqslant n - q$. Now consider the vertex c_e. The spanning tree T must include one of the following edges, since these are the only edges incident to c_e:

$$\{(c_e, x_a), (c_e, y_a), (c_e, x_b), (c_e, y_b)\}.$$

However, recalling the weights of these edges from the construction, the fact that $\rho(c_e) = 0$, and given what we have inferred about the power values on the vertices x_a, y_a, x_b and y_b, it is clear that ρ does not meet the demand of any of these possible edges that is used by T to span the vertex c_e. This is a contradiction, implying that the chosen vertices indeed form an independent set. This argument concludes the proof.   □

Observing that the vertex cover of the reduced graph in the proof of Theorem 2 was bounded linearly in k, and based on the fact that there is no $n^{o(k)}$ algorithm for MCIS unless the ETH is false, we obtain the following consequence.

Corollary 1. *If there exists an algorithm which, given an instance $(G = (V, E), w)$ of MPST where G has a vertex cover of size ℓ, computes an optimal solution in time $|V|^{o(\ell)}$, then the ETH is false. This holds even if all weights are polynomially bounded.*

Our next result shows that when we restrict our attention to unit demands, then the problem of finding a minimum power spanning tree (equivalently, MCST) becomes tractable. Our approach is quite similar to the one used to show that STEINER TREE is FPT parameterized by treewidth, with two main differences: in the Steiner Tree problem, we need to explicitly track the subset of vertices that are "touched" by the solution, which we do not need to do since every vertex is effectively a terminal for the spanning tree problem. On the other hand, for the Steiner Tree problem, the cost of a solution is linked to its size in a rather straightforward manner, while for MCST, the structure of the edges involved in the spanning tree (or, for partial solutions, the spanning forests) is rather relevant. Therefore, we explicitly store not only the partition induced on a bag by a spanning forest, but also guess the specific edges from the bag that participate in the forest. This helps us keep track of the vertex cover as we go along, but it does affect the running time since our dynamic programming (DP) tables have $2^{\theta(w^2)}$ rows in contrast with the $2^{\theta(w \log w)}$ that turn out to be enough for Steiner Tree.

Theorem 3. MCST *admits an algorithm with running time $2^{\theta(w^2)}$ where w denotes the treewidth of the input graph.*

Proof. (Sketch.) Let (G, k) be an instance of MCST with $G = (V, E)$, where $n := |V|$ and $m := |E|$. Let $\mathbb{T} = (T, \{B_t\}_{t \in V(T)})$ be a nice tree decomposition of G (with introduce edge nodes) of width w. Let t be a node of $V(T)$ and B_t be the bag associated with it. Note that $|B_t| \leqslant w + 1$.

We use G_t to denote the graph induced by the vertex set $\bigcup_{t'} B_{t'}$, where t' ranges over all descendants of t, including t. By $E(B_t)$ we denote the edges present in $G[B_t]$. We use H_t to denote the graph on vertex set $V(G_t)$ and the edge set $E(G_t) \setminus E(B_t)$.

Let H be a subgraph of G. For a subset of vertices $X \subseteq V(H)$ and a partition $\mathcal{P}$ of X into s parts $X_1, \ldots, X_s$, we say that a spanning forest F of H is *compatible* with $\mathcal{P}$ if F has exactly s components $C_1, \ldots, C_s$ such that $X_i \subseteq C_i$ for all $i \in [s]$. We are now ready to describe the semantics of our DP table. For each $t \in V(T)$, for all possible partitions $\mathcal{P}$ of B_t, for each $F \subseteq E(B_t)$, $X \subseteq V(B_t)$ and $k \in [n]$, we let:

$$D[t, \mathcal{P}, F, X, k] = \begin{cases} 1 & \text{if there exists a spanning forest } F^* \text{ of } G_t \text{ compatible} \\ & \text{with } \mathcal{P} \text{ with a vertex cover } S \text{ of size at most } k, \\ & \text{such that } F^* \cap E(B_t) = F \text{ and } S \cap B_t = X, \\ 0 & \text{otherwise.} \end{cases}$$

Note that the solution to the problem is the smallest k for which:

$$D[t, \mathcal{P}, F, X, k] = 1,$$

for any choice of F and X, where t is the root node and $\mathcal{P}$ is the partition with exactly one part. Also observe that the size of the DP table is $2^{0^*(w^2)} \cdot n^{0^*(1)}$. The recursive relationships are described in a standard fashion, but we briefly sketch some of the interesting cases.

The calculations for the leaf nodes and the introduce vertex nodes are trivial. When at an introduce edge node, if the introduced edge belongs to F, then it is important to check that X includes one of its endpoints. At a forget node, if v is the forgotten vertex and $\mathcal{P}$ is the partition under consideration, we review all rows in the child bag consistent with $\mathcal{P}$ on the vertices in B_t but where v belongs to one of the given parts, and ignore rows where v is a standalone part. The value of k will also have to be adjusted appropriately depending on whether v was chosen in the vertex cover or not, and in all the cases where v was not in the vertex cover, we need to run a sanity check to ensure that edges incident to v as given by the row of the child node under consideration are covered by the current choice of vertex cover in B_t.

The most non-trivial case is the join nodes. Intuitively, the procedure for the join node involves "patching" the solutions from the two child nodes, while adjusting for the double-counting of the cost of vertex cover vertices in X. However, a straightforward patch may lead to cycles in our combined solution, which might also lead to suboptimal costs for the corresponding vertex covers. It can be verified that in this case, we can follow an approach similar to what is used in the case of Steiner Trees, except that the auxiliary graphs used for "acyclic merges" have to now account for the edges given in F. Due to lack of space, defer a detailed description to a full version of this manuscript.   □

We conclude this section by making some observations about the relationship between spanning trees that have small vertex covers and the vertex covers of the underlying graph.

Observation 1. *Every connected graph* G *admits a spanning tree with a vertex cover of size at most* κ, *where* κ *is the size of a smallest connected vertex cover of* G.

Proof. Let $S \subseteq V(G)$ be a connected vertex cover of size at most κ. Let T be a spanning tree for $G[S]$. For any vertex $v \in V \backslash S$, let s_v denote an arbitrarily chosen vertex from $N(v)$. Note that $N(v) \subseteq S$ (since S is a vertex cover) and $N(v) \neq \emptyset$ (since G is connected). Consider the tree T' obtained from T by adding the edges (v, s_v) for all $v \in V \backslash S$. Observe that T' is a spanning tree and S covers all edges of T, which is $|S| \leqslant \kappa$, as desired. This concludes the proof.   □

Combined with the well-known fact that any connected graph that has a vertex cover of size ℓ also has a connected vertex cover of size at most 2ℓ (for instance, by choosing all the non-leaf vertices of a depth-first search traversal, which is known to be a two-approximate vertex cover), we have the following.

Corollary 2. *Every graph admits a spanning tree with a vertex cover of size at most* 2ℓ, *where* ℓ *is the size of a smallest vertex cover of* G.

Although far from optimal—for instance, consider the complete graphs—the bound in Observation 1 is tight, as witnessed by the graph consisting of two stars on three leaves whose centers are connected by an edge. We remark that the size of the vertex cover of the transformed instance constructed in the proof of Theorem 1 is bounded by $(|R|+1)$, which implies the following consequence in the context of kernelization when parameterized by the size of the vertex cover.

Corollary 3. *When parameterized by the solution size,* MCST *does not admit a polynomial kernel unless* coNP $\subseteq$ NP/poly.

5 Concluding Remarks

We showed that MPST is W[1]-hard when parameterized by the vertex cover of the input graph, and is W[2]-hard when parameterized by the solution size. For the special case of unit demands, however, we demonstrate an FPT algorithm when parameterized by treewidth. We also demonstrated the hardness of polynomial kernelization for MCST when parameterized by the vertex cover. For MPST, the dynamic programming algorithm presented would also give us a FPT algorithm for the combined parameter (w, M), where M is the maximum demand and w is the treewidth. To achieve this, we would have to store all possible power assignments to the vertices in the bags.

Our contributions here for the MCST problem leave open several directions for future work. The most natural question is if we can improve the running time of the dynamic programming algorithm when parameterized by treewidth, possibly using randomized approaches such as "Cut and Count", which have proven to be successful for the related Steiner Tree problem. We also leave open the question of whether there is a deterministic single-exponential algorithm when parameterized by the vertex cover of the input graph. This seems to be an intuitively appealing possibility, and indeed, it is promising to pursue the natural approach where we guess the interaction of a possible optimal solution with a given vertex cover. We also leave open the question of obtaining Turing or lossy kernels for MPST or MCST when parameterized by vertex cover. Finally, we believe that it would be interesting to study these problems on special classes of graphs, especially those relevant to application scenarios, from a parameterized perspective. We note that the reduced instance in Theorem 1 is bipartite.

References

1. Althaus, E., Călinescu, G., Mandoiu, I.I., Prasad, S.K., Tchervenski, N., Zelikovsky, A.: Power efficient range assignment for symmetric connectivity in static ad hoc wireless networks. Wireless Netw. **12**(3), 287–299 (2006)
2. Angel, E., Bampis, E., Chau, V., Kononov, A.: Min-power covering problems. In: Elbassioni, K., Makino, K. (eds.) ISAAC 2015. LNCS, vol. 9472, pp. 367–377. Springer, Heidelberg (2015). https://doi.org/10.1007/978-3-662-48971-0_32
3. Angel, E., Bampis, E., Escoffier, B., Lampis, M.: Parameterized power vertex cover. In: Heggernes, P. (ed.) WG 2016. LNCS, vol. 9941, pp. 97–108. Springer, Heidelberg (2016). https://doi.org/10.1007/978-3-662-53536-3_9
4. Bonsma, P.S., Zickfeld, F.: Improved bounds for spanning trees with many leaves. Discrete Math. **312**(6), 1178–1194 (2012)
5. Calinescu, G., Kortsarz, G., Nutov, Z.: Improved approximation algorithms for minimum power covering problems. In: Epstein, L., Erlebach, T. (eds.) WAOA 2018. LNCS, vol. 11312, pp. 134–148. Springer, Cham (2018). https://doi.org/10.1007/978-3-030-04693-4_9
6. Cygan, M., et al.: Parameterized Algorithms. Springer, Cham (2015). https://doi.org/10.1007/978-3-319-21275-3
7. Diestel, R.: Graph Theory. GTM, vol. 173. Springer, Heidelberg (2017). https://doi.org/10.1007/978-3-662-53622-3
8. Fukunaga, T., Maehara, T.: Computing a tree having a small vertex cover. In: Chan, T.-H.H., Li, M., Wang, L. (eds.) COCOA 2016. LNCS, vol. 10043, pp. 77–91. Springer, Cham (2016). https://doi.org/10.1007/978-3-319-48749-6_6
9. Grandoni, F.: On min-power steiner tree. In: Epstein, L., Ferragina, P. (eds.) ESA 2012. LNCS, vol. 7501, pp. 527–538. Springer, Heidelberg (2012). https://doi.org/10.1007/978-3-642-33090-2_46
10. Ho, J., Lee, D.T., Chang, C., Wong, C.K.: Minimum diameter spanning trees and related problems. SIAM J. Comput. **20**(5), 987–997 (1991)
11. Iglesias, J., Rajaraman, R., Ravi, R., Sundaram, R.: Plane gossip: approximating rumor spread in planar graphs. LATIN 2018. LNCS, vol. 10807, pp. 611–624. Springer, Cham (2018). https://doi.org/10.1007/978-3-319-77404-6_45
12. Kloks, T. (ed.): Treewidth. LNCS, vol. 842. Springer, Heidelberg (1994). https://doi.org/10.1007/BFb0045375

Minimum Conflict Free Colouring
Parameterized by Treewidth

Pradeesha Ashok[1]([⊠]), Rathin Bhargava[1], Naman Gupta[2],
Mohammad Khalid[1], and Dolly Yadav[1]

[1] International Institute of Information Technology Bangalore, Bangalore, India
pradeesha@iiitb.ac.in,
{rathin.bhargava,mohammadkhalid.udayagiri,dolly.yadav}@iiitb.org
[2] Indian Institute of Science Education and Research Mohali, Mohali, India
ms17169@iisermohali.ac.in

Abstract. Conflict free q-Colouring of a graph G refers to the colouring of a subset of vertices of G using q colours such that every vertex has a neighbour of unique colour. In this paper, we study the Minimum Conflict free Q-Colouring problem. Given a graph G and a fixed constant q, Minimum Conflict free Q-Colouring is to find a Conflict free q-Colouring of G that minimises the number of coloured vertices. We study the Minimum Conflict free Q-Colouring problem parameterized by the treewidth of G. We give an FPT algorithm for this problem and also prove running time lower bounds under Exponential Time Hypothesis (ETH) and Strong Exponential Time Hypothesis (SETH).

Keywords: Conflict free colouring of graphs · Parameterized complexity · FPT algorithms · Treewidth · Exponential Time Hypothesis · Strong Exponential Time Hypothesis

1 Introduction

Given a graph $G(V, E)$ a q-colouring refers to a function $c : V \to [q]$, where $[q] = \{1, 2, \ldots, q\}$. A well studied colouring problem in graphs is the *Proper Colouring* problem which is a colouring c with the added constraint that if $(u, v) \in E$ then $c(u) \neq c(v)$. Many other versions of graph colouring are also studied. In this paper, we study the Conflict Free Colouring problem in graphs.

Given a hypergraph $\mathcal{G}(\mathcal{V}, \mathcal{E})$, a Conflict free q-colouring of $\mathcal{G}$ refers to a colouring $c : \mathcal{V} \to [q]$ such that every hyperedge $E \in \mathcal{E}$ has a vertex v with a distinct colour $c(v)$ i.e., no other vertex in E has the colour $c(v)$ under c. Conflict free colouring was initially studied for geometric hypergraphs motivated by the frequency allocation problem in wireless networks [6]. Later, Pach and Tardos [10] studied this problem for hypergraphs induced by graph neighbourhoods. In this version, all vertices of the graph are coloured. Abel et al. [1] studied a closely

© Springer Nature Switzerland AG 2020
M. Changat and S. Das (Eds.): CALDAM 2020, LNCS 12016, pp. 439–450, 2020.
https://doi.org/10.1007/978-3-030-39219-2_35

related problem of colouring only a subset of vertices of G such that for every vertex in V there exists a vertex with a distinct colour in its neighbourhood. They studied algorithmic and combinatorial problems on Conflict free colouring of planar and outerplanar graphs. [1] also studied the bicriteria problem of minimizing the number of coloured vertices in a Conflict free q-colouring of graphs. We study this problem for general graphs.

We now state the problem that we study. Consider a graph $G(V, E)$ and a fixed constant q. Let $N(v)$ denote the open neighbourhood of a vertex v i.e., the set of all vertices u in V such that $(u, v) \in E$ and $N[v]$ denote the closed neighbourhood of v i.e., $N[v] = N(v) \cup \{v\}$. A Closed Neighbourhood Conflict Free q-Colouring is a colouring c of a subset V' of V such that for every vertex $v \in V$, there exists a vertex $u \in N[v]$ such that $c(u) \neq c(u')$ for any vertex $u' \in N[v] \setminus \{u\}$. Similarly, a Open Neighbourhood Conflict Free q-Colouring is a colouring c of a subset V' of V such that for every vertex $v \in V$, there exists a vertex $u \in N(v)$ such that $c(u) \neq c(u')$ for any vertex $u' \in N(v) \setminus \{u\}$. We study the following minimisation problems.

MIN-Q-CNCF: Given a graph $G(V, E)$ and a fixed constant q, find a Closed Neighbourhood Conflict Free q-Colouring that minimises the number of coloured vertices.

MIN-Q-ONCF: Given a graph $G(V, E)$ and a fixed constant q, find a Open Neighbourhood Conflict Free q-Colouring that minimises the number of coloured vertices.

The above problems can be seen as variants of an important problem in graph theory called the MINIMUM DOMINATING SET problem. Specifically, when $q = 1$, MIN-Q-CNCF and MIN-Q-ONCF respectively are the EFFICIENT DOMINATING SET problem and PERFECT DOMINATING SET problem. Therefore MIN-Q-CNCF and MIN-Q-ONCF are NP-hard [7,12].

We study the parameterized complexity of the minimum Conflict Free q-colouring problem when parameterized by the treewidth τ of the graph and prove upper and lower bounds.

1. We show that MIN-Q-CNCF and MIN-Q-ONCF are FPT when parameterized by treewidth. This can also be proved using Courcelle's theorem [3]. We give a constructive proof by giving an algorithm with running time $\mathcal{O}(q^{o(\tau)})$ for both problems.
2. For $q = 1$, we show that an algorithm with running time $\mathcal{O}(2^{o(|V|)})$ cannot exist for MIN-Q-CNCFand MIN-Q-ONCF , under Exponential Time Hypothesis. Since $|V|$ is an upper bound for τ, this also rules out the possibility of algorithms with running time $\mathcal{O}(2^{o(\tau)})$. For $q = 2$, we show that an algorithm with running time $\mathcal{O}(2^{o(|V|)})$ cannot exist for MIN-Q-CNCF and we show that an algorithm with running time $\mathcal{O}(2^{o(\tau)})$ cannot exist for MIN-Q-ONCF.
3. For $q \geq 3$, we show that an algorithm with running time $\mathcal{O}((q - \epsilon)^{o(\tau)})$ cannot exist for MIN-Q-CNCF and MIN-Q-ONCF , under Strong Exponential Time Hypothesis.

2   Preliminaries

In this section, we give definitions and results that will be used in subsequent sections.

Parameterized Complexity: Parameterized complexity was introduced as a technique to design efficient algorithms for problems that are NP-hard. An instance of a parameterized problem is a pair (Π, k) where Π is the input and k is the parameter. A parameter is a positive integer that represents the value of a fixed attribute of the input or output and is assumed to be much smaller than the size of the input, n. A parameterized problem is said to be *fixed parameter tractable* (FPT) if there exists an algorithm that solves it in $f(k)n^{O(1)}$ time, where f is a computable function independent of n. Refer [4,5] for a detailed description of Parameterized Complexity. We denote an FPT running time using the notation $\mathcal{O}(f(k))$ that hides the polynomial functions.

Exponential Time Hypothesis (ETH) [4]: For $q \geq 3$, let δ_q be the infimum of the set of constants of c for which there exists an algorithm solving the n variable q-SAT in time $\mathcal{O}(2^{cn})$. ETH states that $\delta_3 > 0$. **Strong Exponential Time Hypothesis (SETH)** states that $\lim_{q\to\infty} \delta_q = 1$. In other words, ETH implies that 3-SAT cannot be solved faster than $\mathcal{O}(2^{o(n)})$ and SETH implies q-SAT cannot be solved faster than $\mathcal{O}((q-\epsilon)^n)$.

Treewidth [4]: A tree decomposition is a pair $\mathcal{T} = (T, \{X_t\}_{t\in V(T)})$ where T is a tree whose every node t is assigned to a vertex subset, $X_t \subseteq V(G)$, called a bag, such that the following conditions hold.

- $\bigcup_{t\in V(T)} X_t = V(G)$. In other words, every vertex of G is at least in one bag.
- For every $(u, v) \in E(G)$, there exists a node t of T such that bag X_t contains both u and v.
- For every node $u \in V(G)$, the set $T_u = \{t \in V(T) : u \in X_t\}$, i.e. the set of nodes whose corresponding bags contain u, induces a connected subtree of T. The width of the tree decomposition $\mathcal{T}$ is the maximum size of the bag minus 1. The *treewidth* of the graph G, denoted by $\tau(G)$ is the minimum possible width of a tree decomposition of G.

Nice Tree Decomposition: A rooted tree decomposition, $(T, \{X_t\}_{t\in V(T)})$ is nice if

- $X_r = \emptyset$ and $X_l = \emptyset$ where r is the root and l is a leaf of the tree.
- Every other node of T is one of the following
- **Introduce node**: A node t with exactly one child t' such that $X_t = X_{t'} \cup \{v\}$ where $v \notin X_t$
- **Forget node**: A node t with exactly one child t' such that $X_t = X_{t'} \setminus \{w\}$
- **Join node**: A node t with two children t_1, t_2 such that $X_t = X_{t_1} = X_{t_2}$.
- **Introduce edge node**: A node t that introduces the edge (u, v) where $u, v \in X_t$ and has only one child t' such that $X_t = X_{t'}$.

In this variant of the tree decomposition, the total number of the nodes is still $O(\tau n)$. It is known that we can compute a nice tree decomposition $(T, \mathcal{X})$ of G with $|V(T)| \in |V(G)|^{O(1)}$ of width at most 5τ in time $O(2^{O(\tau)}n)$, where τ is the treewidth of G [4].

Positive 1-in-3 SAT Problem: Given a 3-CNF formula ϕ with all positive literals, the POSITIVE 1-IN-3 SAT problem asks whether there exists a truth assignment such that exactly one literal is true in all clauses.

For a given graph G, let $\chi_{CF}(G)$ represent the minimum value of q such that there exists a Conflict free q-colouring of G. In a conflict free colouring c of $V' \subseteq V$, if a vertex v has a neighbour $u \in V'$ such that $c(u)$ is unique in the neighbourhood of v, then v is said to be *conflict-free dominated* by u.

3   FPT Algorithm for Min-q-CNCF

We present an FPT algorithm for the MIN-Q-CNCF problem parameterized by treewidth. Our algorithm uses a popular FPT technique known as Dynamic Programming over Treewidth. Assume a nice tree composition $\mathcal{T}$ of G is given. For a node t in $\mathcal{T}$, let X_t represent the set of vertices in the bag of t. With each node t of the tree decomposition we associate a subgraph G_t of G defined as: $G_t = (V_t, E_t = \{e : e \text{ is introduced in the subtree rooted at } t\})$. Here, V_t is the union of all bags present in the subtree rooted at t.

For every node t, we define colouring functions α, β, f where $\alpha : X_t \to \{c_0, c_1, ..., c_q\}$, $\beta : X_t \to \{c_1, ..., c_q\}$ and $f : X_t \to \{B, W, C, R\}$. Here, c_i represents the ith colour for $1 \le i \le q$ and c_0 denotes a no-colour assignment. $\alpha(u)$ and $\beta(u)$ denotes the colour of the vertex u and the colour it is dominated by respectively. The function f denotes the 'state' of each vertex. We now give a little more insight to what α, β and f represent. For any vertex $u \in X_t$ we have the following.

A *Black* vertex is denoted by $f(u) = B$. Intuitively, a black vertex is coloured and dominated in G_t. A *Cream* vertex is denoted by $f(u) = C$. A cream vertex is coloured but not dominated in G_t. A *White* vertex is denoted by $f(u) = W$ and is not coloured but is dominated in G_t. A *Grey* vertex is denoted by $f(u) = R$. It is not coloured and not dominated in G_t. A tuple $[t, \alpha, \beta, f]$ is *valid* if the following conditions are true for every vertex $u \in X_t$.

- If $f(u) = B$ then $\alpha(u) \ne c_0$.
- If $f(u) = C$ then $\alpha(u) \ne c_0$ and $\alpha(u) \ne \beta(u)$.
- If $f(u) \in \{W, R\}$ then $\alpha(u) = 0$.

A colouring $c : V_t \to \{c_0, c_1, ..., c_q\}$ is said to extend $[t, \alpha, \beta, f]$ if every vertex in $V_t \setminus X_t$ is conflict-free dominated and for every $v \in X_t$, the following is true:

1. $c(v) = \alpha(v)$.
2. if $f(v) \in \{B, W\}$, then v has exactly one neighbour u in G_t such that $c(u) = \beta(v)$.
3. if $f(v) \in \{C, R\}$ then no neighbour of v in G_t is given the colour $\beta(v)$ by c.

We now define sub problems for every node t. Let $dp[t, \alpha, \beta, f]$ denote the minimum number of coloured vertices in any colouring of V_t that extends $[t, \alpha, \beta, f]$. Every tuple, which is either invalid or cannot be extended to a conflict free colouring, corresponds to $dp[t, \alpha, \beta, f] = \infty$.

We define $f_{v \to \gamma}$ where $\gamma \in \{B, W, R, C\}$, as the function where $f_{v \to \gamma}(x) = f(x)$, if $x \neq v$, and $f_{v \to \gamma}(x) = \gamma$, otherwise. Similarly, we define $\alpha_{v \to \gamma}$, for $\gamma \in \{c_0, c_1, \ldots, c_q\}$ and $\beta_{v \to \gamma}$ for $\gamma \in \{c_1, c_2, \ldots, c_q\}$. We now give recursive formulae for $dp[., ., ., .]$.

Leaf Node: In this case $X_t = \phi$. So, $dp[t, \phi, \phi, \phi] = 0$.

Introduce Vertex Node: Let t' be the only child node of t. Then, $\exists\, v \notin X_{t'}$ such that $X_t = X_{t'} \cup \{v\}$.

$$dp[t, \alpha, \beta, f] = \begin{cases} dp[t', \alpha|_{X_{t'}}, \beta|_{X_{t'}}, f|_{X_{t'}}] + 1 & \text{if } f(v) = B \wedge \alpha(v) = \beta(v). \\ dp[t', \alpha|_{X_{t'}}, \beta|_{X_{t'}}, f|_{X_{t'}}] + 1 & \text{if } f(v) = C \wedge \alpha(v) \neq \beta(v). \\ dp[t', \alpha|_{X_{t'}}, \beta|_{X_{t'}}, f|_{X_{t'}}] & \text{if } f(v) = R. \\ \infty & \text{otherwise.} \end{cases}$$

The correctness of the recurrence formula follows from the fact that a vertex is an isolated vertex when it is introduced and can be conflict-free dominated only by itself.

Forget Vertex Node: Let t' be the only child node of t. Then, $\exists v \notin X_t$ such that $X_{t'} = X_t \cup \{v\}$. The vertex v cannot be dominated by a vertex introduced above X_t. Hence $[t, \alpha, \beta, f]$ cannot be extended by a colouring if $f(v) \in \{C, R\}$. Hence we get the following:

$$dp[t, \alpha, \beta, f] = \min_{1 \leq i, j \leq q} \begin{cases} dp[t', \alpha_{v \to c_i}, \beta_{v \to c_j}, f_{v \to B}]. \\ dp[t', \alpha_{v \to c_0}, \beta_{v \to c_i}, f_{v \to W}]. \end{cases}$$

Introduce Edge Node: Let t be an introduce edge node with child node t'. Let (u^*, v^*) be the edge introduced at t. Consider distinct $u, v \in \{u^*, v^*\}$. We decide the value of $dp[t, \alpha, \beta, f]$ based on the following cases.

$$dp[t, \alpha, \beta, f] = \begin{cases} dp[t', \alpha, \beta, f_{u \to C, v \to C}] & ((f(u), f(v)) = (B, B) \wedge (\alpha(u) = \beta(v)) \wedge (\alpha(v) = \beta(u))). \\ dp[t', \alpha, \beta, f_{v \to C}] & (f(u) \in \{B, C\} \wedge f(v) = B \wedge \alpha(u) = \beta(v) \wedge \alpha(v) \neq \beta(u)). \\ dp[t', \alpha, \beta, f_{v \to R}] & (f(u) \in \{B, C\} \wedge f(v) = W \wedge \alpha(u) = \beta(v)). \\ \infty & (f(u) \in \{B, C\} \wedge f(v) \in \{C, R\} \wedge \alpha(u) = \beta(v)). \\ dp[t', \alpha, \beta, f] & \textit{otherwise.} \end{cases}$$

Clearly, the edge (u, v) can only dominate v if u is coloured with $\beta(v)$. If v is conflict-free dominated by u and $f(v) \in \{W, B\}$, then v was not conflict-free dominated in t' under the same colouring functions α and β. Hence if $f(v)$ is black (or white), we set $f(v)$ to cream (or grey) in the child node.

Join Node: Let t be a join node with 2 child nodes t_1, t_2 and $X_t = X_{t_1} = X_{t_2}$. We call tuples $[t_1, \alpha_1, \beta_1, f_1]$ and $[t_2, \alpha_2, \beta_2, f_2]$ as $[t, \alpha, \beta, f]$-consistent if the following conditions hold for all $v \in X_t$.

- $\alpha(v) = \alpha_1(v) = \alpha_2(v)$.
- $\beta(v) = \beta_1(v) = \beta_2(v)$.
- If $f(v) = B$ then $(f_1(v), f_2(v)) = (B, B) \wedge \alpha(v) = \beta(v)$ or $(f_1(v), f_2(v)) \in \{(B, C), (C, B)\} \wedge \alpha(v) \neq \beta(v))$.
- If $f(v) = C$ then $f_1(v) = f_2(v) = C$.
- If $f(v) = R$ then $f_1(v) = f_2(v) = R$.
- If $f(v) = W$ then $(f_1(v), f_2(v)) \in \{(W, G), (G, W)\}$.

All other colourings are not consistent. For example, assume $f_1(v) = f_2(v) = W$ and both $dp[t_1, \alpha_1, \beta_1, f_1]$ and $dp[t_2, \alpha_2, \beta_2, f_2]$ are finite. Then v is conflict free dominated in G_{t_1} and G_{t_2}. By the property of nice tree decomposition, an edge between two vertices in a join node is introduced above the join node. Hence X_t induces an independent set in G_t. Therefore v is conflict free dominated by a vertex outside X_t in both G_{t_1} and G_{t_2}. Now, in G_t, v has two neighbours with colour $\beta(v)$ and hence v cannot be conflict free dominated.

Now we give the recurrence formula for $dp[]$.

$$dp[t, \alpha, \beta, f] = \min\left(dp[t_1, \alpha_1, \beta_1, f_1] + dp[t_2, \alpha_2, \beta_2, f_2] - |f^{-1}(B)| - |f^{-1}(C)|\right)$$

where tuples $[t_1, \alpha_1, \beta_1, f_1]$ and $[t_2, \alpha_2, \beta_2, f_2]$ are $[t, \alpha, \beta, f]$-consistent.

Now $dp[r, \emptyset, \emptyset, \emptyset]$ where r is the root of $\mathcal{T}$ gives the desired solution. Also, it can be seen that all recurrences except those for join nodes, can be computed in $\mathcal{O}((4q^2)^\tau)$ time. For a join node t, two tuples are consistent with $[t, \alpha, \beta, f]$ if (f, f_1, f_2) is in one of 7 forms. Thus, processing a join node can be done in $\mathcal{O}((7q^2)^\tau)$ time. Hence, we get the following result.

Theorem 1. *There exists an FPT algorithm with running time $\mathcal{O}(q^{O(\tau)})$ for* MIN-Q-CNCF *parameterized by the treewidth of the graph.*

We also prove a similar result for MIN-Q-ONCF . The algorithm is very similar to that given above and can be found in the full version of the paper.

Theorem 2. *There exists an FPT algorithm with running time $\mathcal{O}(q^{O(\tau)})$ for* MIN-Q-ONCF *parameterized by the treewidth of the graph.*

4   Lower Bounds

In this section, we give lower bounds that complement the results given in Sect. 3.

4.1   Lower Bounds for Min-q-CNCF

Theorem 3. MIN-1-CNCF *cannot be solved in $\mathcal{O}(2^{o(n)})$ time, unless ETH fails.*

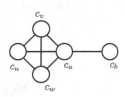

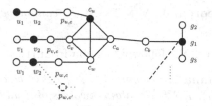

Fig. 1. Clause gadget

Fig. 2. Combination of vertex and clause gadgets

We prove the theorem by giving a linear reduction from the POSITIVE 1-IN-3 SAT problem. It is known that POSITIVE 1-IN-3 SAT cannot be solved in $\mathcal{O}(2^{o(n)})$ time, unless ETH fails [9,11]. Let ϕ be an instance of the POSITIVE 1-IN-3 SAT problem, with n variables and m clauses. We will construct a graph $G(V, E)$ corresponding to ϕ. For every variable u of ϕ, we add two nodes u_1, u_2 to $V(G)$ and the edge (u_1, u_2) to $E(G)$. For every clause c, we add a gadget as shown in Fig. 1.

If a variable u belongs to a clause c, then in G, the vertex u_2 is connected to one of the vertices in $\{c_u, c_v, c_w\}$ in the clause gadget of c, through a connector vertex $p_{u,c}$ as shown in Fig. 2. The vertex c_b in each clause gadget is connected to a *global vertex* g_1. The global vertex also has two neighbours of degree 1, g_2 and g_3. Clearly G has $2n + 8m + 3$ vertices.

Lemma 1. ϕ *is satisfiable if and only if G can be conflict-free coloured using one colour.*

Proof. Assume that ϕ has a satisfying assignment. We now give a valid conflict free 1-colouring of G. Colour the global vertex g. If a variable u is true in the satisfying assignment, then we colour the vertex u_1 of the corresponding variable gadget, otherwise we colour the vertex u_2. Observe that if the vertex u_1 is coloured, then u_2 is conflict-free dominated by u_1 and hence, u_2 cannot be coloured. For the same reason, a connector vertex $p_{u,c}$ that connects u_2 to the clause gadget of clause c cannot be coloured. Therefore, in order to conflict-free dominate the vertex $p_{u,c}$, the vertex c_u of the clause gadget should be coloured. By similar arguments, if the vertex u_1 in variable gadget is uncoloured then the corresponding vertex c_u in clause gadget should also be uncoloured.

Let c be an arbitrary clause in ϕ and let u, v, w be the variables in c. In a satisfying assignment, exactly one among u, v, w is true. Without loss of generality, let u be the variable that is true. Then c_u is coloured and is conflict free dominated by itself. c_v, c_w and c_a are uncoloured but conflict free coloured by c_u. This means that for every clause c in ϕ the vertex c_b of the corresponding clause gadget in G is uncoloured and is conflict-free dominated by g. This gives us a valid 1-conflict free colouring of graph G.

Similarly we can prove that if G has a valid 1-conflict free colouring then ϕ has a satisfying assignment. □

Now we will consider the case $q = 2$. Assume the colours used are red and blue.

Theorem 4. MIN-2-CNCF *cannot be solved in* $\mathcal{O}(2^{o(n)})$ *time, unless ETH fails.*

We use the following lemma from [1].

Lemma 2. *[Lemma 3.2, [1]] Let G be any graph, u, $v \in V(G)$ and $e = (v, u) \in E(G)$. If $N(v)$ contains two disjoint and independent copies of a graph $H = G_q$ with $\chi_{CF}(H) = q$, not adjacent to any other vertex $w \in G$, every q-conflict-free colouring of G colours v. If the same holds for u and in addition, $N_G(u) \cap N_G(v)$ contains two disjoint and independent copies of a graph $J = G_{q-1}$ with $\chi_{CF}(J) = q - 1$, not adjacent to any other vertex $w \in G$, every q-conflict-free colouring of G colours u and v with different colours.*

We're looking at the special case, where $q = 2$. As given in Lemma 2, G_1 is a single vertex. G_2 is $K_{1,3}$ with one edge subdivided by another vertex.

We define a vertex $v' \in V(G)$ as a *special vertex* if $N(v')$ contains two disjoint and independent copies of G_2. We also define an edge gadget between 2 special vertices u' and v' as a *special edge* between u' and v' if $(u', v') \in E(G)$ and $N_G(u) \cap N_G(v)$ contains two disjoint and independent copies of G_1. We note that by Lemma 2, a special vertex in a graph needs to be coloured *red* or *blue* and if there exists a special edge between two special vertices u' and v', then u' and v' needs to be coloured with opposite colours.

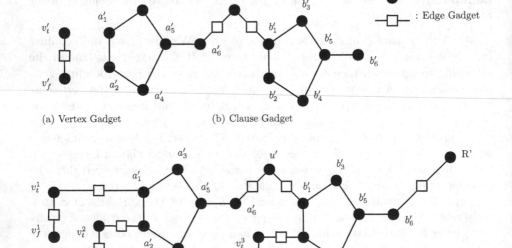

(a) Vertex Gadget   (b) Clause Gadget

(c) Vertex Gadget v_i connected to Clause Gadget C_j connected to the palette vertex R'

Fig. 3. All the gadgets used in this proof

We give a linear reduction from the 3-SAT problem to MIN-2-CNCF . Let ϕ be an instance of the 3-SAT problem with n variables and m clauses. We construct an instance of MIN-2-CNCF , $G(V, E)$, corresponding to ϕ. For every variable v in ϕ, we add a vertex gadget to G which consists of two special vertices v'_t and v'_f connected by a special edge. We define a sub-clause gadget which consists of 6 special vertices denoted as $a'_1, a'_2, a'_3, a'_4, a'_5, a'_6$ where a'_i, $i \in [5]$ form C_5 - $\{a'_1, a'_3, a'_5, a'_4, a'_2, a'_1\}$. The vertex a'_6 is adjacent to a'_5. For every clause c in ϕ, we add a clause gadget to G. A clause gadget is constructed by taking 2 sub-clause gadgets $\{a'_1, \ldots, a'_6\}$ and $\{b'_1, \ldots, b'_6\}$ and connecting them through a vertex u' by adding special edges between a'_6 and u', and b'_1 and u'. Let the variables in the clause c_i be denoted as $v^i_k, k \in [3]$. Then the variable gadgets corresponding to the the variables v^i_1, v^i_2, v^i_3 in G are respectively connected to the vertices a'_1, a'_2, b'_2 of the clause gadget corresponding to c_i, through special edges. If the variable v appears in c_i as a positive literal, then v_f is connected to the clause gadget, otherwise v_t is connected. Finally, there exists a palette vertex R such that there is a special edge between R and b_6 for all clause gadgets. (Refer Fig. 3). Since every vertex gadget and clause gadget has a constant number of vertices, V has $O(n + m)$ vertices. Specifically, V contains $k = 2n + 13m + 1$ special vertices.

Lemma 3. *Let vertices a'_1 and a'_2 have exactly 1 neighbour outside the sub-clause gadget which is coloured the opposite to it and conflict-free dominates itself. Then, given that we colour only special vertices, if a'_1 and a'_2 are both coloured red, a'_6 will be coloured red. If a'_1 or a'_2 is coloured blue, then there exists a colouring where a'_6 will be coloured blue.*

Proof. We first prove that colouring a'_1 and a'_2 red forces a'_6 to be coloured *red*. By construction of the sub-clause gadget, a'_1 and a'_2 are adjacent to each other. It's given that both a'_1 and a'_2 have one blue neighbour outside the gadget. The colour *red* appears twice in the closed neighbourhood of a'_1. Thus, the only colour which can dominate a'_1 is blue. If a'_3 is coloured blue, a'_1 would not be conflict free dominated. Thus, a'_3 is coloured *red*. By a similar argument involving a'_2, we can show that a'_4 has to be coloured *red*. In this way, let a'_5 be coloured *blue*. By contradiction, if a'_5 is coloured *red*, then a'_3 does not have a conflict free neighbour as all its neighbours are *red*. Thus, a'_6 gets coloured red. If it gets coloured blue, then a'_5 will have 2 red and 2 blue neighbours in its closed neighbourhood and cannot get dominated. Thus, a'_6 is coloured *red*. If both a'_1 and a'_2 are coloured blue, then we can prove, in a similar case to the one done above, that a'_6 will be coloured blue.

To prove the second part, let's assume without loss of generality, that a'_1 is coloured *red* and a'_2 is coloured *blue*. Since a'_1 is connected to a blue neighbour outside the sub-clause gadget, a'_3 is forced to be coloured *blue*. Likewise, a'_4 is coloured *red*. To ensure that a'_6 gets a *blue* colour, we need a'_5 to be coloured *blue*. It can be seen that this is a valid colouring. $\square$

Lemma 4. *Any instance ϕ of 3-SAT is satisfiable if and only if G can be conflict free 2-coloured with at most k coloured vertices.*

Proof. Assume there exists a 2-conflict free colouring of G which colours at most k vertices. We will show that ϕ has a satisfying assignment. By Lemma 2, in any conflict free 2-colouring of G every special vertex is coloured. Since we have coloured at most k vertices, only the special vertices are coloured. The palette vertex, R, is coloured since it is a special vertex. Without loss of generality, let it be coloured red. Since b'_6 vertices of all the clause gadgets have special edges to R, they are coloured *blue*. By Lemma 3, we know that at least one of a'_1, a'_2, b'_2 is coloured blue. Let that one vertex be u. Now, u is connected to a corresponding variable gadget. If it is connected to v'_t, then assign that variable v in ϕ false, otherwise assign true. It is easy to see that this is a satisfying assignment. Similarly, we can see that if there exists a satisfying assignment of ϕ, there exists a conflict free 2-colouring of G.   □

Now the theorem follows from ETH.   □

Now we consider $q \geq 3$.

Lemma 5. *For $q \geq 3$, an algorithm with running time $\mathcal{O}((q - \epsilon)^{o(\tau)})$ cannot exist for* MIN-Q-CNCF *, under Strong Exponential Time Hypothesis.*

We reduce PROPER Q-COLOURING to MIN-Q-CNCF. We know from [8] that PROPER Q-COLOURING under SETH, cannot be solved faster than $\mathcal{O}((q-\epsilon)^{\tau(G)})$. As shown in Lemma 3.4 from [1], from any graph G, we can construct G' which can be Conflict free q-coloured if and only if G can be proper q-coloured. From Claim 2, Lemma 7 from [2], we know that the treewidth of G' is $max\{\tau(G), q\}$. where $\tau(G)$ is the treewidth of the graph G. Hence, MIN-Q-CNCF colouring cannot be solved faster than $\mathcal{O}((q - \epsilon)^{max\{\tau(G), q\}})$ and the lemma follows.   □

4.2   Lower Bounds for Min-q-ONCF

Theorem 5. MIN-1-ONCF *cannot be solved in $\mathcal{O}(2^{o(n)})$ time, unless ETH fails.*

Proof. We give a reduction from POSITIVE 1-IN-3 SAT. Let ϕ' be an instance of POSITIVE 1-IN-3 SAT with n variables and m clauses. We will construct a graph $G'(V, E)$ corresponding to ϕ'. For each variable and clause we construct a variable gadget and a clause gadget respectively. The variable gadget for an arbitrary variable u is a path of length 3. The clause gadget is shown in Fig. 4. There is a global gadget which consists of a path of length 4, $g_1 - g_2 - g_3 - g_4$ with a pendant vertex connected at vertex g_3. If a variable u belongs to a clause c, then vertex u_3 is connected to vertex $c_{u,1}$ in G'. Vertex g_4 is connected to vertex c_y of all clause gadgets. Clearly, G' has $8m + 3n + 5$ vertices.

Lemma 6. *ϕ' is satisfiable if and only if G' can be conflict free 1-coloured.*

Proof. We will show that if G' has a valid 1-conflict free colouring then ϕ' has a satisfying truth assignment. We can show the other direction by similar arguments. Observe that the vertices g_2, g_3 should always be coloured and g_1, g_5, g_4

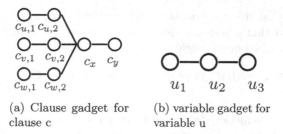

(a) Clause gadget for clause c

(b) variable gadget for variable u

Fig. 4. Clause and variable gadgets

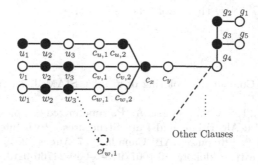

Fig. 5. Combined gadget

should always be uncoloured as it is the only way to conflict free colour the vertices g_1, g_2, g_3, g_4 and g_5. Since g_4 is uncoloured and is dominated by g_3, all its neighbours except g_3 should also be uncoloured and dominated in a valid colouring. Hence, for every clause c, the vertex c_y in G' should be uncoloured and c_x should be coloured. To conflict free dominate c_x, exactly one among the vertices $c_{u,2}$, $c_{v,2}$, $c_{w,2}$ should be coloured. Without loss of generality, assume that the vertex $c_{u,2}$ is coloured. Since $c_{u,2}$ is coloured, the vertex u_1 should be coloured and the vertices v_1, w_1 should be uncoloured. Then, assign true to variable u and false for variables v and w. It can be seen that this gives a satisfying assignment for ϕ' (Fig. 5). $\qquad\square$

Theorem 6. MIN-2-ONCF *cannot be solved in* $\mathcal{O}(2^{o(\tau)})$, *assuming ETH is true.*

Proof. [2] gives a result that a variant of MIN-2-ONCF where every vertex needs to be coloured cannot be solved in time $\mathcal{O}(2^{o(\tau)})$ under ETH. For proving this result, they give a reduction from the 3-SAT problem that reduces an instance of the 3-SAT problem, ϕ, to an instance of the MIN-2-ONCF problem, G, with treewidth a linear function of $|\phi|$. We use the same reduction and modify G by connecting a vertex of degree 1 to every vertex in G. Note that the treewidth of the modified instance, G', is still a linear function of $|\phi|$. Now, we will show that ϕ has a satisfying assignment if and only if G' has a valid conflict free 2-colouring that colours at most $|V(G)|$ vertices. This follows from the result in [2] and the fact that a vertex of degree one can only be conflict free dominated by its neighbour when open neighbourhood is considered.

Lemma 7. MIN-Q-ONCF *cannot be solved in time* $\mathcal{O}((q - \epsilon)^{\tau(G)})$ *under SETH*.

Proof. We give a reduction from the PROPER Q-COLOURING problem to the MIN-Q-ONCF problem. We know from [8] that PROPER Q-COLOURING under SETH, cannot be solved faster than $\mathcal{O}((q - \epsilon)^{\tau(G)})$. We consider the graph G' in lemma 5 in [2] and construct graph G'' by adding a vertex with degree 1 to all the vertices in G'. Now the proof follows as before.

References

1. Abel, Z., et al.: Conflict-free coloring of graphs. SIAM J. Discrete Math. **32**(4), 2675–2702 (2018)
2. Bodlaender, H.L., Kolay, S., Pieterse, A.: Parameterized complexity of conflict-free graph coloring. In: Algorithms and Data Structures - 16th International Symposium, WADS 2019, Edmonton, AB, Canada, 5–7 August 2019, Proceedings, pp. 168–180 (2019). https://doi.org/10.1007/978-3-030-24766-9_13
3. Courcelle, B.: The monadic second-order logic of graphs. I. Recognizable sets of finite graphs. Inf. Comput. **85**(1), 12–75 (1990)
4. Cygan, M., et al.: Parameterized Algorithms, vol. 4. Springer, Cham (2015). https://doi.org/10.1007/978-3-319-21275-3
5. Downey, R.G., Fellows, M.R.: Parameterized Complexity. Springer, Switzerland (2012). https://doi.org/10.1007/978-3-319-21275-3
6. Even, G., Lotker, Z., Ron, D., Smorodinsky, S.: Conflict-free colorings of simple geometric regions with applications to frequency assignment in cellular networks. SIAM J. Comput. **33**(1), 94–136 (2003)
7. Kratochvíl, J., Křivánek, M.: On the computational complexity of codes in graphs. In: Chytil, M.P., Koubek, V., Janiga, L. (eds.) MFCS 1988. LNCS, vol. 324, pp. 396–404. Springer, Heidelberg (1988). https://doi.org/10.1007/BFb0017162
8. Lokshtanov, D., Marx, D., Saurabh, S.: Known algorithms on graphs of bounded treewidth are probably optimal. In: Proceedings of the Twenty-Second Annual ACM-SIAM Symposium on Discrete Algorithms, pp. 777–789. SIAM (2011)
9. Muzi, I., O'Brien, M.P., Reidl, F., Sullivan, B.D.: Being even slightly shallow makes life hard. In: 42nd International Symposium on Mathematical Foundations of Computer Science (MFCS 2017), vol. 83, p. 79. Schloss Dagstuhl-Leibniz-Zentrum fuer Informatik (2017)
10. Pach, J., Tardos, G.: Conflict-free colourings of graphs and hypergraphs. Comb. Probab. Comput. **18**(5), 819–834 (2009)
11. Schaefer, T.J.: The complexity of satisfiability problems. In: Proceedings of the Tenth Annual ACM Symposium on Theory of Computing, pp. 216–226. ACM (1978)
12. Yen, C.C., Lee, R.C.T.: The weighted perfect domination problem. Inf. Process. Lett. **35**(6), 295–299 (1990)

Computational Geometry

Planar Projections of Graphs

N. R. Aravind$^{(\boxtimes)}$ and Udit Maniyar

Indian Institute of Technology Hyderabad, Sangareddy, India
aravind@iith.ac.in

Abstract. We introduce and study a new graph representation where vertices are embedded in three or more dimensions, and in which the edges are drawn on the projections onto the axis-parallel planes. We show that the complete graph on n vertices has a representation in $\lceil \sqrt{n/2}+1 \rceil$ planes. In 3 dimensions, we show that there exist graphs with $6n - 15$ edges that can be projected onto two orthogonal planes, and that this is best possible. Finally, we obtain bounds in terms of parameters such as geometric thickness and linear arboricity. Using such a bound, we show that every graph of maximum degree 5 has a plane-projectable representation in 3 dimensions.

Keywords: Graph drawing · Planarity · Thickness · Planar projections

1 Introduction

In this paper, we consider embeddings of graphs where the vertices are mapped to points in $\mathbb{R}^d$, for $d \geq 3$, and the edges are represented by line-segments on the $\binom{d}{2}$ axis-parallel planes. For example, a 3-dimensional network may be visualized by placing it inside a cube and drawing the edges on the walls of the cube by projecting the points.

One motivation is the connection to two classical parameters, thickness and geometric thickness. The thickness of a graph G, is the smallest number of planar subgraphs into which the edges of G can be decomposed. This was introduced by Tutte in [18]; see also [16] for a survey of thickness. Geometric thickness adds the restriction that all the subgraphs must be embedded simultaneously, that is, with a common embedding of the vertices. This was studied in [4] for complete graphs. The connection between geometric thickness and parameters such as maximum degree and tree-width has been studied in various papers: [2,7,8]. While using the standard co-ordinate planes in high dimensions is more restrictive than thickness, it appears to be less so than geometric thickness (Sect. 3).

Book embeddings, defined by Ollmann in [17], are restrictions of geometric drawings in which the vertices are in convex position. The book thickness of G is the smallest number of subgraphs that cover all the edges of G in such a drawing. This is also known as stack number, and is studied in [6]. Also see [5]

© Springer Nature Switzerland AG 2020
M. Changat and S. Das (Eds.): CALDAM 2020, LNCS 12016, pp. 453–462, 2020.
https://doi.org/10.1007/978-3-030-39219-2_36

for a survey. More generally, a survey on simultaneous embedding of graphs may be found in [3].

In [14], the authors showed that n-vertex graphs of geometric thickness 2 can have at most $6n - 18$ edges. Such graphs can also be represented as projections in two orthogonal planes; orthogonal planes appear to allow a greater degree of freedom, as we give a construction of graphs with $6n - 15$ edges. We also note that a plane-projectable construction with $6n - 17$ edges was given in [15].

1.1 Preliminaries

For a point q in $\mathbb{R}^d$, we denote by $\pi_{i,j}(q)$, the projection of q on the plane $\{x \in \mathbb{R}^d \mid x_i = x_j = 0\}$ formed by the ith and jth co-ordinate axes.

Definition 1. *Given a graph $G = (V, E)$, we say that an injective map $\pi : V \to \mathbb{R}^d$ is a* **plane-projecting map** *of G if there exists a decomposition $E = \cup_{1 \le i < j \le d} E_{i,j}$ such that the projection $\pi_{i,j}$ is injective and induces a straight-line planar embedding of the subgraph $(V, E_{i,j})$.*

We define the **plane-projecting dimension** of a graph G to be the smallest integer d for which a plane-projecting map in $\mathbb{R}^d$ exists for G. We denote this by $pdim(G)$.

If $pdim(G) \le d$, we shall say that G is d-**dimensionally projectable** or **plane-projectable in d-dimensions**.

We note the following connection between the plane-projecting dimension and the two thickness parameters of a graph.

Observation 1. *Let G have thickness $\theta(G) = r$ and geometric thickness $\bar{\theta}(G) = s \ge r$. Then we have:*

(i) $\dfrac{\sqrt{8r + 1} + 1}{2} \le pdim(G) \le 2r.$

(ii) $pdim(G) \le 2\lceil \sqrt{s} \rceil.$

Proof. The first inequality in (i) follows from the observation that $r \le \binom{pdim(G)}{2}$; the second inequality is easy to see. For (ii), we let $k = \lceil \sqrt{s} \rceil$. For a vertex v, let (a, b) be its position in an optimal geometric thickness representation of G. Then the map $f(v) = (a, a, \dots, a, b, b, \dots, b)$ (with number of a's and b's each equal to k), is a plane-projecting map, with the edge sets $\{E_{i,j} : 1 \le i \le k, k + 1 \le j \le 2k\}$ corresponding to the subgraphs of the geometric thickness representation, and $E_{i,j}$ drawn on the plane with $x_i = x_j = 0$. $\square$

In [7], the author obtained a bound of $O(\log n)$ on the geometric thickness of graphs with arboricity two; thus as a corollary, we obtain a bound of $O(\sqrt{\log n})$ on the plane-projecting dimension of such graphs.

1.2   Our Results

In Sect. 2, we obtain an upper bound of $\sqrt{n/2} + O(1)$ on the plane-projecting dimension of K_n.

In Sect. 3, we give a construction of graphs having n vertices and $6n - 15$ edges that can be projected on two orthogonal planes, and further show that this is tight. We also obtain an upper bound on the maximum number of edges of a n-vertex graph that is plane-projectable in 3 dimensions.

In Sect. 4, we show that every graph of maximum degree five is plane- projectable in three dimensions, by obtaining an upper bound in terms of the linear arboricity of G (which is the minimum number of linear forests that partition the edges of G). We also obtain a general upper bound of $\Delta(G)\left(\dfrac{1}{2} + o(1)\right)$ on $pdim(G)$. Note that an upper bound of $\Delta(G)+1$ follows from Observation 1 and a result of [13], which states that the thickness of a graph of maximum degree Δ is at most $\lceil\dfrac{\Delta}{2}\rceil$.

2   Plane-Projecting Dimension of Complete Graphs

In the paper [4], the authors show that the geometric thickness of K_n is at most $\lceil n/4 \rceil$. Combining this with Observation 1, we get $pdim(K_n) \leq \lceil\sqrt{n}\rceil$.

The thickness of K_n is known to be 1 for $1 \leq n \leq 4$, 2 for $5 \leq n \leq 8$, 3 for $9 \leq n \leq 10$, and $\lceil\dfrac{n+2}{6}\rceil$ for $n > 10$. Thus, for $n > 10$, we get $pdim(K_n) \geq \sqrt{n/3}$.

By using the construction of [4] in a more direct way, we obtain the following improved upper bound.

Theorem 2. $pdim(K_n) \leq \lceil\dfrac{\sqrt{2n+7}+1}{2}\rceil$.

To prove Theorem 2, we shall use the following lemma, which we first state and prove.

Lemma 1. *For every natural number $d \geq 2$ and every natural number n, there exist n points in $\mathbb{R}^d$ such that for every $1 \leq i < j \leq d$, the projections of these points to the (i,j)-plane are in convex position, and in the same order on the convex hull.*

Proof (of Lemma 1). We consider the point-set $P_i = (\cos(a_i + ib), \cos(a_i + (i + 1)b), \ldots, \cos(a_i + (i + d - 1)b))$ for some suitable b and a_is. For given $i, j \in \{1, 2, \ldots, d\}$, the projection of these points in the (i, j) plane lie on an ellipse. $\square$

We now prove Theorem 2.

Proof (of Theorem 2). Let d be such that $\dbinom{d}{2} \geq \lceil\dfrac{n}{4}\rceil$. We assume that $n = 2k$, where k is even, and find sets S, T of $n/2$ points each in $\mathbb{R}^d$ such that the projections of S and T to two given axis-parallel planes lie on an ellipse, with the ellipse corresponding to S always contained inside and congruent to the ellipse corresponding to T, as shown in Fig. 1 (right).

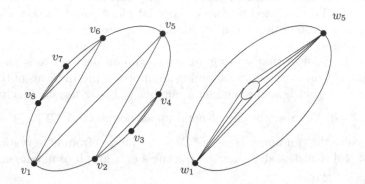

Fig. 1. Left: Path in S/T; Right: Edges between diametrically opposite vertices of T and the vertices of S

The drawing of edges is now the same as in [4], which we explain for the sake of completeness.

We decompose the complete graph on each of S, T into $k/2$ disjoint paths and draw their edges in $k/2$ planes such that each path contains exactly one pair of diametrically opposite vertices that are adjacent. Here, we use the phrase "diametrically opposite" for a pair of vertices if the points corresponding to them have exactly $k/2 - 1$ other points between them on the convex hull. This is illustrated in Fig. 1(a), where v_1, v_5 are the diametrically opposite pair which are adjacent. Other diametrically opposite pairs are $\{v_2, v_6\}, \{v_3, v_7\}$ etc., each of which shall be adjacent in a different plane.

Finally, we add edges between every vertex of S and the diametrically opposite pair of T. That this can be done is shown in [4], by showing that there exist a set of parallel lines each passing through one point in S, and arguing by continuity that if the diametrically opposite pair is far enough, they may be joined to the points of S without intersections. This is illustrated in Fig. 1(b). □

3   Plane-Projectable Graphs in $\mathbb{R}^3$

In this section, we focus on $\mathbb{R}^3$, and ask the following two extremal questions.

Q1. What is the maximum number of edges of a n-vertex graph with plane-projecting dimension three?

Q2. What is the maximum number of edges of a n-vertex graph whose edges can be projected into two orthogonal planes in $\mathbb{R}^3$?

We shall answer Question 1 partially by giving an upper bound of $9n - 24$, and Question 2 completely, by giving matching upper and lower bounds.

As mentioned earlier, any graph with geometric thickness two can be projected in two of the co-ordinate planes. In [14], it was shown that a n-vertex graph of geometric thickness two can have at most $6n - 18$ edges and that $6n - 20$ edges is achievable. This was improved in [9], where it was shown that for every $n \geq 9$, $6n - 19$ edges is achievable.

By modifying their construction, we show the following:

Theorem 3. *For every $n \geq 14$, there exists a graph G_n on $6n - 15$ edges with an embedding in $\mathbb{R}^3$ such that the restriction to two of the planes form planar straight-line embeddings of two graphs whose edge-union is equal to G_n.*

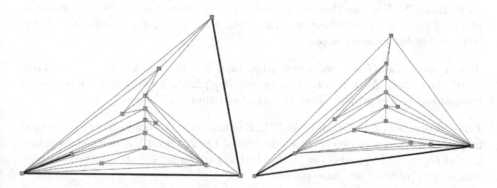

Fig. 2. On left: H_{14}, On right: M_{14}; G_{14} has $6 \times 14 - 15 = 69$ edges. The dark edges are common to both planes. Also the exact vertex positions are not important, but the ordering of the vertices on the common axis should be the same.

Proof. Let H_n, M_n be the projection of G_n on XY, YZ planes respectively. Observe that if we fix the embedding of vertices of G in H, then in M we would have the freedom to move vertices along Z axis because z coordinate of vertices have not been fixed yet.

Figure 2 gives the construction of a graph G_{14} with 14 vertices and $6 \times 14 - 15 = 69$ edges.

Let us assume that we are given H_k, M_k which are the planar projections of G_k on XY, YZ planes respectively, such that $|E_k| = 6k - 15$.

We now show that we can add a vertex v_{k+1} with 6 neighbors, such that three of the new edges are added to H_k to obtain H_{k+1} and the other three are added to M_k to obtain M_{k+1}.

In H_k, we place v_{k+1} inside a triangle whose vertices are disjoint from the vertices present on the convex hull of M_k namely v_1, v_2, v_k. Now the (x, y) coordinates of v_{k+1} are fixed we can take the z coordinate of v_{k+1} to be a value strictly greater than the z coordinate of v_k. Now in M_k (YZ plane) we can add v_{k+1} by connecting v_{k+1} to v_1, v_2, v_k.

We can always find a triangle whose vertices should not contain any one of v_1, v_2, v_k. If k is odd we add v_{k+1} inside the triangle v_6, v_4, v_{k-1} in H_k and if k is even we add v_{k+1} inside the triangle v_7, v_8, v_k in H_k.

Since we have added 6 edges to the graph G_k the new graph G_{k+1} contains $6k - 15 + 6 = 6(k + 1) - 15$. The vertex v_{n+1} is also being mapped to a suitable point in $\mathbb{R}^3$.

Inductively using the above process, we can generate G_n for all n such that $|E_n| = 6n - 15$.

We will now show that the above upper bound is in fact tight; we first need the following definition.

Definition 2. *We say that an embedding of a planar graph G is maximally planar if no non-adjacent pair of vertices can be joined by a line-segment without intersecting the existing edges.*

Theorem 4. *Let G be a connected graph on $n \geq 3$ vertices having an embedding in $\mathbb{R}^3$ such that the restriction to two of the planes form straight-line planar embeddings of two graphs. Then G has at most $6n - 15$ edges.*

Proof. Consider an embedding of $G(V, E)$ such that the edges of G are covered by planar drawings in two (projected) planes. Let XY and YZ be the two planes and let $G_1 = (V, E_1)$ and $G_2 = (V, E_2)$ be the two planar sub-graphs respectively, which are projected on these planes.

Clearly we can assume that the embeddings of both G_1 and G_2 are maximally planar.

Let

A to be the vertex with lowest y coordinate value,

B to be the vertex with highest y coordinate value,

C to be the vertex with second lowest y coordinate value,

D to be the vertex with second highest y coordinate value.

Claim 5. *Both AC and BD belong to $G_1 \cap G_2$.*

Proof of Claim 5: Suppose for contradiction that $AC \notin G_1$. Since G_1 is maximally planar, there must be an edge uv that intersects the line-segment joining AC. Therefore the y co-ordinate of u or the y co-ordinate of v must lie between the y co-ordinates of A and C, which contradicts the choice of A, C. The proof for BD is identical.

We now consider two cases.

Case 1: $|E_1| < 3n - 6$ or $|E_2| < 3n - 6$. In this case, we have: $|E_1 \cup E_2| = |E_1| + |E_2| - |E_1 \cap E_2| \leq 6n - 13 - 2 = 6n - 15$.

Case 2: $|E_1| = |E_2| = 3n - 6$. In this case, we show that in addition to AC and BD, the edge AB also belongs to both E_1 and E_2, which shows that $|E_1 \cup E_2| \leq 6n - 15$.

Since G_1 has $3n - 6$ edges, its convex hull is a triangle. By the definition of A, B, we see that AB should be on the convex hull, and hence is an edge of G_1. Similarly, AB belongs to G_2 as well.

This completes the proof of Theorem 4. □

We now give an upper bound on graphs with plane-projectable dimension three.

Theorem 6. *Let G be a connected graph on $n \geq 3$ vertices having an embedding in $\mathbb{R}^3$ such that the restriction to the three planes form straight-line planar embeddings of three graphs. Then G has at most $9n - 24$ edges.*

Proof. Consider an embedding of $G(V, E)$ such that the edges of G are covered by planar drawings in three (projected) planes. Let XY, YZ and ZX be the two planes and let $G_1 = (V, E_1)$, $G_2 = (V, E_2)$ and $G_3 = (V, E_3)$ be the three planar sub-graphs respectively, which are projected on these planes.

We may assume that G_1, G_2, G_3 are maximally planar.

Here we have to consider few cases:

Case 1: $|E_1| = |E_2| = |E_3| = 3n - 6$. In this case we use the same argument as Theorem 4, we get $|E_2 \setminus E_1| = 3n - 9$, $|E_3 \setminus E_1| = 3n - 9$.

$|E_1 \cup E_2 \cup E_3| \leq |E_1| + |E_2 \setminus E_1| + |E_3 \setminus E_1|$.

$\implies |E_1 \cup E_2 \cup E_3| \leq (3n - 6) + (3n - 9) + (3n - 9) = 9n - 24$.

Case 2: $|E_1| = |E_2| = 3n - 6, |E_3| \leq 3n - 7$. In this case if we use the same argument as Theorem 4 we get $|E_2 \setminus E_1| = 3n - 9$.

Since G_1, G_3 are maximally planar using Claim 5, we get $|E_3 \setminus E_1| \leq 3n - 7 - 2 = 3n - 9$.

$\implies |E_1 \cup E_2 \cup E_3| \leq (3n - 6) + (3n - 9) + (3n - 9) = 9n - 24$.

Case 3: $|E_1| = 3n - 6, |E_2| \leq 3n - 7, |E_3| \leq 3n - 7$.

Since G_1, G_2, G_3 are maximally planar using Claim 5, we get $|E_3 \setminus E_1| \leq 3n - 7 - 2 = 3n - 9$.

$|E_2 \setminus E_1| \leq 3n - 7 - 2 = 3n - 9$.

$\implies |E_1 \cup E_2 \cup E_3| \leq (3n - 6) + (3n - 9) + (3n - 9) = 9n - 24$.

Case 4: $|E_1| \leq 3n - 7, |E_2| \leq 3n - 7, |E_3| \leq 3n - 7$.

Since G_1, G_2, G_3 are maximally planar, using Claim 5, we get $|E_3 \setminus E_1| \leq |E_2 \setminus E_1| \leq 3n - 7 - 2 = 3n - 9$. Similarly $|E_3 \setminus E_1| = \leq 3n - 9$;

$\implies |E_1 \cup E_2 \cup E_3| \leq (3n - 7) + (3n - 9) + (3n - 9) = 9n - 25$.

This completes the proof of Theorem 6.

4   Relation with Linear Arboricity and Maximum Degree

A linear forest is a forest in which every tree is a path. The linear arboricity of a graph G is the minimum number of linear forests into which the edges of G can be decomposed.

We have the following.

Proposition 1. *If G has linear arboricity at most k, then there is an embedding of G in $\mathbb{R}^k$ such that the edges of G can be drawn on the following k standard planes: for $i = 1, \ldots, k-1$, the ith plane is $\{x \in \mathbb{R}^k : x_j = 0 \forall j \notin \{1, i\}\}$, and the kth plane is $\{x \in \mathbb{R}^k : x_j = 0 \forall j \notin \{2, 3\}\}$. In particular, $pdim(G) \leq k$.*

In [10], it was shown that graphs of maximum degree 5 have linear arboricity at most 3. Thus, we get the following.

Corollary 1. *Any graph of maximum degree 5 is plane-projectable.*

In [1], Alon showed that a graph of maximum degree Δ has linear arboricity at most $\dfrac{\Delta}{2} + o(\Delta)$. Thus, we have: $pdim(G) \leq \Delta(G)\left(\dfrac{1}{2} + o(1)\right)$.

We shall actually prove a stronger form of Proposition 1, in which we replace linear arboricity by caterpillar arboricity, which we define below.

A caterpillar tree is a tree in which all the vertices are within distance 1 of a central path. A caterpillar forest is a forest in which every tree is a caterpillar tree. The *caterpillar arboricity* of a graph G is the minimum number of caterpillar forests into which the edges of G can be decomposed. This has been studied previously in [12].

The main idea behind Proposition 1 is the following.

Lemma 2. *Given a caterpillar tree G with vertex set $V = \{v_1, v_2, \ldots, v_n\}$, and $n \geq 2$ distinct real numbers $y_1, \ldots, y_n$, there exist n real numbers $x_1, \ldots, x_n$ such that G has a straight-line embedding with the vertex v_i mapped to (x_i, y_i).*

Proof. Let $w_1, w_2, \ldots w_k$ be the vertices on the central path such that w_i has an edge with w_{i-1} and w_{i+1}. All the indices are taken modulo k. Also, let L_i denote the set of leaf vertices adjacent to w_i.

We set the x co-ordinate of w_i to be i, and the x co-ordinate of every vertex of L_i to be $i + 1$. Clearly the edges of the caterpillar drawn with the above embedding are non-crossing. □

We remark that the above result and concept have also been studied as "unleveled planar graphs" in [11].

Lemma 3. *Given a cycle G with vertex set $V = \{v_1, v_2, \ldots, v_n\}$, and $n \geq 2$ distinct real numbers $y_1, \ldots, y_n$, there exist n real numbers $x_1, \ldots, x_n$ such that G has a straight-line embedding with the vertex v_i mapped to (x_i, y_i).*

Proof. Without loss of generality, let $v_1, v_2, \ldots v_n$ be the vertices on the cycle such that v_i has an edge with v_{i-1} and v_{i+1} and v_1 be the vertex with smallest y coordinate value. All the indices are taken modulo n.

We first remove the edge between v_1 and v_n so that the remaining graph is a path for which we use the previous lemma to construct the embedding.

Now to add back the edge $v_1 v_n$, we have to make sure that none of the other edges intersect with the edge between v_1 and v_n. Let *slope*$_i$ to be the slope between v_1 and v_i, and note that this is positive for all i, since v_1 has the lowest y coordinate. Let $m = min_i\{slope_i\}$. We now draw a line L through v_1 with slope less than m and place the vertex v_n on L, as illustrated in the figure below (Fig. 3).

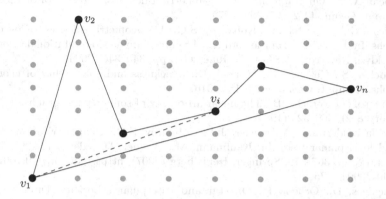

Fig. 3. Cycle graph with given y co-ordinates

The proposition below also follows from an application of Lemma 2.

Proposition 2. *Let G be the edge-union of a planar graph and d paths. Then* $pdim(G) \leq d + 2$.

5   Open Problems

1. What is the plane-projecting dimension of K_n?
2. Find tight bounds for the maximum number of edges in a n-vertex graph that is plane-projectable in $\mathbb{R}^3$.
3. Is every graph of maximum degree 6 plane-projectable in three dimensions?
4. Is $pdim(G) = O(\sqrt{\Delta(G)})$?
5. Is it true that $pdim(G)$ is at most the smallest d such that $\binom{d}{2} \geq \bar{\theta}(G)$?

References

1. Alon, N.: The linear arboricity of graphs. Isr. J. Math. **62**(3), 311–325 (1988)
2. Barát, J., Matoušek, J., Wood, D.R.: Bounded-degree graphs have arbitrarily large geometric thickness. Electron. J. Comb. **13**(1), 3 (2006)
3. Bläsius, T., Kobourov, S.G., Rutter, I.: Simultaneous embedding of planar graphs. In: Handbook on Graph Drawing and Visualization, pp. 349–381. Chapman and Hall/CRC (2013)

4. Dillencourt, M.B., Eppstein, D., Hirschberg, D.S.: Geometric thickness of complete graphs. J. Graph Algorithms Appl. **4**(3), 5–17 (2000)
5. Dujmovic, V., Wood, D.R.: On linear layouts of graphs. Discrete Math. Theor. Comput. Sci. **6**(2), 339–358 (2004)
6. Dujmovic, V., Wood, D.R.: Stacks, queues and tracks: layouts of graph subdivisions. Discrete. Math. Theor. Comput. Sci. **7**(1), 155–202 (2005)
7. Duncan, C.A.: On graph thickness, geometric thickness, and separator theorems. Comput. Geom. **44**(2), 95–99 (2011)
8. Duncan, C.A., Eppstein, D., Kobourov, S.G.: The geometric thickness of low degree graphs. In: Proceedings of the 20th ACM Symposium on Computational Geometry, Brooklyn, New York, USA, 8–11 June 2004, pp. 340–346 (2004)
9. Durocher, S., Gethner, E., Mondal, D.: Thickness and colorability of geometric graphs. Comput. Geom. **56**, 1–18 (2016)
10. Enomoto, H., Péroche, B.: The linear arboricity of some regular graphs. J. Graph Theory **8**(2), 309–324 (1984)
11. Estrella-Balderrama, A., Fowler, J.J., Kobourov, S.G.: Characterization of unlabeled level planar trees. In: Kaufmann, M., Wagner, D. (eds.) GD 2006. LNCS, vol. 4372, pp. 367–379. Springer, Heidelberg (2007). https://doi.org/10.1007/978-3-540-70904-6_35
12. Gonçalves, D., Ochem, P.: On star and caterpillar arboricity. Discrete Math. **309**(11), 3694–3702 (2009)
13. Halton, J.H.: On the thickness of graphs of given degree. Inf. Sci. **54**(3), 219–238 (1991)
14. Hutchinson, J.P., Shermer, T.C., Vince, A.: On representations of some thickness-two graphs. Comput. Geom. **13**(3), 161–171 (1999)
15. Malviya, P.: Graph visualization. Master's thesis, Indian Institute of Technology Hyderabad (2016)
16. Mutzel, P., Odenthal, T., Scharbrodt, M.: The thickness of graphs: a survey. Graphs Comb. **14**(1), 59–73 (1998)
17. Ollmann, T.: On the book thickness of various graphs. In: Proceedings of the 4th SouthEastern Conference on Combinatorics, Graph Theory and Computing, vol. VIII, p. 459 (1973)
18. Tutte, W.: The thickness of a graph. Indag. Math. **25**, 567–577 (1963)

New Algorithms and Bounds for Halving Pseudolines

Sergey Bereg[✉] and Mohammadreza Haghpanah

University of Texas at Dallas, Richardson, TX 75080, USA
{besp,Mohammadreza.Haghpanah}@utdallas.edu

Abstract. Let P be a set of points in general position in the plane. A *halving line* of P is a line passing through two points of P and cutting the remaining $n - 2$ points in a half (almost half if n is odd). Generalized configurations of points and their representations using allowable sequences are useful for bounding the number of halving lines.

We study a problem of finding generalized configurations of points maximizing the number of halving pseudolines. We develop algorithms for optimizing generalized configurations of points using the new notion of *partial allowable sequence* and the problem of computing a partial allowable sequence maximizing the number of k-transpositions. It can be viewed as a sorting problem using transpositions of adjacent elements and maximizing the number of transpositions at position k.

We show that this problem can be solved in $O(nk^n)$ time for any $k > 2$, and in $O(n^k)$ time for $k = 1, 2$. We develop an approach for optimizing allowable sequences. Using this approach, we find new bounds for halving pseudolines for even n, $n \leq 100$.

1 Introduction

Let S be a set of n points in the plane in general position. A *halving line* of S is a line passing through two points of S and

(i) cutting the remaining points in a half, if n is even, or
(ii) having $(n - 1)/2$ and $(n - 3)/2$ points of S on each side, if n is odd.

The problem of finding $h(n)$, the maximum number of halving lines for a set of n points, is one of the important open problems in the field of discrete geometry. Erdős, Lovász, Simmons Straus [13, 19] raised this problem for first time.

This problem is extended from the real plane $\mathbb{R}^2$ to the real projective plane $\mathbb{P}^2$. A *generalized configuration of points* consists of n distinct points in the projective plane and an arrangement of $\binom{n}{2}$ pseudolines crossing from each pair of points and intersect each other exactly once. Halving lines in $\mathbb{R}^2$ can be similarly extended to *halving pseudolines* [15] for a generalized configuration of points in $\mathbb{P}^2$ and define $\widetilde{h}(n)$ is the maximum number of halving pseudolines.

The research is supported in part by NSF award CCF-1718994.

M. Changat and S. Das (Eds.): CALDAM 2020, LNCS 12016, pp. 463–475, 2020.
https://doi.org/10.1007/978-3-030-39219-2_37

A part of extensive research on finding bounds on $h(n)$ and $\widetilde{h}(n)$ can be found in [1,7,12,24].

Goodman and Pollack [17] introduced *allowable sequences of permutations* (allowable sequence for short) which are useful for encoding configurations of points in $\mathbb{P}^2$. Allowable sequence is a doubly infinite sequence and half-period of it can be represented by a sequence $\Pi = (\pi_0, \pi_1, \ldots, \pi_{\binom{n}{2}})$ of permutations on n elements such that:

(1) Any permutation $\pi_i, i \geq 1$ can be obtained from the previous permutation π_{i-1} by a transposition of two adjacent elements.
(2) Every two elements are transposed exactly one time.

A transposition between elements at positions k and $k + 1$ is called a k-*transposition*. We denote by $\tau(k, \Pi)$ the number of k-transpositions in Π.

Allowable sequences are one of most the important tools in proving bounds for many problems of discrete geometry including bounds on the number of k-sets, halving lines, halving pseudolines, also problems of finding rectilinear crossing number of graph K_n and pseudolinear crossing number of K_n, see [1–3,6,9]. For example, the number of $(\leq k)$-sets of a set of n points in the plane in general position were studied in [11,20] using allowable sequences.

Most known upper bounds for $h(n)$ use the upper bound

$$h(n) \leq \widetilde{h}(n), \tag{1}$$

where $\widetilde{h}(n)$ is the maximum number of halving pseudolines. The definition of $\widetilde{h}(n)$ is based on generalized configurations of points in $\mathbb{P}^2$. For the sake of simplicity, we define it using allowable sequences as follows. Let Π be an allowable sequence of permutations on $[n]$. In this paper, we denote by $[n]$ the set $\{1, 2, \ldots, n\}$. First, we define $\widetilde{h}(\Pi)$ using two cases. If n is even then $\widetilde{h}(\Pi) = \tau(n/2, \Pi)$. If $n \geq 3$ is odd then $\widetilde{h}(\Pi) = \tau(\frac{n-1}{2}, \Pi) + \tau(\frac{n+1}{2}, \Pi)$. Then $\widetilde{h}(n)$ is the maximum value of $\widetilde{h}(\Pi)$ over all allowable sequences of permutations on $[n]$.

The bound (1) is used to show the tight bounds for the halving numbers by proving upper bounds for $\widetilde{h}(n)$ matching the lower bounds for $h(n)$. The tight bounds $h(n) = \widetilde{h}(n)$ are known for all $n \leq 27$ [4]. Inequality (1) can be viewed as the lower bound for $\widetilde{h}(n)$. Current lower bounds of $\widetilde{h}(n)$ and $h(n)$, for small n, are mostly can be attained by point configurations of Aichholzer's construction [5]. Can the bound (1) be improved?

In this paper, we propose to study lower bounds for $\widetilde{h}(n)$ using allowable sequences. This can be viewed as the problem of finding an allowable sequence Π maximizing $\widetilde{h}(\Pi)$ for a given n. The problem is known to be difficult for large n. Checking all possible configurations is computationally expensive as the number of simple arrangements of n pseudolines B_n, grows exponentially in n. Best upper bound for B_n is found by Felsner and Valtr [16] and best lower bound of B_n is provided by Dumitrescu and Mandal [10]. (see also [14,18,21])

$$\Omega(2^{0.2053n^2}) = B_n = O(2^{0.6571n^2}).$$

We propose an approach using *partial allowable sequences* described in Sect. 2. An interesting problem in our approach is the following sorting problem.

MAX-k Sorting Problem. Given a permutation π on $[n]$ and an integer $1 \leq k < n$, sort π using transpositions to $(n, n-1, n-2, \ldots, 1)$ such that

(1) The number of k-transpositions is maximized, and
(2) Every pair (i, j) can transposed at most one time.

Our Results. In this paper, we show that MAX-k sorting problem can be solved in $O(nk^n)$ time for any k and in $O(n^k)$ time for $k = 1, 2$. We develop an approach for optimizing allowable sequences and use it to find new bounds for halving pseudolines for even n, $n \leq 100$.

2   Transforming Allowable Sequences

Allowable sequences are very flexible and can be modified to increase $\tilde{h}(\Pi)$. For example, if the transpositions in permutations π_i and π_{i+1} are non-overlapping, say transpositions at positions $j, j+1$ in π_i and positions $j', j'+1$ in π_{i+1} with $|j - j'| \geq 2$, then the transpositions in π_i and in π_{i+1} can be exchanged.

We define a *push operation* as follows. Consider a permutation π_j of an allowable sequence $\Pi = (\pi_0, \pi_1, \ldots, \pi_{\binom{n}{2}})$. Consider two elements $\pi_j(i) = a$ and $\pi_j(i+1) = b$. If $a > b$ then the transposition of a and b is at some permutation $\pi_{j'}$ before π_j, i.e. $j' < j$. We can push the transposition of a and b down from the j'th permutation to the jth permutation. Thus, a and b should be exchanged in all permutations between these two permutations, see an example at Fig. 1. We call this operation *push-down*.

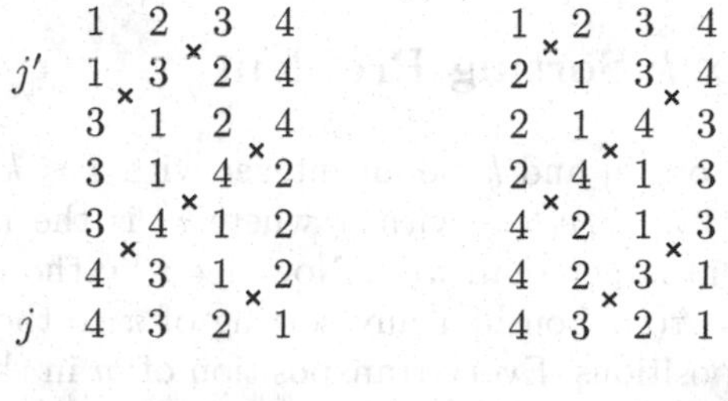

$$
\begin{array}{cccc}
 & 1 & 2 & 3 & 4 \\
j' & 1 & 3 & 2 & 4 \\
 & 3 & 1 & 2 & 4 \\
 & 3 & 1 & 4 & 2 \\
 & 3 & 4 & 1 & 2 \\
 & 4 & 3 & 1 & 2 \\
j & 4 & 3 & 2 & 1 \\
\end{array}
\qquad
\begin{array}{cccc}
1 & 2 & 3 & 4 \\
2 & 1 & 3 & 4 \\
2 & 1 & 4 & 3 \\
2 & 4 & 1 & 3 \\
4 & 2 & 1 & 3 \\
4 & 2 & 3 & 1 \\
4 & 3 & 2 & 1 \\
\end{array}
$$

Fig. 1. Pushing down the transposition of 2 and 3 from permutation $\pi_{j'}$ to π_j.

Now suppose that $a < b$ and the transposition of a and b is at some permutation $\pi_{j'}$ after π_j, i.e. $j' > j$. We can push the transposition of a and b up from the j'th permutation to the jth permutation. Thus, a and b should be exchanged in all permutations between these two permutations. We called the operation *push-up*.

Proposition 1. *For any two allowable sequences* Π, Π' *of permutations on* n *elements, there exists a sequence of push operations transforming* Π *into* Π'. *It holds even if the operations are restricted to push-down (or push-up) operations.*

We describe another tool for transforming allowable sequences. Let $\Pi = (\pi_0, \pi_1, \ldots, \pi_{\binom{n}{2}})$ be an allowable sequence. A *partial allowable sequence* $\Pi_{i,j}$ is a subsequence of consecutive permutations of Π, i.e. $\Pi_{i,j} = (\pi_i, \pi_{i+1}, \ldots, \pi_j)$. One way to optimize $\widetilde{h}(\Pi)$ is to choose a partial allowable sequence $\Pi_{i,j}$ and find another partial allowable sequence $\Pi'_{i,j} = (\pi'_i, \pi'_{i+1}, \ldots, \pi'_j)$ such that

(1) $\pi'_i = \pi_i, \pi'_j = \pi_j$, and
(2) the partial allowable sequence $\Pi_{i,j}$ can be transformed to $\Pi'_{i,j}$ using push operations within $\Pi_{i,j}$.

In many cases, the permutations of $\Pi_{i,j}$ have a common prefix and a common suffix. By removing them and renumbering the elements of the permutations, we reduce the size of the problem. Suppose that n is even. Then the halving transpositions correspond to k-transpositions for some value of k in the reduced problem. The elements of the reduced partial allowable sequence can be renumbered such that the last permutation is $(m, m-1, \ldots, 2, 1)$ for some $m \leq n$. Then the problem of optimizing the halving transpositions can be viewed as MAX-k sorting problem. In this paper, we mostly focus on this problem.

For odd n, the problem is to maximize the sum $\tau(k, \Pi) + \tau(k + 1, \Pi)$. This problem will be discussed in Sect. 7.

The output of MAX-k sorting problem is a partial allowable sequence $\Pi_{i,j} = (\pi_i, \pi_{i+1}, \ldots, \pi_j)$. It can be represented by a *transcript* which is a sequence of integers $k_1, k_2, \ldots, k_{j-i}$ such that $\pi_s, (1 \leq s \leq j - i)$ is obtained from π_{s-1} by a k_s-transposition.

3   General MAX-k Sorting Problem

Let π a permutation on $[n]$ and k be an integer with $1 \leq k < n$. We associate a vector $v(\pi) = (v_1, v_2, \ldots, v_{n-1})$ with π where v_i is the number of elements $\pi(1), \pi(2), \ldots, \pi(i - 1)$ larger than $\pi(i)$. Now, let i be the largest integer such that $v_i \geq k$. Let $m = \pi(i)$. Consider any sorting of π to the reverse of identity using adjacent transpositions. Every transposition of m in the sorting will be at positions $j, j + 1$ where $j \geq k$. Thus, m will never be used in a k-transposition; therefore, we can remove it from π. By renumbering the permutation elements, we reduce π to a permutation on $[n-1]$. If $v_i < k$ for all i, we call the permutation π k-*bounded*. By repeating the above process, a permutation π can be reduced to a unique k-bounded permutation, say σ. We called $v(\sigma)$ a k-*vector* of π.

Let $\lambda_k(\pi)$ be the maximum number of k-transpositions in a sorting of permutation π. The use of vectors of k-bounded permutations is motivated by the following proposition.

Proposition 2. *Let π_1, and π_2 be two permutation on $[n]$. If the k-vectors of π_1 and π_2 are equal then $\lambda_k(\pi_1) = \lambda_k(\pi_2)$.*

Lemma 1. *The number of k-bounded permutations on $[n]$ is $(k-1)!k^{n-k+1}$.*

Theorem 3. MAX-k *sorting problem can be solved in $O(nk!k^{n-k})$ time.*

Proof. First, we will explain the main idea of the algorithm. Then, we show the improvement of the algorithm efficiency by changing the indexing process of storing array. We employ dynamic programming and compute arrays A_m for $m = 1, 2, \ldots, n$. An array A_m stores the maximum number of k-transpositions for all k-bounded permutations of size m. We use k^m entries in A_m since the number of k-bounded permutations of size m is $(k-1)!k^{m-k+1} \leq k^m$. A k-bounded permutation π of size m corresponds to $A_m[j]$ where $j = \sum_{i=1}^{m} v_i k^{i-1}$ and $v = v(\pi)$.

For a given vector v_π, the corresponding k-bounded permutation π can be computed in linear time following the proof of Proposition 1. Then the maximum number of k-transpositions in a sorting of π can be computed as follows. Apply one transposition and find the corresponding k-bounded permutation. Its length is either m or $m-1$. So, it is stored in A_m or A_{m-1}. There are at most $m-1$ possible transposition for π. We find the maximum value for them in $O(m)$ time. The running time for computing each A_m is $O(m^2 k^m)$. Then the total running time is $\sum_{m \leq n} cm^2 k^m = O(n^2 k^n)$.

We modify the previous approach to improve both the running time and the space. The size of each array A_m can be reduced using Lemma 1. We change indexing for k-bounded permutations and introduce a one-to-one map T from set of k-bounded permutations of size m to $\{0, 1, \ldots, (k-1)!k^{m-k+1} - 1\}$ [1]. Let π be a k-bounded permutations of size m, and $v_\pi = (v_1, v_2, \ldots, v_m)$ be the associated vector with π. We define $T_a^b(\pi) = \sum_{i=a}^{b} \delta_k(i)v_i$ where $1 \leq a \leq b \leq m$, and $T(\pi) = T_1^m(\pi)$ where

$$\delta_k(i) = \begin{cases} 1 & i = 1 \\ \delta_k(i-1) \cdot i & 1 < i < k \\ \delta_k(i-1) \cdot k & i \geq k. \end{cases}$$

The vector $\delta_k = (\delta_k(1), \ldots, \delta_k(m))$ can be computed once at the beginning. Given an index $t = T(\pi)$ of a permutation π, π can be computed in $O(m)$ time. For a permutation π of size m, the values of $T_t^m(\pi)$, and $T_1^t(\pi)$ for $t \in [m]$ can be computed in $O(m)$ time. Let π' be the permutation obtained from π by applying one transposition, say p-transposition. The permutations π and π' are different only at positions p and $p+1$, the vector $v_{\pi'}$ may be different from v_π only at positions p and $p+1$. Specifically, if $v = v(\pi)$ and $u = v(\pi')$, then $u_p = v_{p+1}, u_{p+1} = v_p + 1$ and $u_i = v_i, i = 1, \ldots, p-1, p+2, \ldots, m$.

---

[1] This improves the space by a factor of $k^{k-1}/(k-1)!$. For example, if $k = 5$ this a factor of 26.041.

If $u_{p+1} < k$ then permutation π' is k-bounded and

$$T(\pi') = T(\pi) + \delta_k(p)v_{p+1} + \delta_k(p+1)(v_p+1) - \delta_k(p)v_p - \delta_k(p+1)v_{p+1}.$$

Suppose that $u_{p+1} \geq k$. Then $p \geq k$ and $\pi'(p+1)$ will be deleted and the elements of π' will be renumbered. Then π' corresponds to an entry of array A_{m-1} and

$$T(\pi') = T_1^p(\pi) + T_{p+2}^m(\pi)/k.$$

Notice that the computation of the final π' takes $O(m)$ time whereas $T(\pi')$ can be computed in $O(1)$ time. We avoid the computation of π' in this case and use $A_{m-1}[T(\pi')]$ directly.

The runnig time for computing each A_m is $O(m(k-1)!k^{m-k+1})$. Then the total running time is $O(nk!k^{n-k})$.   □

4   MAX-1 Sorting Problem

We show that MAX-1 sorting problem can be solved by a greedy algorithm. Let π be a permutation on $[n]$. The following algorithm has two steps. The first step will maximize the number of 1-transposition and step 2 will complete the sorting of the permutation.

Step 1. While $\pi(1) \neq n$, pick the smallest i such that $\pi(i) > \pi(1)$ and move it to the first position, i.e. by swapping $\pi(i)$ with $\pi(i-1), \pi(i-2), \ldots, \pi(1)$.

Step 2. While $\pi(i) < \pi(i+1)$ for some i, swap $\pi(i)$ and $\pi(i+1)$.

To compute the maximum number of 1-transpositions for π, one can apply Step 1 without doing swaps. This can be done by computing the longest increasing sequence in π where start at first position.

Proposition 4. MAX-*1 sorting problem can be solved in linear time, i.e. the maximum number of 1-transpositions to sort π can be found in $O(n)$ time and the corresponding transcript can be found in time $O(n+I)$ where I is the size of the transcript.*

5   MAX-2 Sorting Problem

First, we solve MAX-2 sorting problem in a special case where the input permutation is $1, 2, \ldots, n$. Define a function $f : \mathbb{Z}_+ \to \mathbb{Z}_+$

$$f(n) = \begin{cases} 3n/2 - 3, & \text{if } n \text{ is even,} \\ 3(n-1)/2 - 1, & \text{if } n \text{ is odd.} \end{cases}$$

Theorem 5. *For any $n \geq 3$, the maximum number of 2-transpositions in a sorting of $1, 2, \ldots, n$ is $f(n)$.*

Proof. We omit the base case $M(3) \leq 2$ and $M(4) \leq 3$ due to the lack of space.

Inductive step. First, we show that, for any $n \geq 3$, $M(n) \leq f(n)$ implies $M(n + 2) \leq f(n + 2)$.

Let Π be an allowable sequence of size $n + 2$. We transform it by pushing operators as follows. Consider elements $n, n + 1$, and $n + 2$. There are three transpositions in Π between these elements, and at most two of them are 2-transpositions. We push them down in the following order: transposition of n and $n + 1$, transposition of n and $n + 2$, and transposition of $n + 1$ and $n + 2$. The number of 2-transpositions in Π will not decrease. This is shown in Fig. 2 (the last four lines).

Now, $n + 2$ is in third position, and it was never swapped to this position before. Therefore we can push-down the transpositions of $n + 2$ with $n - 1, n - 2, \ldots, 1$, see Fig. 2. Similarly, we can push-down the transpositions of $n + 1$ with $n - 1, n - 2, \ldots, 1$ as shown in Fig. 2. Let Π' be the allowable sequence of the remaining permutations after removing $n + 1$ and $n + 2$ from them. Let t and t' be the number of 2-transpositions in Π and Π', respectively. Then $t \leq t' + 3$. By induction hypothesis, $t' \leq f(n)$. Then $t \leq f(n) + 3 = f(n + 2)$.

1	2	3	4	$\cdots$	n	$n+1$	$n+2$
$\cdots$			$\cdots$				
n	$n-1$	$n-2$	$n-3$	$\cdots$	1	$n+1$	$n+2$
n	$n-1$	$n-2$	$n-3$	$\cdots$	$n+1$	1	$n+2$
n	$n-1$	$n-2$	$n+1$	$\cdots$	2	1	$n+2$
n	$n-1$	$n+1$	$n-2$	$\cdots$	2	1	$n+2$
n	$n+1$	$n-1$	$n-2$	$\cdots$	2	1	$n+2$
n	$n+1$	$n-1$	$n-2$	$\cdots$	2	$n+2$	1
n	$n+1$	$n-1$	$n-2$	$\cdots$	$n+2$	2	1
n	$n+1$	$n-1$	$n+2$	$\cdots$	3	2	1
n	$n+1$	$n+2$	$n-1$	$\cdots$	3	2	1
n	$n+2$	$n+1$	$n-1$	$\cdots$	3	2	1
$n+2$	n	$n+1$	$n-1$	$\cdots$	3	2	1
$n+2$	$n+1$	n	$n-1$	$\cdots$	3	2	1

Fig. 2. An allowable sequence of size $n + 2$.

The above argument can be used to construct an allowable sequence with $f(n)$ 2-transpositions for any $n \geq 5$, see Fig. 3 for an example. $\qquad\square$

We show how to solve MAX-2 sorting problem for a given permutation π on n elements. As in Sect. 3, consider the vector $v_\pi = (v_1, v_2, \ldots, v_{n-1})$ where v_i is the number of elements $\pi(1), \pi(2), \ldots, \pi(i - 1)$ larger than $\pi(i)$. First, we observe that an element $\pi(i)$ can be removed from π if $v_i \geq 2$. We remove all elements $\pi(i)$ from π with π if $v_i \geq 2$. Without loss of generality, we assume that

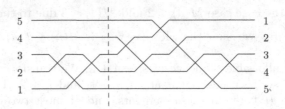

Fig. 3. The inductive step. The wiring diagrams for $n = 5$ constructed from the wiring diagrams for $n = 3$. The corresponding transcript of it is $(2, 1, 2, 3, 2, 4, 3, 2, 1, 2)$.

all $v_i < 2$ in the vector v for π. Then vector v is simply a binary sequence. We define an *i-block*, $i = 0, 1$, as a maximal subsequence of v of consecutive i. Then vector v is a sequence of alternating blocks

$$v = B_1^0 B_1^1 B_2^0 B_2^1 \ldots B_k^0 B_k^1, \qquad (2)$$

where B_j^i is an i-block and block B_k^1 may be not present. Note that the first block of vector v must be a 0-block since $v_1 = 0$. Thus, the first block in (2) is B_1^0. We employ dynamic programming and compute $m_j^i, i = 0, 1, j = 1, 2, \ldots, k$, the maximum number of 2-transpositions in a sorting of $\pi(1), \pi(2), \ldots, \pi(l)$ where $l = |B_1^0| + |B_1^1| + |B_2^0| + |B_2^1| + \cdots + |B_j^i|$. Initially $m_1^0 = f(|B_1^0|)$ if $|B_1^0| \geq 3$; otherwise $m_1^0 = 0$. If $|B_1^0| = 1$ then $m_1^1 = |B_1^1| - 1$; otherwise $m_1^1 = m_j^0 + |B_1^1|$. To make a recursive formula we consider a pair (j', j) such that $1 \leq j' < j \leq k$. Let α be the permutation obtained from sequence $\pi(1), \pi(2), \ldots, \pi(l)$ by

(i) sorting first $|B_1^0| + |B_1^1| + \cdots + |B_{j'}^1|$ elements, and then
(ii) deleting elements of π corresponding to blocks $B_{j'+1}^1, B_{j'+2}^1, \ldots, B_{j-1}^1$.

Let $g(j', j)$ be the maximum number of 2-transpositions in a sorting of permutation α. Then,

$$m_j^0 = \max(f(l_j), \max_{1 \leq j' < j} (m_{j'}^1 + g(j', j))), \qquad (3)$$

$$m_j^1 = \begin{cases} m_j^0 + |B_j^1|, & \text{if } j > 1 \text{ or } (j = 1 \text{ and } |B_1^0| > 1) \\ |B_j^1| - 1, & \text{otherwise.} \end{cases} \qquad (4)$$

where $l_j = |B_1^0| + |B_2^0| + \cdots + |B_j^0|$.

Computing $g(j', j)$. Let a be the total length of blocks $B_1^0, B_1^1, \ldots, B_{j'}^1$ and let b be the total length of blocks $B_{j'+1}^0, B_{j'+2}^0, \ldots, B_j^0$. Then vector α (after relabeling) is

$$\alpha = a, a - 1, \ldots, 1, a + 1, a + 2, \ldots, a + b$$

and $g(j', j)$ can be computed as

$$g(j', j) = \begin{cases} 3b/2, & \text{if } b \text{ is even,} \\ 3(b - 1)/2 + 1, & \text{if } b \text{ is odd.} \end{cases} \qquad (5)$$

This can be shown similar to the proof of Theorem 5. In the base case, $b = 1, 2$ and α can be sorted using one and three 2-transpositions, respectively. The inductive case is similar to Fig. 3 and three 2-transpositions can be added.

The maximum number of 2-transpositions in a sorting of π is m_k^0 or m_k^1 if block B_k^1 exists. The corresponding transcript can be computed as follows. First, we modify the dynamic program and, for each $j = 1, 2, \ldots, k$, we store the value of j' that is used in computing m_j^0. If m_j^0 is computed as $f(l_j)$ in Eq. (3) then we store $j' = -1$. We denote by t_j^i the transcript corresponding to m_j^i, i.e. the transcript for sorting $\pi(1), \pi(2), \ldots, \pi(l)$ where $l = |B_1^0| + |B_1^1| + |B_2^0| + |B_2^1| + \cdots + |B_j^i|$. Then transcript t_j^1 is the transcript t_j^0 followed by inserting the elements of π corresponding to B_j^1 as in the insertion sort.

The transcript t_j^0 can be computed as follows. If $m_j^0 = f(l_j)$ then the transcript t_j^0 is obtained by using Theorem 5, see Fig. 3 for an example. Suppose that $m_j^0 = m_{j'}^1 + g(j', j)$ for some $1 \leq j' < j$. Then the transcript t_j^0 is the transcript $t_{j'}^1$ followed by the transcript obtained using the proof of Eq. 5.

Theorem 6. *For any permutation π on $[n], n \geq 3$, a transcript maximizing* MAX-*2 sorting problem can be solved in $O(n^2)$ time.*

6 Improving Lower Bounds for $\tilde{h}(n)$

In this section we use algorithms developed in previous sections to improve lower bounds for $\tilde{h}(n)$ for even n. Since n is even, we use allowable sequences maximizing the number of $\frac{n}{2}$-transpositions. We can directly apply the MAX-k sorting algorithm from Theorem 3 for $k = \frac{n}{2}$ but this is infeasible for large n. Instead, we devise a heuristic approach to the problem using a local optimization by

(i) creating a block and an allowable sequence and
(ii) solving the corresponding MAX-k sorting problem for small k.

Let $\Pi = (\pi_0, \pi_1, \ldots, \pi_{\binom{n}{2}})$ be an allowable sequence. We define a (l, r)-*window* or simply a *window* for a permutation π_i of Π as the sequence $\pi_i(l), \pi_i(l+1), \ldots, \pi_i(r)$, (see Fig. 4(a)). In general, we define a *block*[2] of Π as a sequence of (l, r)-windows in consecutive permutations $\pi_i, \pi_{i+1}, \ldots, \pi_j$ of Π such that each window of the block has the same set of elements and every two consecutive windows are different, see Fig. 4 for examples. For a permutation in an allowable sequence, we call the $\frac{n}{2}$-th position of the permutation, the *halving position*.

Let B be a block in Π which consists of (l, r)-windows of length $r - l + 1 = t$ on permutations $\pi_i, \pi_{i+1}, \ldots, \pi_j$. Suppose that there is at least one l-transposition and at least one $(r-1)$-transposition in the block. We also assume that block B overlaps with the halving position in Π. Consider the partial allowable sequence $\Pi_{i,j}$. Note that the optimization of this partial allowable sequence (as described in Sect. 2) corresponds to MAX-k sorting problem derived from the block B as

---

[2] The blocks in this section are different from alternating blocks used in the proof of Theorem 6.

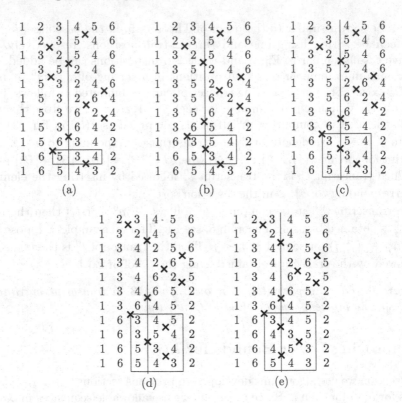

Fig. 4. (a) Initial window. (b–d) Extension of the window using push-up/push-down. (e) Applying the MAX-1 sorting algorithm.

follows. There is a bijection $\alpha : \{\pi_j(l), \pi_j(l+1), \ldots, \pi_j(r)\} \rightarrow [t]$ such that $\alpha(\pi_j(x)) = t - (x - l)$, $l \leq x \leq r$. This map transforms the last window of B into sequence $(t, t-1, t-2, \ldots, 1)$. Then the permutation for MAX-k sorting problem is $\alpha(\pi_i(l)), \alpha(\pi_i(l+1)), \ldots, \alpha(\pi_i(r))$ and the value of k corresponds to the halving position in the block. The solution of an algorithm for MAX-k sorting problem can be used to increase $\tau(\frac{n}{2}, \Pi)$ if the block in Π is replaced by $\alpha^{-1}(\Pi')$ where Π' is the output of the algorithm (a sequence of permutations).

A block B that allows to increase $\tau(\frac{n}{2}, \Pi)$ may not exist in the allowable sequence Π. We create new blocks using push operations as follows. First, we choose a (l, r)-window in a permutation of Π such that it includes the halving position We consider it as the initial block B_0. To construct a new block $B_i, i > 0$, we consider the permutation π_j containing the lower window w_j of B_{i-1}. Take two adjacent elements $\pi_j(t)$ and $\pi_j(t+1)$ in w_j such that $\pi_j(t) < \pi_j(t+1)$ and the corresponding transposition of $\pi_j(t)$ and $\pi_j(t+1)$ in Π is not a $\frac{n}{2}$-transposition. Apply the push operation (push-up) to this transposition such that it is a transposition between π_j and π_{j+1}. Then the block B_i is the extension of B_{i-1} using the (l, r)-window in permutation π_{j+1}. Note that this extension preserves the block property that every two consecutive windows are different.

The same process can be done to extend the block upward by pushing down transpositions (see Fig. 4(b)).

Let B be the block constructed by the above procedure. The block contains the halving position, say at its k-th position. By applying a MAX-k sorting algorithm for block B we may increase $\tau(k, B)$. This will increase $\tau(\frac{n}{2}, \Pi)$, see Fig. 4(e)) for an example. We run this algorithm all even n up to 100 for $k = 1$. The running time for each n is between two hours and two days. The results are shown in Table 1. Many of them improve known lower bounds for $\widetilde{h}$. The transcripts of these allowable sequences are available at http://www.utdallas.edu/~besp/soft/pseudo/halving/even.zip.

7 Future Work

In this paper, we used the algorithm discussed in Sect. 6 combined with the MAX-1 sorting algorithm from Sect. 4 on allowable sequences of even length to achieve the results shown in Table 1. For odd n, the problem of maximizing $\widetilde{h}(\Pi)$ for an

Table 1. Bounds for the $\widetilde{h}(n)$. **LB/UB** denotes the current lower bound/upper bound for $\widetilde{h}(n)$. The bounds in bold are obtained in this paper. **Prev** denotes the previous lower bound for $h(n)$ if they are improved in this paper. The numbers in this column are from [5,24].

n	LB	UB	Prev	n	LB	UB	Prev
28	63	64	-	66	**202**	236	197
30	69	72	-	68	**207**	246	203
32	74	79	-	70	**226**	257	211
34	**81**	86	79	72	**229**	268	211
36	**88**	94	84	74	**237**	279	228
38	**97**	102	94	76	**253**	291	237
40	**104**	110	103	78	**262**	303	242
42	111	119	-	80	**265**	315	250
44	**117**	127	112	82	**268**	327	261
46	**126**	136	122	84	**282**	339	264
48	**133**	146	129	86	**292**	351	276
50	**141**	155	139	88	**297**	363	282
52	**146**	164	143	90	**312**	376	290
54	**153**	174	152	92	**317**	388	300
56	**163**	183	158	94	**326**	401	309
58	**169**	193	165	96	**345**	414	308
60	**177**	204	172	98	**338**	427	320
62	**187**	214	180	100	**366**	440	328
64	**195**	225	187				

allowable sequence Π uses two halving positions. A heuristic solution is to use a MAX-k sorting algorithm on a block for each of halving positions separately. Another approach is to provide an algorithm for the problem of maximizing total number of k-transpositions and $(k+1)$-transpositions in sorting of a permutation. We will explore this approach for odd values of n in the future.

Every allowable sequence from Table 1 corresponds to an arrangement of pseudolines. If an arrangement of n pseudolines is stretchable, i.e., is isomorphic to an arrangement of n straight lines, then one can find a set of n points in the plane providing a new bound for $h(n)$. The problem of determining whether a pseudoline arrangement is stretchable is NP-hard, see [22,23,25]. It would be interesting to explore the stretchability of the pseudoline arrangements from Table 1, perhaps using the heuristic method by Bokowski [8].

References

1. Ábrego, B.M., Balogh, J., Fernández-Merchant, S., Leaños, J., Salazar, G.: An extended lower bound on the number of ($\leq k$)-edges to generalized configurations of points and the pseudolinear crossing number of K_n. J. Comb. Theory Ser. A **115**(7), 1257–1264 (2008)
2. Ábrego, B.M., Cetina, M., Fernández-Merchant, S., Leaños, J., Salazar, G.: On $\leq$ k-edges, crossings, and halving lines of geometric drawings of k_n. Discrete Comput. Geom. **48**(1), 192–215 (2012)
3. Ábrego, B.M., Fernández-Merchant, S.: A lower bound for the rectilinear crossing number. Graphs Combin. **21**(3), 293–300 (2005)
4. Ábrego, B.M., Fernández-Merchant, S., Leaños, J., Salazar, G.: The maximum number of halving lines and the rectilinear crossing number of kn for n $\leq$ 27. Electron. Notes Discrete Math. **30**, 261–266 (2008)
5. Aichholzer, O.: On the rectilinear crossing number. http://www.ist.tugraz.at/staff/ aichholzer/research/rp/triangulations/crossing/
6. Alon, N., Györi, E.: The number of small semispaces of a finite set of points in the plane. J. Comb. Theory Ser. A **41**(1), 154–157 (1986)
7. Beygelzimer, A., Radziszowski, S.: On halving line arrangements. Discrete Math. **257**(2–3), 267–283 (2002)
8. Bokowski, J.: On heuristic methods for finding realizations of surfaces. In: Bobenko, A.I., Sullivan, J.M., Schröder, P., Ziegler, G.M. (eds.) Discrete Differential Geometry. Oberwolfach Seminars, vol. 38, pp. 255–260. Springer, Basel (2008). https:// doi.org/10.1007/978-3-7643-8621-4_13
9. Cetina, M., Hernández-Vélez, C., Leaños, J., Villalobos, C.: Point sets that minimize ($\leq k$)-edges, 3-decomposable drawings, and the rectilinear crossing number of K_{30}. Discrete Math. **311**(16), 1646–1657 (2011)
10. Dumitrescu, A., Mandal, R.: New lower bounds for the number of pseudoline arrangements. In: Proceedings of 13th Symposium on Discrete Algorithms, pp. 410–425 (2019)
11. Edelsbrunner, H., Hasan, N., Seidel, R., Shen, X.J.: Circles through two points that always enclose many points. Geometriae Dedicata **32**, 1–12 (1989)
12. Eppstein, D.: Sets of points with many halving lines. Technical Report ICS-TR92-86, University of California, Irvine, Department of Information and Computer Science, August 1992

13. Erdős, P., Lovász, L., Simmons, A., Straus, E.: Dissection graphs of planar point sets. In: Srivastava, J.N. (ed.) A Survey of Combinatorial Theory, pp. 139–154. North-Holland, Amsterdam (1973)
14. Felsner, S.: On the number of arrangements of pseudolines. Discrete Comput. Geom. **18**, 257–267 (1997)
15. Felsner, S., Goodman, J.E.: Pseudoline arrangements. In: Handbook of Discrete and Computational Geometry, pp. 125–157. Chapman and Hall/CRC (2017)
16. Felsner, S., Valtr, P.: Coding and counting arrangements of pseudolines. Discrete Comput. Geom. **46**(3), 405–416 (2011)
17. Goodman, J.E., Pollack, R.: Semispaces of configurations, cell complexes of arrangements. J. Comb. Theory Ser. A **37**(3), 257–293 (1984)
18. Knuth, D.E. (ed.): Axioms and Hulls. LNCS, vol. 606. Springer, Heidelberg (1992). https://doi.org/10.1007/3-540-55611-7
19. Lovász, L.: On the number of halving lines. Annal. Univ. Scie. Budapest. de Rolando Eötvös Nominatae, Sectio Math. **14**, 107–108 (1971)
20. Lovász, L., Vesztergombi, K., Wagner, U., Welzl, E.: Convex quadrilaterals and k-sets. In: Pach, J. (ed.) Towards a Theory of Geometric Graphs, pp. 139–148. Contemporary Mathematics, American Mathematical Society (2004)
21. Matoušek, J.: Lectures on Discrete Geometry, vol. 212. Springer-Verlag, New York (2002). https://doi.org/10.1007/978-1-4613-0039-7
22. Mnëv, N.: On manifolds of combinatorial types of projective configurations and convex polyhedra. Soviet Math. Doklady **32**, 335–337 (1985)
23. Mnev, N.E.: The universality theorems on the classification problem of configuration varieties and convex polytopes varieties. In: Viro, O.Y., Vershik, A.M. (eds.) Topology and Geometry — Rohlin Seminar. LNM, vol. 1346, pp. 527–543. Springer, Heidelberg (1988). https://doi.org/10.1007/BFb0082792
24. Rodrigo, J., López, M.D.: An improvement of the lower bound on the maximum number of halving lines in planar sets with 32 points. Electr. Notes in Discr. Math. **68**, 305–310 (2018)
25. Shor, P.: Stretchability of pseudolines is NP-hard. Applied Geometry and Discrete Mathematics-The Victor Klee Festschrift (1991)

Algorithms for Radon Partitions with Tolerance

Sergey Bereg[✉] and Mohammadreza Haghpanah

University of Texas at Dallas, Richardson, TX 75080, USA
{besp,Mohammadreza.Haghpanah}@utdallas.edu

Abstract. Let P be a set n points in a d-dimensional space. Tverberg theorem says that, if n is at least $(k-1)(d+1)$, then P can be partitioned into k sets whose convex hulls intersect. Partitions with this property are called *Tverberg partitions*. A partition has tolerance t if the partition remains a Tverberg partition after removal of any set of t points from P. A tolerant Tverberg partition exists in any dimensions provided that n is sufficiently large. Let $N(d, k, t)$ be the smallest value of n such that tolerant Tverberg partitions exist for any set of n points in $\mathbb{R}^d$. Only few exact values of $N(d, k, t)$ are known.

In this paper, we study the problem of finding Radon partitions (Tverberg partitions for $k = 2$) for a given set of points. We develop several algorithms and found new lower bounds for $N(d, 2, t)$.

Keywords: Tverberg's theorem · Linear classifiers · Tolerance

1   Introduction

Tverberg's theorem and Tverberg partitions are of crucial importance in combinatorial convexity and stands on the intersection of combinatorics, topology and linear algebra. Tverberg partitions with tolerance showed importance in these fields, years after the main theorem.

Theorem 1 (Tverberg [32]). *For any set $P \subset \mathbb{R}^d$ of at least $(k-1)(d+1)+1$ points, there exists a partition of $P \subset \mathbb{R}^d$ into k sets $P_1, P_2, \ldots, P_k$ such that their convex hulls intersect*

$$\bigcap_{i=1}^{k} \operatorname{conv}(P_i) \neq \emptyset. \tag{1}$$

This is a generalization of Radon's theorem from 1921 [26] which provides a partition of at least $d+1$ points for $k = 2$. We call a partition satisfying Equation (1) a *Tverberg partition*. If $k = 2$, we call it a *Radon partition*.

The computational complexity of finding a Tverberg partition according to Theorem 1 is not known. Teng [31] showed that testing whether a given point is in the intersection of convex hulls of a partition is coNP-complete. On the

© Springer Nature Switzerland AG 2020
M. Changat and S. Das (Eds.): CALDAM 2020, LNCS 12016, pp. 476–487, 2020.
https://doi.org/10.1007/978-3-030-39219-2_38

other hand, such a point can be computed in n^{d^2} time if d is fixed [2]. Mulzer and Werner [24] found an approximation algorithm for Tverberg partitions with linear running time.

A Tverberg partition has *tolerance* t if after removing t points from P it still remains Tverberg partition.

Definition 2. (*t-tolerant Tverberg partition*) Let P be set of point in $\mathbb{R}^d$. $\Pi = \{P_1, P_2, \ldots, P_k\}$ be a partition of size k of P. A partition $\Pi = \{P_1, P_2, \ldots, P_k\}$ of P is called *t-tolerant* if for every $C \subset P$ with $|C| \leq t$

$$\bigcap_{i=1}^{k} \mathsf{conv}(P_i \setminus C) \neq \emptyset.$$

In 1972, Larman [21] proved that every set of size $2d+3$ admits a 1-tolerant Tverberg partition into two sets, i.e., a 1-tolerant Radon partition. García-Colín [13] proved the existence of a Radon partition for any tolerance t, see also [14]. Soberon *et al.* [30] proved if $|P| > (t+1)(k-1)(d+1)$ then P has a t-tolerant Tverberg partition. Examples of a Tverberg partition (with tolerance 0) and tolerant Tverberg partitions are shown in Fig. 1.

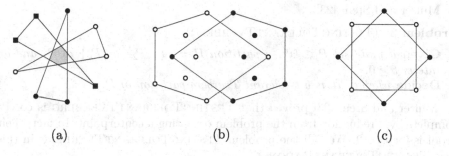

|(a)|(b)|(c)|

Fig. 1. (a) A Tverberg partition for $k = 3$ (the intersection of 3 convex hulls is shaded). Points from the same set of the partition have the same shape (disk, circle or square). (b) A 1-tolerant Tverberg partition for $k = 2$. (c) A 2-tolerant Tverberg partition for $k = 2$.

The problem of finding Tverberg partitions with tolerance seems more difficult. For example, Tverberg's theorem provides a tight bound for the number of points. On the other hand, only a few tight bounds are known for Tverberg partitions with tolerance. Let $N(d, k, t)$ be minimum number such that every set of points $P \subset \mathbb{R}^d$ with $|P| \geq N(d, k, t)$ has a t-tolerant Tverberg partition. For fixed t and d, Garcia *et al.* showed [15], $N(d, k, t) = kt + o(t)$ using a generalization of the Erdos-Szekeres theorem for cyclic polytopes in $\mathbb{R}^d$. Soberon [29] improved the bound to $N(d, k, t) = kt + O(\sqrt{t})$. Mulzer and Stein [23] provided an algorithm for finding a t-tolerant Tverberg partition of size k for a set $P \subset \mathbb{R}^d$ in $O(2^{d-1}dkt + kt \log t)$ time. The algorithm by Mulzer and Stein [23] for finding a t-tolerant Tverberg partition uses large number of points. A natural question is to find algorithms when n is relatively small. In this paper, we consider the following problems.

Problem 1. (COMPUTINGTOLERANTPARTITION)

> **Given** *a finite set* $P \subset \mathbb{R}^d$ *and an integer* t.
> **Compute** *a t-tolerant Tverberg partition for P if it exists.*

Our motivation for this problem is to construct sets of points in $\mathbb{R}^d$ with large tolerance t to find new lower bounds for $N(d, t, k)$ for some small values of d and t. One approach to this problem is to check all possible partitions and solve the following problem for them. This is possible in practice only for relatively small n, k, and d.

Problem 2. (COMPUTINGMAXTOLERANCE)

> **Given** *a finite set* $P \subset \mathbb{R}^d$, *a Tverberg partition* $\Pi = \{P_1, P_2, \ldots, P_k\}$ *of P.*
> **Compute** *the largest t such that Π is t-tolerant Tverberg partition.*

For a Tverberg partition Π, we say that the *tolerance* of Π is t and write $\tau(\Pi) = t$ if partition Π is t-tolerant but not $(t + 1)$-tolerant. Thus, the problem COMPUTINGMAXTOLERANCE is to compute the tolerance of a Tverberg partition. The decision problem of COMPUTINGMAXTOLERANCE has been studied by Mulzer and Stein [23].

Problem 3. (TESTINGTOLERANTTVERBERG)

> **Given** *a finite set* $P \subset \mathbb{R}^d$, *a partition* $\Pi = \{P_1, P_2, \ldots, P_k\}$ *of P, and an integer* $t \geq 0$.
> **Decide** *whether Π is a t-tolerant Tverberg partition of P.*

Mulzer and Stein [23] proved that TESTINGTOLERANTTVERBERG is coNP-complete by a reduction from the problem of testing a centerpoint. In fact, their proof is for $k = 2$. We call the problem TESTINGTOLERANTTVERBERG in this case TESTINGTOLERANTRADON.

In this paper we study algorithms for problems COMPUTINGTOLERANTPAR-TITION, COMPUTINGMAXTOLERANCE and TESTINGTOLERANTTVERBERG aiming to compute point configurations in d dimensions with high tolerance. These problems are hard even for $k = 2$, i.e., for Radon Partitions. In this paper we focus on Radon Partitions only. We use $M(d, t) = N(d, 2, t)$ for simplicity. They would provide new lower bounds for $M(d, t)$. We are not aware of any program supporting lower bounds for $M(d, t)$. Our results can be summarized as follows.

1. We found that the problem COMPUTINGMAXTOLERANCE for $k = 2$ (Radon partitions) is related to linear classifiers with outliers which is a well-known classification problem in machine learning and statistics. The literature on linear classifiers is vast, see for example [3, 9, 16, 18, 25, 27, 33]. This classification problem is also known as the *weak separation problem* [5, 10, 12, 19, 22] and *linear programming with violations* [7]. In fact, our Theorem 3 states that two optimization problems are equivalent (COMPUTINGMAXTOLERANCE and *optimal weak separation problem*).

2. The relation between COMPUTINGTOLERANTPARTITION and the classification problem can be used to solve COMPUTINGTOLERANTPARTITION more efficiently. We provide three algorithms for the problem COMPUTINGTOLER-ANTPARTITION. The first algorithm is simple and easy to implement. The second algorithm improved testing separable partition (Step 3) in the first algorithm by using BFS and hamming distances. As a result, the second algorithm is faster. To provide a more memory efficient algorithm, we used gray code in the last algorithm.
3. Using the algorithms for COMPUTINGRADONPARTITION, we established new lower bounds on $M(d, t)$. For this purpose, we design algorithms for generating sets of points and improving them. The bounds computed by the program are shown in Table 1 (different algorithms for different pairs of d and t). Because of the efficiency of these algorithms, we could solve COMPUTINGRADONPAR-TITION for set of points as large as 26.

Following [5, 19, 20], in this paper we assume that the points of set P are in general position. In Sect. 2 we show that Radon partitions and linear classifiers are related. In Sect. 3 we discuss algorithms for Radon partitions. In Section 4 we discuss experiments and lower bounds for $M(d, t)$.

2   Radon Partitions and Linear Classifiers

In this section we show a relation between the problem COMPUTINGMAXTOL-ERANCE for two sets (i.e. the problem of computing the maximum tolerance of a Radon partition in d dimensions) and linear classifiers with outliers which is a well-known classification problem in machine learning and statistics, see for example [3, 25, 27, 33]. Outlier detection algorithms are often computationally intensive [27] (Fig. 2).

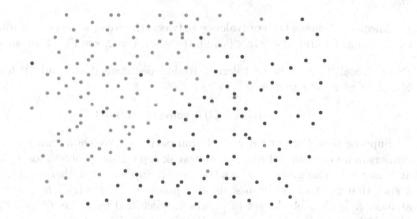

Fig. 2. Example of a classification problem in the plane that can be solved with a linear classifier and few outliers.

This classification problem is also known as the *weak separation problem* [5,10,12,19,22] and can be defined as follows. Let P be a bicolored set of points in $\mathbb{R}^d$, i.e. $P = R \cup B$ where R be a set red points in $\mathbb{R}^d$ and B is a set blue points. Let h be a hyperplane $a_1x_1 \cdots + a_dx_d = a_0$. Let h^+ be the halfspace that contains the points satisfying $a_1x_1 \cdots + a_dx_d \leq a_0$ and let h^- be the halfspace that contains the points satisfying $a_1x_1 \cdots + a_dx_d \geq a_0$. If h were a separator (or classifier) of P, we would have $R \subset h^+$ and $B \subset h^-$. A red point x is an *outlier*, if $x \notin h^+$. A blue point x is an outlier, if $x \notin h^-$. The weak separation problem is to find a hyperplane h minimizing the number of misclassified points (outliers)

$$mis(h) = |R \setminus h^+| + |B \setminus h^-|.$$

The weak separation problem in the plane is well studied. Gajentaam and Overmars [12] showed the weak separation problem is 3Sum-hard by reducing the point covering problem to it. An algorithm with $O(n^2)$ time complexity is provided for the weak separation problem by Houle [19]. Cole *et al.* [8] presented an $O(N_k(n) \log^2 k + n \log n)$ time algorithm to compute the k-hull of n points in the plane where $N_k(n)$ is the maximum number of k-sets for a set of n points. This algorithm can be used to compute the space of all classifiers misclassifying up to k points in the plane in $O(nt \log^2 t + n \log n)$ time [5]. Thus, a t-weak separator can be computed within the same time. A better algorithm with $O(nt \log t + n \log n)$ time have been found by Everett *et al.* [10]. In higher dimensions, Aronov *et al.* [5] proved that the weak separation problem can be solved using duality in $O(n^d)$ time.

The connection of tolerant Radon partitions and the weak separations is established in the next theorem.

Theorem 3. *Let $\Pi = \{P_1, P_2\}$ be a Radon partition of a set $P \subset \mathbb{R}^d$ (i.e. $\mathsf{conv}(P_1) \cap \mathsf{conv}(P_2) \neq \emptyset$). The tolerance of partition Π is t if and only if the number of outliers in an optimal solution for the weak separation problem for P_1 and P_2 is $t + 1$.*

Theorem 3 shows the equivalence between the weak separation problem and the problem COMPUTINGMAXTOLERANCE for $k = 2$, see Fig. 3 for an example.

Proof. Recall that Π is a t-tolerant Radon partition if and only if for any set $C \subset P$ of at most t points

$$\mathsf{conv}(P_1 \setminus C) \cap \mathsf{conv}(P_2 \setminus C) \neq \emptyset.$$

Suppose that Π is a t-tolerant Radon partition. We show that the number of outliers in an optimal solution for the weak separation problem for P_1 and P_2 is at least $t + 1$. The proof is by contradiction. Suppose that there is a hyperplane h such that the number of misclassified points $mis(h) = |P_1 \setminus h^+| + |P_2 \setminus h^-|$ is at most t. Let C_h be the set of points misclassified by h, i.e. $C_h = (P_1 \setminus h^+) \cup (P_2 \setminus h^-)$. Then $P_1 \setminus C_h \subset h^+$ and $P_2 \setminus C_h \subset h^-$. Therefore

$$\mathsf{conv}(P_1 \setminus C_h) \cap \mathsf{conv}(P_2 \setminus C_h) \subset h.$$

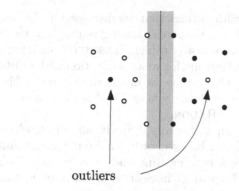

outliers

Fig. 3. A linear classifier with 2 outliers (and maximum margin) corresponding to the 1-tolerant Radon partition in Fig. 1(b).

The hyperlane h contains at most d points of $P_1 \cup P_2$ (due to general position). Then

$$\mathrm{conv}(P_1 \cap h) \cap \mathrm{conv}(P_2 \cap h) = \emptyset.$$

Therefore

$$\mathrm{conv}(P_1 \setminus C_h) \cap \mathrm{conv}(P_2 \setminus C_h) = \mathrm{conv}(P_1 \cap h) \cap \mathrm{conv}(P_2 \cap h) = \emptyset.$$

and Π is not t-tolerant. Contradiction.

Now, suppose that Π is not a $(t+1)$-tolerant Radon partition. We show that the number of outliers in an optimal solution for the weak separation problem for P_1 and P_2 is at most $t + 1$. There is a set C of size at most $t + 1$ such that

$$\mathrm{conv}(P_1 \setminus C) \cap \mathrm{conv}(P_2 \setminus C) = \emptyset.$$

By Minkowski hyperplane separation theorem [6, Section 2.5.1][1] there is a separating hyperplane h for $\mathrm{conv}(P_1 \setminus C)$ and $\mathrm{conv}(P_2 \setminus C)$, i.e., $\mathrm{conv}(P_1 \setminus C) \subset h^+$ and $\mathrm{conv}(P_2 \setminus C) \subset h^-$. Then, the number of misclassified points $mis(h) = |P_1 \setminus h^+| + |P_2 \setminus h^-|$ is at most $|C| \leq t + 1$.

Therefore, if the tolerance of partition Π is t, then the number of outliers in an optimal solution for the weak separation problem for P_1 and P_2 is $t + 1$. The converse is true and the theorem follows.   □

3   Algorithms for Tolerant Radon Partitions

In this section, we design several algorithms for the problem COMPUTING TOLER-ANTPARTITION in order to find new lower bounds for tolerant Radon partitions. In this section, we assume that given points are in general position. The first idea is based on the connection of the problem COMPUTING MAXTOLERANCE for

---

[1] See also https://en.wikipedia.org/wiki/Hyperplane_separation_theorem.

$k = 2$ to linear classifiers with outliers that we discussed in the previous section. We can iterate through all possible partitions of given set into two sets and check if the partition is t-tolerant or not (Problem TESTINGTOLERANTRADON). The test can be done using an algorithm for weak separation with $O(nt \log t + n \log n)$ time for the plane [10] and $O(n^d)$ time for higher dimensions. This approach has $O(2^n T_d(n))$ running time complexity where $T_d(n)$ is the time complexity of the problem TESTINGTOLERANTRADON.

Since the problem is computationally difficult, all our algorithms have exponential running time and can be used only for bounded n, t and d. However, the algorithms have different running time and space bounds. This allows us to obtain lower bounds for n up to 27 in Sect. 4. We assume in this section that $t = O(1)$ is a constant.

The above approach uses TESTINGTOLERANTRADON with $O(n^d)$ running time which is not easy to implement. Our first algorithm is simpler. The algorithm uses separable partitions. A partition of P into k subsets is *seprable* [20] if their convex hulls are pair-wise disjoint. The number of separable partitions for $k = 2$ is well-known *Harding number* $H(n, d)$ [17]. Harding proved

$$H(n, d) = \sum_{j=0}^{d} \binom{n-1}{j} = \Theta(n^d).$$

Hwang and Rothblum [20] provided a method for enumerating separable 2-partitions in $O(nH(n, d))$ time. It is based on the following recursive formula

$$H(n, d) = H(n-1, d) + H(n-1, d-1).$$

Algorithm 1

1. Let $P = \{p_1, p_2, \ldots, p_n\}$. Construct $\mathcal{S}$, the set of all separable partitions using the enumeration from [20]. For any separable partition $P = P_1 \cup P_2$, we assume that $p_1 \in P_1$ and we encode the partition with a binary code $b_1 \ldots b_n$ where $b_i = j - 1$ if $p_i \in P_j$. Note hat $|\mathcal{S}| = H(n, d)$.
2. Construct $\mathcal{P}$, the set of all partitions of $P = P_1 \cup P_2$ with $p_1 \in P_1$. Encode the partitions as in Step 1.
3. For every binary code $b = (b_1, b_2, \ldots, b_n) \in \mathcal{S}$ and every $C \subset [n]$ with $|C| \leq t$, we make a b' by flipping b_i for all $i \in C$ and remove b' from $\mathcal{P}$.
4. Return any remaining partition in $\mathcal{P}$ as tolerant partition of points P.

Running Time Analysis. The partitions of $\mathcal{S}$ and their correspondent binary codes can be computed in $O(n^{d+1})$. The number of all binary codes in $\mathcal{P}$ is 2^{n-1}, and creating each of them takes $O(n)$. Therefore, Step (2) takes $O(n2^n)$ time. Step (3) searches $|\mathcal{S}|n^t = O(n^{d+t})$ binary codes in $\mathcal{P}$. Thus, Step (3) takes $O(n^{d+t+1})$ time. Each binary code contains n bits. The total time for Algorithm 1 is $O(n2^n + n^{d+t+1})$.

Correctness. We prove the following for the correctness of the algorithm.

(1) Every binary code deleted from S is not t-tolerant,
(2) Every binary code remained in S is t-tolerant.

Clearly, every binary code b deleted from $\mathcal{P}$ can be transformed into a binary code in S by flipping at most t bits. Since a binary code in S corresponds to a separable partition, the partition of b is not t-tolerant.

For the second part, suppose that a binary code b in $\mathcal{P}$ corresponds to a partition that is not t-tolerant. Then, it can be transformed to a separable code by flipping at most t bits. Therefore, b must be deleted from $\mathcal{P}$.

Algorithm 1 can be improved using the fact that it tries to delete the same binary code a multiple numbers of times. The second algorithm avoids it.

Algorithm 2

1. Construct S and $\mathcal{P}$ as in Algorithm 1.
2. Let $S_0 = S$ and remove S_0 element form $\mathcal{P}$. For each $i \in [t]$ compute S_i as follows.
2a. For each $b \in S_{i-1}$ and each position $j \in [n]$, we change j-th position of b and call it b'. Then, if b' is in $\mathcal{P}$, we remove it from $\mathcal{P}$ and add it to S_i.
3. Return remaining elements in $\mathcal{P}$ as tolerant partition of points P.

We show that Algorithm 2 is correct. First if binary code b' is removed from $\mathcal{P}$ in Step (2a) there is a binary code b in S such that hamming distance between b' and b is at most t. So the partition corresponding to b' is not t-tolerant.

It remains to show that every binary code $b \in \mathcal{P}$ corresponding to a partition Π with $\tau(\Pi) < t$ is removed from $\mathcal{P}$. Let $t_1 = \tau(\Pi)$. There exists $b'' \in S$ such that hamming distance between b and b'' is t_1. Algorithm 2 will change t_1 bits in b'' in Step (2a), and create binary code b. Then binary code b will be removed from $\mathcal{P}$. Only codes correspondent to t-tolerant partition will be left in $\mathcal{P}$.

Since both S and $\mathcal{P}$ contain at most 2^{n-1} binary codes, Algorithm 2 takes only $O(n2^n + n^{d+1})$ time. Using Algorithm 2 we were able to obtain more bounds on $M(d, t)$ (the bounds are shown in Sect. 4).

We also develop Algorithm 3, which is slower than Algorithm 2 but it is memory efficient. The idea is to apply gray code to enumerate all binary codes for $\mathcal{P}$. Let $hd(u, v)$ be the hamming distance between binary codes u and v (the number of positions where u and v are different). For each binary code b, we compute a *hamming vector* $v_b = (v_1, v_2, \ldots, v_N)$ where

- $N = |S| = H(n, d)$ is the size of S,
- $v_i = hd(b, s_i)$, $i = 1, 2, \ldots, N$ and
- s_i is the binary code of the i-th partition of S.

Algorithm 3

1. Construct S as in Algorithm 1.
2. For each binary code $b \in \mathcal{P}$, generated by the gray code, compute the hamming vector v_b as follows.

2.1 For the first binary code b, compute v_b directly by computing every $v_i = hd(b, s_i)$ in $O(n)$ time.

2.2 For every other binary code b following binary code b', b and b' are different in only one position. Then $hd(b, s_i) = hd(b', s_i) \pm 1$ and the hamming vector v_b can be computed in $O(N)$ time.

2.3 If all entries of v_b are greater than t, then the Radon partition corresponding to binary code b is t-tolerant and the algorithms stops.

3. If the algorithm does not stop in Step (2.3), then P does not admit a t-tolerant Radon partition.

For the correctness of Algorithm 3 it is sufficient to proof following lemma.

Lemma 1. *Let Π be a Radon partition, and b is binary representation of it. All entries of v_b are greater than k if and only if Π is t-tolerant.*

Proof. It follows as a sequence of equivalences. All entries of v_b are greater than $k \iff hd(s_i, b) > k$ for every $i \in N \iff$ for every separable partition $s_i \in S$ there is at least $k + 1$ outliers $\iff \Pi$ is a t-tolerant Radon partition. $\square$

Running Time Analysis. Step (2.1) calculates v_b in $O(nN)$ time, and it only happens one time through the algorithm. Step (2.2) takes $O(2^n N)$ time. So, the time complexity of above algorithm is $O(2^n n^d)$.

4   Experimental Results

There have been some known lower bound for $M(d, t)$ which are listed as follows. Larman [21] proved for $M(d, 1) \geq 2d + 3$ for d = 2, 3. Forge *et al.* [11] proved $M(4, 1) \geq 11$. Ramírez-Alfonsí [4] proved that, for any $d \geq 4$,

$$M(d, 1) \geq \left\lceil \frac{5d}{3} \right\rceil + 3. \tag{2}$$

García-Colín and Larman [14] proved

$$M(d, t) \geq 2d + t + 1. \tag{3}$$

Soberón [28] proved a lower bound for N

$$N(d, k, t) \geq k(t + \lfloor d/2 \rfloor + 1). \tag{4}$$

As we concern about lower bound of M, $M(d, t) \geq 2t + d$ for odd d, and $M(d, t) \geq 2t + d + 1$ for even d.

To improve a lower bound on $M(d, t)$ for a pair of d and t, it is sufficient to find a set of points in $\mathbb{R}^d$ which its size is larger than previous lower bound on $M(d, t)$ such that every partition of it into two sets is not t-tolerant. One approach finding such a set of points is as follows. For a given number of points n, we start with initial points set P computed randomly. We can use one of

Table 1. Lower bounds on $M(d,t)$ using point configurations computed by algorithms from Sect. 2 for tolerance $t \leq 10$ and dimension $d = 2, 3, 4$. We omit bounds for $d = 1$ and $t = 1$ because the tight bounds are known [4, 21, 23].

t	$d = 2$	$d = 3$	$d = 4$
2	10	11	13
3	12	14	16
4	14	16	18
5	17	18	21
6	19	20	23
7	21	-	-
8	23	-	-
9	25	-	-
10	27	-	-

the algorithms for problem COMPUTINGTOLERANTPARTITION from the previous section. There are two possible outcomes. If a t-tolerant Radon partition for P does not exists, then $M(d,t) \geq n+1$, which a lower bound for M. Otherwise, the algorithm output a t-tolerant Radon partition for P, say $\Pi = \{P_1, P_2\}$. Since P is t-tolerant Radon partition every classifier of Π has at least $t + 1$ outliers. We compute all classifiers of Π, and choose a classifier c which has the minimum number of misclassification. We want to decrease the number of misclassifications of c by moving one of the points of P. Therefore, we compute the distance of c and all outliers of c and pick one of the outliers p which has the minimum distance to c. Finally, we move p to the other side of c randomly and continue this process with the new set of points.

Table 1 shows the lower bounds we obtained by using of mentioned algorithms in this paper. Using Algorithm 1, we have achieved new lower bounds for $M(2,5)$ and $M(2,6)$; however, it is slow for larger t in the plane. The results in the Table for $d = 3$ and $d = 4$ are computed by Algorithm 2. Algorithm 3 is more memory efficient than Algorithm 2 and it is used larger number of points in the plane, including $M(2,10) \geq 27$. In higher dimensions, Algorithm 2 performed better than others since it has less dependency on the dimension of points than other Algorithms.

We provide a website with the point sets corresponding to the lower bounds in Table 1 at [1]. A point set providing a lower bound for $M(d,t)$ must have points in general position and there must be no t-tolerant Radon partition for it. The following basic tests can be used for verification.

1. Test whether $d + 1$ points lie on the same hyperplane,
2. Given a point p and a hyperplane π such that $p \notin \pi$, test whether $p \in \pi^+$ or $p \in \pi^-$.

Both tests can be done using the determinant of the following matrix defined by points $p_1, p_2, \ldots, p_{d+1} \in \mathbb{R}^d$

$$
\begin{bmatrix}
p_{1,1} & p_{2,1} & p_{3,1} & \cdots & p_{d+1,1} \\
p_{1,2} & p_{2,2} & p_{3,2} & \cdots & p_{d+1,2} \\
\vdots & \vdots & \vdots & \ddots & \vdots \\
p_{1,d} & p_{2,d} & p_{3,d} & \cdots & p_{d+1,d} \\
1 & 1 & 1 & \cdots & 1
\end{bmatrix}.
$$

If the determinant is equal to 0, then the points lie on the same hyperplane. Otherwise, let π be the hyperplane passing through the points $p_1, p_2, \ldots, p_d$. Then the sign of the determinant corresponds to one of the cases $p_{d+1} \in \pi^+$ or $p_{d+1} \in \pi^-$. The points in our sets have integer coordinates and the determinant can be computed without rounding errors.

References

1. Point sets. http://www.utdallas.edu/~besp/soft/NonTolerantRadon.zip
2. Agarwal, P.K., Sharir, M., Welzl, E.: Algorithms for center and Tverberg points. ACM Trans. Algorithms **5**(1), 5:1–5:20 (2008)
3. Aggarwal, C.C., Sathe, S.: Theoretical foundations and algorithms for outlier ensembles. SIGKDD Explor. **17**(1), 24–47 (2015)
4. Alfonsín, J.R.: Lawrence oriented matroids and a problem of mcmullen on projective equivalences of polytopes. Eur. J. Comb. **22**(5), 723–731 (2001)
5. Aronov, B., Garijo, D., Rodríguez, Y.N., Rappaport, D., Seara, C., Urrutia, J.: Minimizing the error of linear separators on linearly inseparable data. Discrete Appl. Math. **160**(10–11), 1441–1452 (2012)
6. Boyd, S., Vandenberghe, L.: Convex optimization. Cambridge University Press, New York (2004)
7. Chan, T.M.: Low-dimensional linear programming with violations. SIAM J. Comput. **34**(4), 879–893 (2005)
8. Cole, R., Sharir, M., Yap, C.K.: On k-hulls and related problems. SIAM J. Comput. **16**, 61–77 (1987)
9. Corrêa, R.C., Donne, D.D., Marenco, J.: On the combinatorics of the 2-class classification problem. Discrete Optim. **31**, 40–55 (2019)
10. Everett, H., Robert, J., van Kreveld, M.J.: An optimal algorithm for the ($\leq k$)-levels, with applications to separation and transversal problems. Int. J. Comput. Geom. Appl. **6**(3), 247–261 (1996)
11. Forge, D., Las Vergnas, M., Schuchert, P.: 10 points in dimension 4 not projectively equivalent to the vertices of a convex polytope. Eur. J. Comb. **22**(5), 705–708 (2001)
12. Gajentaan, A., Overmars, M.H.: On a class of $O(n^2)$ problems in computational geometry. Comput. Geom. **5**(3), 165–185 (1995)
13. García-Colín, N.: Applying Tverberg type theorems to geometric problems. Ph.D. thesis, University College of London (2007)
14. García-Colín, N., Larman, D.G.: Projective equivalences of k-neighbourly polytopes. Graphs Comb. **31**(5), 1403–1422 (2015)

15. García-Colín, N., Raggi, M., Roldán-Pensado, E.: A note on the tolerant Tverberg theorem. Discrete Comput. Geom. **58**(3), 746–754 (2017)
16. Hamel, L.H.: Knowledge Discovery with Support Vector Machines. Wiley-Interscience, New York (2009)
17. Harding, E.F.: The number of partitions of a set of n points in k dimensions induced by hyperplanes. Proc. Edinb. Math. Soc. **15**(4), 285–289 (1967)
18. Hastie, T., Tibshirani, R., Friedman, J.H.: The Elements of Statistical Learning: Data Mining, Inference, and Prediction. Springer Series in Statistics, 2nd edn. Springer, New York (2009). https://doi.org/10.1007/978-0-387-84858-7
19. Houle, M.F.: Algorithms for weak and wide separation of sets. Discrete Appl. Math. **45**(2), 139–159 (1993)
20. Hwang, F.K., Rothblum, U.G.: On the number of separable partitions. J. Comb. Optim. **21**(4), 423–433 (2011)
21. Larman, D.G.: On sets projectively equivalent to the vertices of a convex polytope. Bull. London Math. Soc. **4**(1), 6–12 (1972)
22. Matouvsek, J.: On geometric optimization with few violated constraints. Discrete Comput. Geom. **14**(4), 365–384 (1995)
23. Mulzer, W., Stein, Y.: Algorithms for tolerant Tverberg partitions. Int. J. Comput. Geom. Appl. **24**(04), 261–273 (2014)
24. Mulzer, W., Werner, D.: Approximating Tverberg points in linear time for any fixed dimension. Discrete Comput. Geom. **50**(2), 520–535 (2013)
25. Niu, Z., Shi, S., Sun, J., He, X.: A survey of outlier detection methodologies and their applications. In: Deng, H., Miao, D., Lei, J., Wang, F.L. (eds.) AICI 2011, Part I. LNCS (LNAI), vol. 7002, pp. 380–387. Springer, Heidelberg (2011). https://doi.org/10.1007/978-3-642-23881-9_50
26. Radon, J.: Mengen konvexer Körper, die einen gemeinsamen Punkt enthalten. Math. Ann. **83**, 113–115 (1921)
27. Sathe, S., Aggarwal, C.C.: Subspace histograms for outlier detection in linear time. Knowl. Inf. Syst. **56**(3), 691–715 (2018)
28. Soberón, P.: Equal coefficients and tolerance in coloured Tverberg partitions. Combinatorica **35**(2), 235–252 (2015)
29. Soberón, P.: Robust Tverberg and colourful Carathéodory results via random choice. Comb. Probab. Comput. **27**(3), 427–440 (2018)
30. Soberón, P., Strausz, R.: A generalisation of Tverberg's theorem. Discrete Comput. Geom. **47**(3), 455–460 (2012)
31. Teng, S.-H.: Points, Spheres, and Separators: a unified geometric approach to graph partitioning. Ph.D. thesis, School of Computer Science, Carnegie-Mellon University (1990). Report CMU-CS-91-184
32. Tverberg, H.: A generalization of Radon's theorem. J. London Math. Soc. **1**(1), 123–128 (1966)
33. Yuan, G., Ho, C., Lin, C.: Recent advances of large-scale linear classification. Proceedings of the IEEE **100**(9), 2584–2603 (2012)

Author Index